Petroleum Industry Wastewater
Advanced and Sustainable Treatment Methods

Petroleum Industry Wastewater
Advanced and Sustainable Treatment Methods

Edited by

Muftah H. El-Naas
Gas Processing Center, Qatar University, Doha, Qatar

Aditi Banerjee
KTH Royal Institute of Technology, Stockholm, Sweden

ELSEVIER

Elsevier
Radarweg 29, PO Box 211, 1000 AE Amsterdam, Netherlands
The Boulevard, Langford Lane, Kidlington, Oxford OX5 1GB, United Kingdom
50 Hampshire Street, 5th Floor, Cambridge, MA 02139, United States

Notices

Knowledge and best practice in this field are constantly changing. As new research and experience broaden our understanding, changes in research methods, professional practices, or medical treatment may become necessary.

Practitioners and researchers must always rely on their own experience and knowledge in evaluating and using any information, methods, compounds, or experiments described herein. In using such information or methods they should be mindful of their own safety and the safety of others, including parties for whom they have a professional responsibility.

To the fullest extent of the law, neither the Publisher nor the authors, contributors, or editors, assume any liability for any injury and/or damage to persons or property as a matter of products liability, negligence or otherwise, or from any use or operation of any methods, products, instructions, or ideas contained in the material herein.

ISBN: 978-0-323-85884-7

For Information on all Elsevier publications
visit our website at https://www.elsevier.com/books-and-journals

Publisher: Susan Dennis
Editorial Project Manager: Susan Ikeda
Production Project Manager: Paul Prasad Chandramohan
Cover Designer: Matthew Limbert

Typeset by MPS Limited, Chennai, India

Contents

List of contributors

Saba Abdolalian Civil and Environmental Engineering, Faculty of Civil Engineering, Babol Noshirvani University of Technology, Babol, Iran

Riya Agarwal Amity Institute of Biotechnology, Amity University Kolkata, Kolkata, India

Ahmed M.D. Al ketife Department of Chemical Engineering, Qatar University, Doha, Qatar; Faculty of Engineering, University of Thi-Qar, Nasiriyah, Iraq

Taghreed Al-Khalid Department of Chemical and Petroleum Engineering, College of Engineering, UAE University, Al-Ain, United Arab Emirates

Fares Almomani Department of Chemical Engineering, Qatar University, Doha, Qatar

Abdelrahman M. Awad Gas Processing Center, Qatar University, Doha, Qatar

Muneer M. Ba-Abbad Gas Processing Center, College of Engineering, Qatar University, Doha, Qatar

Fawzi Banat Center for Membranes and Advanced Water Technology (CMAT), Department of Chemical Engineering, Khalifa University of Science and Technology, SAN Campus, Abu Dhabi, United Arab Emirates

Aditi Banerjee KTH Royal Institute of Technology, Stockholm, Sweden

Abdelbaki Benamor Gas Processing Center, College of Engineering, Qatar University, Doha, Qatar

S. Bhatia Department of Chemistry, Isabella Thoburn P. G. College, Lucknow, Uttar Pradesh, India

Yuh Nien Chow River Engineering and Urban Drainage Research Center (REDAC), Engineering Campus, Universiti Sains Malaysia, Penang, Malaysia

Jhanvi Desai School of Petroleum Technology, Pandit Deendayal Energy University, Gandhinagar, India

Muftah H. El-Naas Gas Processing Center, Qatar University, Doha, Qatar

Keng Yuen Foo River Engineering and Urban Drainage Research Center (REDAC), Engineering Campus, Universiti Sains Malaysia, Penang, Malaysia

Shadi W. Hasan Center for Membranes and Advanced Water Technology (CMAT), Department of Chemical Engineering, Khalifa University of Science and Technology, SAN Campus, Abu Dhabi, United Arab Emirates

Ibnelwaleed A. Hussein Gas Processing Center, Qatar University, Doha, Qatar

Mohamed H. Ibrahim Gas Processing Center, Qatar University, Doha, Qatar

Rem Jalab Gas Processing Center, Qatar University, Doha, Qatar

S. Joshi Department of Chemistry, Isabella Thoburn P.G. College, Lucknow, Uttar Pradesh, India

Omar Khalifa Center for Membranes and Advanced Water Technology (CMAT), Department of Chemical Engineering, Khalifa University of Science and Technology, SAN Campus, Abu Dhabi, United Arab Emirates

Arindam Kushagra Amity Institute of Nanotechnology, Amity University Kolkata, Kolkata, India

Hamish R. Mackey Division of Sustainable Development, College of Science and Engineering, Hamad Bin Khalifa University, Qatar Foundation, Doha, Qatar

Ebrahim Mahmoudi Department of Chemical and Process Engineering, Faculty of Engineering and Built Environment, Universiti Kebangsaan Malaysia, Bangi, Selangor, Malaysia

Abdul Wahab Mohammad Department of Chemical and Process Engineering, Faculty of Engineering and Built Environment, Universiti Kebangsaan Malaysia, Bangi, Selangor, Malaysia

Mustafa S. Nasser Gas Processing Center, Qatar University, Doha, Qatar

Abdullah Omar Department of Chemical Engineering, Qatar University, Doha, Qatar

Snigdhendubala Pradhan Division of Sustainable Development, College of Science and Engineering, Hamad Bin Khalifa University, Qatar Foundation, Doha, Qatar

Farhad Qaderi Civil and Environmental Engineering, Faculty of Civil Engineering, Babol Noshirvani University of Technology, Babol, Iran

Kritica Rani Amity Institute of Biotechnology, Amity University Kolkata, Kolkata, India

Naim Rashid Division of Sustainable Development, College of Science and Engineering, Hamad Bin Khalifa University, Qatar Foundation, Doha, Qatar

Mohammad-Hossein Sarrafzadeh UNESCO Chair on Water Reuse, School of Chemical Engineering, College of Engineering, University of Tehran, Tehran, Iran

Manan Shah Department of Chemical Engineering, School of Technology, Pandit Deendayal Energy University, Gandhinagar, India

Vrutang Shah School of Petroleum Technology, Pandit Deendayal Energy University, Gandhinagar, India

Puja Singh Amity Institute of Biotechnology, Amity University Kolkata, Kolkata, India

Riham Surkatti Gas Processing Center, College of Engineering, Qatar University, Doha, Qatar

Maryam Taghizadeh Civil and Environmental Engineering, Faculty of Civil Engineering, Babol Noshirvani University of Technology, Babol, Iran

K.L. Tan River Engineering and Urban Drainage Research Center (REDAC), Universiti Sains Malaysia, Nibong Tebal, Malaysia

Chapter 1

Treatment of petroleum industry wastewater: current practices and perspectives

Mohamed H. Ibrahim[1], Aditi Banerjee[2] and Muftah H. El-Naas[1]

[1]*Gas Processing Center, Qatar University, Doha, Qatar,* [2]*KTH Royal Institute of Technology, Stockholm, Sweden*

Introduction

Oil and gas industries are evolving to meet the ever-growing energy demand that will lead to more crude oil extraction in the foreseeable future. Over 50,000 oil and gas fields are established across the globe which can damage the surrounding and extended environments. Generally, oil and gas extraction operations are carried out using extraction platforms where the extracted oil is accompanied by water that is called produced water (PW). PW consists of different chemicals such as polyaromatic hydrocarbons, phenols, benzene, toluene, ethylbenzene, and heavy metals. These pollutants can have a detrimental effect on the environment; nonetheless, the industry continues to generate billions of gallons of wastewater annually. In addition, these chemicals can lead to the accumulation of petroleum hydrocarbons and other contaminants in different natural habitats. Typical wastewater treatment of oily wastewater involves several physical treatment steps, such as oil and water separators, dissolved or induced air flotation systems. Secondary treatment is represented in biological processes, such as activated sludge systems, oxidation lagoons, or trickling filters. The biological treatment is followed by a tertiary treatment which is considered to be a polishing step that usually utilizes activated carbon or sand filters in addition to different filtration technologies and other advanced treatment techniques such as electrodialysis (Treviño-Reséndez, de, Medel, & Meas, 2021). In this chapter, different methods for treating petroleum wastewater are discussed. More details about these technologies are provided in the next chapters.

Electrochemical methods

Several electrochemical techniques are discussed in the literature to treat petroleum wastewater including electrocoagulation (EC), electrooxidation, and electro-fenton. EC is an electrochemical technique that has been applied on bench and pilot scale to remove a wide range of pollutants in petroleum wastewater. The technique utilizes sacrificial electrodes that are usually made of aluminum or iron which go through electrodissolution to create flocs of metal hydroxides. The generated hydroxides lead to the effective separation of the oil contaminants in the water due to several mechanisms, such as charge neutralization, entrapment, and adsorption. These are followed by the precipitation or flotation of the generated flocs which facilitates the separation of the oil from water (Treviño-Reséndez et al., 2021). Moreover, the generated hydroxides can promote the removal of other contaminants such as heavy metals through adsorption or precipitation through adsorption which can lead to an overall decrease in the total dissolved solids in the wastewater. The performance of the EC process can be measured through several parameters such as chemical and biological oxygen demand (COD and BOD), total petroleum hydrocarbons, and total organic carbons. The change in these parameters mainly represents the efficiency of a certain process in removing pollutants. Nonetheless, measuring the content of phenols, benzene, toluene, and xylene could give a better insight into the removal of various chemical organic pollutants from the wastewater. It is also worth noting that the EC mechanism that leads to the removal of suspended solids also contributes to the removal of organic and inorganic pollutants during the process.

Petroleum Industry Wastewater. DOI: https://doi.org/10.1016/B978-0-323-85884-7.00015-1

Most of the EC work in the literature focus on finding a feasible solution for the PW generated through the fracking process that will enable the recycling of the vast quantities of PW during the extraction process. The vast majority of these studies have been carried out on a lab-scale with few studies that evaluated the technology at a pilot scale. The technology readiness level of EC to treat petroleum wastewater is estimated to be in the range of 5−6. Hence, extra efforts are required to increase the applicability of the technology through the implementation of more prototypes and pilot-scale units. This will definitely lead to an increase in the viability of EC compared to the currently used technologies. El-Naas et al. designed a three-step pilot plant with a capacity of 1 m^3 of wastewater per hour. The pilot was evaluated for the treatment of highly contaminated refinery wastewater (El-Naas, Alhaija, & Al-Zuhair, 2014). The target of the study was to decrease the contraction of the pollutants in the wastewater in accordance with the UAE's discharge limits (COD: 150 mg L^{-1} and phenols: 0.1 mg L^{-1}). The three-step process consisted of an EC reactor, a biomass reactor, and an activated carbon adsorption column. The pilot plant was operated continuously for 24 h, and it reached a complete COD and phenol removal after 8 h only. In addition, the reported results were in line with the laboratory scale result reported in a different study. Overall, the EC process is considered to be the most application-ready approach for the treatment of several types of petroleum wastewater. It is recommended that more attention is paid to the utilization of other electrochemical technologies and assess their performance under real field conditions and larger scales (pilot scale). Although there is no universal electrochemical technology for the treatment of highly contaminated wastewater from the petroleum industry, coupling electrochemical technology with other treatment options (physicochemical and biological) may prove to be effective and more work is needed in this direction (Treviño-Reséndez et al., 2021).

Membrane-based methods

Recently, membrane-based technology has emerged as a promising alternative for the treatment of wastewater (Tanudjaja, Hejase, Tarabara, Fane, & Chew, 2019). With their rapid growth and development, membrane-based technologies are expected to become the leading technology for the treatment of oily wastewater. Membranes are primarily classified as organic (polymeric) and inorganic (ceramic, metal, glass, etc.) depending on the base materials used for their production (Gao & Xu, 2019). The possibility of using a wide range of materials for membranes synthesis provides further options for their customization depending on the required application and process conditions. These features render the application of membrane-based technology to be more competitive compared to the other available oily wastewater treatment technologies (Xu, Jiang, Wei, Chen, & Jing, 2018).

Membrane fouling is one of the major challenges for the implementation of membrane-based technology. Fouling can lead to loss of performance of the membranes over time and its remedy requires the use of chemicals for cleaning which in turn complicates its operation and ultimately increases its operating and capital cost (Jepsen, Bram, Pedersen, & Yang, 2018). Organic membranes are susceptible to fouling during the treatment of oily wastewater owing to the hydrophobic nature of the polymers present within the membranes that readily interact with the oil molecules (Zuo et al., 2018). Therefore studies are now focused on converting the surface of the organic membranes hydrophilic to counter this interaction (Jepsen et al., 2018).

Several review studies have been conducted on the treatment of oily wastewater focusing on the technological improvements of membrane-based technologies. Dickhout et al. (2017) reviewed from a colloid perspective emphasizing the operating parameters and modifications on membrane surface wetting for the treatment of PW using membranes. The performance and developments of thin-film composite membranes for the treatment of oil wastewater were reviewed by Ahmad, Goh, Karim, and Ismail (2018). Moreover, the different operating modes, modules, and economics of membranes were reviewed in the work of Tanudjaja et al. (2019). Additional studies have been conducted addressing the advancements in materials for polymeric and ceramic membranes as well as on surface modification for the treatment of oily wastewater (Zoubeik, Ismail, Salama, & Henni, 2018). However, to rationalize the approach for research and development of membranes for oily wastewater treatment, further studies need to be conducted on the membrane materials emphasizing membrane surface chemistry, membrane surface patterning, and hydrodynamics.

Biomaterial-based methods

There are numerous studies where cellulose and its derivatives are applied in the petroleum industry. This can be done through cellulose-derived materials such as chitin. Chitin is the second most used biodegradable biopolymer in the globe which is extracted from crustaceans and some arthropods. Chitosan is a derivative of chitin that is synthesized through alkaline hydrolysis treatment using 40%−50% NaOH (Negi, Verma, & Singh, 2021). Sun et al. (2008) utilized a modified chitosan-based polymeric surfactant for the removal of oil from wastewater. The chitosan was chemically modified using carboxymethyl which achieved an oil removal efficiency of 99% at a pH of approximately 7. It is worth noting that the chemical composition of chitosan consists of amine functional groups that can readily form a bond with the oil residues in the wastewater. The authors reported that increasing the temperature above 50°C can destroy the formed bonds and recorded an optimum operating temperature between 40°C and 50°C. Similarly, Bratskaya, Avramenko, Schwarz, and Philippova (2006) utilized modified chitosan in the application of removing diesel from wastewater. Chitosan was utilized to synthesize an adsorbent in the form of flakes to treat oily water generated from the biodiesel industry (Pitakpoolsil & Hunsom, 2014). The study reported that increasing the dosage of the adsorbent led to an increase in the removal percentage. In addition, at a dose of 3.5 g of adsorbent, the removal reached 95.8% of oil and grease, 93.6% of BOD, and 97.6% of COD. In a different study, chitosan in a powder form was synthesized to remove oil from saline water (Vidal, Desbrières, Borsali, & Guibal, 2020). The study reported a removal efficiency of 95% was achieved and the efficiency increases at lower concentrations of the pollutants.

Biological methods

Biological treatment methods have proved to be very effective in the area of wastewater treatment. Compared to other treatment methods, biological treatment is environment-friendly, inexpensive, and can lead to the complete conversion of complex organic pollutants. Biological treatment has been widely applied to remove organics, metals, oil, and grease from wastewater using several microorganisms including bacteria, microalgae, fungus, and yeasts (Di Caprio, Altimari, & Pagnanelli, 2015; Li, Sun, & Li, 2005; Qiu, Zhong, Li, Bai, & Li, 2007). These microorganisms have the ability to degrade several components under aerobic, anaerobic, and anoxic conditions (Sharma & Philip, 2016).

Biotreatment systems are often classified into three categories: suspended growth, attached growth, and hybrid. In suspended growth, microorganisms are suspended in a liquid medium under aerobic or anaerobic conditions, whereas attached growth is formed by granulation of activated sludge or attachment of the biomass as biofilms (Tziotzios, Teliou, Kaltsouni, Lyberatos, & Vayenas, 2005). This technique can result in a high concentration of biomass within the biotreatment system and is often used in several bioreactors, such as granular sludge reactors, packed bed reactors, fluidized bed bioreactors, spouted bed bioreactors, rotating biological contactors, and biological activated filters (Banerjee & Ghoshal, 2017; El-Naas, Surkatti, & Al-Zuhair, 2016; Rana, Gupta, & Rana, 2018). The attached-growth approach creates a surface that is necessary for the development of a biofilm structure that can achieve higher biomass concentration and allow the microorganisms to remain in the reactor for a long residence time leading to better biodegradation performance (Tziotzios et al., 2005). Hybrid systems involve the combination of both suspended- and attached-growth systems. For example, this may include the combination of activated sludge with biofilters and submerged membrane bioreactors (Sharma & Philip, 2016). Refinery wastewater using biological treatment methods has been investigated by a number of researchers. Yamaguchi, Ishida, and Suzuki (1999) reported a 65% removal of organic pollutants in petroleum wastewater using *Prototheca zopfii* in a rotating biological reactor. Sanghamitra, Mazumder, and Mukherjee (2020) reported a removal percentage between 54.6 and 80 of oil from synthetic oily wastewater by utilizing aerobic batch reactor. In addition, the authors reported a complete reduction in BOD, COD, and oil and grease content in the wastewater. Raghukumar, Vipparty, David, and Chandramohan (2001) reported an overall removal of crude oil in wastewater in the range of 45%–55% using *Oscillatoria salina*, *Aphanocapsa* sp., and *Plectonema terebrans* gathered from a local marine environment. Moreover, it was reported that using enzymes in biological treatment proved to be effective in removing the oil from wastewater (Bhumibhamon, Koprasertsak, & Funthong, 2002). The biological treatment of refinery wastewater using immobilized bacterial cells in polyvinyl alcohol gel was also investigated. El-Naas, Al-Muhtaseb, and Makhlouf (2009) reported an efficient degradation of phenol using immobilized *Pseudomonas putida* in spouted bed bioreactor. The authors reported that complete removal of phenol was achieved in less than 5 h. Moreover, the same biological treatment, utilizing the same biomass, was also tested for the removal of cresols from synthetic wastewater in batch and continuous modes. The results indicated that *P. putida* had high potential for the biodegradation of p-cresol at concentrations up to 200 mg L^{-1} in continuous mode achieving more than 85% removal efficiency (Surkatti & El-Naas, 2014). In addition, the biological treatment was tested in the pilot scale for the treatment of real refinery wastewater by El-Naas et al. (2016). The treatment process was combined with other pre- and posttreatment processes. Results showed that the bioreactor achieved complete removal of phenol and o-cresols within 8 h of operation, and both m- and p-cresols had significant removal. The biological treatment of refinery wastewater was also investigated on a pilot scale of a fixed-film bioreactor system. The reactor was operated using immobilized biomass in polyurethane foam (Jou & Huang, 2003). The system showed high performance in the removal of organic pollutants from refinery wastewater using an 8-h hydraulic retention time and achieved 85%–90% COD reduction and complete phenol reduction.

Life cycle assessment

The life cycle technique is a methodology for the economic and environmental assessment of products, services, or processes. Life cycle assessment (LCA) and life cycle cost analysis (LCCA) are used to conduct environmental and economic evaluations. Both these methodologies are used to evaluate and measure the effectiveness of every stage of product, process, or service (environmentally and economically). LCCA and LCA were originally employed during the late 1900s in the field of wastewater treatment. Since then, the number of publications examining the environmental and economic analyses of technologies, facilities, and processes has been expanding during the past two decades. Several approaches and frameworks were used to examine the environmental and economic implications. LCCA is a helpful tool to estimate the larger economic effect of the design, construction, and operational decisions; nonetheless, it is necessary to fully examine the various approaches and frameworks used in LCCA to find best practices and the latest state of the art.

The petroleum refining process consumes a considerable amount of water. The typical European refinery utilizes 5.9 m^3 per ton of feed (Barthe, Chaugny, Roudier, & Sancho, 2010). The water is used for different processes, such as distillation, cracking, steam generation, and process cooling purposes. Therefore substantial amounts of wastewater are generated. It is estimated that a single refinery that consumes 400–1000 m^3 h^{-1} of raw water can generate 200–600 m^3 h^{-1} of wastewater that needs to be treated or discharged (Nabzar, 2011). The generated water consists of a complex mixture of different pollutants such as aromatic hydrocarbons, emulsified oil, and grease as well as inorganic substances, including ammonia, sulfides, and cyanides (Taghreed Al-Khalid & Muftah, 2017). As discussed earlier in this chapter, there are different treatment methods that can be used to treat the generated wastewater. However, there are numerous drivers that are pushing the petroleum refining industry from the treatment approach to the reuse approach such as government regulations and water scarcity. The regulations can be different depending on the region; nonetheless, there is a general trend of having more strict legislation that advocates for water reuse (Muñoz et al., 2020). In addition, the countries that have the highest oil production happen to be located in water-stressed regions which further necessities the need for water reuse. Hence, selecting the most appropriate technology or combination of technologies is paramount to successful water reuse. This requires a comprehensive selection criterion that can assess the technical performance and environmental sustainability of the desired technology. The latter is often determined by the means of LCA that can provide a complete assessment of the direct and indirect possible environmental effects of a certain technology. LCA examines the life cycle of the process in a comprehensive way. Nonetheless, LCA work in the area of the oil and gas industry is considered to be rather limited and scarce.

Vlasopoulos, Memon, Butler, and Murphy (2006) discussed the environmental impacts of 20 oily water treatment technologies that can treat the water to be suitable for reuse in other industries or for agricultural purposes. It is worth noting that the study mainly focused on PW rather than petroleum refinery wastewater. Moreover, Ronquim et al. optimized the performance of a reverse osmosis system to treat and then reuse the refinery wastewater in cooling towers. Nonetheless, the study did not consider the raw refinery wastewater but rather already treated water that can be discharged (Ronquim, Sakamoto, Mierzwa, Kulay, & Seckler, 2020). Muñoz et al. carried out an LCA on wastewater reuse in a petroleum refinery in Turkey. The study aim was to determine if refinery wastewater reuse in several utilities in the industry such as cooling water can lead to overall economic benefit in addition to the environmental one (Muñoz et al., 2020).

Summary and future perspectives

Wastewater treatment processes and methodologies need to be innovative to maintain efficient treatment. Combining various technologies will considerably improve the performance of the treatment process. While combining different technologies may be quite productive, more studies and real field implementations are needed to address the scale and effectiveness of the integration of multiple technologies. Most of the recent research on wastewater treatment has been conducted at the laboratory scale with a limited number of studies at the bench or pilot scale. In addition, the vast majority of the research work used simulated water rather than real wastewater. Selecting a certain treatment technology mainly depends on the composition and properties of the wastewater. For example, if the wastewater contains highly resistive contaminants, a biological treatment technique would be more suitable. This can be followed by a tertiary/polishing step to ensure the complete removal of the contaminants. There are several techniques that proved to be effective for the treatment of highly contaminated petroleum wastewater including electrochemical, biological, and membrane-based techniques. Despite their effectiveness in the treatment process, they still suffer from drawbacks that must be addressed in future research.

Electrochemical techniques consist of different technologies such as electrooxidation and EC. EC is considered to be the most application-ready technology for the treatment of refinery wastewater (Moussa, El-Naas, Nasser, & Al-Marri, 2017). Nonetheless, it can suffer from several disadvantages such as electrode passivation. This can be mitigated by the frequent cleaning of electrodes or using different electrode designs that can reduce passivation (Ibrahim, Moussa, El-Naas, & Nasser, 2020). Combining several electrochemical treatment techniques is the most used approach in the literature to treat refinery wastewater. Membrane-based technologies have the potential to be the most successful technology for refinery wastewater treatment. Membranes are versatile and can be fabricated with specific properties that can suit different compositions of wastewater. Nonetheless, membrane fouling is a major challenge for the technology to overcome. Fouling can be mitigated by using cleaning chemicals but at the expense of operating costs. Therefore the research direction in the literature is to mitigate membrane fouling by modifying the membrane surface. Cellulose-derived materials are also investigated in the literature for treating petroleum wastewater. This is mainly due to their ability to form a bond with the oily contaminants which facilitates their removal. Biological treatment techniques have been very effective in dealing with highly contaminated petroleum wastewater and have been characterized by their simplicity, low operating and maintenance cost, and ability to degrade complex hydrocarbons. However, biotreatment is often considered to be slow and relatively sensitive to operating conditions. LCA is rather an important tool in evaluating the long-term sustainability of the treatment process. Nonetheless, its use in the area of petroleum refinery wastewater is often limited. Hence, more LCA is required for better long-term evaluation of specific treatment technology.

References

Ahmad, N. A., Goh, P. S., Karim, Z. A., & Ismail, A. F. (2018). Thin film composite membrane for oily waste water treatment: Recent advances and challenges. *Membranes, 8*(4). Available from https://doi.org/10.3390/membranes8040086.

Banerjee, A., & Ghoshal, A. K. (2017). Biodegradation of an actual petroleum wastewater in a packed bed reactor by an immobilized biomass of *Bacillus cereus. Journal of Environmental Chemical Engineering, 5*(2), 1696−1702. Available from https://doi.org/10.1016/j.jece.2017.03.008.

Barthe, P., Chaugny, M., Roudier, S., & Sancho. (2010). *Best available techniques (BAT) reference document for the refining of mineral oil and gas.* Industrial Emissions Directive. Available from https://doi.org/10.2791/010758.

Bhumibhamon, O., Koprasertsak, A., & Funthong, S. (2002). Biotreatment of high fat and oil wastewater by lipase producing microorganisms. *Agriculture and Natural Resources, 36*(3), 261−267.

Bratskaya, S., Avramenko, V., Schwarz, S., & Philippova, I. (2006). Enhanced flocculation of oil-in-water emulsions by hydrophobically modified chitosan derivatives. *Colloids and Surfaces A: Physicochemical and Engineering Aspects, 275*(1−3), 168−176. Available from https://doi.org/10.1016/j.colsurfa.2005.09.036.

Di Caprio, F., Altimari, P., & Pagnanelli, F. (2015). Integrated biomass production and biodegradation of olive mill wastewater by cultivation of *Scenedesmus* sp. *Algal Research, 9*, 306−311. Available from https://doi.org/10.1016/j.algal.2015.04.007.

Dickhout, J. M., Moreno, J., Biesheuvel, P. M., Boels, L., Lammertink, R. G. H., & de Vos, W. M. (2017). Produced water treatment by membranes: A review from a colloidal perspective. *Journal of Colloid and Interface Science, 487*, 523−534. Available from https://doi.org/10.1016/j.jcis.2016.10.013.

El-Naas, M. H., Alhaija, M. A., & Al-Zuhair, S. (2014). Evaluation of a three-step process for the treatment of petroleum refinery wastewater. *Journal of Environmental Chemical Engineering, 2*(1), 56−62. Available from https://doi.org/10.1016/j.jece.2013.11.024.

El-Naas, M. H., Al-Muhtaseb, S. A., & Makhlouf, S. (2009). Biodegradation of phenol by *Pseudomonas putida* immobilized in polyvinyl alcohol (PVA) gel. *Journal of Hazardous Materials, 164*(2−3), 720−725. Available from https://doi.org/10.1016/j.jhazmat.2008.08.059.

El-Naas, M. H., Surkatti, R., & Al-Zuhair, S. (2016). Petroleum refinery wastewater treatment: A pilot scale study. *Journal of Water Process Engineering, 14*, 71−76. Available from https://doi.org/10.1016/j.jwpe.2016.10.005.

Gao, N., & Xu, Z. K. (2019). Ceramic membranes with mussel-inspired and nanostructured coatings for water-in-oil emulsions separation. *Separation and Purification Technology, 212*, 737−746. Available from https://doi.org/10.1016/j.seppur.2018.11.084.

Ibrahim, M. H., Moussa, D. T., El-Naas, M. H., & Nasser, M. S. (2020). A perforated electrode design for passivation reduction during the electrochemical treatment of produced water. *Journal of Water Process Engineering, 33*. Available from https://doi.org/10.1016/j.jwpe.2019.101091.

Jepsen, K. L., Bram, M. V., Pedersen, S., & Yang, Z. (2018). Membrane fouling for produced water treatment: A review study from a process control perspective. *Water (Switzerland), 10*(7). Available from https://doi.org/10.3390/w10070847.

Jou, C. J. G., & Huang, G. C. (2003). A pilot study for oil refinery wastewater treatment using a fixed film bioreactor. *Advances in Environmental Research, 7*(2), 463−469. Available from https://doi.org/10.1016/S1093-0191(02)00016-3.

Li, B., Sun, Y. L., & Li, Y. Y. (2005). Pretreatment of coking wastewater using anaerobic sequencing batch reactor (ASBR). *Journal of Zhejiang University: Science, 6*(11), 1115−1123. Available from https://doi.org/10.1631/jzus.2005.B1115.

Moussa, D. T., El-Naas, M. H., Nasser, M., & Al-Marri, M. J. (2017). A comprehensive review of electrocoagulation for water treatment: Potentials and challenges. *Journal of Environmental Management, 186*, 24−41. Available from https://doi.org/10.1016/j.jenvman.2016.10.032.

Muñoz, I., Aktürk, A. S., Ayyıldız, Ö., Çağlar, Ö., Meabe, E., Contreras, S., . . . Jiménez-Banzo, A. (2020). Life cycle assessment of wastewater reclamation in a petroleum refinery in Turkey. *Journal of Cleaner Production, 268*, 121967. Available from https://doi.org/10.1016/j.jclepro.2020.121967.

Nabzar, L. (2011). Panorama 2011: Water in fuel production Oil production and refining; Panorama 2011: L'eau dans la production de carburants. Production petroliere et raffinage. France.

Negi, H., Verma, P., & Singh, R. K. (2021). A comprehensive review on the applications of functionalized chitosan in petroleum industry. *Carbohydrate Polymers*, *266*, 118125. Available from https://doi.org/10.1016/j.carbpol.2021.118125.

Pitakpoolsil, W., & Hunsom, M. (2014). Treatment of biodiesel wastewater by adsorption with commercial chitosan flakes: Parameter optimization and process kinetics. *Journal of Environmental Management*, *133*, 284−292. Available from https://doi.org/10.1016/j.jenvman.2013.12.019.

Qiu, X., Zhong, Q., Li, M., Bai, W., & Li, B. (2007). Biodegradation of *p*-nitrophenol by methyl parathion-degrading *Ochrobactrum* sp. B2. *International Biodeterioration and Biodegradation*, *59*(4), 297−301. Available from https://doi.org/10.1016/j.ibiod.2006.09.005.

Raghukumar, C., Vipparty, V., David, J. J., & Chandramohan, D. (2001). Degradation of crude oil by marine cyanobacteria. *Applied Microbiology and Biotechnology*, *57*(3), 433−436. Available from https://doi.org/10.1007/s002530100784.

Rana, S., Gupta, N., & Rana, R. S. (2018). Removal of organic pollutant with the use of rotating biological contactor. *Materials Today: Proceedings*, *5*(Issue 2), 4218−4224. Available from https://doi.org/10.1016/j.matpr.2017.11.685.

Ronquim, F. M., Sakamoto, H. M., Mierzwa, J. C., Kulay, L., & Seckler, M. M. (2020). Eco-efficiency analysis of desalination by precipitation integrated with reverse osmosis for zero liquid discharge in oil refineries. *Journal of Cleaner Production*, *250*. Available from https://doi.org/10.1016/j.jclepro.2019.119547.

Sanghamitra, P., Mazumder, D., & Mukherjee, S. (2020). A study on aerobic biodegradation of oil and grease containing wastewater. *Journal of the Indian Chemical Society*, *97*(5), 819−822. Available from https://indianchemicalsociety.com/portal/uploads/journal/2020_05_17_Extended_1601005922.pdf.

Sharma, N. K., & Philip, L. (2016). Combined biological and photocatalytic treatment of real coke oven wastewater. *Chemical Engineering Journal*, *295*, 20−28. Available from https://doi.org/10.1016/j.cej.2016.03.031.

Sun, G., Chen, X., Li, Y., Zheng, B., Gong, Z., Sun, J., ... Lin, W. (2008). Preparation of *H*-oleoyl-carboxymethyl-chitosan and the function as a coagulation agent for residual oil in aqueous system. *Frontiers of Materials Science in China*, *2*(1), 105−112. Available from https://doi.org/10.1007/s11706-008-0019-3.

Surkatti, R., & El-Naas, M. H. (2014). Biological treatment of wastewater contaminated with p-cresol using *Pseudomonas putida* immobilized in polyvinyl alcohol (PVA) gel. *Journal of Water Process Engineering*, *1*, 84−90. Available from https://doi.org/10.1016/j.jwpe.2014.03.008.

Taghreed Al-Khalid & Muftah H. El-Naas (2017). Organic contaminants in refinery wastewater: Characterization and novel approaches for biotreatment, recent insights in Petroleum Science and Engineering, Mansoor Zoveidavianpoor, *IntechOpen*, Available from: https://doi.org/10.5772/intechopen.72206. https://www.intechopen.com/chapters/58096.

Tanudjaja, H. J., Hejase, C. A., Tarabara, V. V., Fane, A. G., & Chew, J. W. (2019). Membrane-based separation for oily wastewater: A practical perspective. *Water Research*, *156*, 347−365. Available from https://doi.org/10.1016/j.watres.2019.03.021.

Treviño-Reséndez, J., de, J., Medel, A., & Meas, Y. (2021). Electrochemical technologies for treating petroleum industry wastewater. *Current Opinion in Electrochemistry*, *27*, 100690. Available from https://doi.org/10.1016/j.coelec.2021.100690.

Tziotzios, G., Teliou, M., Kaltsouni, V., Lyberatos, G., & Vayenas, D. V. (2005). Biological phenol removal using suspended growth and packed bed reactors. *Biochemical Engineering Journal*, *26*(1), 65−71. Available from https://doi.org/10.1016/j.bej.2005.06.006.

Vidal, R. R. L., Desbrières, J., Borsali, R., & Guibal, E. (2020). Oil removal from crude oil-in-saline water emulsions using chitosan as biosorbent. *Separation Science and Technology (Philadelphia)*, *55*(5), 835−847. Available from https://doi.org/10.1080/01496395.2019.1575879.

Vlasopoulos, N., Memon, F. A., Butler, D., & Murphy, R. (2006). Life cycle assessment of wastewater treatment technologies treating petroleum process waters. *Science of the Total Environment*, *367*(1), 58−70. Available from https://doi.org/10.1016/j.scitotenv.2006.03.007.

Xu, Z., Jiang, D., Wei, Z., Chen, J., & Jing, J. (2018). Fabrication of superhydrophobic nano-aluminum films on stainless steel meshes by electrophoretic deposition for oil-water separation. *Applied Surface Science*, *427*, 253−261. Available from https://doi.org/10.1016/j.apsusc.2017.08.189.

Yamaguchi, T., Ishida, M., & Suzuki, T. (1999). Biodegradation of hydrocarbons by *Prototheca zopfii* in rotating biological contactors. *Process Biochemistry*, *35*(3−4), 403−409. Available from https://doi.org/10.1016/S0032-9592(99)00086-2.

Zoubeik, M., Ismail, M., Salama, A., & Henni, A. (2018). New developments in membrane technologies used in the treatment of produced water: A review. *Arabian Journal for Science and Engineering*, *43*(5), 2093−2118. Available from https://doi.org/10.1007/s13369-017-2690-0.

Zuo, J. H., Cheng, P., Chen, X. F., Yan, X., Guo, Y. J., & Lang, W. Z. (2018). Ultrahigh flux of polydopamine-coated PVDF membranes quenched in air via thermally induced phase separation for oil/water emulsion separation. *Separation and Purification Technology*, *192*, 348−359. Available from https://doi.org/10.1016/j.seppur.2017.10.027.

Chapter 2

Chemistry of petroleum wastewater

Farhad Qaderi and Maryam Taghizadeh

Civil and Environmental Engineering, Faculty of Civil Engineering, Babol Noshirvani University of Technology, Babol, Iran

Introduction

Water plays an essential role in human life. The industrial revolution and economic development have led to population growth and increased urbanization (Chen, 2018; Li & Yu, 2011; Zafra, Moreno-Montaño, Absalón, & Cortés-Espinosa, 2015; Zhang et al., 2015). The development of communities, consumerism, and the expansion of urbanization has led to an alarming rate in waste/wastewater production around the world, and this waste is disposed of without proper management and treatment (Chen, 2018; Li & Yu, 2011; Zafra et al., 2015; Zhang et al., 2015). Today, the oil industry is the most important source of energy production in the world, but this industry is one of the main causes of environmental problems due to the production of large volumes of wastewater.

The word Petroleum originated from the Latin roots Petra (rock) and oleum (oil) (Jafarinejad, 2016). Petroleum is formed from the accumulation of hydrocarbons. Hydrocarbons accumulate naturally, thousands of feet below the surface of the Earth, from the decomposition of organic materials like plants and marine animals that died millions of years ago. It is a naturally occurring fluid found in rock formations. In other words, vast quantities of the remains of decomposed organic materials settled into sea or lake bottoms and are mixed with sediments buried in layers of clay, silt, and sand. As further layers settled into the bed, in the lower regions, heat and pressure began to rise. This process causes the organic matter to change, first into a waxy material, which is found in various oil shales around the world, and then with more heat into liquid and gaseous hydrocarbons. Oil or natural gas formation is related to the amount of pressure, the degree of heat, and the type of biomass. It is believed that More heat is believed to produce lighter oil, even higher heat or biomass made predominantly of plant material produced by natural gas. The petroleum industry generates a large amount of oily waste due to upstream and downstream operations (Abdulredha, Siti Aslina, & Luqman, 2020; Li et al., 2014; Varjani, Kumar, & Rene, 2019). This oily waste can be divided into a solid or liquid. The upstream process includes extracting, transporting, and storing crude oil, and the downstream process includes refining crude oil (Al-Futaisi, Jamrah, Yaghi, & Taha, 2007; Hu, Li, & Zeng, 2013; Thakur, Srivastava, Mall, & Hiwarkar, 2018). Based on the amount of water in solids, this waste is categorized into simple sewage crude oil and sludge.

Crude oil consists of approximately 10−14 wt.% hydrogen and 83−87 wt.% carbon. Oxygen (0.05−1.5 wt.%), sulfur (0.05−6 wt.%), nitrogen (0.1−2 wt.%), and metals such as vanadium, nickel, iron, and copper (nickel and vanadium <1000 ppm) are some impurities found in crude oil. Crude oil is not a uniform material, and its exact molecular and fractional composition varies widely with the formation of oil, location, age of the oil field, and the depth of the individual well. Crude oils obtained from different oil reservoirs have widely different characteristics. Many oil reservoirs contain live bacteria. Some crude oils are black, heavy, and thick like tar, while others are brown or nearly clear with low viscosity and low specific gravity. Usually, four different types of hydrocarbon molecules (alkanes or paraffins (15%−60%), naphthenes, or cycloparaffins (30%−60%), aromatics or benzene molecules (3%−30%), and asphaltics (remainder)) appear in crude oil. The relative percentage of each component, which varies from oil to oil, determines the properties of the oil. (Thakur et al., 2018).

Oilfield wastewater is composed of produced water, used drilling fluid, and wastewater that is produced after washing the drilling/extraction equipment and storage tanks (Bakke, Klungsøyr, & Sanni, 2013; Ottaviano, Cai, & Murphy, 2014).

Oil and gas production is usually accompanied by the production of large quantities of water. The amount of produced water depends on the extraction technology, the reservoir characteristics, and the rate of oil extraction. In different sites, The ratio of produced water to oil may vary from to 2 to 5.

Petroleum Industry Wastewater. DOI: https://doi.org/10.1016/B978-0-323-85884-7.00007-2

Petroleum refinery effluents are generated in oil refinery processes that convert crude oil into numerous refined products, such as liquefied petroleum gas, fuels, lubricants, and petrochemical intermediates. These processes consume about $0.2 \, m^3 \, t^{-1} \, L$ to $25 \, m^3 \, t^{-1} \, L$ of water, which has decreased in European refineries due to water reuse and recycling (Barthe et al., 2015). The amount and composition of the refinery wastewater vary considerably depending on different factors such as plant configuration, crude oil characteristics, and process designs. According to a study, regardless of the configuration, the main wastewater is generated through a number of processes such as distillation, hydrotreating, cooling system, and desalting (Diya'Uddeen, Daud, & Abdul Aziz, 2011). The water is used or produced? directly with hydrocarbon compounds in desalination, distillation, and cracking or washing and cleaning operations, and leads to increasing the organic load. On the other hand, the biggest part of the wastewater effluent is water used for cooling systems and boilers, while this effluent is usually less polluted.

Petroleum refineries process raw crude oil into three categories of products enumerated as follows:

1. *Fuel products*: Gasoline, distillate fuel oil, jet fuels, residual fuel oil, liquefied petroleum gases, refinery fuel, coke, and kerosene.
2. *Nonfuel products*: Asphalt and road oil, lubricants, naphtha solvents, waxes, nonfuel coke, and miscellaneous products.
3. *Petrochemicals and petrochemical feedstock*: Naphtha, ethane, propane, butane, ethylene, propylene, butylene, and BTEX compounds (benzene, toluene, ethylbenzene, and xylene).

Petroleum industry wastewater consists of oil impurities, BOD and COD are not contaminants, they are measure of contamination. high total solids, hydrocarbons, and other waste such as oily sludge, heavy metals, volatile organic compounds, ammonia, waste catalyst, high total dissolved, salts, nitrates, sulfides, oil, and grease content, etc. (Jasmine & Mukherji, 2015). Oily wastewater is divided into four major types of petroleum hydrocarbons, namely, aliphatic, aromatic, asphaltenes, and compounds containing oxygen, nitrogen, and sulfur (Honse, Ferreira, Mansur, Lucas, & González, 2012; Thakur et al., 2018; Varjani & Upasani, 2017).

These petroleum compounds consist of three main hydrocarbon groups, namely, paraffins (straight-chain n-alkanes from C1 to C40 and branched isoalkanes), naphthenes or cycloparaffins (naphthene rings contain typically 5 or 6 carbon atoms, more condensed dicyclonaphthenes C8 and C9 are present in addition to monocyclonaphthenes), and aromatics (having the benzene ring comprising alternate double and single bonds with adjacent carbon atoms, monocyclic, and polycyclic aromatics). In Naphthenic acids (NAs) (a mixture of alkyl-substituted acyclic, monocyclic, and polycyclic carboxylic acids with an aliphatic chain of 9 to 20 carbon atoms), which are known to cause toxic effects, are also difficult to remove from refinery wastewater (Wang, Ghimire, Xin, Janka, & Bakke, 2017). Other pollutants found in wastewater include hydrogen sulfide, ammonia, phenols, benzene, cyanides and suspended solids containing metals and inorganic compounds (e.g., halides, sulfates, phosphates, sulfides) (Barthe et al., 2015).

These pollutants are easily dispersed or dissolved in oily wastewater. In general, about 80% of petroleum hydrocarbons in oily wastewater are aromatic and aliphatic compounds. (Jasmine & Mukherji, 2015; Perera et al., 2012; Varjani & Chaithanya Sudha, 2018; Ward, Singh, & Van Hamme, 2003).

Every year, in the petroleum, transportation, and other high-consumption industries, agriculture, and urban management produce large volumes of hydrocarbon wastewater. This wastewater is very complex and varies depending on the raw materials and technologies used. They refers to hydrocarbon wastewater are the main source of organic pollution of the environment and contain toxic, mutagenic, and carcinogenic substances.

The metallurgical industry is another major source of oily wastewater. Most of the oily waste comes from metal-forming operations or metal-working, such as coke quenching, steel rolling, solvent extraction, and electroplating (Wu, Jiang, He, & Song, 2017). Oil-in-water and water-in-oil emulsions of different compositions, ranging from a trace of oil in water to a trace of water in oil, are used as cooling and lubricating agents. They also provide corrosion protection for machined parts and machining tools. Such spent oily water emulsion can be highly viscous and often severely hamper wastewater treatment plant capabilities, thus causing increased maintenance costs and energy consumption. Also, a distinctive feature of metallurgical process effluent is the use of large amounts of toxic organic matter in extraction and electro-deposition processes. These toxic substances are used as extractants, diluents, matrix modifiers, flocculants, brightening agents, acid fog inhibitors.

Emulsified oil, emulsifiers, degreasing agents, surfactants, solvents, suspended solids, metals, and acids/alkalines are the primary components of metallurgical oily wastewater (Wu et al., 2017). This wastewater, due to the paragenetic relation of these elements with rare earth metals, contains radioactive metals, such as thorium, radium, and uranium.

Advances in the transportation industry have led to a considerable increase in the use of motor fuels and oils that are the main sources of hydrocarbon pollution in wastewater. Furthermore, the car wash, engine wash, paint spraying workshop, or petrol station import a large number of hydrocarbon wastes to the urban wastewater system.

About 10−50 mg L^{-1} of typical domestic wastewater consists of oil and grease (O&G). In the raw water resources of European areas, more than 6000 organic compounds have been identified that have entered these resources due to human activities. While some of these are highly persistent, others are easily biodegradable. These organic compounds include aliphatic and aromatic hydrocarbons, polyaromatic hydrocarbons (PAHs), fatty acids, ketones, phthalate esters, plasticizers, and other polar compounds. Solvent extractable organics are dominated by petroleum hydrocarbons, which arise from motor oil leaks, degraded asphalt, and worn tires from the roads. Municipal wastewater contains large amounts of toxic and nondegradable organic pollutants such as PAHs, polychlorinated biphenyls (PCBs), di-(2-ethylhexyl) phthalate (DEHP), liner alkyl benzene sulphonates (LASs), nonylphenol ethoxylates (NPEs), dioxins (PCDD), and furans (PCDF). Also, advances in technology have led to the emergence of micropollutants such as pharmaceuticals, personal care compounds, flame-retardants, biocides, and pesticides in municipal wastewater (Visser & Aloisi de Larderel, 1997).

Petroleum wastewater characteristics

Wastewater produced in refineries includes soluble and insoluble pollutants. The types of pollutants in petroleum wastewater can be classified as follows (Visser & Aloisi de Larderel, 1997):

- Total hydrocarbon content (THC);
- Total petroleum hydrocarbon index (TPH-index);
- Biochemical oxygen demand (BOD);
- Chemical oxygen demand (COD);
- Total organic carbon (TOC);
- Ammoniacal nitrogen;
- Total nitrogen;
- Total suspended solids (TSS);
- Total metals;
- Cyanides;
- Fluorides phenols;
- Phosphates;
- Special metals such as Cd, Ni, Hg, Pb, and vanadium;
- Benzene, Toluene, ethylbenzene, and xylene (BTEX);
- PH (acids, alkalis);
- Sulfides;
- Other micropollutants.

The main sources of wastewater from petroleum industry activities are produced water, cutting, cooling water, drilling fluids, well treatment chemicals, process, drainage and wash, spill and leakage, domestic wastewater, and sewage. The volumes of wastewater are minimal during seismic operations and often relate to camp and vessel activities.

The primary discharge from production operations is produced water and the major wastewater effluents from exploratory drilling are drilling fluids and cuttings. The volumes of produced water depend on the type of production (gas and oil), geographical, the lifetime of a field, field, and the level of activity (Visser & Aloisi de Larderel, 1997; Congress & Office of Technology Assessment, 1992). The characteristics of effluent from different sections of the petroleum industry process are summarized in Table 2.1.

Petroleum industry wastewater contains many aromatic organic compounds such as PAHs and phenolic substances. These substances decompose in nature with difficulty. Therefore, these are a serious threat to the environment. Organic pollutants in industrial wastewater have carcinogenic or mutagenic properties and the US Environmental Protection Agency (EPA) has considered these compounds among the priority pollutants.

These contaminants accumulate in human and animal tissues. Some of these compounds are soluble in water and have been detected in concentrations ranging from a few milligrams per liter to a maximum of 7000 mg L^{-1} in petroleum wastewater (Waseem et al., 2019).

TABLE 2.1 The characteristics of effluent from different sections of the petroleum industry process.

Main sources	Environmentally significant components
Produced water, Process water, for example, engine cooling water, brake cooling water, wash water, Ballast water, Hydro-test fluids, Contaminated rain/drainage water, Drilling fluid, chemicals Spent stimulation or fracturing fluids, Spent completion fluids, Waste lubricants, Water-based (include brine), muds and cuttings, Oil-based muds and cuttings, Mercury Dehydration and sweetening wastes, Domestic sewage	Hydrocarbons, inorganic salts, heavy metals, solids, organics, sulfides, corrosion inhibitors, biocides, phenols, BOD, benzene, organo-halogens, PAHs, radioactive material Inorganic salts, heavy metals, solids, organics, BOD, sulfides, corrosion inhibitors, biocides, demulsifiers, wax inhibitors, detergents, hydrocarbons Hydrocarbons, phenols, PAHs, Solids, corrosion inhibitors, biocides, BOD, dyes, oxygen scavengers Inorganic salts, Heavy metals solids, organics, BOD, sulfides, corrosion inhibitors, scale inhibitors, detergents, hydrocarbons Metals, salts, organics, pH, surfactants, biocides, emulsifiers, viscosifiers Inorganic acids (HCL, HF), hydrocarbons, methanol, corrosion inhibitors, oxygen scavengers, formation fluids, naturally occurring radioactive materials (NORM), geling agents Hydrocarbons, corrosion inhibitors, inorganic salts Organics, heavy metals High pH, inorganic salts, hydrocarbons, solids/cutting, drilling fluid chemicals, heavy metals Hydrocarbons, solids/cutting, heavy metals, inorganic salts, drilling fluid chemical Mercury Amines, glycols, filter sludges, metal sulfides, H_2S, metals, benzene BOD, solids, detergents, coliform bacteria

From Praveen, P., & Loh, K. C. (2013). Simultaneous extraction and biodegradation of phenol in a hollow fiber supported liquid membrane bioreactor. *Journal of Membrane Science, 430*, 242−251. https://doi.org/10.1016/j.memsci.2012.12.021.
Diya'uddeen, B.H., Daud, W.M.A.W., & Abdul Aziz, A.R. (2011).Treatment technologies for petroleum refinery effluents: A review. *Process Safety and Environmental Protection.* 89(2):95−105.
Raza, W., Lee, J., Raza, N., Luo, Y., Kim, K.H., & Yang. J. (2018). Removal of phenolic compounds from industrial waste water based on membrane-based technologies. Journal of Industrial and Engineering Chemistry (In Press. https://doi.org/10.1016/j.jiec.2018.11.024).

Under favorable conditions and through various reactions like methylation and chlorination, these compounds can produce toxic and dangerous substances such as cresols and chlorophenols (Jasmine & Mukherji, 2015; Perera et al., 2012; Varjani & Chaithanya Sudha, 2018; Ward et al., 2003).

Thus, global concerns about the direct discharge of petroleum wastewater without removing or reducing toxic pollutants into the environment are growing. For this reason, industries are required to treat their wastewater before discharging it into the ecosystem (Jasmine & Mukherji, 2015; Perera et al., 2012; Varjani & Chaithanya Sudha, 2018; Ward et al., 2003).

Wastewater from the petroleum industry such as oil refineries, petrochemical products, and transportation have a large number of toxic substances (Viggi et al., 2015). These wastewaters contain various types of organic and inorganic pollutants such as phenol, sulfide, BTEX, hydrocarbon, and heavy metals (He & Jiang, 2008; Usman, Faure, Hanna, Abdelmoula, & Ruby, 2012; Varjani & Upasani, 2017; Waseem et al., 2019).

Research has shown that most of this wastewater is composed of aliphatic hydrocarbons and aromatic compounds such as benzene, toluene, and ethylbenzene. Table 2.2 shows the different types of organic compounds in petroleum wastewater.

Petroleum wastewater contains complex compounds of organic pollutants and often contains oils and greases, which clog pipes and cause corrosion and unpleasant odors (Xu & Zhu, 2004). Phenolic pollutants are also a serious threat to the environment due to their high persistence in the environment (Abdelwahab, Amin, & El-Ashtoukhy, 2009; Kavitha & Palanivelu, 2004; Lathasree, Rao, Sivasankar, Sadasivam, & Rengaraj, 2004; Pardeshi & Patil, 2008; Yang, 2008).

NAs in petroleum wastewater cause toxic effects and form a large volume of wastewater, and their removal from petrochemical effluents is a critical challenge (Wang et al., 2017). showed that the percentage of aromatic NAs was about 2.1%−8.8% in a refinery wastewater treatment plant.

The following general results can be expressed from reports made by various researchers as shown in Table 2.2.

Based on the studies, a wide range of pollutants was identified and a large difference was seen in the characteristics of wastewater due to differences in wastewater source, crude oil quality, and operating conditions.

Petroleum wastewater compounds are mainly soluble in the form of organic and inorganic compounds, with organic compounds including hydrocarbons (PAH) and inorganic compounds containing heavy metals.

Sour water stream (SWS) has a high concentration of sulfide and complex chemical compositions such as oil, phenols, sulfides, mercaptans, ammonia, cyanides, and other micropollutants (El-Naas, Alhaija, & Al-Zuhair, 2014).

TABLE 2.2 Different types of organic compounds in petroleum wastewater reported by various researchers.

Parameters references	pH	COD (mg L^{-1})	BOD (mg L^{-1})	TSS (mg L^{-1})	Phenols (mg L^{-1})	Turbidity (UTN)	Oil & grease (mg L^{-1})	TDS (mg L^{-1})
Vendramel, Bassin, Dezotti, & Sant' Anna (2015)	8.3	1250	–	150	–	–	–	–
Aljuboury, Palaniandy, Aziz, & Feroz (2014)	6.5–9.5	550–1600	–	–	–	–	–	1200–1500
Saber, Hasheminejad, Taebi, & Ghaffari (2014)	6.7	450	174	150	–	–	870	–
Gasim, Kutty, Hasnain-Isa, & Alemu (2013)	8.48	7896	3378	–	–	–	–	–
Tony, Purcell, & Zhao (2012)	7.6	364	–	105	–	42	946	–
Hasan, Abdul Aziz, & Daud (2012)	7	1343	846	74	–	83	240	–
Farajnezhad and Gharbani (2012)	7.5	1120	–	110	–	–	–	–
Abdelwahab, Amin, & El-Ashtoukhy (2009)	8	80–120	40.25	22.8	13	–	–	–
El-Naas, Al-Zuhair, & Alhaija (2010)	9.5	4050	–	80	–	–	–	–
Altas Buyukgungor (2008)	7.19–9.22	220	–	–	–	–	–	–
Dincer, Karakaya, Gunes, & Gunes (2008)	2.5	21000	8000	2580	–	–	1140	37000
Zeng, Yang, Zhang, & Pu (2007)	6.5–6.8	500–1000	–	90–300	10–20	150–350	400–1000	3000–5000
Demirci, Erdogan, & Ozcimder (1998)	6.5–8.5	800	350	100	8	–	3000	–

From El-Naas, M. H., Alhaija, M. A., & Al-Zuhair, S. (2014). Evaluation of a three-step process for the treatment of petroleum refinery wastewater. *Journal of Environmental Chemical Engineering, 2*(1), 56–62. https://doi.org/10.1016/j.jece.2013.11.024.
Farajnezhad, H., & Gharbani, P. (2012). Coagulation treatment of wastewater in petroleum industry using poly aluminum chloride and ferric chloride. *International Journal of Research and Reviews in Applied Sciences, 13*, 306–310.
Altas, L., & Buyukgungor, H. (2008). Sulfide removal in petroleum refinery wastewater by chemical precipitation. *Journal of Hazardous Materials, 153*, 462–469.

TABLE 2.3 Minimum standard discharge limits for refinery effluents.

pH	Composition (mg L^{-1})								References
	COD	BOD	TDS	SS	TOC	Ammonia	Phenols	Sulfides	
6–9	150	30	–	30	–	–	–	1	Environmental Health Safety Guidelines

From IFC: Environmental, Health and Safety (EHS) Guidelines, General EHS Guidelines, April 30 2009.

Various environmental protection agencies have set a maximum discharge limit to prevent the entry of large amounts of hazardous petroleum compounds into the environment as described in Table 2.3. Also, most of the hydrocarbons of undegraded petroleum hydrocarbons are fuel additives such as dichloroethane (DCE), dichloromethane (DCM), and t-butyl methyl ether (tBME), which are carcinogenic.

The methods for measuring the main compounds in petroleum wastewater

PH measurements

The pH has a great effect on wastewater treatment. It affects chlorination and coagulation efficiency. The pH meter includes a glass electrode, a potentiometer, a reference electrode, and a temperature compensating device. Modern pH meters use a single composite electrode to measure and buffer solutions for calibration with pH 4, 7, and 10. The pH meter can detect these setpoints automatically in calibration mode. Measurements should be taken immediately after collection, but if samples must be stored, they should be kept at 4°C and should be measured no later than 6 h after collection. If pH values vary widely, standardize each sample with a buffer solution having a pH within 1–2 pH units of the sample (American Society for Testing & Materials ASTM, 2018).

Total residual chlorine

The chlorination of the treated effluent serves primarily to destroy or deactivate disease-causing microorganisms. Chlorination may produce adverse effects. Potentially carcinogenic chloro-organic compounds such as chloroform may be formed. To fulfill the primary purpose of chlorination and minimize any adverse effects, proper testing procedures must be used. Several methods for measurement of total residual chlorine are available, including iodometric methods, amperometric titration methods, and N,N-diethyl-phenylenediamine (DPD) methods. The following operating procedure will discuss a DPD Colorimetric Method. First, select the volume of the sample that will need not more than 20 mL 0.01 N $Na_2S_2O_3$ and not less than 0.2 mL are needed for the starch iodide until the end of the reaction. Use a 500 mL sample for a chlorine range of 1 to 10 mg L^{-1}, if the amount of chlorine is above 10 mg L^{-1}, use proportionately less sample. Then reach pH by adding acetic acid to about 3–4 in a flask. Add about 1 g KI estimated on a spatula. Add the sample to it and mix it with a stirrer and perform the titration process away from direct sunlight. And then from a burette, add 0.025 N or 0.01 N $Na_2S_2O_3$ until the yellow color of the liberated iodine is almost discharged. Add 1 mL starch solution and titrate until the disappearance of the blue color (American Society for Testing & Materials ASTM, 2018).

Total suspended solids

TSS of a wastewater sample is determined by pouring a carefully measured volume of wastewater through a pre-weighed filter of specified pore size, then weighing the filter again after the drying process that removes all water on the filter. Filters for TSS measurements are typically composed of glass fibers. The gain in weight is a dry weight measure of the particulates present in the water sample expressed in units derived or calculated from the volume of water filtered (typically milligrams per liter or mg L^{-1}).

If the wastewater contains an appreciable amount of dissolved substances (as certainly would be the case when measuring TSS in seawater), these will add to the weight of the filter as it is dried. Therefore, it is necessary to "wash" the filter and sample with deionized water after filtering the sample and before drying the filter. Failure to add this step is a fairly common mistake made by inexperienced laboratory technicians working with seawater samples, and it will completely invalidate the results as the weight of salts left on the filter during drying can easily exceed that of the suspended particulate matter.

Total suspended solids (TSSs) are measured in laboratories by filtering a known volume of a sample, drying the filter and captured solids, then weighing the filter to determine the weight of the captured suspended solids in the sample. TSS is calculated as follows:

$$TSS \ (mg \ L^{-1}) = (W_{fss} - W_f)/V_s$$

where W_{fss} is the weight of filter with suspended solids, W_f is the weight of the filter, and V_s is the volume of sample.

The entire process takes about 2 h or more and does not lend itself to instantaneous, continuous measurements. See ASTM D5907, EPA Method 10.2, Standard Methods 2540D or similar gravimetric method for details of the lab method ("2540 SOLIDS", Standard Methods for the Examination of Water and Wastewater, n.d; American Society for Testing & Materials ASTM, 2018). TSS samples can be collected in plastic or glass containers. Sample analysis should be performed immediately after sampling; otherwise, it should be stored on ice or in the refrigerator to reduce the activity of microorganisms, in which case the analysis should be performed for a maximum of 7 days.

Oil and grease

The high volume of oil and grease generated during the extraction of oil and gas is a serious challenge in the petroleum industry. The chemical nature (toxicity and solubility) of the contaminants in the wastewater is another challenge. In petroleum wastewater, oil can be in three forms: free, dispersed, emulsion, or soluble (Demirci, Erdogan, & Ozcimder, 1998; Zeng, Yang, Zhang, & Pu, 2007). In petroleum wastewater, total oil and grease (TOG) have different concentrations (100−2000 ppm). In monitoring oil pollutants in petroleum wastewater, two issues of measuring their concentration and analyzing the chemical transformations of petroleum hydrocarbons in the environment should be considered. In general, the TOG in petroleum wastewater can be measured by gravimetry and infrared spectroscopy (Karyakin & Galkin, 1995; National Environmental Council (CONAMA), 1986; Vorontsov, Nikanorova, & Dorokhov, 1998). Gravimetric methods, such as Method 413.1 and Standard Method 5520B from the US EPA, involve liquid−liquid extraction (LLE) from oily water samples using 1,1,2-trichloro1,2,2-trifluoroethane (CFC-113), a compound that severely depletes ozone in the atmosphere (Cirne, Boaventura, Guedes, & Lucas, 2016).

A new gravimetric method, Method 1664, has been developed by the EPA specifically to measure oil and grease in oil wastewater. In this method, hexane is used as an alternative extraction solvent. This method is less expensive and less harmful to the environment than the LLE for determining oil and grease in wastewater samples. In this method, the use of different concentrations of solvent and extraction methods such as solid-phase extraction (SPE) is allowed. In the SPE method, faster analysis is done, even though other steps are necessary to prevent the clogging of SPE cartridges when used on aqueous samples containing high levels of particles. Therefore, the successful application of Method 1664 depends on satisfying many quality control parameters, which are hard to achieve (Cirne et al., 2016).

The method for measurement of TOG by infrared spectrometry typically uses a Horiba OCMA-350 analyzer. In this method, hydrocarbons in wastewater are used to measure contamination (Gonzalez, Teixeira, & Lucas, 2004). This instrument operates on the technique of energy absorption in the infrared spectrum, with a wavelength range from 3.4 to 3.5 μm. Because in this device, the amount of energy absorbed is related to the concentration of hydrocarbons present in the wastewater samples, so it only measures C−H bonds. Because water can absorb energy in the range of 3.4−3.5 μm, we must use a water-insoluble solvent to extract hydrocarbons and also does not absorb energy wavelengths in this range.

Researchers have suggested the S-316 solvent (an oligomer based on poly (trifluoroethylene)), which has the disadvantage of being very expensive. Of course, carbon tetrachloride can be used instead of S-316, which is cheaper but very toxic. Also, to achieve reliable results, the device must be calibrated. Therefore, a calibration solution should be prepared according to the standard formula of the manufacturer and the calibration range should be defined (0−50 or 0−200 $mg \ L^{-1}$). Therefore, it is important to evaluate other methods to determine TOG concentration (Cirne et al., 2016).

Crude oil contains a wide range of contaminants, so there are several other methods for measuring TOG in petroleum wastewater, including colorimetry, spectrofluorimetry, gas chromatography (GC) combined with mass spectrometry, and high-performance liquid chromatography (HPLC). Each of these methods can detect all or some of the contaminants in petroleum wastewater (Minty, Ramsey, & Davies, 2000).

GC is an analytical technique used to separate the chemical components of a sample mixture and then detect them to determine their presence or absence and/or how much is present. These chemical components are usually organic molecules or gases. The separation is based on the differences in the speed with which these components are carried in the fixed phase by the flow of a mobile liquid or gas phase. The stationary phase in GC is the part of the system where the mobile phase will flow and distribute the solutes between the phases. The stationary phase plays a vital role in determining the selectivity and retention of solutes in a mixture. To monitor the outflow from the column, a detector can be used. When each chemical compound leaves the column, it is detected and recognized electronically, and the quantity can also be determined. Generally, the substances are identified (qualitatively) by the order in which they emerge from the column and by the retention time of the column (Cirne et al., 2016).

In colorimetry or visible spectrophotometry, a solution is used to measure the absorption of electromagnetic radiation in the ultraviolet (UV)-visible region. Based on this, useful information can be obtained regarding the concentration of chemical species in the effluent sample.

The spectrophotometer is an instrument that measures the amount of light that a sample absorbs. The spectrophotometer works by passing a light beam through a sample to measure the light intensity of a sample. Photometers employ a filter, together with a transducer, to select the wavelength with suitable radiation. Spectrophotometers have a significant advantage because the wavelength can be changed continuously, making it possible to record an absorption spectrum. In addition, photometers are very simple, robust, and cheap. The spectrophotometers work in the UV-visible region, whereas photometers are used in the visible region. Photometers are widely used as detectors for electrophoresis, chromatography, immunoassays, or continuous flow analysis (Aversa, Queiros, & Lucas, 2014; Queiros, Clarisse, & Oliveira, 2006).

Nitrate

The highest oxidation state of nitrogen is nitrate. It may occur in wastewater due to the oxidation of organic nitrogen to ammonia and then to nitrate and then to nitrate. Nitrate at levels above 105 mg L^{-1} as N can cause an illness. Nitrogen is also one of the fertilizing essential elements of the growth of algae. Among the existing technologies for nitrate removal, physicochemical denitrification, biological and chemical reductions have been noticeable. Samples with a volume of at least 100 mL, are collected in glass or plastic bottles. Samples should be analyzed as soon as possible after collection (within 1 h), but maybe stored at 4°C for up to 24 h. For storage, up to 7 days, preserve the samples with 2 mL concentrated sulfuric acid per liter and store them at 4°C (Perera et al., 2012).

Biochemical oxygen demand

Determination of BOD is extremely difficult because there are so many variable factors that affect the outcome. The BOD distilled water must be aerated and noncontaminated and the pH environment must be acceptable (6.5−7.5). In the BOD test, several important factors such as temperature, the activity of microorganisms, and the concentration of waste material should be considered. Microorganisms must be alive, incubation temperature must be constant and controlled, and waste concentration must be acceptable. Wastewater samples must be free of toxic substances. The strength of the effluent sample should be such as to reduce at least 2 mg L^{-1} of dissolved oxygen, but at least 1 mg L^{-1} should remain in the sample. As a result, the BOD test must provide the ideal conditions for the growth of microorganisms so that microorganisms can effectively consume digestible organic matter. In the BOD test, microorganisms consume all organic matter. In a BOD bottle, the organic matter in the sample is added to dilute water containing nutrients, oxygen, and microorganisms, then the sample is kept at 20°C for 5 days. Initially, the number of microorganisms is small, but due to the excellent environmental growth conditions, their population increases rapidly and they enter the logarithmic reproductive stage and consume all organic matter. Due to the need for microorganisms to consume oxygen, the amount of oxygen consumption also increases. If the BOD bottle is closed tightly, then the microorganisms use only the oxygen inside the bottle. Therefore, within a few days, the growth of microorganisms enters the logarithmic stage and with the digestion of organic matter, a large amount of oxygen is consumed. On the fourth day of the BOD test, the microorganisms enter a stable growth phase because the amount of organic matter is limited, and they can hardly maintain the population of microorganisms. By reducing the amount of organic matter, the population of microorganisms decreased, so the rate of oxygen consumption also decreases. At this stage, by reducing the organic matter, microorganisms enter the endogenous stage. They use their internal reserves and their population is declined. The BOD test expresses the total amount of organic matter consumed in 5 days. Complete stabilization of a wastewater sample requires 20 days since most of the organic matter is consumed in the first 5 days, so the latter is defined as the standard incubation period (Dincer et al., 2008; Perera et al., 2012).

Chemical oxygen demand

COD is an important wastewater quality parameter because similar to BOD, it provides an index to assess the effect discharged wastewater will have on the receiving environment. Higher COD levels mean a greater amount of oxidizable organic material in the sample, which will reduce dissolved oxygen (DO) levels. A reduction in DO can lead to anaerobic conditions, which is deleterious to higher aquatic life forms. COD is a test that measures the amount of oxygen required to chemically oxidize the organic material and inorganic nutrients, such as ammonia or nitrate in the wastewater. It is often used as an alternative to BOD due to the shorter length of testing time. The most important disadvantages of oxygen testing are the presence of hazardous and toxic chemicals. COD testing is often used to measure contaminants in wastewater and water samples. In municipal wastewater, the BOD content of the sample represents 50%−60% of the COD content. The presence of toxic substances or industrial wastewater can change this ratio. Similar to the BOD test, this test uses oxygen to oxidize organic matter into water and carbon dioxide, but the COD test uses chemically bound oxygen in the potassium dichromate compound. By using dichromate, Cr^{3+} ions are produced. The amount of Cr^{3+} ions produced depends on the amount of organic matter in the wastewater sample. The traditional COD analysis method is the wet chemistry method. This involves 2-h digestion at high heat under acidic conditions in which potassium dichromate acts as the oxidant for any organic material present in a wastewater sample. Silver sulfate is present as the catalyst and mercuric sulfate acts to complex out any interfering chloride. Following the digestion, the extent of oxidation is measured through indirect measurement of oxygen demand via electrons consumed in the reduction of Cr^{6+} to Cr^{3+}. This can be done by titration or spectrophotometry. Collect the samples in glass bottles, if possible. The use of plastic containers is permissible if it is known that no organic contaminants are present in the containers. Biologically active samples should be tested as soon as possible. Samples containing settleable material should be well mixed, preferably homogenized, to permit the removal of representative aliquots. Samples should be preserved with sulfuric acid to a pH < 2 and maintained at 4°C until analysis (Dincer et al., 2008).

Polyaromatic hydrocarbon determination of PAHs in the liquid phase is performed using three common extraction methods, such as liquid−liquid extraction, solid-phase microextraction, and solid-phase extraction. In recent years, methods such as Headspace solvent microextraction, dispersive liquid−liquid microextraction, cloud point extraction, have been used.

Various solvents have been used in various studies for the extraction of liquid−liquid polynuclear aromatic hydrocarbons from petroleum wastewater. These solvents include hexane, dichloromethane, carbon tetrachloride, and cyclohexane. According to recent studies, hexane has been identified as the most suitable solvent (Tony, Purcell, & Zhao, 2012). Hexane, a highly volatile substance, has low solubility in water (13 mg L_{21} at 20C) and a density of 0.6548 g mL_{21}, and it is compatible with the technique of analysis.

The analytical techniques based on liquid or gas chromatography are used to determine and detect the amount of PAHs. PAHs are much more suitable for liquid chromatography because they have less solubility in water and their volatility decreases with increasing molecular weight (Gocan & Cobzac, 2006).

Total organic carbon

There are several methods to determine TOC. The two main components required to measure TOC are to convert the organic carbon into carbon dioxide (CO_2) and the means to detect CO_2. The three primary oxidation methods most commonly used are chemical agents, high-temperature combustion, and photocatalytic. All three methods have provided acceptable results.

The chemical agent method is known as the persulfate or nondispersive infrared (NDIR) sensor process, which can be accomplished through either persulfate oxidation with UV light and irradiation activation, or through the alternative process, which is known as heated persulfate oxidation.

The second method to determine TOC is known as the high-temperature combustion (catalytic) process. This process measures TOC by heating a sample in a high-temperature furnace with a cobalt or platinum catalyst.

The third method to determine TOC is photocatalytic or UV light. The UV light process oxidizes the carbon into CO_2 gas. There are multiple manufacturers of TOC analyzers using all three of these measurement methods. All the products have their advantages and disadvantages depending on the specific application, location of the instrument, etc.

A TOC analyzer, for example, the one shown before, utilizes the UV persulfate oxidation method, which detects generated CO_2 using its NDIR detector for analysis. This method and the analyzer conform to standards set by US EPA, DIN, CE, ASTM, and NAMUR regulations, as well as ISO.

To use this type of TOC analyzer, the water sample first is acidified and then sparged to remove inorganic carbon. The remaining liquid is mixed with sodium persulfate and digested by two high-performance reactors. The resulting

CO_2 is then stripped from the liquid and, after drying, its concentration is measured by the NDIR analyzer. The analyzer measures TOC ranging from $0-5$ to 20,000 mg L^{-1} (Aljuboury, Palaniandy, Aziz, & Feroz, 2014).

References

"2540 SOLIDS", Standard Methods for the Examination of Water and Wastewater. (n.d). Available from https://doi.org/10.2105/SMWW.2882.030.

Abdelwahab, O., Amin, N. K., & El-Ashtoukhy, E. S. Z. (2009). Electrochemical removal of phenol from oil refinery wastewater. *Journal of Hazardous Materials*, *163*(2−3), 711−716. Available from https://doi.org/10.1016/j.jhazmat.2008.07.016.

Abdulredha, M. M., Siti Aslina, H., & Luqman, C. A. (2020). Overview on petroleum emulsions, formation, influence and demulsification treatment techniques. *Arabian Journal of Chemistry*, *13*(1), 3403−3428. Available from https://doi.org/10.1016/j.arabjc.2018.11.014.

Al-Futaisi, A., Jamrah, A., Yaghi, B., & Taha, R. (2007). Assessment of alternative management techniques of tank bottom petroleum sludge in Oman. *Journal of Hazardous Materials*, *141*(3), 557−564. Available from https://doi.org/10.1016/j.jhazmat.2006.07.023.

Altas, L., & Buyukgungor, H. (2008), Sulfide removal in petroleum refinery wastewater by chemical precipitation. *Journal of Hazardous Materials*, *153*, 462−469.

Aljuboury, D. D. A., Palaniandy, P., Aziz, A., & Feroz, H. B. (2014). Organic pollutants removal from petroleum refinery wastewater with nanotitania photo-catalyst and solar irradiation in Sohar oil refinery. *Journal of Innovative Engineering*, *2*(3), 1−12.

American Society for Testing and Materials (ASTM), D5907-18, Standard Test Methods for Filterable Matter (Total Dissolved Solids) and Nonfilterable Matter (Total Suspended Solids) in Water, ASTM International, West Conshohocken, PA, 2018, www.astm.org.

Aversa, T., Queiros, Y., & Lucas, E. (2014). Synthesis and sulfonation of macroporous polymeric resins and their assessment for removal of oil and aniline from. *Water.Polimeros*, *24*, 45−51. Available from http://dx.doi.org/10.4322/polimeros.2013.048.

Bakke, T., Klungsøyr, J., & Sanni, S. (2013). Environmental impacts of produced water and drilling waste discharges from the Norwegian offshore petroleum industry. *Marine Environmental Research*, *92*, 154−169. Available from https://doi.org/10.1016/j.marenvres.2013.09.012.

Barthe, P., Chaugny, M., Roudier, S., & Sancho, L. D. (2015). *EU best available techniques (BAT): Reference document for the refining of mineral oil and gas.* Luxembourg.: Publications Office of the European Union.

Chen, Y. C. (2018). Evaluating greenhouse gas emissions and energy recovery from municipal and industrial solid waste using waste-to-energy technology. *Journal of Cleaner Production*, *192*, 262−269. Available from https://doi.org/10.1016/j.jclepro.2018.04.260.

Cirne, I., Boaventura, J., Guedes, Y., & Lucas, E. (2016). Methods for determination of oil and grease contents in wastewater from the petroleum industry. *Chemistry and Chemical Technology*, *10*(4), 437−444. Available from https://doi.org/10.23939/chcht10.04.437.

Demirci, S., Erdogan, B., & Ozcimder, R. (1998). Wastewater treatment at the petroleum refinery, Kirikkale, Turkey using some coagulants and Turkish clays as coagulant aids. *Water Research*, *32*(11), 3495−3499. Available from https://doi.org/10.1016/S0043-1354(98)00111-0.

Dincer, A. R., Karakaya, N., Gunes, E., & Gunes, Y. (2008). Removal of COD from oil recovery industry wastewater by the advanced oxidation processes (AOP) based on H2O2. *Global Nest Journal*, *10*(1), 31−38. Available from http://ast.gnest.org/Journal/Vol10_No1/31-38_479_DINCER_10-1.pdf.

Diya'Uddeen, B. H., Daud, W. M. A. W., & Abdul Aziz, A. R. (2011). Treatment technologies for petroleum refinery effluents: A review. *Process Safety and Environmental Protection*, *89*(2), 95−105. Available from https://doi.org/10.1016/j.psep.2010.11.003.

El-Naas, M. H., Al-Zuhair, S., & Alhaija, M. A. (2010). Reduction of COD in refinery wastewater through adsorption on Date-Pit activated carbon. *Journal of Hazardous Materials*, *173*(1−3), 750−757.

El-Naas, M. H., Alhaija, M. A., & Al-Zuhair, S. (2014). Evaluation of a three-step process for the treatment of petroleum refinery wastewater. *Journal of Environmental Chemical Engineering*, *2*(1), 56−62. Available from https://doi.org/10.1016/j.jece.2013.11.024.

Farajnezhad, H., & Gharbani, P. (2012). Coagulation treatment of wastewater in petroleum industry using poly aluminum chloride and ferric chloride. *International Journal of Research and Reviews in Applied Sciences, 13*, 306−310.

Gasim, H. A., Kutty, S. R. M., Hasnain-Isa, M., & Alemu, L. T. (2013). Optimization of anaerobic treatment of petroleum refinery wastewater using artificial neural networks. *Research Journal of Applied Sciences, Engineering and Technology*, *6*(11), 2077−2082.

Gocan, S., & Cobzac, S. (2006). *Modern method of organic samples' processing.* Cluj-Napoca, Romania: Risoprint.

Gonzalez, G., Teixeira, C., & Lucas, E. (2004). The use of polymers in the treatment of produced oily waters. *Journal of Applied Polymeric Science*, *94*, 1473−1479.

Hasan, D. U. B., Abdul Aziz, A. R., & Daud, W. M. A. W. (2012). Oxidative mineralisation of petroleum refinery effluent using Fenton-like process. *Chemical Engineering Research and Design*, *90*(2), 298−307.

He, Y., & Jiang, Z. W. (2008). Technology review: Treating oilfield wastewater. *Filtration and Separation*, *45*(5), 14−16. Available from https://doi.org/10.1016/S0015-1882(08)70174-5.

Honse, S. O., Ferreira, S. R., Mansur, C. R. E., Lucas, E. F., & González, G. (2012). Separation and characterization of asphaltenic subfractions. *Quimica Nova*, *35*(10), 1991−1994. Available from https://doi.org/10.1590/S0100-40422012001000019.

Hu, G., Li, J., & Zeng, G. (2013). Recent development in the treatment of oily sludge from petroleum industry: A review. *Journal of Hazardous Materials*, *261*, 470−490. Available from https://doi.org/10.1016/j.jhazmat.2013.07.069.

Jafarinejad, S. (2016). Control and treatment of sulfur compounds specially sulfur oxides (SOx) emissions from the petroleum industry: A review. *Chem International*, *2*(4), 242−253.

Jasmine, J., & Mukherji, S. (2015). Characterization of oily sludge from a refinery and biodegradability assessment using various hydrocarbon degrading strains and reconstituted consortia. *Journal of Environmental Management*, *149*, 118−125. Available from https://doi.org/10.1016/j.jenvman.2014.10.007.

Karyakin, A. V., & Galkin, A. V. (1995). Fluorescence of water-soluble components of oils and petroleum products forming the oil pollution of waters. *Journal of Analytical Chemistry (Transl. of Zh. Anal. Khim.)*, *50*, 1078−1080.

Kavitha, V., & Palanivelu, K. (2004). The role of ferrous ion in Fenton and photo-Fenton processes for the degradation of phenol. *Chemosphere*, *55* (9), 1235−1243. Available from https://doi.org/10.1016/j.chemosphere.2003.12.022.

Lathasree, S., Rao, A. N., Sivasankar, B., Sadasivam, V., & Rengaraj, K. (2004). Heterogeneous photocatalytic mineralisation of phenols in aqueous solutions. *Journal of Molecular Catalysis A: Chemical*, *223*(1−2), 101−105. Available from https://doi.org/10.1016/j.molcata.2003.08.032.

Li, W. W., & Yu, H. Q. (2011). From wastewater to bioenergy and biochemicals via two-stage bioconversion processes: A future paradigm. *Biotechnology Advances*, *29*(6), 972−982. Available from https://doi.org/10.1016/j.biotechadv.2011.08.012.

Li, X., Cao, X., Wu, G., Temple, T., Coulon, F., & Sui, H. (2014). Ozonation of diesel-fuel contaminated sand and the implications for remediation end-points. *Chemosphere*, *109*, 71−76. Available from https://doi.org/10.1016/j.chemosphere.2014.03.005.

Minty, B., Ramsey, E., & Davies, I. (2000). Development of an automated method for determining oil in water by direct aqueous supercritical fluid extraction coupled on-line with infrared spectroscopy. *Analyst*, *125*, 2356−2363.

National Environmental Council (CONAMA). (1986). *(Conselho Nacional do Meio Ambiente)—Resolution number, Vol. 20*.

Ottaviano, J. G., Cai, J., & Murphy, R. S. (2014). Assessing the decontamination efficiency of a three-component flocculating system in the treatment of oilfield-produced water. *Water Research*, *52*, 122−130. Available from https://doi.org/10.1016/j.watres.2014.01.004.

Pardeshi, S. K., & Patil, A. B. (2008). A simple route for photocatalytic degradation of phenol in aqueous zinc oxide suspension using solar energy. *Solar Energy*, *82*(8), 700−705. Available from https://doi.org/10.1016/j.solener.2008.02.007.

Perera, F. P., Tang, D., Wang, S., Vishnevetsky, J., Zhang, B., Diaz, D., . . . Rauh, V. (2012). Prenatal polycyclic aromatic hydrocarbon (PAH) exposure and child behavior at age 6−7 years. *Environmental Health Perspectives*, *120*(6), 921−926. Available from https://doi.org/10.1289/ehp.1104315.

Queiros, Y., Clarisse, M., Oliveira, R., et al. (2006). Oily water treatment using polymeric material: Use, saturation and regeneration. *Polimeros*, *16*, 224−229.

Raza, W., Lee, J., Raza, N., Luo, Y., Kim, K.H., & Yang. J. (2018). Removal of phenolic compounds from industrial waste water based on membrane-based technologies. Journal of Industrial and Engineering Chemistry. (In Press. https://doi.org/10.1016/j.jiec.2018.11.024).

Saber, A., Hasheminejad, H., Taebi, A., & Ghaffari, G. (2014). Optimization of Fenton-based treatment of petroleum refinery wastewater with scrap iron using response surface methodology. *Applied Water Science*, *4*, 283−290.

Thakur, C., Srivastava, V. C., Mall, I. D., & Hiwarkar, A. D. (2018). Mechanistic study and multi-response optimization of the electrochemical treatment of petroleum refinery wastewater. *Clean - Soil, Air, Water*, *46*(3). Available from https://doi.org/10.1002/clen.201700624.

Tony, M. A., Purcell, P. J., & Zhao, Y. (2012). Oil refinery wastewater treatment using physicochemical, Fenton and Photo-Fenton oxidation processes. *Journal of Environmental Science and Health - Part A Toxic/Hazardous Substances and Environmental Engineering*, *47*(3), 435−440. Available from https://doi.org/10.1080/10934529.2012.646136.

U.S. Congress, Office of Technology Assessment, (1992). Managing Industrial Solid Wastes from Manufacturing, Mining, Oil and Gas Production, and Utility Coal Combustion-Background Paper, OTA-BP-O-82 (Washington, DC: U.S. Government Printing Office, February 1992).

Usman, M., Faure, P., Hanna, K., Abdelmoula, M., & Ruby, C. (2012). Application of magnetite catalyzed chemical oxidation (Fenton-like and persulfate) for the remediation of oil hydrocarbon contamination. *Fuel*, *96*, 270−276. Available from https://doi.org/10.1016/j.fuel.2012.01.017.

Varjani, S. J., & Chaithanya Sudha, M. (2018). Treatment technologies for emerging organic contaminants removal from wastewater. In S. Bhattacharya, A. Gupta, & A. Pandey (Eds.), *Water Remediation* (pp. 91−115). Springer. Available from https://doi.org/10.1007/978-981-10-7551-3_6.

Varjani, S. J., & Upasani, V. N. (2017). A new look on factors affecting microbial degradation of petroleum hydrocarbon pollutants. *International Biodeterioration and Biodegradation*, *120*, 71−83. Available from https://doi.org/10.1016/j.ibiod.2017.02.006.

Varjani, S., Kumar, G., & Rene, E. R. (2019). Developments in biochar application for pesticide remediation: Current knowledge and future research directions. *Journal of Environmental Management*, *232*, 505−513. Available from https://doi.org/10.1016/j.jenvman.2018.11.043.

Vendramel, S., Bassin, J. P., Dezotti, M., & Sant' Anna, G. L., Jr (2015). Treatment of petroleum refinery wastewater containing heavily polluting substances in an aerobic submerged fixedbed reactor. *Environ Technol*, *36*, 2052−2205. Available from https://doi.org/10.1080/09593330.2015.1019933.

Viggi, C. C., Presta, E., Bellagamba, M., Kaciulis, S., Balijepalli, S. K., Zanaroli, G., & Aulenta, F. (2015). The Oil Spill Snorkel: an innovative bioelectrochemical approach to accelerate hydrocarbons biodegradation in marine sediments. *Front Microbiol*, *6*(881), 1−11. Available from https://doi.org/10.3389/fmicb.2015.00881.

Visser, J. P., & Aloisi de Larderel, J. (1997). *Environmental Management in Oil and Gas Exploration and Production, an Overview of Issues and Management Approaches*. Joint E&P Forum.

Vorontsov, A., Nikanorova, M., Dorokhov, A., et al. (1998). Assay methods for monitoring the contaminants of natural waters by petroleum products. *Journal of Optical Technology*, *65*, 335−340.

Wang, S., Ghimire, N., Xin, G., Janka, E., & Bakke, R. (2017). Efficient high strength petrochemical wastewater treatment in a hybrid vertical anaerobic biofilm (HyVAB) reactor: A pilot study. *Water Practice and Technology*, *12*(3), 501−513. Available from https://doi.org/10.2166/wpt.2017.051.

Ward, O., Singh, A., & Van Hamme, J. (2003). Accelerated biodegradation of petroleum hydrocarbon waste. *Journal of Industrial Microbiology and Biotechnology*, *30*(5), 260−270. Available from https://doi.org/10.1007/s10295-003-0042-4.

Waseem, R., Jechan, L., Nadeem, R., Yiwei, L., Ki-Hyun, K., & Jianhua, Y. (2019). Removal of phenolic compounds from industrial waste water based on membrane-based technologies. *Journal of Industrial and Engineering Chemistry*, 1−18. Available from https://doi.org/10.1016/j.jiec.2018.11.024.

Wu, P., Jiang, L. Y., He, Z., & Song, Y. (2017). Treatment of metallurgical industry wastewater for organic contaminant removal in China: Status, challenges, and perspectives. *Environmental Science: Water Research and Technology, 3*(6), 1015−1031. Available from https://doi.org/10.1039/c7ew00097a.

Xu, X., & Zhu, X. (2004). Treatment of refectory oily wastewater by electro-coagulation process. *Chemosphere, 56*(10), 889−894. Available from https://doi.org/10.1016/j.chemosphere.2004.05.003.

Yang, X. (2008). Sol -gel synthesized nanomaterials for environmental applications. PhD Thesis, Kansas State University.

Zafra, G., Moreno-Montaño, A., Absalón, Á. E., & Cortés-Espinosa, D. V. (2015). Degradation of polycyclic aromatic hydrocarbons in soil by a tolerant strain of *Trichoderma asperellum*. *Environmental Science and Pollution Research, 22*(2), 1034−1042. Available from https://doi.org/10.1007/s11356-014-3357-y.

Zeng, Y., Yang, C., Zhang, J., & Pu, W. (2007). Feasibility investigation of oily wastewater treatment by combination of zinc and PAM in coagulation/flocculation. *Journal of Hazardous Materials, 147*(3), 991−996. Available from https://doi.org/10.1016/j.jhazmat.2007.01.129.

Zhang, X., He, W., Ren, L., Stager, J., Evans, P. J., & Logan, B. E. (2015). COD removal characteristics in air-cathode microbial fuel cells. *Bioresource Technology, 176*, 23−31. Available from https://doi.org/10.1016/j.biortech.2014.11.001.

Chapter 3

Concomitant degradation of petroleum products and microplastics in industrial wastewater using genetically modified microorganisms

Kritica Rani[1], Puja Singh[1], Riya Agarwal[1] and Arindam Kushagra[2]

[1]Amity Institute of Biotechnology, Amity University Kolkata, Kolkata, India, [2]Amity Institute of Nanotechnology, Amity University Kolkata, Kolkata, India

Introduction

Pollution has tremendously increased owing to the stepping urbanization that has slowly and steadily overpowered conventional means of resources that once was the reason for clear blue sky. Water pollution is one such aspect of pollution with the major contributors being anthropogenous substances that subsequently turn dangerous to humans and wildlife. Similar to mankind, these substances are highly diverse with petroleum, agrochemicals, microplastics including their by-products being the major perpetrators.

Petroleum is a fossil fuel, composed of rock minerals, found in deposits beneath the surface of the earth (Petroleum, Pollution Issues, n.d.). Like coal and natural gas, petroleum is formed by the decomposition of organic matters, such as large deposits of dead organisms—primarily algae and zooplanktons, under the sedimentary rocks over the million years, subjected to intense pressure and heat (Chen, 2020). In the present day, petroleum is found and are extracted from vast petroleum reservoirs where ancient seas were located, with giant drills in the form of crude oils, which usually shows a variety of colors like, black or dark brown, and can also be in reddish, yellowish, tan, or greenish color (Turgeon & Morse, 2018). The petrochemicals mainly consist of high molecular weight hydrocarbons. These hydrocarbons are made up of aliphatic, aromatic, and heterocyclic compounds. Petroleum hydrocarbons are usually divided into four groups namely- the saturates, the aromatics, the asphaltenes (phenols, fatty acids, ketones, esters, and porphyrins), and the resins (pyridines, quinolines, carbazoles, sulfoxides, and amides) (Colwell, Walker, & Cooney, 1997) that together contribute to the toxicity of petroleum hydrocarbons. Petroleum extraction and processing, as well as availability, plays a very significant role in the world's economy and geopolitics (Chen, 2020). Companies extract and process petroleum to produce various other kinds of petroleum-based products, which include automobiles, plastics, fertilizers, and many more, thus providing various sources of petroleum pollution such as accidental release, the result of exploration, production, maintenance, transportation, storage, etc. (Das & Chandran, 2011). It has been found that approx. 5.74 million tons of oils were lost due to tanker incidents from 1970 to 2016 (Stewart, Lipton, Celentano, & Reed, 1992) as depicted in Fig. 3.1. These incidents contaminated the natural environment.

Petroleum as a source of energy as well as a primary raw material for the chemical industries has always increased its worth in the world, which resulted in the increased world production from 29 to 2400 million metric tons per year over the 20th century (Gutnick & Rosenberg, 1977). The real problem is the fact the major oil-producing countries are not the major oil consumers, which follows another problem of the massive movement of petroleum across countries overseas (Gutnick & Rosenberg, 1977). This usually gives the oil a way to enter the sea, usually through accidental oil spills, but also through deliberate discharge of wash waters and ballast from the oil tankers, which has drastic effects on aquatic lives as well as on the lives of organisms that depend on it. Petroleum refineries are generally considered as one

Petroleum Industry Wastewater. DOI: https://doi.org/10.1016/B978-0-323-85884-7.00013-8

19

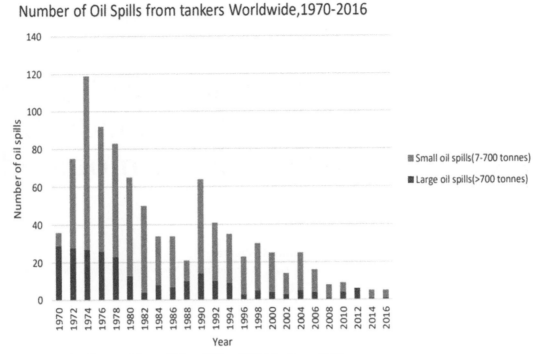

FIGURE 3.1   Number of oil spills around the world from 1970 to 2016 (Max Roser, 2013).

of the major sources of the air, water, and soil pollutants in regions where they are located (Environmental Impact of the Petroleum Industry, Hazardous Substance Research Centers/South, & Southwest Outreach Program). These pollutants are found to contain mutagenic, carcinogenic and growth-inhibiting chemicals, and also small quantities ($5-100 \, \mu g \, L^{-1}$) of certain petroleum can destroy microalgae and juvenile forms of many marine organisms (Gutnick & Rosenberg, 1977).

The next most contagious pollutant is microplastics in an aqueous ecosystem. In the recent era, these have been increased substantially with outgrowing industrialization and urbanization. Microplastics are small pieces of plastic that are less than 5 mm in size (Arthur, Baker, & Bamford, 2009; Collignon, Hecq, Galgani, Collard, & Goffart, 2014). They enter the ecosystem as a result of industrial processes and the use of cosmetics, clothes, plastic bottles, cars and truck tires, etc. They can be primary or secondary depending on the nature of the occurrence. Primary microplastics are naturally 5 mm or less before they barge into the environment whereas secondary microplastics are formed due to the weathering of larger macro plastics already present in the environment (Cole et al., 2013; Conkle, Báez Del Valle, & Jeffrey, 2018). Both the types of microplastics escape the natural process of degradation and persist in the aquatic or marine ecosystems thereby causing bioaccumulation and further biomagnification. Poly(ethylene terephthalate) (PET), a common durable plastic used in bottled water, juice, and soda leaches antimony in amounts that exceed US safety guidelines under high temperature. Polyvinyl chloride (PVC) is used to wrap meat and sandwiches, floats in the tub in the form of bath toys, makes for stylish jackets and household plumbing, is responsible for leaching toxic chemicals when in contact with water. Polystyrene, a common type of plastic used in packaging for takeout food and the fishing industry, releases carcinogens when in contact with hot beverages. Bisphenol A (BPA), an ingredient used to harden plastic also causes a wide range of disorders like cardiovascular disease, type 2 diabetes, and abnormalities in liver enzymes (Thompson, Moore, Vom Saal, & Swan, 2009).

Thinking of various ways of reclaiming oil spilled water bodies, we found that natural attenuation through bioremediation can bring great results. Also, the use of chemical oxidants in the remediation of oil-contaminated water bodies has gained prominence over the last few decades. In addition to this, the energy recovering process has been sought as a way to reduce plastic accumulation. Incineration of plastic wastes can help in turning the loss of energy into the atmosphere in landfills into energy that can be used for further use. However, as opposed to recycling, this method does not diminish the amount of plastic material that is produced. Therefore, recycling plastics is considered a more efficient solution (Thompson et al., 2009). Bioremediation is another possible solution for the degradation of large amounts of microplastic waste here as well (Thompson et al., 2009).

Bioremediation is a process in which the exploitation of microorganism's aid in the consumption and decomposition of products from the petroleum industry along with synthetic polymers using enzymes and diverse metabolic capabilities that augment the removal and degradation of these environmental pollutants. Thus, bioremediation is presently the need of the hour, since it is cost-effective and a noninvasive technique best suited for eliminating these pollutants (April, Foght, & Currah, 2000). A viable remedial technology requires microorganisms to quickly adapt and efficiently use pollutants to give favorable results (Leahy & Colwell, 1990). Microorganisms are influenced by many external factors such as pH, temperature, and availability of nitrogen and phosphorus sources, which determine the rate and the extent of degradation (Atlas & Bartha, 1992; Atlas, 1981, 1984; Foght & Westlake, 1987; Zobell, 1946). Genetically modified organisms (GMOs) have been highly efficient in the degradation process. These genetically modified bacteria have shown a positive result in the degradation process pertaining to all the criteria are fulfilled. However, due to ecological and regulatory constraints, there are still many problems in field testing (Shukhla, Singh, & Sharma, 2010).

In this chapter, we are trying to make in-depth understanding of genetically modified bacteria for the remediation of etiological pollutants mainly Petroleum compounds and Microplastics. This will also be helpful to understand the importance of genetically modified microorganism (GEM) in the remediation process.

Interaction of petroluem hydrocarbons with microorganisms

Crude oil is an extremely complex naturally occurring mixture of hydrocarbons along with small amounts of oxygen, sulfur, and nitrogen-containing compounds and some trace quantities of metallic constituents (Gutnick & Rosenberg, 1977). One important requirement for bioremediation success is the presence of microbes with the required metabolic capabilities, in optimal conditions of nutrients, pH, and other physical factors, along with optimal growth rates (Das & Chandran, 2011). Essential characteristics defining the hydrocarbon-degrading microorganisms are: (1) membrane-bound, group-specific dioxygenases and (2) mechanisms for optimizing interactions between the microbes and the water-soluble hydrocarbons (Rosenberg, Navon-Venezia, Zilber-Rosenberg, & Ron, 1998). Two main approaches can be used for oil spill remediation: (1) bioaugmentation, in which known oil-degrading microbes are added to supplement the already existing bacterial population present in the oil spilled area and (2) biostimulation, in which nutrients are added to stimulate the growth of indigenous microbial population already present in the area (Das & Chandran, 2011). The susceptibility of hydrocarbons to biodegradation differs and can be ranked as linear alkanes > branched alkanes > small aromatics > cyclic alkanes (Perry, 1984; Ulrici, 2000). Some compounds may not be degraded at all, like the high molecular weight polycyclic aromatic hydrocarbons (PAHs) (Atlas & Bragg, 2009).

We know that, for millions of years, hydrocarbons have been present in water and on land as a result of plant synthesis and natural oil seeps, thus, it is not surprising that hydrocarbon oxidizers can be found in almost all natural environments (Rosenberg, 2012). Hydrocarbons present in the environment are found to be biodegradable by bacteria, fungi, and yeast, with the recorded efficiency range from 6% to 82% for soil fungi, 3.13% to 50% for soil bacteria, and 0.003% to 100% for marine bacteria (Das & Chandran, 2011). The processes of oxidation-reduction, adsorption, ion exchange, and chelation reactions are some of the processes found to be involved in biodegradation (Xenia & Refugio, 2016).

The oxygenases in the hydrocarbon-degrading microbes are always membrane-bound, never extracellular or intracellular, thus it is very important that they must make direct contact with their substrate (Rosenberg et al., 1998). This is a challenge faced by the microorganisms themselves as well as the researchers working in this field, because of the chemical heterogeneity and the water insolubility of the crude oils (Gutnick & Rosenberg, 1977). To understand how the cell-surface properties of the bacteria interact with the hydrocarbons of crude oils, we need to understand the dynamics of crude oil in different natural environments. After the seepage of oil into the sea, the oil rises to the surface and comes in contact with the atmosphere where some of the lower molecular weight hydrocarbons become volatile and gets evaporated (Rosenberg et al., 1998). Also, in an aqueous environment, some of the hydrocarbons get dissolved which are rapidly degraded by the microorganisms present naturally, and the remaining major part of the oil remains dispersed in the oil-in-water form (Gutnick & Rosenberg, 1977). Also, as the microbial degradation and the evaporation of the easily available hydrocarbons proceeds, the remaining oil phase tends to solidify to form tar balls, leaving a solid phase as well (Gutnick & Rosenberg, 1977). Two biological strategies can be used to increase contact between microbes and the hydrocarbons: (1) mechanisms of specific adhesion, or desorption, and (2) emulsification of the hydrocarbons (Rosenberg et al., 1998). There is another fact that the oxygenases are group-specific, which means, that some works on some fractions of alkanes, others will work on different aromatic or cyclic hydrocarbons, thus it is concluded that only mixtures of different microbes can efficiently work for the degradation of petroleum oil fractions (Rosenberg et al., 1998).

Biodegradation of petroleum hydrocarbons by microorganisms

Hydrocarbons can be easily degraded by microorganisms aerobically or anaerobically. Bacteria, fungi, algae, archaea all have the tendency to break down complex hydrocarbons inducing enhanced bioremediation. While fungi and algae can degrade these pollutants aerobically, bacteria and archaea can work in both aerobic and anaerobic conditions (Haritash & Kaushik, 2009; Weelink, van Eekert, & Stams, 2010). Owing to the relatively reduced nature of most hydrocarbons, they are typically the energy source or electron donors for microbial metabolism, and their oxidation must be coupled to the reduction of a suitable electron acceptor. Aerobic degradation occurs with molecular oxygen (O_2) as both a reactant to oxidize the substrate and an electron acceptor for microbial respiration. In contrast, anaerobic degradation uses different biotransformation pathways that don't depend on oxygen, coupled to microbial respiration of a variety of electron acceptors. Under anaerobic conditions, biodegradation often results from the stepwise concerted action of many different microbes in a process called syntropy. Other means of degrading organic compounds include the use of biochemical methods like the action of enzymes on complex hydrocarbon molecules, the use of biosurfactants for emulsification into smaller droplets or via chemical methods like oxidation of petroleum waste via Fenton's reaction. The section also deals with the importance of GMOs in playing the role.

Aerobic degradation

For aerobic degradation of hydrocarbons, the availability of oxygen often turns out to be the rate-limiting step. Oxygen availability is dependent on the diffusion of oxygen from the affected site of the environment as well as uptake by the microorganism. High amounts of oxygen can speed up the degradation process of several manifolds (Atlas & Philp, 2005). The degradation rates also depend on the complexity of hydrocarbon, environmental parameters, and the age of spill. The degradation occurs in the order of alkenes > short-chain alkanes > branched alkanes > aromatics (Xue, Yu, Bai, Wang, & Wu, 2015). The most rapid and complete degradation of the majority of organic pollutants is brought about under aerobic conditions. Fig. 3.2 shows the main principle of aerobic degradation of hydrocarbons (Fritsche & Hofrichter, 2000). The initial intracellular attack of organic pollutants is an oxidative process and the activation, as well as the incorporation of oxygen, is the enzymatic key reaction catalyzed by oxygenase and peroxidases. These enzymes add molecular oxygen to the hydrocarbon molecules that lead to the formation of alcohols that are subsequently

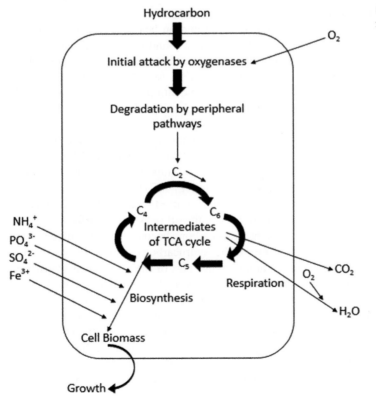

FIGURE 3.2 Aerobic degradation of hydrocarbons by microorganisms.

TABLE 3.1 Table 1: Enzymes involved in biodegradation of petroleum hydrocarbons.

Enzymes	Substrates	Microorganisms	References
Soluble methane monooxygenases	C1–C8 alkanes alkenes and cycloalkanes	*Methylococcus Methylosinus Methylocystis Methylomonas Methylocella*	McDonald, Miguez, and Rogge (2006)
Particulate methane monooxygenases	C1–C5 (halogenated) alkanes and cycloalkanes	*Methylobacter Methylococcus Methylocystis*	McDonald et al. (2006)
AlkB-related alkane hydroxylases	C5–C16 alkanes, fatty acids, alkylbenzenes, cycloalkanes and so forth	*Pseudomonas Burkholderia Rhodococcus Mycobacterium*	Jan, Beilen, and Neuenschwunder (2003)
Eukaryotic P450	C10–C16 alkanes, fatty acids	*Candida maltose Candida tropicalis Yarrowia lipolytica*	Iida, Sumita, Ohta, and Takagi (2008)
Bacterial P450 oxygenase system	C5–C16 alkanes, cycloalkanes	*Acinetobacter Caulobacter Mycobacterium*	Van Beilen, Funhoff, and Van Loon (2006)
Dioxygenases	C10–C30 alkanes	*Acinetobacter sp.*	Maeng, Sakai, Tani, and Kato (1996)

oxidized to fatty acids (Abbasian, Lockington, Mallavarapu, & Naidu, 2015). Peripheral degradation pathways convert organic pollutants step by step into intermediates of the central intermediary metabolism, for example, the tricarboxylic acid cycle. Biosynthesis of cell biomass occurs from the central precursor metabolites, like acetyl-CoA, succinate, pyruvate. Sugars required for various biosynthesis and growth are generated by gluconeogenesis (Das & Chandran, 2011; Fritsche & Hofrichter, 2000). The degradation of petroleum hydrocarbons can be mediated by specific enzyme systems like the Cytochrome P450 family as illustrated in Table 3.1. Cytochrome P450 enzyme systems were found to be involved in the biodegradation of petroleum hydrocarbons. Fig. 3.3 shows the initial attack on xenobiotics by oxygenases (Das & Chandran, 2011).

Anaerobic degradation

Although anaerobic degradation has a slower rate of decomposition than aerobic means yet facultative and obligatory bacteria and archaea are known to develop at hydrocarbon-impacted sites deficit of oxygen and readily degrade the organic compounds within days or months. In this form of degradation addition of an oxidized functional group to activate the molecules in an anoxic environment is the rate-limiting step (Meckenstock & Mouttaki, 2011). Despite being slow, this process results in the complete degradation of pollutants even in the absence of oxygen.

Anaerobic microbes use terminal electron acceptors (TEAs) unlike, aerobic microbes that use oxygen for respiration, including compounds such as nitrate, sulfate, carbon dioxide, oxidized metals, or even certain organic compounds (Atlas & Philp, 2005). In a few cases, specific species of denitrifying or sulfate-reducing microorganisms have been known to metabolize certain hydrocarbons completely to CO_2 and water. However, anaerobic degradation of hydrocarbons mostly occurs via syntrophy, where the degradation of a substrate by one microbe is dependent on the activity of another microbe maintaining low concentrations of intermediate products like hydrogen and format that drive otherwise thermodynamically unfavorable reactions. Syntrophy is more common under anaerobic conditions because the use of oxygen as a TEA is more energetically favorable (Gieg, Fowler, & Berdugo-Clavijo, 2014). Syntrophic processes are important for the complete degradation of methane and carbon dioxide since methanogens metabolize simple substrates like acetate and hydrogen. Multiple syntrophic relationships may be present in any given environment, based on available substrates and conditions as shown in Fig. 3.4 (Gieg et al., 2014). In environments rich in methanogens, when all other electron acceptors are used up, primary degraders such as Peptococcaceae and Clostridium degrade hydrocarbons to intermediates like H_2 and acetate, which are later consumed by methanogens (Gieg et al., 2014).

Degradation using biosurfactants

Biosurfactants are extracellular are membrane-bound surface-active compounds produced by microorganisms like bacteria, fungi, yeast. They can be classified as glycolipids, phospholipids, neutral peptides, lipopeptides, fatty acids based

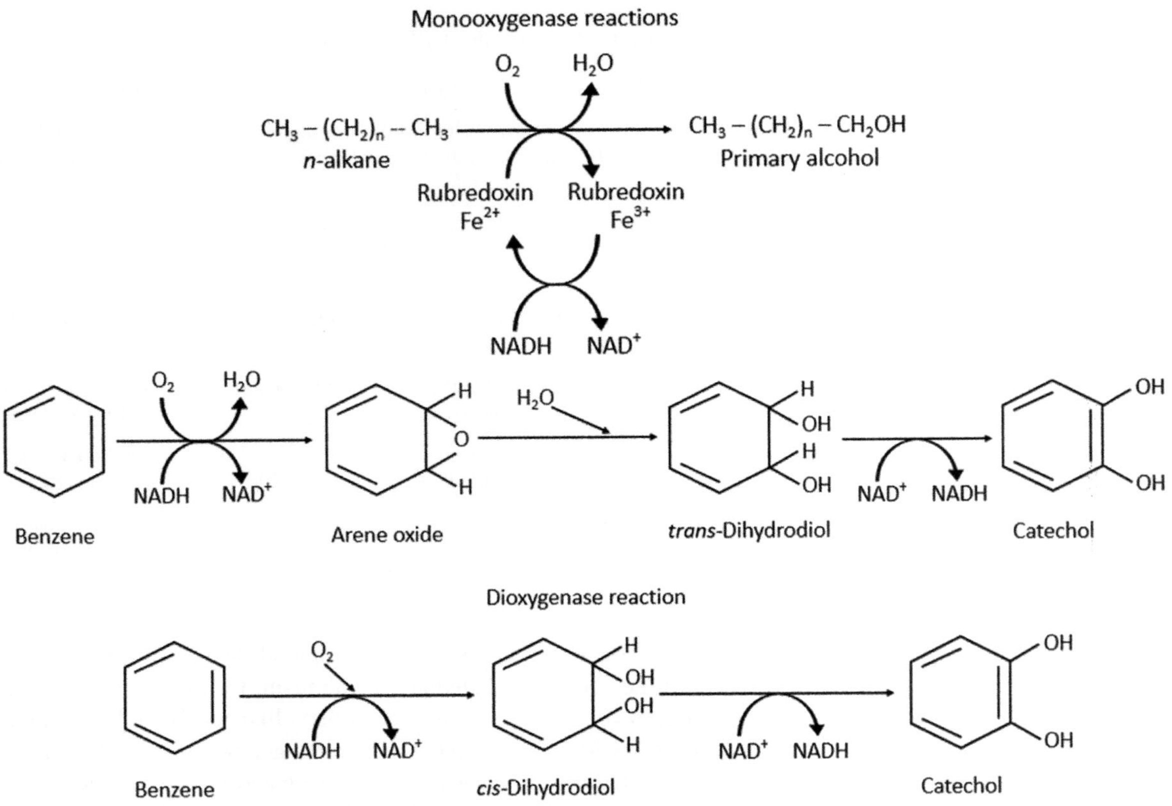

Monooxygenase reactions

FIGURE 3.3   Enzymatic reactions involved in the breakdown of petroleum hydrocarbons.

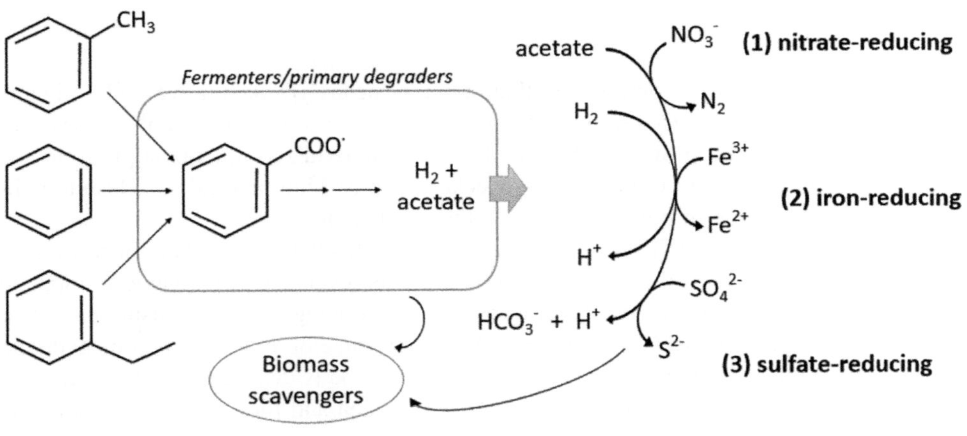

FIGURE 3.4 Conceptual model for syntrophic anaerobic degradation of benzene and alkylbenzenes (Gieg et al., 2014).

on their chemical nature. These biosurfactants tend to decrease the surface and Interfacial tension in solids, liquids, gases that increase their solubilization in liquid solutions. They are found to be more stable at extreme temperatures, pH, and salt concentrations (Musale & Thakar, 2015). Owing to the better performance and excellent characteristics of biosurfactants they have found applications in oil recovery, environmental bioremediation, food processing, and pharmaceutical industries over synthetic surfactants. Table 3.2 summarizes the variety of biosurfactants produced by different microorganisms (Das & Chandran, 2011). Among all the biosurfactants used for the purpose, rhamnolipids have been seen to have shown the most productive result in the bioremediation of hydrocarbons (Musale & Thakar, 2015). Belonging to a group of surfactants, besides reducing the surface tension of the crude oil, they behave as potential emulsifying agents too that help in the formation of micelles. These micelles in the form of microdroplets are encapsulated around hydrophobic microbial cell surface and engulfed inside and degraded (Das & Chandran, 2011). Fig. 3.5 shows

TABLE 3.2 Table 2: List of biosurfactants and the microorganisms responsible for their production.

Biosurfactants	Microorganisms	References
Sophorolipid	*Candida bombicola*	Daverey and Pakshirajan (2009)
Rhamnolipids	*Pseudomonas aeruginosa*	Kumar, Léon, De Sisto Materano, Ilzins, and Luis (2008)
Glycolipid	*Aeromonas sp.*	Ilori, Amobi, and Odocha (2005)
Surfactin	*Bacillus subtilis*	Youssef, Simpson, and Duncan (2007)
Rhamnolipids	*Pseudomonas fluorescens*	Mahmound, Aziza, Abdeltif, and Rachida (2008)
Lipomannan	*Candida tropicalis*	Muthusamy, Gopalakrishnan, Ravi, and Sivachidambaram (2008)
Glycolipid	*Bacillus* sp.	Tabatabaee, Assadi, Noohi, and Sajadian (2005)

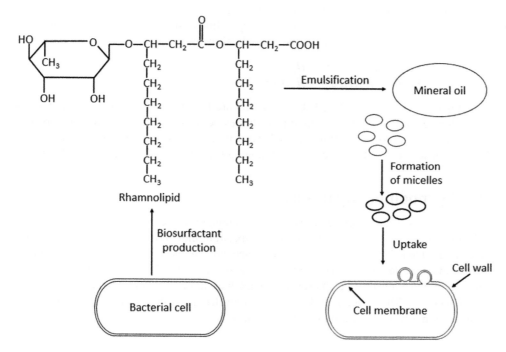

FIGURE 3.5 Degradation of hydrocarbons by biosurfactant (rhamnolipid) produced from *Pseudomonas* sp. (Das & Chandran, 2011).

the production of rhamnolipid from *Pseudomonas* sp. and the formation of micelles that are uptaken and degraded by microorganisms (Das & Chandran, 2011).

Degradation of hydrocarbons by chemical oxidation

The use of chemical compounds in the remediation of hydrocarbons has shown great efficiency in the last few decades. The reason might be partly due to its effectiveness over a broad range of contaminants and partly because of its quick mode of action with some areas remediating within hours (Goi, Trapido, & Kulik, 2009; Karpenko, Karpenko, Lubenets, & Novikov, 2009; Ojinnaka, Osuji, & Achugasim, 2011). As we know, crude oil is a product of a collection of organic compounds the most prominent among them being hydrocarbons. These hydrocarbons can be efficiently degraded by the use of chemical compounds that can also act as oxidants. Some of the chemical oxidants serving the purpose include hydrogen peroxide, Fenton's reagent, sodium or potassium persulfate, etc.

Hydrogen peroxide has been widely used to remove hydrocarbons and hydrocarbon derivatives from contaminated soils (Goi et al., 2009; Mahmoud, McCormick, Dicki, McClayment, & Stoke-Rees, 2000; Mohamed, Achami, & Mahmound, 2002; Ojinnaka et al., 2011; Stuart, Tai, & John, 2001). The compound can be used in direct or indirect oxidation. The direct method involves the breakdown of the hydrogen peroxide into water and oxygen with oxygen participating in autooxidation with the contaminants. The indirect methods make use of a catalyst that transfers one-electron to generate hydroxyl free radical OH, a powerful oxidizing agent with a standard electrode potential of +2.8 V. When hydrogen peroxide is used in tandem with Fe (II) salts, it is known as Fenton's reagent and the reaction is called Fenton's reaction (Ojinnaka et al., 2011; Watts, 1992).

$$\left(Mn^{n+}\right) + H_2O_2 \left(Mn^{n+1}\right) + OH^- + OH$$

$$\text{Eg. } Fe^{2+} + H_2O_2 Fe^{3+} + OH^- + OH \tag{3.1}$$

Step 1 above is the initiation step whereas the propagation steps follow as shown below:

$$OH + H_2O_2 OH^- + H_2O \tag{3.2}$$

$$Fe^{3+} + HO_2 Fe^{2+} + H^+ + O_2 \tag{3.3}$$

$$Fe^{2+} + HO_2 Fe^{3+} + HO_2^- \tag{3.4}$$

$$Fe^{2+} + HO Fe^{3+} + HO^- \tag{3.5}$$

Steps (2) and (3) are propagation steps while steps (4) and (5) are termination steps.

The generation of the hydroxyl radicals (HO) takes place at a low pH range while at the neutral pH range, the ferryl ion (FeO^{2+}) mechanism predominates (Kaysztof, 2009). It follows the nonradical pathway shown below:

$$Fe^{2+} + H_2O_2 FeO^{2+} + H_2O$$
$$FeO^{2+} + H_2O_2 Fe^{2+} + H_2O + O_2$$

The two most reactive intermediates formed above, that is, FeO^{2+} and HO are the ones responsible for attacking or oxidizing the hydrocarbon molecules thus leading to their degradation. In studies performed by various researchers, it is estimated that the highest efficiency of hydrocarbon reduction was achieved in 7 days where HO played the role of reactive intermediate at acidic pH medium whereas FeO^{2+} overpowered at neutral pH levels (Ojinnaka et al., 2011).

Degradation using genetically modified organisms

GEMs are the future of various research projects that are directed towards the cleaning of the ecosystem. Owing to their fast mode of reproduction, ease of availability, higher degradative capacity, broad-spectrum they have found applications in the removal of hazardous waste from the environment. Table 3.3 enlists several genetically engineered bacteria exploited for the breakdown of diverse petroleum hydrocarbon waste and their contaminants (Das & Chandran, 2011). Evidence and studies of the application of GEMs in the testing of contaminants like chlorinated compounds, aromatic hydrocarbons, and nonpolar toxicants have been recorded (Das & Chandran, 2011). One such GEM capable of degrading petroleum hydrocarbons augmented with the breakdown of microplastics has been discussed in detail in the next section.

TABLE 3.3 Table 3: Genetic engineering of bacterial species for biodegradation of contaminants.

Microorganism	Modification	Contaminants	References
Pseudomonas putida	Pathway	4-Ethylbenzoate	Ramos, Wasserfallen, Rose, and Timmis (1987)
Pseudomonas putida KT2442	Pathway	Toluene/benzoate	Panke, Sanchez-Romero, and De Lorenzo, (1998)
Pseudomonas sp. FRI	Pathway	Chloro-, methylbenzoates	Rojo, Pieper, Engesser, Knackmuss, and Timmis (1987)
Comamonas. testosteroni VP44	Substrate specificity	o-, p-Monochlorobiphenyls	Hrywna, Tsoi, Maltseva, Quensen, and Tiedje (1999)
Pseudomonas sp. LB400	Substrate specificity	PCB	Erickson and Mondello, (1993)
Pseudomonas pseudoalcaligenes KF707-D2	Substrate specificity	TCE, toluene, benzene	Suyama et al. (1996)

TABLE 3.4 Table 4: Different hydrocarbon degrading microorganisms, their substrates and mode of degradation.

Kingdom	Species	Substrate	Mode of degradation	References
Bacteria, Fungi, Algae, Archaea	*(Aerobes) Pseudomonas sp., Rhodococcus sp., Burkholderia. Mycobacterium. Arthrobacter. Oleispira antarctica. Halobacterium (Halorubrum) distributum. VKM B-1916 Haloferax sp., Halobacterium sp., Halococcus sp., Haloarcula EH4, Haloferax sp., Cladosporium sp., Prototheca zopfii, Candida tropicalis*	n-Alkanes, PAHs, cyclohexane monoaromatic, PAHs, heterocyclic compounds. monoaromatics, aromatic hydrocarbonn-alkanes (C10–C18), n-alkanes (C10-C30), n-alkanes (C10-C34), anthracene PAHs, aromatic and aliphatic hydrocarbon, aromatic hydrocarbon	Degradation by surfactant production, enzymatic degradation, degradative catabolic pathway, enzymatic degradation, metabolic degradation, metabolic degradation in saline medium, metabolic degradation	Daverey and Pakshirajan (2009), Iida et al. (2008), Morya, Salvachúa, and Shekhar (2020), Maeng et al. (1996), Arigar, Mahesh, Nagenahalli, and Yun (2006), Gentile et al. (2016), Kulichevskaya, Milekhina, Borzenkov, Zvyagintseva, and Belyaev (1991), Al-Mailem, Sorkhoh, Al-Awadhi, Eliyas, and Radwan (2010), Bertand, Almallah, Acquaviva, and Mille (1990), Birollia, Santosa, Anushka, Luciane, and Porto (2017), Semple, Cain, and Schmidt (1998), Farag and Soliman (2011)
Bacteria	*(Facultative anaerobes) Dietzia sp. Pseudomonas sp. BS2201, BS2203 Bre6ibacillus sp. BS2202*	n-Alkanes (C6-C40), alkanes, PAHs	Enzymatic degradation	Wang et al. (2011), Grishchenkov et al. (2000)
Bacteria, Archaea	*(Anaerobes) Geobacter metallireducens (GS-15) Dechloromonas sp. (RCB, JJ)*	Aromatic hydrocarbon	Metabolic degradation	Lovely et al. (1993), Coates et al. (2001), Coates et al. (2001)

Different kingdoms of microorganisms (bacteria, fungi, algae, archaea) have their specific mode of degradation of scavenging their respective substrates. Table 3.4 enlists the classification of several aerobic, anaerobic and facultative anaerobic microorganism species that are utilized for the bioremediation of environmental pollutants.

Degradation of contaminants through genetically modified synthetic microorganisms

In the chapter above, we discussed numerous ways of degrading environmental waste using chemical and biological means. All these processes although successful, are time-consuming. Many indigenous microorganisms already present in water are capable of decomposing the contaminants. Also, the commercially available bioremediating agents augment the task. However, all these wild strains are target-specific. They remain restricted within their natural ecosystems and machinery.

With the help of Recombinant DNA Technology, it has been possible to manipulate the genomic structure of microorganismal species to modify certain aspects of these organisms according to our desire. We can then, target a broad spectrum of contaminants with an explicit approach of scavenging these pollutants manifolds. Using this technology, we can also produce an indefinite number of such GMOs which can fasten the process unlike the conventional approaches hired for the same.

This section deals with the designing of a genetically modified bacteria that aims at the removal of petroleum hydrocarbons in addition to the microplastics that are elevating the attention of various environmentalists, researchers and scientists. The primary function of this synthetic bacteria is the degradation of organic pollutants that are depriving the aqueous ecosystems of its essential oxygen thus leading to their death. Secondly, they attack the minute particles of plastics that unlike their size has occupied vast areas of land, soil, water bodies, etc. with far more potential threats than any other pollutant recorded so far. The important components and their role in the bioremediation is discussed in further detail below.

Geobacter sp

For this purpose, we chose to deploy *Geobacter* sp., an obligate anaerobe that has a prominent utility in bioremediation. The capability of oxidizing organic pollutants, metals like iron, and heavy compounds including petroleum hydrocarbons into environmentally safe carbon dioxide and water has made them even more productive for the bioremediation of oil spills, radioactive waste, and oil-based contaminants (Childers, 2002). *Geobacter* species were discovered with a highly conductive pilus that helped them to use metal ions in the form of TEAs (Reguera et al., 2005). This property of *Geobacter* assisted in a variety of extracellular electron transport mechanisms with the outside world (Tan et al., 2016).

In terms of biodegradation of oil-covered contaminants and petroleum hydrocarbons in aqueous ecosystems into carbon dioxide as a waste by-product and mineralizing complex metal compounds into simple ones (Fig. 3.6), *Geobacter* species like *Geobacter metallireducens*, *G. sulfurreducens* use electrical pili between themselves and the pollutant as a channel to accept electrons for their energy source (Cologgi, 2014). The genome of *G. metallireducens* have been notified with reductive dehalogenases and iron-reducing genes that make them compatible for our use with an additional advantage of breaking down halogenated compounds along with hydrocarbons as well (Heider & Rabus, 2008). *Geobacter* sp. invariably degrades acetate, short-chain fatty acids with addition to mono-aromatic compounds like benzoate, toluene, phenol to name a few, into carbon dioxide with ferric ions as their sole electron acceptor. These species reduce environmental ferric ions (Fe^{3+}) into ferrous ions (Fe^{2+}) that can perform Fenton's reaction in the presence of hydrogen peroxide with the release of nascent hydroxyl radicals (OH) (refer to Section 3.4). This reactive intermediate has the potential to degrade the organic contaminants into carbon dioxide and water.

Pollutants + OH CO_2 + H_2O

For microplastic degradation, with the help of rDNA technology, we can produce recombinants of *Geobacter* sp. by taking genes from a plastic degrading microorganism, *Ideonella sakaiensis* 201-F6.

Ideonella sakaiensis 201-F6

I. sakaiensis 201-F6 (*Is*206-F6), a Gram-negative, aerobic, rod-shaped, motile with a polar flagellum, nonspore-forming bacterium, first isolated from a natural microbial consortium collected in Sakai city, Japan, was discovered with the ability to use low-crystallinity plastic, especially PET films (1.9%), as their carbon and energy source (Tanasupawat, Takehana, Yoshida, Hiraga, & Oda, 2016; Yoshida et al., 2016). It was found (Fig. 3.7) that their microbial culture was

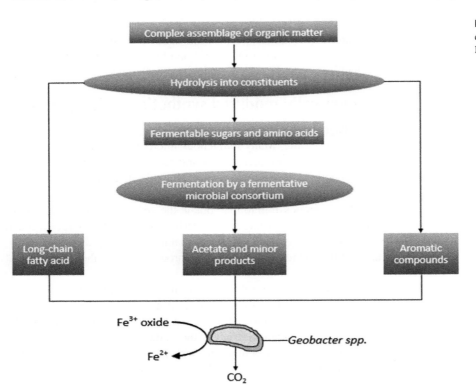

FIGURE 3.6 Degradation of organic contaminants using *Geobacter* sp. via Fenton's reaction (Mostofa et al., 2012).

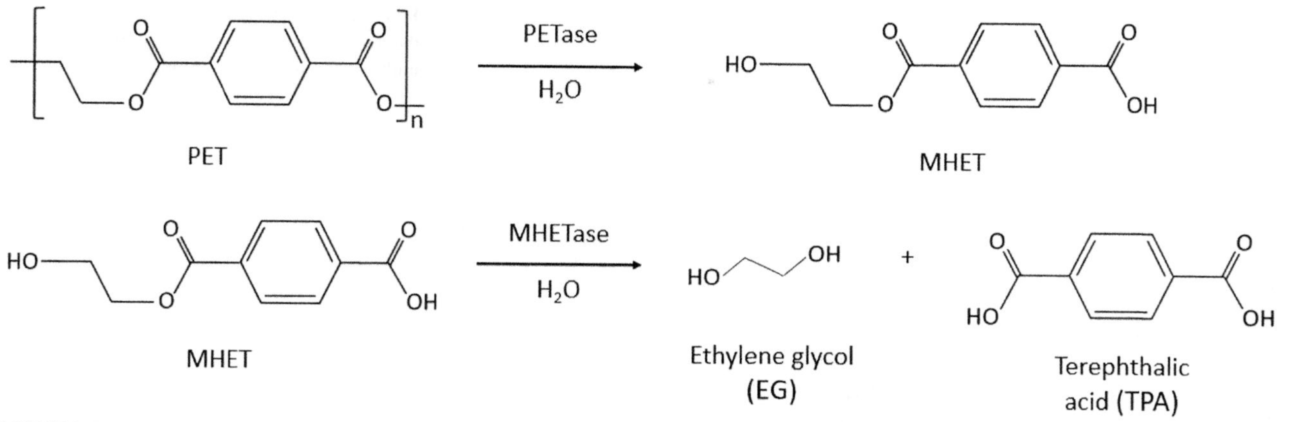

FIGURE 3.7 PETase and MHETase degrade PET to terephthalic acid and ethylene glycol (Palm et al., 2019).

able to degrade the PET films at the rate of 0.13 mg cm^{-2} day^{-1} at 30°C and 75% of the PET carbon was catabolized into CO_2 at 28°C (Yoshida et al., 2016). The strain is found to contain cytochrome oxidase and catalase that can thrive well within the pH range of 5.5–9.0 (optimally at pH 7–7.5) and at 15°C–42°C (optimally at 30°C–37°C) (Tanasupawat et al., 2016).

*Is*206-F6 can produce two α/β hydrolase fold enzymes (α/β hydrolases), PETase and MHETase, both works together to degrade PET in two steps (Palm et al., 2019) as shown in Fig. 3.7. The first α/β hydrolase fold enzyme, PETase, breaks PET to mono-(2-hydroxyethyl) terephthalate (MHET), and then the second α/β hydrolase fold enzyme, MHETase, converts MHET into simpler products terephthalate (TPA) and ethylene glycol (EG) (Palm et al., 2019). It is found that PETase doesn't degrade aliphatic polyesters, which suggests that it is an aromatic polyesterase, and with certain improvements, PETase can also degrade another semi-aromatic polyester, polyethylene-2,5-furandicarboxylate (PEF), which is an emerging, bio-based PET replacement based on sugar-derived 2,5-furandicarboxylic acid (FDCA), having improved gas barrier properties over PET (Austin et al., 2018).

Vector

An ideal vector should have certain characteristics that make it suitable for transformation and selection in the host (Nora et al., 2019). The first basic element is the origin of replication (ori) which is responsible for the copy number of a given plasmid in the cell and will be recognized by cellular replication machinery. The replication mechanism that can be recognized by a specific set of organisms is called narrow host range vectors and the origin of replication that replicate in more than one species or genus is called broad-host-range vectors, the protein encoded recognizes their replication origin inside the plasmid (Durland, Toukdarian, Fang, & Helinski, 1990). Another important element for a vector is Selection markers. This allows the selection of positive transformants stretching from auxotroph to drug resistance (Gnugge & Rudolf, 2017). Multiple cloning site is also an important component for the cloning of desired DNA (Nora et al., 2019).

The vector utilized for our cloning is RK2 vector (Fig. 3.8). It is a broad host range vector belonging to the IncP incompatibility group (Figurski, Pohlman, Bechhofer, Prince, & Kelton, 1982). It has been found suitable for replication in *G. sulfurreducens* and was maintained for over 15 generations in it (Chan et al., 2015). RK2 is about 60 kb long and carries the gene for replication, conjugation, maintenance, and antibiotic resistance. The resistance genes provide resistance to the antibiotics kanamycin, ampicillin, and tetracycline (Ingram, Richmond, & Sykes, 1973). RK2 contains *kil* gene (lethal) and *kor* gene which acts as a repressor for *kil* gene. These genes are suspected for the broad host range behavior of RK2 vectors (Kornacki, Chang, & Figurski, 1993).

Mechanism

The synthesis of the GM bacteria begins with the isolation of PETase and MHETase encoding genes from *I. sakaiensis* 201-F6 using restriction enzymes. The nicked genes are then ligated into the vector RK2 using T4 DNA ligase. The recombinant vector is then inserted into the host *G. sulfurreducens* through electroporation. The vector plasmid takes over the machinery of the host with the production of PET degrading PETase enzyme that when comes in contact with

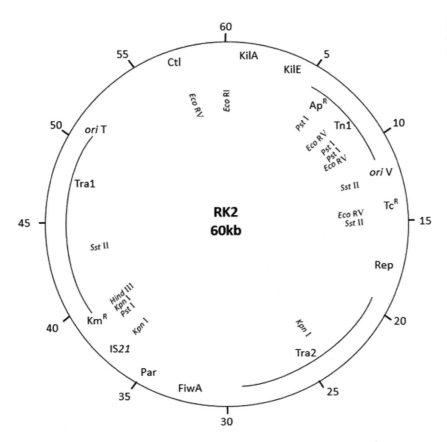

FIGURE 3.8   Structure of RK2 vector containing the resistance genes, origin of replication (ori), multiple cloning sites (Virginia, 1999).

debris rich in microplastics breaks it down to MHET which is subsequently decomposed into its final end products viz. Terephthalic acid (TPA) and Ethylene glycol (EG) under the action of MHETase. This completes the mechanism for microplastic bioremediation. For the bioremediation of organic compounds, the host bacterium can oxidize such contaminants with Fe^{3+} as electron acceptors. They reduce ferric ions to ferrous ions in the presence of H_2O_2 with the release of either of the two compounds, OH or FeO^{2+} according to the pH levels. Both of these reactive intermediates can degrade hydrocarbon pollutants into carbon dioxide and water in acidic and neutral pH mediums respectively. The overall representation of the working mechanism of the designed GMO is depicted through a flowchart (Fig. 3.9).

Fate of the end products

Terephthalic acid released as an end product can be used in pharmaceutical industries as a raw material for drugs, as a carrier in paints, hot melt adhesives, manufacture of polyester coatings resins. Ethylene glycol (EG) can find commercial and industrial applications as an antifreeze and cooling agent. They can also be used in the manufacture of capacitors and as a dehydrating agent. Water and carbon dioxide gas released as end products of hydrocarbon bioremediation directly escape into the environment. Thus, we can acknowledge that the GMO poses no danger to our nature and that its synthesis shall prove beneficial in getting our surroundings rid of toxic pollutants.

Conclusions

Eradicating our environment from hazardous pollutants has been the need of the hour for ages. Although, the chemical heterogeneity and the water insolubility of crude oil is a major challenge, with some improvements along with the use of GMOs we can pave the way for a research frontier with broader implications. However, insufficient knowledge about the potential microorganisms and their bioremediation mechanisms has been delaying the action. This chapter covers in numerous ways of eliminating organic compounds from our water bodies enlisting possible physical, chemical, and biological means of accomplishing the task. The most efficient among all these methods was the use of genetically

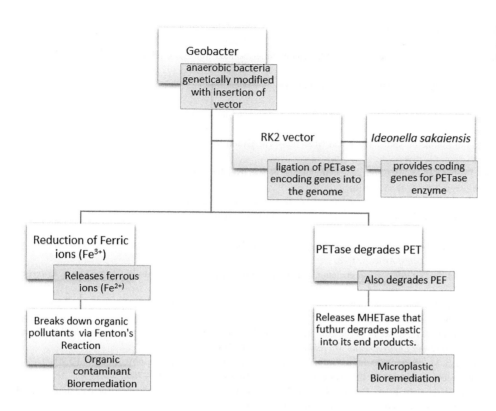

FIGURE 3.9 Flow chart depicting the overall working of the recombinant *Geobacter* sp.

modified microorganisms that were designed and synthesized in the pursuit of cleaning up the aqueous ecosystems of its unwanted guests.

Thus, according to the present review, it can be deduced that GMOs play a significant role in petroleum hydrocarbon degradation with no less effect on the overpiling microplastics endangering our homeland. Therefore, they can be considered as potential candidates for the remediation strategies undertaken in the future.

Future prospects

The use of GMOs as a key component in the bioremediation of pollutants is ground-breaking and shall pave the way for future endeavors adopted for the same cause. The world of bioinformatics has also made it easy to design in silico models of the desired organism. With the knowledge of microbial hydrocarbon degradation process supplemented with the in-depth understanding of ecological relationships, biological mechanisms, and field engineering techniques it shall be possible to develop models that can predict the fate of these organic pollutants and help plan strategies of utilizing them for the removal of organic contaminants.

It has been found that *Geobacter* species communicate between themselves using nanowires that are made of protein having metal-like conductivity (Malvankar et al., 2015). Evidences that several *Geobacter sp.* work in symphony for the degradation of diverse contaminants that they are unable to process alone can be useful in targeting more than one pollutant thereby increasing the spectrum of these host organisms (Williams, 2011). *Geobacter sp.* ought to show biofilm conductivity which helps in the generation of electricity. The presence of pili and nanowires increases the electric current generation through effective electron transfer (Reguera et al., 2006). This property of microbial electrogenesis can be useful in developing *Geobacter* models that can scavenge pollutants with an additional quality of generating electricity.

GEMs are cheaper as compared to other means of remediation. Future research much focus on the large-scale field testing of these engineered microbes and the associated risk factors. Owing to the diversity in the nature and function of microorganisms there also lay a variety of novel approaches that can be sought for tackling future challenges.

References

Abbasian, F., Lockington, R., Mallavarapu, M., & Naidu, R. (2015). A comprehensive review of aliphatic hydrocarbon biodegradation by bacteria. *Applied Biochemistry and Biotechnology, 176*(3), 670–699.

Al-Mailem, D. M., Sorkhoh, N. A., Al- Awadhi, H., Eliyas, M., & Radwan, S. S. (2010). Biodegradation of crudeoil and pure hydrocarbons by extreme halophilic archaea from hypersaline coasts of the Arabian Gulf. *Extremophiles: Life Under Extreme Conditions, 14*, 321−328.

April, T. M., Foght, J. M., & Currah, R. S. (2000). Hydrocarbondegrading filamentous fungi isolated from flare pit soils in northern and western Canada. *Canadian Journal of Microbiology, 46*(1), 38−49.

Arigar, C. K., Mahesh, A., Nagenahalli, M., & Yun, D. J. (2006). *Phenol degradation by immobilized cells of Arthrobacter citreus, 17,* 47−55, Springer.

Arthur, C., Baker, J., & Bamford, H. (2009). Proceedings of the international research workshop on the occurrence. *Effects and Fate of Microplastic Marine Debris,* NOAA Technical Memorandum.

Atlas, R., & Bragg, J. (2009). Bioremediation of marine oil spills: When and when not—The Exxon Valdez experience. *Microbial Biotechnology, 2* (2), 213−221.

Atlas, R. M. (1981). Microbial degradation of petroleum hydrocarbons: an environmental perspective. *Microbiological Reviews, 45*(1), 180−209.

Atlas, R. M. (Ed.), (1984). *Petroleum microbiology.* New York: Macmillion.

Atlas, R. M., & Bartha, R. (1992). Hydrocabon biodegradation and oil spill bioremediation. *Advances in Microbial Ecology, 12,* 287−338.

Atlas, R. M., & Philp, J. (2005). Bioremediation: Applied microbial solutions for real-world environmental cleanup. *American Association for Microbiology,* 1−366.

Austin, H. P., Allen, M. D., Donohoe, B. S., Rorrer, N. A., Kearns, F. L., Silveira, R. L., Pollard, B. C., et al. (2018). Characterization and engineering of a plastic-degrading aromatic polyesterase. *PNAS Latest Articles,* 1−8.

Bertand, J. C., Almallah, M., Acquaviva, M., & Mille, G. (1990). Biodegradation of hydrocarbons by extremely halophilic archaebacterium. *Letters in Applied Microbiology, 11,* 260−263.

Birollia, W. G., Santosa, D. A., Anushka, N. A. A., Luciane, C. F. S. G., & Porto, A. L. M. (2017). *Biodegradation of anthracene and several PAHs by the marine-derived fungus* Cladosporium sp. Elsevier, pp. 13563−120.

Chan, C. H., Levar, C. E., Zacharoff, L., Badalamenti, J. P., Bond, D. R., & Loffler, F. E. (2015). Scarless genome editing and stable inducible expression vectors for geobacter sulfurreducens. *Applied and Environmental Microbiology, 81,* 7178−7186.

Chen, Y. (2020). Petroleum, *Investopedia,* Dotdash, March 29, 2020, https://www.investopedia.com/terms/p/petroleum.asp Accessed 17.05.20.

Childers, S. (2002). Geobacter metallireducens accesses insoluble Fe (III) oxide by chemotaxis. *Nature, 416*(6882), 767−769.

Coates, J. D., Chakraborty, R., Lack, J. G., Connor, S., Cole, K. A., Bender, K. S., & Achenbach, L. A. (2001). Anaerobic benzene oxidation coupled to nitrate reduction in pure culture by two strains of Dechloromonas. *Nature, 411,* 1039−1043.

Cole, M., Lindeque, P., Fileman, E., Halsband, C., Goodhead, R., Moger, J., & Galloway, T. S. (2013). Microplastic Ingestion by Zooplankton. *Environmental Science & Technology, 47*(12), 6646−6655.

Collignon, A., Hecq, J.-H., Galgani, F., Collard, F., & Goffart, A. (2014). Annual variation in neustonic micro- and meso-plastic particles and zooplankton in the Bay of Calvi (Mediterranean−Corsica). *Marine Pollution Bulletin, 79*(1−2), 293−298.

Cologgi, D. (2014). Enhanced uranium immobilization and reduction by Geobacter sulfurreducens biofilms. *Applied and Environmental Microbiology, 80*(21), 6638−6646.

Colwell, R. R., Walker, J. D., & Cooney, J. J. (1997). Ecological aspects of microbial degradation of petroleum in the marine environment. *Critical Reviews in Microbiology, 5*(4), 423−445.

Conkle, J. L., Báez Del Valle, C. D., & Jeffrey, W. T. (2018). Are we underestimating microplastic contamination in aquatic environments? *Environmental Management, 61*(1), 1−8.

Das, N., & Chandran, P. (2011). Microbial degradation of petroleum hydrocarbon contaminants: An overview. *Biotechnology Research International, 1,* 941810.

Daverey, A., & Pakshirajan, K. (2009). Production of sophorolipids by the yeast *Candida bombicola* using simple and low cost fermentative media. *Food Research International, 42*(4), 499−504.

Durland, R. H., Toukdarian, A., Fang, F., & Helinski, D. R. (1990). Mutations in the trfA replication gene of the broadhost-range plasmid RK2 result in elevated plasmid copy numbers. *Journal of Bacteriology, 172,* 3859−3867.

Environmental Impact of the Petroleum Industry, Hazardous Substance Research Centers/South & Southwest Outreach Program, June 2003. Website: https://cfpub.epa.gov/ncer_abstracts/index.cfm/fuseaction/display.files/fileID/14522.

Erickson, B. D., & Mondello, F. J. (1993). Enhanced biodegradation of polychlorinated biphenyls after site-directed mutagenesis of a biphenyl dioxygenase gene. *Applied and Environmental Microbiology, 59*(11), 3858−3862.

Farag, S., & Soliman, N. A. (2011). Biodegradation of crude petroleum oil and environmental pollutants by Candida tropicalis strain. *Brazilian Archives of Biology and Technology, 54.*

Figurski, D. H., Pohlman, R. F., Bechhofer, D. H., Prince, A. S., & Kelton, C. A. (1982). Broad host range plasmid RK2 encodes multiple kil genes potentially lethal to *Escherichia coli* host cells. *Genetics, 79,* 1935−1939.

Foght, J. M., & Westlake, D. W. S. (1987). *Biodegradation of hydrocarbons in freshwater, oil in freshwater: chemistry, biology. Countermeasure technology* (pp. 217−230). New York: Pergamon Press.

Fritsche, W., & Hofrichter, M. (2000). Aerobic degradation by microorganisms, environmental processes-soil decontamination, biotechnolgy (pp. 146−155). Weinheim, Germany: Wiley-VCH.

Gentile, G., Bonsignore, M., Santisi, S., Catalfamo, M., Giuliano, L., Genovese, L., Yakimov, M. M., Denaro, R., Genovese, M., & Cappello, S. (2016). *Biodegradation potentiality of psychrophilic bacterial strain Oleispira antarctica RB-8T.* Elsevier.

Gieg, L. M., Fowler, S. J., & Berdugo-Clavijo, C. (2014). Syntrophic biodegradation of hydrocarbon contaminants. *Current Opinion in Biotechnology, 27,* 21−29.

Gnugge, R., & Rudolf, F. (2017). A shuttle vector series for precise genetic engineering of *Saccharomyces* cerevisae. *Yeast (Chichester, England)*, *26*, 545–551.

Goi, A., Trapido, M., & Kulik, N. (2009). Contaminated soil remediation with hydrogen peroxide oxidation. *Chemical and Materials Engineering*, *2*(3), 144–148.

Grishchenkov, V. G., Townsend, R. T., McDonald, T. J., Autenrieth, R. L., Bonner, J. S., & Boronin, A. M. (2000). *Degradation of petroleum hydrocarbons by facultative anaerobic bacteria under aerobic and anaerobic conditions. Elsevier Process Biochemistry* (pp. 896–899).

Gutnick, D. L., & Rosenberg, E. (1977). Oil tankers and pollution: A microbiological approach. *Annual Reviews of Microbiology*, *31*, 379–396.

Haritash, A. K., & Kaushik, C. P. (2009). Biodegradation aspects of polycyclic aromatic hydrocarbons (PAHs): A review. *Journal of Hazardous Materials*, *169*(1), 1–15.

Heider, J., & Rabus, R. (2008). *Genomic insights in the anaerobic biodegradation of organic pollutants. Microbial biodegradation: Genomics and molecular biology.* Caister Academic Press, ISBN 978-1-904455-17-2.

Hrywna, Y., Tsoi, T. V., Maltseva, O. V., Quensen, J. F., III, & Tiedje, J. M. (1999). Construction and characterization of two recombinant bacteria that grow on ortho- and parasubstituted chlorobiphenyls. *Applied and Environmental Microbiology*, *65*(5), 2163–2169.

Iida, T., Sumita, T., Ohta, A., & Takagi, M. (2008). The cytochrome P450ALK multigene family of an n-alkane-assimilating yeast, *Yarrowia lipolytica*:C and characterization of genes coding for new CYP52 family members. *Yeast (Chichester, England)*, *16*(12), 1077–1087.

Ilori, M. O., Amobi, C. J., & Odocha, A. C. (2005). Factors affecting biosurfactant production by oil degrading Aeromonas sp. isolated from a tropical environment. *Chemosphere*, *61*(7), 985–992.

Ingram, L. C., Richmond, M. H., & Sykes, R. B. (1973). Molecular characterization of the R factors implicated in the carbenicillin resistance of a sequence of *Pseudomonas aeruginosa* strains isolated from burns. *Antimicrobial Agents and Chemotherapy*, 279–288.

Jan, B., Beilen, V., Neuenschwunder, M., et al. (2003). Rubredoxins involved in alkane degradation. *The Journal of Bacteriology*, *184*(6), 1722–1732.

Karpenko, O., Karpenko, E., Lubenets, V., & Novikov, V. (2009). Chemical oxidants for remediation of contaminated soil and water: A review. *Chemistry & Chemical Technology*, *3*(1), 41–45.

Kaysztof, B. (2009). Fenton's reaction—Controversy concerning the chemistry. *Ecological Chemistry and Engineering*, *16*(3), 347–358.

Kornacki, J. A., Chang, C. H., & Figurski, D. H. (1993). kil-kor regulon of promiscuous plasmid RK2: Structure, products, and regulation of two operons that constitute the kil E locus. *Journal of Bacteriology*, *175*(16), 5078–5090.

Kulichevskaya, I. S., Milekhina, E. I., Borzenkov, I. A., Zvyagintseva, I. S., & Belyaev, S. S. (1991). Oxidation of petroleum hydrocarbons by extremely halophilic archaebacteria. *Microbiology (Eng)*, *60*, 596–601.

Kumar, M., Léon, V., De Sisto Materano, A., Ilzins, O. A., & Luis, L. (2008). Biosurfactant production and hydrocarbon degradation by halotolerant and thermotolerant *Pseudomonas* sp. *World Journal of Microbiology and Biotechnology*, *24*(7), 1047–1057.

Leahy, J. G., & Colwell, R. R. (1990). Microbial degradation of hydrocarbons in the environment. *Microbiological Reviews*, *54*(3), 305–315.

Lovely, D. R., Giovannoni, S. J., White, D. C., Champine, J. E., Phillips, E. J., Gorby, Y. A., & Goodwin, S. (1993). *Geobacter* metallireducens gen. nov. sp. nov., a microorganism capable of coupling the complete oxidation of organic compounds to the reduction of iron and other metals. *Archives of Microbiology*, *159*, 336–344.

Maeng, J. H. O., Sakai, Y., Tani, Y., & Kato, N. (1996). Isolation and characterization of a novel oxygenase that catalyzes the first step of n-alkane oxidation in *Acinetobacter* sp. strain M-1. *Journal of Bacteriology*, *178*(13), 3695–3700.

Mahmoud, M., McCormick, M., Dicki, E. W., McClayment, G., & Stoke-Rees, P. (2000). Hydrogen peroxide in cleaning of residual hydrocarbons adjacent to structures with restricted access. *Proceedings of the Canadian Geotechnical Conference, Montreal*, *1*, 575–584.

Mahmound, A., Aziza, Y., Abdeltif, A., & Rachida, M. (2008). Biosurfactant production by Bacillus strain injected in the petroleum reservoirs. *Journal of Industrial Microbiology & Biotechnology*, *35*, 1303–1306.

Malvankar, N., Vargas, M., Nevin, K., Tremblay, P.-L., Evans-Lutterodt, K., Nykypanchuk, D., Martz, E., Tuominen, M. T., & Lovley, D. R. (2015). Structural basis for metallic-like conductivity in microbial nanowires. *mBio*, *6*(2).

Max Roser. (2013) Oil spills. Published online at ourworldindata.org. Retrieved from (https://ourworldindata.org/oil-spills) [Online Resource] Accessed 20.05.20.

McDonald, I. R., Miguez, C. B., Rogge, G., et al. (2006). Diversity of soluble methane monooxygenase-containing methanotrophs isolated from polluted environments. *FEMS Microbiology Letters*, *255*(2), 225–232.

Meckenstock, R. U., & Mouttaki, H. (2011). Anaerobic degradation of nonsubstituted aromatic hydrocarbons. *Current Opinion in Biotechnology*, *22*(3), 406–414.

Mohamed, H., Achami, G., & Mahmound, M. (2002). Hydrocarbon peroxide remediation of diesel contaminated sand: Impact on volume change. *Proceedings of the Canadian Geotechnical Conference Niagara Falls*, Canada.

Morya, R., Salvachúa, D., & Shekhar, I. (2020). Burkholderia: An untapped but promising bacterial genus for the conversion of aromatic compounds. *Cell Press Reviews*.

Mostofa, K. M. G., Liu, C.-Q., Minakata, D., Wu, F., Vione, D., Abdul, M., Takahito, M., & Sakugawa, Y. H. (2012). *Photoinduced and microbial degradation of dissolved organic matter in natural waters*, *1*(4), 273–364.

Musale, V., & Thakar, S. B. (2015). Biosurfactant and hydrocarbon degradation. *Life Science Informatics Publications*, *1*(1), 2.

Muthusamy, K., Gopalakrishnan, S., Ravi, T. K., & Sivachidambaram, P. (2008). Biosurfactants: Properties, commercial production and application. *Current Science*, *94*(6), 736–747.

Nora, L. C., Westmann, C. A., Santana, L. M., Alves, L. F., Monteiro, L. M. O., Guazzaroni, M.-E., & S-Rocha, R. (2019). The art of vector engineering: Towards the construction of next-generation genetic tools. *Microbial Biotechnology (Reading, Mass.)*, 125–147.

Ojinnaka, C., Osuji, L., & Achugasim, O. (2011). *Remediation of hydrocarbons in crude oil-contaminated soils using Fenton's reagent*, 6527–6540.

Palm, G. J., Reisky, L., Böttcher, D., Müller, H., Michels, E. A. P., Walczak, M. C., Berndt, L., Weiss, M. S., Bornscheuer, U. T., & Weber, G. (2019). Structure of the plastic-degrading Ideonella sakaiensis MHETase bound to a substrate. *Nature Communications, 10*(1717).

Panke, S., Sanchez-Romero, J. M., & De Lorenzo, V. (1998). Engineering of quasinatural Pseudomonas putida strains toluene metabolism through an ortho-cleavage degradation pathway. *Applied and Environmental Microbiology, 64*(2), 748–751.

Perry, J. J. (1984). *Microbial metabolism of cyclic alkanes. Petroleum microbiology* (pp. 61–98). New York: Macmillan.

Petroleum, Pollution Issues, pollutionissues.com/Na-Ph/Petroleum. (n.d.).

Ramos, J. L., Wasserfallen, A., Rose, K., & Timmis, K. N. (1987). Redesigning metabolic routes: manipulation of TOL plasmid pathway for catabolism of alkylbenzoates. *Science (New York, N.Y.), 235*(4788), 593–596.

Reguera, G., McCarthy, K. D., Mehta, T., Nicoll, J. S., Tuominen, M. T., & Lovley, D. R. (2005). Extracellular electron transfer via microbial nanowires. *Nature, 435*(7045), 1098–1101.

Reguera, G., Nevin, K. P., Nicoll, J. S., Covalla, S. F., Woodard, T. L., & Lovley, D. R. (2006). Biofilm and nanowire production leads to increased current in geobacter sulfurreducens fuel cells. *Applied and Environmental Microbiology, 72*(11), 7345–7348.

Rojo, F., Pieper, D. H., Engesser, K.-H., Knackmuss, H.-J., & Timmis, K. N. (1987). Assemblage of ortho cleavage route for simultaneous degradation of chloro- and methylaromatics. *Science (New York, N.Y.), 238*(4832), 1395–1398.

Rosenberg, E. (2012). *The hydrocarbon-oxidizing bacteria. The Prokaryotes.* New York: Springer. (Chapter 5).

Rosenberg, E., Navon-Venezia, S., Zilber-Rosenberg, I., & Ron, E. Z. (1998). *11 rate-limiting steps in the microbial degradation of petroleum hydrocarbons* (pp. 159–172). Springer-Verlag Berlin Heidelberg.

Semple, K. T., Cain, R. B., & Schmidt, S. (1998). Biodegradation of aromatic compounds by microalgae. *FEMS Microbiology Letters*, 291–300.

Shukhla, K. S., Singh, N. K., & Sharma, S. (2010). Bioremediation: Developments, current practices and perspectives. *GEBJ.*

Stewart, W. F., Lipton, R. B., Celentano, D. D., & Reed, M. L. (1992). Prevalence of migraine headache in the United States: Relation to age, income, race, and other sociodemographic factors. *Journal of the American Medical Association, 267*(1), 64–69.

Stuart, M. O., Tai, T. W., & John, G. A. (2001). *Proceedings of the 54th Canadian geotechnical conference* (pp. 1170–1177). Calgary.

Suyama, A., Iwakiri, R., Kimura, N., Nishi, A., Nakamura, K., & Furukawa, K. (1996). Engineering hybrid pseudomonads capable of utilizing a wide range of aromatic hydrocarbons and of efficient degradation of trichloroethylene. *Journal of Bacteriology, 178*(14), 4039–4046.

Tabatabaee, A., Assadi, M. M., Noohi, A. A., & Sajadian, V. A. (2005). Isolation of biosurfactant producing bacteria from oil reservoirs. *Iranian Journal of Environmental Health Science & Engineering, 1*, 6–12.

Tan, Y., Adhikari, R. Y., Malvankar, N. S., Ward, J. E., Nevin, K. P., Woodard, T. L., Smith, J. A., Snoeyenbos-West, O. L., & Franks, A. E. (2016). The low conductivity of geobacter uraniireducens pili suggests a diversity of extracellular electron transfer mechanisms in the genus *Geobacter*. *Frontiers in Microbiology, 7*, 980.

Tanasupawat, S., Takehana, T., Yoshida, S., Hiraga, K., & Oda, K. (2016). Ideonella sakaiensis sp. nov., isolated from a microbial consortium that degrades poly(ethylene terephthalate). *International Journal of Systematic and Evolutionary Microbiology, 66*, 2813–2818.

Thompson, R. C., Moore, C. J., Vom Saal, F. S., & Swan, S. H. (2009). Plastics, the environment and human health: Current consensus and future trends, philosophical transactions of the royal society B. *Biological Sciences, 364*(1526), 2153–2166.

Turgeon, A., & Morse, E. (2018). Petroleum. *National Geographic.* https://www.nationalgeographic.org/encyclopedia/petroleum/ Accessed 17.05.20.

Ulrici, W. (2000). Contaminant soil areas, different countries and contaminant monitoring of contaminants. *Environmental Process II. Soil Decontamination Biotechnology, 11*, 5–42.

Van Beilen, J. B., Funhoff, E. G., Van Loon, A., et al. (2006). Cytochrome P450 alkane hydroxylases of the CYP153 family are common in alkane-degrading eubacteria lacking integral Biotechnology Research International 11 membrane alkane hydroxylases. *Applied and Environmental Microbiology, 2*(1), 59–65.

Virginia, 1 (1999). *Waters, conjugative transfer in the dissemination of beta-lactam and aminoglycoside resistance.*

Wang, X. B., Chi, C., Nie, Y., Tang, Y., Tan, Y., Wu, G., & Wu, X. L. (2011). *Degradation of petroleum hydrocarbons (C6–C40) and crude oil by a novel Dietzia strain. Elsevier Bioresource Technology* (pp. 7755–7761).

Watts, R. J. (1992). Hydrogen peroxide for physicochemically degrading petroleum-contaminated soils. *Remediation Table 6 Journal, 2*(4), 413–421.

Weelink, S. A., van Eekert, M. H., & Stams, A. J. (2010). Degradation of BTEX by anaerobic bacteria: Physiology and application. *Reviews in Environmental Science and Bio/Technology, 9*(4), 359–385.

Williams, C. (2011). Who are you calling simple? *New Scientist, 211*(2821), 38–41.

Xenia, M. E., & Refugio, R. V. (2016). Microorganisms metabolism during bioremediation of oil contaminated soils. *Journal of Biodegradation and Bioremediation, 7*(2).

Xue, J., Yu, Y., Bai, Y., Wang, L., & Wu, Y. (2015). Marine oil-degrading microorganisms and biodegradation process of petroleum hydrocarbon in marine environments: A review. *Current Microbiology, 71*(2), 220–228.

Yoshida, S., Hiraga, K., Takehana, T., Taniguchi, I., Yamaji, H., Maeda, Y., Toyohara, K., Miyamoto, K., Kimura, Y., & Oda, K. (2016). A bacterium that degrades and assimilates poly(ethylene terephthalate). *Science (New York, N.Y.), 351*(6278), 1196–1199.

Youssef, N., Simpson, D. R., Duncan, K. E., et al. (2007). In situ biosurfactant production by Bacillus strains injected into a limestone petroleum reservoir. *Applied and Environmental Microbiology, 73*(4), 1239–1247.

Zobell, C. E. (1946). Action of microorganisms on hydrocarbons. *Bacteriological Reviews, 10*, 1–49.

Chapter 4

Constraints and advantages of bacterial bioremediation of petroleum wastewater by pure and mixed culture

Vrutang Shah[1], Jhanvi Desai[1] and Manan Shah[2]

[1]School of Petroleum Technology, Pandit Deendayal Energy University, Gandhinagar, India, [2]Department of Chemical Engineering, School of Technology, Pandit Deendayal Energy University, Gandhinagar, India

Introduction

Population increase and extensive economic activities lead to the greater demand for energy; looking at the current tread, global energy obligation may double over the consecutive three decades, making the petroleum wastewater production and discharge an essential issue to overcome. Water pollution by chemical and biological activity is the principal concern for various public and social authorities, especially for many industries (Crini & Lichtfouse, 2019). The crude oil fulfills one-third of the global energy requirement, which requires significant refining operations, making a tremendous amount of wastewater, before extracting to the end product (Pal, Banat, Almansoori, & Abu Haija, 2016). With increased emphasis on stream pollution prevention, the petroleum industry is faced with ever-increasing demands for greater efficiency in wastewater treatment. Additionally, the current significant economic growth rate results in greater energy demand and clean water crisis with increased energy production and wastewater. It is only a matter of time before most refineries will be producing wastewater of a quality high enough to permit their partial reuse (McKinney, 1967).

In the oil industry, petroleum wastewater is an ingrained part as it comprises various inorganic and organic components. The discharge of these harmful components needs to be treated to remove the toxic constituents before entering the receiving waters. Different methods are amalgamated for this treatment as the wastewater is a complex compound, and also, there are rigid limits for its discharge (Ghimire & Wang, 2019). The disposal of wastewater generated from the oil and gas industry is challenging because of the increased toxicity levels and generated volume while the environmentally accepted values are much lower. The strict global standards demand novel facilities and existing facilities to recover more water for reuse by accomplishing higher wastewater treatment levels. For these concerns, the treatment of petroleum wastewater becomes a widely scrutinized area of study by many researchers.

Compared to various chemical and physical processes, bioremediation processes for treating wastewater are considered cheaper and greener sources in treating effluents (Pal et al., 2016). Due to biological processes' economic viability, considering the investment and operations cost, it is preferred in any treatment plant for wastewater over physical and chemical processes. For a century, the aerobic activated sludge process is utilized to treat the effluents in biological methodology (Mittal, 2011).

Petroleum wastewater disposal is an environmentally unsuitable and costly problem (Sawyer, McCarty, & Parkin, 2003). Generally, refineries have an onsite facility to treat this wastewater and make the pollutants level suitable for disposal according to local and national norms (Tchobanoglous, Burton, & Stensel, 2003). About 70% of wastewater consists of dissolved solids and to precipitate these solids biochemical methods are considered potent. When physical solids are discharged into the environment, they can lead to the development of the anaerobic condition and sludge deposits (Maiti, 2004). Wastewater is typically gray having a foul odor (Tajrishy & Abrishamchi, 2005). However, petroleum wastewater hazardous characteristics can vary based on the operation processes, refining crude, various intermediate products, and manufacturing lubricants from which it is separated; and are a chief source of aquatic

Petroleum Industry Wastewater. DOI: https://doi.org/10.1016/B978-0-323-85884-7.00004-7

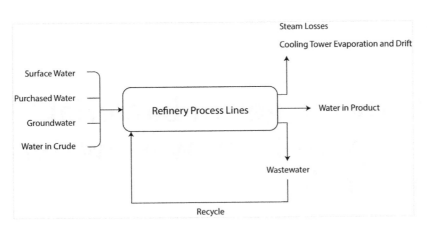

FIGURE 4.1 Refinery water process lines.

environmental pollution (Harry, 1995; Saien & Nejati, 2007; Wake, 2005). The ratio of wastewater generated to the crude processed during the refining process is about 0.4−1.6 (Coelho, Castro, Dezotti, & Sant'Anna, 2006). Significant concentrations of organic compounds like aliphatic and aromatic hydrocarbon are present during the oil processing of wastewater. A direct discharge will create a toxic effect on biological creatures and groundwater. To mitigate these toxic properties of the wastewater, the refinery must treat it well before discharging it; this biological treatment can be beneficial. Fig. 4.1 shows the typical water process lines in petroleum refinery.

Depending on the operating condition and the crude oil quality, the effluent composition varies (Benyahia, Abdulkarim, & Embaby, 2006). In refinery water is required for various process, and also crude oil contains traces of the water. While required hydrocarbon is separated as the crude is broken down, a large amount of water is produced containing the organic pollutants and recalcitrant compounds; and it is obligatory to remove these pollutants as per the practiced standards (Asatekin & Mayes, 2009; Farajnezhad & Gharbani, 2012; Rasheed, Pandian, & Muthukumar, 2011; Vendramel, Bassin, Dezotti, & Sant'Anna, 2015). There arises a problem of decreased level of dissolved oxygen, <2 mg L^{-1}, required for aquatic life by discharging the wastewater, containing the higher organic matter, into aquatic environment (Attiogbe, Glover-Amengor, & Nyadziehe, 2007). These create the anaerobic system and products of bio-chemical reactions produce the contaminated displeasing colored water. Therefore, oxygen availability is necessary to mitigate this situation.

Due to rigid regulations, researchers are motivated to design a better-equipped treatment facility providing cost-effective and highly efficient treatments. For making the water characteristics suitable for effective bioremediation methods, primary physicochemical treatment is essential. Biological process facilities primarily include anaerobic reactors, trickling filters, and activated sludge processes (Kulkarni & Kherde, 2015; Kulkarni, 2015; Marques, Souza, Souza, & Rocha, 2008). Aerobic and anaerobic treatment methods can carry out biological treatment of effluent. In situ biological treatment mainly includes methods of Bioremediation and Phytoremediation. The bioremediation method constitutes aerobic and anaerobic methods of treatment. Aerobic treatment, which uses oxygen for stabilizing organic matter, can be used for biological treatment with the disadvantage of the high pumping cost. Most of the treatment facility uses the activated sludge method for secondary treatment. While in the aerobic method, breaking of waste contaminants into simple compounds using acidogenesis, acetogenesis, hydrolysis, or methanogenesis. There are higher concentrations of methane, COD, and BOD present in the petroleum wastewater. A higher concentration of methane is being used as a fuel. A high concentration of wastewater can be treated with biological treatment as the most suitable method.

According to CONCAWE report number 2/11 (CONCAWE, 2011), in 2008, 111 of 125 refineries apply a biological treatment to their wastewater before discharge, for which 103 sites apply a three-stage biological system. Most refineries have an aerated activated sludge reactor as the biological unit (Santo et al., 2013). Also, various researchers have performed investigations for the biological treatments for petroleum wastewater.

Petroleum wastewater

In the petroleum, sector wastewater is usually termed petroleum wastewater. Various sources for petroleum wastewater exist like oilfield production, processing plants of olefin, petroleum refineries, and others. The toxicity and degradability of wastewater vary based on their originating source (Gutiérrez et al., 2007; Llop, Pocurull, & Borrull, 2009).

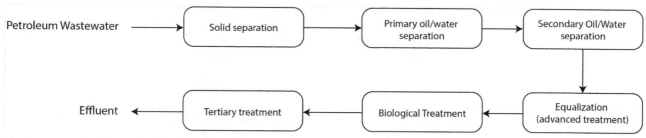

FIGURE 4.2   Different treatment stages of wastewater.

In the case of wastewater produced from the oil fields, artificial surfactants and emulsified crude are contained in large concentrations characterized by low biodegradability and high COD values (Xiao-ling, 2015). During the extraction of oil, they are generated and consist of recalcitrant organic complex pollutants such as a polymer, radioactive substances, surfactants benzenes, hummus, phenols, polycyclic aromatic hydrocarbons, and different kinds of heavy mineral oil (Tong, Zhang, Liu, Ye, & Chu, 2013; Zhao, Wang, Ye, Borthwick, & Ni, 2006b). Effluent processing of produced water is a useful option for produced water handling. Water generated from wastewater treatment is nontoxic and thus could be re-sed. The fundamental aim for processing generated water for the operators is Deoiling, Disinfection, Sand and Suspended Solid removal, Soluble organics removal, Dissolved gas removal, Desalination, Softening, and removing NORM (Arthur, Langhus, & Patel, 2005). Refinery processes, producing a surplus of 2500 products, generate petroleum wastewater. Different refinery configurations alter crude into a different ratio of utility products like Kerosene and Petrol. Units like Hydro-cracking, Hydro-skimming, Condensate, Hydro-cracker flare, Hydro-skimmer flare, Condensate flare, sour water, and the desalter accounted for the yielding of wastewater. Here, Fig. 4.2 shows different stages of wastewater in the refinery.

Additionally, in the refinery plants for wastewater, main phenol sources are obtained from neutralized spent caustic, the desalter effluent, and the tank water drain found in the receiving waste streams (Al Hashemi, Maraqa, Rao, & Hossain, 2015). Alkanes, aromatics, and polycyclic hydrocarbons were major pollutants in the refinery wastewater (Shokrollahzadeh, Azizmohseni, Golmohammad, Shokouhi, & Khademhaghighat, 2008). Also, varying concentrations of ammonia, sulfide, phenols and Benzo are generally present (Benyahia et al., 2006; Santo et al., 2013). The wastewater treatment plants in refineries comprise the following steps for oil/water separation:

a. Pretreatment/primary separation.
b. Secondary separation.

These separation processes are followed by biological treatment and in some cases, tertiary treatment. Wastewater treatment equipment is designed based on the quantity and rate at which water needs to be handled and the substance with which the inner body expose. Irrespective of equipment size, keeping the output flow rate constant and sudden change input flow leads to the lower efficiency of the treating equipment; the biological system does not respond immediately to the flow change. Thus, in petroleum refineries, biological treatment is mostly preferred wastewater treatment technology for eliminating dissolved organic compounds.

Petroleum wastewater characteristics

The first step in planning a biological treatment process in a refinery is to identify the petroleum wastewater characteristics. Refinery waters have been segregated into three types:

1. *Clean water*: It includes cooling water, boiler blowdown, and stormwater. Clean waters, often collected in large ponds to allow the solids to settle before discharge, contain considerable organic matter and little organic matter.
2. *Oily water*: It includes tank drainage, pump leakage, and all other oil-containing waters. In most cases, they are passed through separators before mixing them with process waters, but in some cases, the separation is not feasible. Thus, both waste streams are passed through oil separators. Generally, oily waters have low water-soluble organics, but the growing requirements for wastewater treatment make it a preliminary necessity to treat oily wastewater and process wastewater.
3. *Process wastewater*: It is produced by catalytic cracking and desulfurization. A primary cause of water pollution is process wastewater due to its high concentration of phenolics, ammonia, hydrogen sulfide, and mercaptans (McKinney, 1963).

Table 4.1 depicts the types of inorganic and organic materials found in petroleum. High concentrations of various contaminants in petroleum wastewater are termed environmentally hazardous (Mrayyan & Battikhi, 2005; 2010; Wake, 2005). Wastewater identification of organic pollutants exhibited that most compounds present were several fractions of (a) Aliphatic hydrocarbons up to C10, (b) Aromatic compounds like benzene, toluene, and ethyl-benzene (Saien & Nejati, 2007). The behavior of the petroleum wastewater characteristics depends on production/storage, chemicals used in the processing, and the operational conditions. Also, their composition varies by their order of magnitude. However, it is similar to oil and gas production in quality (Fillo, Koraido, & Evans, 1992). Petroleum wastewater compounds majorly include dissolved formation minerals, dissolved gases, production chemical compounds, production solids, and dissolved and dispersed oil compounds (Veil, Puder, Elcock, & Redweik, 2004). Due to the limited dissolution capacity of oil in water, mainly oil is dispersed in water (Ekins, Vanner, & Firebrace, 2007); Amount of oil dissolved or suspended in the petroleum wastewater is depends on various factors (dos Santos, Bezerra Rocha, de Araújo, de Moura, & Martínez-Huitle, 2014; Fakhru'l-Razi et al., 2009; Hansen, 1994). Employing the physical pretreatment process, maximum COD removal is about $1400-1500$ mg L^{-1}, and of BOD is about $25-35$ mg L^{-1}, (Seif, 2001). Seif (2001) conclude that separation and individual treatment for each resource was a great option against full treatment volume following blending of different resources.

Removal of naphthenic acids (NAs) from the wastewater is quite difficult. In a wastewater treatment plant of the refinery, the concentration of NAs having toxic effects is estimated to be about $2\%-9\%$ (Beili et al., 2015). Table 4.1 bridges produced water characteristics ranges from various oil fields.

Minimum discharge limits have been set by the Environmental Protection Agency (EPA) of the USA and the Central Pollution Control Board (CPCB) of the Government of India for each waste component as shown in Table 4.2

TABLE 4.1 Wastewater parameter form oilfield production (Fakhru'l-Razi et al., 2009).

Parameter	Values	Heavy metal	Values (mg L^{-1})
Density (kg m^{-3})	1014–1140	Calcium	13–25,800
Surface Tension (dynes/cm)	43–78	Sodium	132–97,000
TOC (mg L^{-1})	0–1500	Potassium	24–4300
COD (mg L^{-1})	1220	Magnesium	8.0–6000
TSS (mg L^{-1})	1.2–1000	Iron	<0.1–100
pH	4.3–10	Aluminum	310–410
Total Oil (IR; mg L^{-1})	2–565	Boron	5.0–95
Volatile (BTX; mg L^{-1})	0.39–35	Barium	1.3–650
Base/neutrals (mg L^{-1})	<140	Cadmium	<0.005–0.2
(Total nonvolatile oil and grease by GLC/Ms) base (g L^{-1})	275	Chromium	0.02–1.1
Chloride (mg L^{-1})	80–200,000	Copper	<0.002–1.5
Bicarbonate (mg L^{-1})	77–3990	Lithium	3.0–50
Sulfate (mg L^{-1})	<2–1650	Manganese	<0.004–175
Ammoniacal nitrogen (mg L^{-1})	10–300	Lead	0.002–8.8
Sulfite (mg L^{-1})	10	Strontium	0.02–1000
Total polar (mg L^{-1})	9.7–600	Titanium	<0.01–0.7
Higher acids (mg L^{-1})	<1–63	Zinc	0.01–35
Phenols (mg L^{-1})	0.009–23	Arsenic	<0.005–0.3
VFA's (volatile fatty acids) (mg L^{-1})	2.0–4900	Mercury	<0.001–0.002
		Silver	<0.001–0.15
		Beryllium	<0.001–0.004

From Fakhru'l-Razi, A., Pendashteh, A., Abdullah, L. C., Biak, D. R. A., Madaeni, S. S. & Abidin, Z. Z. (2009). Review of technologies for oil and gas produced water treatment. *Journal of Hazardous Materials, 170*(2–3), 530–551. https://doi.org/10.1016/j.jhazmat.2009.05.044.

TABLE 4.2 Environmental standards for the minimum discharge of petroleum oil refineries (Doltade et al., 2019).

Sr. No.	Parameter	As per EPA norms	As per CPCB norms	Unit
1	pH	6−8.5	6−8.5	
2	Oil and Grease	5	5	$mg\,L^{-1}$
3	BOD (3 days 27°C)	15	15	$mg\,L^{-1}$
4	COD	125	125	$mg\,L^{-1}$
5	Suspended solids	20	20	$mg\,L^{-1}$
6	Phenols	0.35	0.35	$mg\,L^{-1}$
7	Sulfides	0.5	0.5	$mg\,L^{-1}$
8	Ammonical nitrogen	15	15	$mg\,L^{-1}$

From Doltade, S. B., Dastane, G. G., Jadhav, N. L., Pandit, A. B., Pinjari, D. V., Somkuwar, N. & Paswan, R. (2019). Hydrodynamic cavitation as an imperative technology for the treatment of petroleum refinery effluent. Journal of Water Process Engineering, 29. https://doi.org/10.1016/j.jwpe.2019.02.008.

Out of the total undegraded petroleum hydrocarbon, the carcinogenic fuel additives such as dichloroethane, dichloromethane, and t-butyl methyl ether, were considered mostly (Aljuboury, Palaniandy, Abdul Aziz, & Feroz, 2017; Diya'Uddeen, Daud, & Abdul Aziz, 2011; Squillace, Zogorski, Wilber, & Price, 1996).

Pretreatment process for biological stabilization

Petroleum wastewater contains different chemicals, whose treatment process depends on wastewater chemistry, discharge requirement, and treatment efficiency. Various factors inhibit aerobic and anaerobic digestion process in biological treatment: the ability of mixed and pure culture to degrade contaminants, presence of macromolecules inside sludge, biological flocks aggregation, presence of inhibitory compounds. Therefore, the application of pretreatment can mitigate these problems and improve the efficiency of the pure and mixed culture bacteria to treat petroleum wastewater. Additionally, pretreatment applied to wastewater increases the treatment efficiency of the biological treatment (Santo et al., 2013). Primary treatment includes Flow Equalization, eliminating free oil and gross solids; eradication of dispersed oil and solids particles by flocculation/ Coagulation, flotation, sedimentation, and filtration Equalization, micro electrolysis, and others; improve biodegradability, acid and alkaline neutralization (Prabu, Suriyaprakash, Kandasamy, & Rathinasabapathy, 2015). An overview of this pretreatment process is given below.

Physical treatment

The physical pretreatment process for petroleum wastewater usually removes unwanted materials, either floating or settled, using the physical operations and separates oil, water, and solid particles. Additionally, these treatments are essential for making the wastewater suitable for bacterial (pure or mixed cultured) activities, purifying, or making the water suitable for further activities. Physical treatment includes various treatments such as screening process, evaporation, dissolved air flotation, sedimentation, microfiltration, ultrafiltration, and reverse osmosis. Initially, large suspended and floating materials are removed from the wastewater through the screening process, which can decrease the biological process's effectiveness. Then depending on the impurities or the process to implement, other physical treatments were decided (Ahmadun Fakhru'l-Razi et al., 2009; dos Santos et al., 2014; Hansen, 1994).

When the wastewater contains oil residual and salt content, then evaporation is usually prepared. However, if suspended solids were there along with oil and salt, then the dissolved air flotation method is used. If the particles are large enough to get settled by gravitational force, they are allowed to, and this process is known as sedimentation, here, settling time oil and grease can also form an insoluble layer above the water level (Abd El-Gawad, 2014). To remove the grease and oil content from the wastewater, simple skimming methods are used. The flotation method can also be implementable for separating suspended solids and liquid contaminants. If the solute molecules are of the same size as the dissolved solid or the metal content in the water, then Reverse Osmosis can be the better alternative to remove it, but it requires the pretreatment of microfiltration and ultrafiltration.

Chemical treatment

Chemical treatment is used to make wastewater more adaptable with the bioremediation agent to be used in the future. Depending on the bioremediation agent, an adequate environment is created using this treatment. Generally, chemical pretreatment includes neutralization, coagulation-flocculation, micro aeration, ozonation, and other treatment depending on the chemical composition.

Neutralization of the wastewater, depending on the bioremediation agents for main treatment, is done to make the suitable pH environment flourish biological activity. If the water is comparatively acidic for effective biological activity, then it is neutralized using the basic agent. Moreover, if the water is more basic comparatively, then wastewater is neutralized by adding acid. The neutralization process also helps in the effective coagulation-flocculation process.

Coagulation-flocculation method is used to remove the extra charge on the suspended particle and agglomerate the destabilized particle (Verma, Prasad, & Mishra, 2010). Following the coagulation-flocculation process, it is advisable to use microfiltration and ultrafiltration physical processes to remove the suspended solids effectively. The coagulation-flocculation process has proved an efficient process to remove turbidity from industrial wastewater effluents (Hossam, Emad, & Waid, 2011).

Readily biodegradable organic compounds are generated from the microaerated breaking of hydrocarbon from wastewater. Hydrolysis of water organic content increases at dissolved oxygen (DO) concentration of 0.2 to 0.3 mg L^{-1}. Reduction of the sulfate compounds is achieved by increasing the BOD/COD ratio. These, in turn, decrease the concentration of H$_2$S gas which inhibits the biological process. These also improve the biological process by increasing benzene ring organics concentration (Wu, Zhou, Wang, & Guo, 2015).

The ozonation of wastewater containing phenol or benzoic compounds can increase the BOC/COD demand for the bioremediation agent from about 20% to 30% (Lin, Tsai, Liu, & Chen, 2001).

In situ biological treatment of petroleum wastewater

In situ biological treatments are employed to eliminate physical, chemical, and biological substances from petroleum wastewater, which are disadvantageous for the ecosystem and further use. Compare to a physical and chemical process for wastewater treatment, the biological process required lower pressure and temperature, which potentially makes it the cost and energy-effective technology. Technological advancement in this field makes this technology cost-effective for generating water parameters accepted by the environment and public legislation (Fulekar, 2010). shows the overall biological technologies available for remediation of wastewater. Fig. 4.3 shows the overall biological technologies for the remediation of wastewater.

The chemical and biological parameters of petroleum wastewater are complex. This complex nature leads to many challenges despite the technologically advanced methods. Complex parameters include various aromatic, heterocyclic

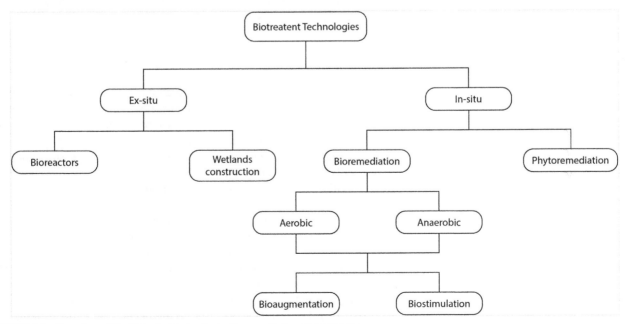

FIGURE 4.3 Overview of the biological technologies for remediation of wastewater.

and polycyclic ringed compounds, which are not easily degraded by biological degradation methods (Dai, Chen, Yan, Chen, & Guo, 2016). However, many researchers focus on making this method efficient and cost-efficient along with an eco-friendly manner. Additionally, for effluent treatment, various bioremediation methods are in commercialization (Prasad & Freitas, 2003).

Bioremediation

Studies show that certain organisms have the potential for in situ degradation of hydrocarbon from the water (Haritash & Kaushik, 2009). Researchers are trying to investigate various microbes for their feasibility of removing the hydrocarbon using the bench-scale treatment in a laboratory (Affandi, Suratman, Abdullah, Ahmad, & Zakaria, 2014).

Potential for in situ biodegradation depends on the nature of hydrocarbon, the microbes' potential, the amount of hydrocarbon and microbes, and the environment where the process takes place (Das & Chandran, 2011). This hydrocarbon acts as a source of food and energy in both aerobic and anaerobic conditions for microbes. This biological attack's potential or susceptibility depends on the type of hydrocarbon; and less branched and aliphatic compounds degrade faster than the complex aromatics compounds (Mohee & Mudhoo, 2012).

Anaerobic process

Recovery of biomethane and biohydrogen by anaerobic digestion from petroleum wastewater, as a part of the energy-recovery method, gained popularity recently (Brentner, Jordan, & Zimmerman, 2010; Rafael et al., 2008; Rahman, Khan, Khan, & Halder, 2018). This method proved to be an environment-friendly, profitable and feasible strategy (Elreedy & Tawfik, 2015; Elreedy, Fujii, & Tawfik, 2019). For biomethane recovery (Siddique, Munaim, & Ab. Wahid, 2016) achieved desired results from petrochemical wastewater. Moreover, the studies showed that pretreatment of wastewater using ultrasonic or microwave enhances anaerobic codigestion and aids in yielding biomethane of 53% and 25%, respectively (Siddique, Munaim, & Wahid, 2017). The removal efficiency depends on the chemical constituents, operating conditions, reactor type, and wastewater sources (Ji, Sun, Ni, & Tong, 2009). Anaerobic digesters have certain advantages: uniform reaction environment suitable for various wastewater ranges, efficient in handling suspended solids and having performance independent of sludge settleability, and large reactor volume that dilutes the inhibitors. However, various drawbacks include requiring a separate mixing mechanism and larger reactor volume, poor effluent quality resulting from nondegradable organic matters, and more significant anaerobic organism generation.

Performance of the various system is measured by measuring COD removing efficiency. (Ghimire & Wang, 2019) show that crude oil extraction from light, medium, heavy petroleum wastewater using different anaerobic systems at thermophilic or mesophilic conditions. Thermophilic condition removes almost 56%−71% COD and same in mesophilic can be about 93% using UASB. This accounts for a probable conclusion that extraction of light petroleum wastewater is degradable easily compared to medium and heavy oil wastewater extraction. One reason for the enhanced efficiency for light oil wastewater by UASB is the granular sludge and plug flow pattern. Compared to 20%−30% from UASB, batch systems provide improved efficiency of about 50%−60% for medium and heavy oil wastewater. Low efficiency can be due to the toxic chemicals present in wastewater and large organic molecules' high concentration.

Various advantages of UASB and AF include being mechanically simple, compact, and producing high-quality effluent, and high biomass concentration is attainable. However, certain drawbacks include reactor configuration based on experience, performance dependent on a solid phase, and less process control. Hybrid UASB/AF process removes the media cost and support required in the conventional AF process.

The bioaccumulation process is dependent on growth as compared to biosorption. Bioremediation is the most acceptable technology as it is based on natural attenuation. Most of the bioremediation methods fall under aerobic processes, and some processes containing recalcitrant molecules fall under anaerobic processes as they allow the microorganisms to degrade those hazardous molecules (Glick, Pasternak, & Patten, 2007; Ke, Bao, Chen, Wong, & Tam, 2009). A study suggested crystallization of minerals like Fe_2O_3, FeS, and $CaCO_3$ in the anaerobic sludge. Thus, mineral precipitation could hinder treatment of produced waters by anaerobic reactors (Ji et al., 2009). Another study showcased periodic purging with nitrogen gas to control hydrogen sulfide by altering pH (Vieira, Sérvulo, & Cammarota, 2005). The variation of treatment depends on the characteristics of produced water and studies of (Rincón et al., 2003) yielded different removal rates for light (87%), medium (20%), and heavy (37%) oil separation from produced water.

Energy production from the anaerobic treatment is a surplus benefit of this wastewater treatment process. Degradation of most hydrocarbons from the produced water is done aerobically as the salinity from produced water acts as an inhibitor, but some recalcitrant halogenated aromatics have to be degraded anaerobically (Xiao & Roberts, 2010).

Aerobic process

Anaerobic treatment is followed by aerobic treatment. The aerobic process has broad application in petroleum wastewater treatment due to its feasibility, more straightforward operation, the enhanced growth rate of organisms, and less sensitivity toward toxic effects than anaerobic processes. It is also called a secondary treatment process as anaerobic treatment is preferred at a primary stage when there is a high concentration of untreated wastewater. Aerobic treatment consists of:

- Activated sludge process.
- Aerated Lagoon.

The addition of a step to moving bed bioreactors (MBBRs) in the active sludge treatment causes enhanced tolerance against toxins and fluctuations. These aerobic processes account for higher chemical and COD removal efficiencies comparatively as it converts the wastewater organic compounds and recalcitrant components into carbon dioxide, water, and solid biological products. It has also increased tolerance toward organic shock load and lower biomass loss levels (Satyawali & Balakrishnan, 2008). TOC removal (78%) and oil removal (94%) from petroleum wastewater are achieved through aerobic reactor immobilization with microorganisms (Zhao et al., 2006b).

1. *The activated sludge process*: The activated sludge process (Fig. 4.4) is usually preferred out of all other biological processes available as it is most effective. It is a compact process, yet it needs considerable construction of aeration tanks and sludge settlers. Most refineries globally use this process as it accounts for reliability in biological processes. The trapped dissolved organic and inorganic materials and suspended solids in wastewater are a continuous suspension of aerobic biological growth. Organic material is used as energy and a carbon source by microorganisms for the microbial growth and conversion of food into tissue cells, water, and oxidized products like carbon dioxide.

   Microorganisms are exposed to organic contaminants of the wastewater when wastewater enters the aeration tank. To maintain the aerobic nature of sludge and solids in suspension air is injected continuously into the system. Typically biomass, in the mixture of wastewater and sludge, comprises 70%−90% organic portions and other inert solids. Typically, the conventional activated sludge process has certain advantages such as predictable and well-characterized performance, process and design well known, and beneficial for various applications. However, this process's main disadvantage is excess sludge production, longer aeration time, more considerable energy demand, and moderate operating and capital expenses (Renou, Givaudan, Poulain, Dirassouyan, & Moulin, 2008).

2. *Aerated lagoons*: Usually, biological wastewater treatment is carried out in aerated lagoons or activated sludge reactors for petroleum wastewater. At times, in an effluent treatment plant, as aerated lagoons require larger areas for their biomass concentrations, they do not meet the discharge requirements (Ma, Guo, Zhao, Chang, & Cui, 2009). Aerated lagoon promotes biological activity by mechanically injecting aeration unit in the underground basin.

There are usually two types of aerated lagoons:

1. *Aerobic-anaerobic/facultative lagoons*: In facultative lagoons, organic material is stabilized by both aerobic and anaerobic and helps to remove nitrogen and phosphorus from the wastewater. This amalgamated system consists of two chief compartments. Initially, suspended solids progress downwards via the aerobic (upper) compartment in the existence of oxygen. These solids settle in an anaerobic (lower) compartment on account of gravity and encounter

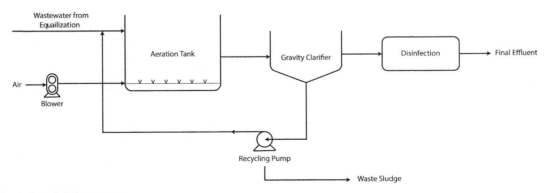

FIGURE 4.4 Activated sludge system.

anaerobic decomposition. There are certain advantages of using a facultative lagoon system such as: simple construction and operation system with pathogen destruction and periodic solid management, and lower capital, maintenance, and operation cost. However, there are certain drawbacks including vast land requirements, poor process control, and performance with higher odor potential.

2. *Aerobic lagoons*: Oxygen existing at all stages makes the settlement of suspended solids feasible at any stage. For the separation to occur, baffles are incorporated in the system or by using sludge settling. Removal sludge settling is necessary at predecided intervals, and hence sludge removal/disposal facilities are essential. Certain benefits of aerobic lagoon system include: simple construction and operation with little odor production, production of low sludge and high effluent quality, and lower capital cost. However, operation cost is higher and the land required for aerobic lagoons is significant.

The main drawback of the aerated lagoon system is extensive land, less process efficacy, and strict discharge government policies; therefore, it is less often utilized in refineries.

Integrated biological process

Individual aerobic and anaerobic processes show good potential in petroleum wastewater treatment with better efficiencies. Some chemicals cannot be degraded easily using a single process. In such cases, the advantage of both systems can be beneficial. This can be achieved by amalgamating both processes and achieving enhanced removal efficiency from petroleum wastewater. The process has some disadvantages like high sludge production of aerobic process and sensitivity toward toxic materials in the anaerobic process. In the anaerobic process, sulfur content is a fundamental disadvantage as one of the end products is hydrogen sulfide (Lettinga, Field, Sierra-Alvarez, Van Lier, & Rintala, 1991). A high amount of sulfur in the anaerobic process has a detrimental effect on the contaminant removal capacity at low pH levels (Ashrafi, Yerushalmi, & Haghighat, 2015).

An integrated biological treatment process is required to take advantage of these treatment processes (aerobic and anaerobic or physicochemical and biological processes). This process should be carried out under favorable environmental conditions to achieve the desired result (Kumar, Saha, & Sharma, 2015). Integrated biological is useful for the removal of pollutants, from petroleum wastewater, such as COD, NH_4^+-N, Oil, PAHs, and Ammonia with the efficiency ranging from 75% to 100%. The performance of properly implemented integrated systems can be more compared to individual aerobic and anaerobic treatment. Additionally, the integrated system uses an anaerobic stage to effectively remove the impurities of degradable COD, while producing biogas; and the aerobic system stage degrades other pollutants such as sulfide, ammonia, and others. Various examples of the integrated biological process include: upflow sludge blanket reactor coupled with immobilized biological aerated filters; upflow sludge blanket reactor coupled with biological aerated filters; integration of hydrolysis, moving bed biofilm reactor, ozonation unit, and biologically active carbon reactor; integration of upflow sludge blanket reactor with aerobic packed bed biofilm reactor (Liu, Ye, Tong, & Zhang, 2013; Lu, Gu, & Xu, 2013; Nasirpour, Mousavi, & Shojaosadati, 2015; Wang, Ghimire, Xin, Janka, & Bakke, 2017; Zhao et al., 2006b; Zheng, 2015; Zou, 2015).

Phytoremediation

A process where plants are used to extricate or detoxify pollutants at in situ conditions is known as Phytoremediation. Here, the plants can be used individually or with their microorganisms (Epuri & Sorensen, 1997). It is also referred to as phytotransformation as it includes the role of the internal mechanisms of plants to remove contaminants from the surfaces. These mechanisms include hydraulic control, phytovolatilization, and rhizoremediation (Newman & Reynolds, 2004).

The Alabama Department of Environmental Management provided an area of approximately 1500 cubic meters of soil, of which 70% of the original samples contained more than 100 parts per million petroleum hydrocarbons. After one year of vegetation, approximately 83% of the samples contained less than 10 ppm hydrocarbon. Removal of petroleum hydrocarbons from a range of deposits contaminated with crude oil, diesel, or refinery waste has also been investigated at initial TPH concentrations ranging from 1700 to 16000 mg kg^{-1}. The presence of certain species has resulted in higher oil losses than in soils devoid of other species of vegetation, kou (*Cordia subcordata*), milo (*Thespesia populnea*) and kiave (*Prosopis pallida*), as well as the native naupaka shrub (*Scaevola serica*), tolerate environmental conditions well and promote the restoration of diesel-contaminated soils. A huge number of fine roots in the soil have been found to bind and convert hydrophobic pollutants such as TPH, BTEX, and PAH. Grasses such as Fescue

(*Vulpia myuros*), rye (*Elymus* sp.), Clover (*Trifolium* sp.), and reed canaries (*Phalaris arundinacea*) have been used successfully in many places, especially those polluted with petrochemical waste (Das & Chandran, 2011).

Phytoremediation uses green plants to eliminate pollutants and is useful mainly in eliminating hydrocarbon and heavy metals toxins from soil and groundwater. Phytoremediation is an easily implementable, cost-effective, ecologically friendly method that virtually eliminates or detoxifies unwanted material. The fundamental limitation of using this method is its higher contaminant removing time and environmental dependent nature.

There are negligible information and data available on the contaminant removal rates and the transformation pathway at field condition (Das & Chandran, 2011). The residents lack awareness about phytoremediation, which is a significant drawback that limits the large-scale implementation of bacterial treatment by phytoremediation for the organics and metal pollutants. The environmental effect, regulatory rules, and conversion efficiency are considered while identifying the appropriate bacteria.

Commercially available bioremediation agents

(Ammar, Wang, & Yang, 2013; Brown, Gunasekera, Striebich, & Ruiz, 2016; Duan et al., 2018; Eskandari et al., 2017; Hara, Syutsubo, & Harayama, 2003; Hedlund, Geiselbrecht, Bair, & Staley, 1999; Huo et al., 2018; Jahromi, Fazaelipoor, Ayatollahi, & Niazi, 2014; Kasai, Kishira, & Harayama, 2002; Ma et al., 2015; Mielczarek, Kragelund, Eriksen, & Nielsen, 2012; Thevenieau, Fardeau, Ollivier, Joulian, & Baena, 2007; Venkateswaran, Hoaki, Kato, & Maruyama, 1995; Wang et al., 2011; Yakimov et al., 2003; Yuan, Chen, & Chang, 2011; Zeppilli et al., 2015; Zhukov, Murygina, & Kalyuzhnyi, 2007). The list of commercially available bacteria, the impurity which it can remove, and the key functions are shown in Table 4.3.

Advantages and disadvantages of biological process (using both pure or mixed culture)

Overall advantages and disadvantages of the biological processes, such as bioreactor, biologically activated sludge, lagoons, and other microbiological treatments, using both pure and mixed biological cultures are as follows:

Advantages of the biological process:

- Biodegradation using microorganisms is a simple and economically feasible process.
- Also, it is well accepted by the public.
- Various extracellular enzymes are produced by Pure cultures having high biodegradability capacity.
- Organic matter, NH3, $NH4^+$, iron can be eliminated efficiently using biodegradable wastewater treatment.
- In situ biological treatment attenuates color well.
- Exhibits enhanced removal rates of BOD and suspended solids (BAS).
- Future technologies use microbiological processes to remove the emergent pollutants from petroleum wastewater.

Disadvantages of the biological process:

- Biological treatment to remove contaminants from petroleum wastewater faces many challenges as wastewater is very complex.
- Biological degradation is restricted by the complex aromatic, polycyclic, and heterocyclic ring structured chemicals present in the wastewater (Dai et al., 2016).
- It is inefficient due to the large requirement for the maintenance and management of microorganisms.
- It is necessary to create an optimally favorable environment that is challenging to provide by biological treatment processes.
- Only some biological processes are commercially viable due to the limited research available on these processes.
- Also, it requires physical or chemical pretreatment as it is not efficient in the case of toxic or nondegradable components.
- It is a slow process treatment.
- Specific molecules like dyes have low biodegradability
- Biological wastewater treatment through BAS has poor decolorization.
- BAS also accounts for the probability of bulking of sludge and foaming.
- The biological sludge generated and degradation of products in an uncontrolled manner is a significant drawback.
- During the decomposition process, the composition of the mixed cultures tends to change.
- Complex microbiological mechanisms are one of the root causes for the failure of biological treatment.

TABLE 4.3 Petroleum hydrocarbon-degrading microorganism and their preferred degradation substrates.

Impurity	Potential bacteria/ Bacteria affiliation	Key function	Reference
n-alkanes (C6-C40)	Dietzia sp	Aliphatic	Wang et al. (2011)
n-alkanes (C10−C18)	Oleispira antarctica	Aliphatic	Yakimov et al. (2003)
n-alkanes(C13−C17)	Rhodococcus ruber	Aliphatic	Zhukov et al. (2007)
Branched and normal alkanes	Alcanivorax sp. Gordonia sihwensis	Aliphatic	Brown et al. (2016); Hara et al. (2003)
Phenolic Hydroxyl	Proteiniphilum acetatigenes Propionibacterium acidipropionici	Phenolic Hydroxyl removed after alkyl chain firstly was oxidized and then intermediates mineralized into CO_2	Duan et al. (2018)
Tetrachlorobisphenol-A	Desulfobacter Desulfofustis Desulfomicrobium	Tetrachlorobisphenol-A dechlorinated with sulfate-reducing bacteria	Thevenieau et al. (2007); Yuan et al. (2011)
Benzene	Chlorella sp.	Aromatic pollutants, including benzene, were almost completely removed	Huo et al. (2018)
Polyaromatics	Cycloclasticus Neptunomonas naphthovoran Bacillus Licheniformis Bacillus Mojavensis	Especially for aromatic pollutants	Eskandari et al. (2017); Hedlund et al. (1999); Kasai et al. (2002)
Macromolecular organics	Chloroflexi	Degrading macromolecular organics	Mielczarek et al. (2012)
Organic matters	Proteobacteria	degrading or mineralizing organic matters	Mielczarek et al. (2012)
protein and carbohydrate	Bacteroidetes Actinobacteria	degrading protein and carbohydrate for producing acetic and propionic acids	Ammar et al. (2013); Zeppilli et al. (2015)
Aromatic compounds, chlorinated hydrocarbons	Syntrophorhabdus	Degrading aromatic compounds, PAHs, chlorinated hydrocarbons and other toxic substances	Ma et al. (2015)
Asphaltenes	Citrobacter sp. Enterobacter sp. Staphylococcus sp. Lysinibacillus sp. Bacillus sp. Pseudomonas sp.	Asphaltenes degradation	Jahromi et al. (2014)
Resins	Pseudomonas sp.	Resin Degradation	Venkateswaran et al. (1995)

Conclusions

A holistic perspective should be approached in this technologically advanced century to develop a cheaper and effective method for wastewater treatment. Since water pollution is becoming a significant concern, wastewater' biological treatment shall be considered a preliminary requirement. Numerous factors challenge biodegradation underneath the natural state. Some of these include the microorganisms competing among themselves, shortfall of biologically available pollutants, deficiency in essential substrates supply, and critical external state like aeration, pH, moisture content, and temperature that can be unfavorable at times. Microbes never reside in isolation but in communities. Therefore, it is necessary to know the microbial interaction and communication amongst microorganisms themselves to mitigate bioremediation failures.

References

Abd El-Gawad, H. S. (2014). Oil and grease removal from industrial wastewater using new utility approach. *Advances in Environmental Chemistry*, 1–6. Available from https://doi.org/10.1155/2014/916878.

Affandi, I. E., Suratman, N. H., Abdullah, S., Ahmad, W. A., & Zakaria, Z. A. (2014). Degradation of oil and grease from high-strength industrial effluents using locally isolated aerobic biosurfactant-producing bacteria. *International Biodeterioration and Biodegradation*, *95*, 33–40. Available from https://doi.org/10.1016/j.ibiod.2014.04.009.

Al Hashemi, W., Maraqa, M. A., Rao, M. V., & Hossain, M. M. (2015). Characterization and removal of phenolic compounds from condensate-oil refinery wastewater. *Desalination and Water Treatment*, *54*(3), 660–671. Available from https://doi.org/10.1080/19443994.2014.884472.

Aljuboury, D. A. D. A., Palaniandy, P., Abdul Aziz, H. B., & Feroz, S. (2017). Treatment of petroleum wastewater by conventional and new technologies — A review. *Global Nest Journal*, *19*(3), 439–452. Available from https://doi.org/10.30955/gnj.002239.

Ammar, E. M., Wang, Z., & Yang, S. T. (2013). Metabolic engineering of Propionibacterium freudenreichii for n-propanol production. *Applied Microbiology and Biotechnology*, *97*(10), 4677–4690. Available from https://doi.org/10.1007/s00253-013-4861-6.

Arthur, J., Langhus, B., & Patel, C. (2005). Technical summary of oil and gas produced water treatment technologies. *ALL Consulting, LLC*. Available from http://www.all-llc.com/publicdownloads/ALLConsulting-WaterTreatmentOptionsReport.pdf.

Asatekin, A., & Mayes, A. M. (2009). Oil industry wastewater treatment with fouling resistant membranes containing amphiphilic comb copolymers. *Environmental Science and Technology*, *43*(12), 4487–4492. Available from https://doi.org/10.1021/es803677k.

Ashrafi, O., Yerushalmi, L., & Haghighat, F. (2015). Wastewater treatment in the pulp-and-paper industry: A review of treatment processes and the associated greenhouse gas emission. *Journal of Environmental Management*, *158*, 146–157. Available from https://doi.org/10.1016/j.jenvman.2015.05.010.

Attiogbe, F. K., Glover-Amengor, M., & Nyadziehe, K. T. (2007). Correlating biochemical and chemical oxygen demand of effluents, a case study of selected industries in Kumasi, Ghana, W. Afr. *The Journal of Applied Ecology*, *11*, 110–118.

Beili, W., Yi, W., Yingxin, G., Guomao, Z., Min, Y., Song, W., & Jianying, H. (2015). Occurrences and behaviors of naphthenic acids in a petroleum refinery wastewater treatment plant. *Environmental Science & Technology*, 5796–5804. Available from https://doi.org/10.1021/es505809g.

Benyahia, F., Abdulkarim, M., & Embaby, A. (2006). *Refinery wastewater treatment: a true technological challenge*.

Brentner, L. B., Jordan, P. A., & Zimmerman, J. B. (2010). Challenges in developing biohydrogen as a sustainable energy source: Implications for a research agenda. *Environmental Science and Technology*, *44*(7), 2243–2254. Available from https://doi.org/10.1021/es9030613.

Brown, L. M., Gunasekera, T. S., Striebich, R. C., & Ruiz, O. N. (2016). Draft genome sequence of Gordonia sihwensis strain 9, a branched alkane-degrading bacterium. *Genome Announcements*, *4*(3). Available from https://doi.org/10.1128/genomeA.00622-16.

Coelho, A., Castro, A. V., Dezotti, M., & Sant'Anna, G. L. (2006). Treatment of petroleum refinery sourwater by advanced oxidation processes. *Journal of Hazardous Materials*, *137*(1), 178–184. Available from https://doi.org/10.1016/j.jhazmat.2006.01.051.

CONCAWE. (2011). Trends in oil discharged with aqueous effluents from oil refineries in Europe 2005 and 2008 survey data. https://www.concawe.eu/publication/report-no-211/.

Crini, G., & Lichtfouse, E. (2019). Advantages and disadvantages of techniques used for wastewater treatment. *Environmental Chemistry Letters*, *17*(1), 145–155. Available from https://doi.org/10.1007/s10311-018-0785-9.

Dai, X., Chen, C., Yan, G., Chen, Y., & Guo, S. (2016). A comprehensive evaluation of re-circulated bio-filter as a pretreatment process for petroleum refinery wastewater. *Journal of Environmental Sciences (China)*, *50*, 49–55. Available from https://doi.org/10.1016/j.jes.2016.05.022.

Das, N., & Chandran, P. (2011). Microbial degradation of petroleum hydrocarbon contaminants: an overview. *Biotechnology Research International*, 1–13. Available from https://doi.org/10.4061/2011/941810.

Diya'Uddeen, B. H., Daud, W. M. A. W., & Abdul Aziz, A. R. (2011). Treatment technologies for petroleum refinery effluents: a review. *Process Safety and Environmental Protection*, *89*(2), 95–105. Available from https://doi.org/10.1016/j.psep.2010.11.003.

dos Santos, E. V., Bezerra Rocha, J. H., de Araújo, D. M., de Moura, D. C., & Martínez-Huitle, C. A. (2014). Decontamination of produced water containing petroleum hydrocarbons by electrochemical methods: A minireview. *Environmental Science and Pollution Research*, *21*(14), 8432–8441. Available from https://doi.org/10.1007/s11356-014-2780-4.

Duan, X., Wang, X., Xie, J., Feng, L., Yan, Y., Wang, F., & Zhou, Q. (2018). Acidogenic bacteria assisted biodegradation of nonylphenol in waste activated sludge during anaerobic fermentation for short-chain fatty acids production. *Bioresource Technology*, *268*, 692–699. Available from https://doi.org/10.1016/j.biortech.2018.08.053.

Ekins, P., Vanner, R., & Firebrace, J. (2007). Zero emissions of oil in water from offshore oil and gas installations: Economic and environmental implications. *Journal of Cleaner Production*, *15*(13–14), 1302–1315. Available from https://doi.org/10.1016/j.jclepro.2006.07.014.

Elreedy, A., Fujii, M., & Tawfik, A. (2019). Psychrophilic hydrogen production from petrochemical wastewater via anaerobic sequencing batch reactor: Techno-economic assessment and kinetic modelling. *International Journal of Hydrogen Energy*, 5189–5202. Available from https://doi.org/10.1016/j.ijhydene.2018.09.091.

Elreedy, A., & Tawfik, A. (2015). Effect of hydraulic retention time on hydrogen production from the dark fermentation of petrochemical effluents contaminated with ethylene glycol. In. *Energy Procedia, 74*, 1071−1078. Available from https://doi.org/10.1016/j.egypro.2015.07.746.

Epuri, V., & Sorensen. (1997). *Benzo(a)pyrene and hexachlorobiphenyl contaminated soil: phytoremediation potential.*

Eskandari, S., Hoodaji, M., Tahmourespour, A., Abdollahi, A., Baghi, T., Eslamian, S., & Ostad-Ali-Askari, K. (2017). Bioremediation of polycyclic aromatic hydrocarbons by bacillus licheniformis ATHE9 and bacillus mojavensis ATHE13 as newly strains isolated from oil-contaminated soil. *Journal of Geography, Environment and Earth Science International*, 1−11. Available from https://doi.org/10.9734/JGEESI/2017/35447.

Fakhru'l-Razi, Ahmadun., Pendashteh, A., Abdullah, L. C., Biak, D. R. A., Madaeni, S. S., & Abidin, Z. Z. (2009). Review of technologies for oil and gas produced water treatment. *Journal of Hazardous Materials, 170*(2−3), 530−551. Available from https://doi.org/10.1016/j.jhazmat.2009.05.044.

Farajnezhad, H., & Gharbani, P. (2012). Coagulation treatment of wastewater in petroleum industry using poly aluminum chloride and ferric chloride. *International Journal of Research and Reviews in Applied Sciences, 13*, 306−310.

Fillo, J. P., Koraido, S. M., & Evans, J. M. (1992). Sources, characteristics, and management of produced waters from natural gas production and storage operations. In J. P. Ray, & F. R. Engelhardt (Eds.), *Environmental science research*. Boston, MA: Springer. Available from https://doi.org/10.1007/978-1-4615-2902-6_12.

Fulekar, M. H. (2010). *Bioremediation Technology*. Dordrecht: Springer. Available from https://doi.org/10.1007/978-90-481-3678-0.

Ghimire, N., & Wang, S. (2019). Biological treatment of petrochemical wastewater. *Petroleum Chemicals - Recent Insight*. Available from https://doi.org/10.5772/intechopen.79655.

Glick, B. R., Pasternak, J. J., & Patten, C. L. (2007). *Bioremediation and biomass utilization. Molecular Biotechnology* (4th ed., pp. 378−415).

Gutiérrez, E., Caldera, Y., Fernández, N., Blanco, E., Paz, N., & Mármol, Z. (2007). Thermophilic anaerobic biodegradability of water from crude oil production in batch reactors. *Revista Tecnica de La Facultad de Ingenieria Universidad Del Zulia, 30*(2), 111−117. Available from https://produccioncientificaluz.org/index.php/tecnica/article/view/6137.

Hansen, B. R. (1994). Review of potential, technologies for the removal of dissolved components from produced water. *Chemical Engineering Research and Design, 72*(2), 176−188.

Hara, A., Syutsubo, K., & Harayama, S. (2003). Alcanivorax which prevails in oil-contaminated seawater exhibits broad substrate specificity for alkane degradation. *Environmental Microbiology, 5*(9), 746−753. Available from https://doi.org/10.1046/j.1468-2920.2003.00468.x.

Haritash, A. K., & Kaushik, C. P. (2009). Biodegradation aspects of polycyclic Aromatic hydrocarbons (PAHs): A review. *Journal of Hazardous Materials, 169*(1−3), 1−15. Available from https://doi.org/10.1016/j.jhazmat.2009.03.137.

Harry, M. F. (1995). *Industrial pollution handbook.*

Hedlund, B. P., Geiselbrecht, A. D., Bair, T. J., & Staley, J. T. (1999). Polycyclic aromatic hydrocarbon degradation by a new marine bacterium, *Neptunomonas naphthovorans* gen. nov., sp. nov. *Applied and Environmental Microbiology, 65*(1), 251−259. Available from https://doi.org/10.1128/aem.65.1.251-259.1999.

Hossam, A., Emad, E., & Waid, O. (2011). Pretreatment of wastewater streams from petroleum/petrochemical industries using coagulation. *Advances in Chemical Engineering and Science*, 245−251. Available from https://doi.org/10.4236/aces.2011.14035.

Huo, S., Zhu, F., Zou, B., Xu, L., Cui, F., & You, W. (2018). A two-stage system coupling hydrolytic acidification with algal microcosms for treatment of wastewater from the manufacture of acrylonitrile butadiene styrene (ABS) resin. *Biotechnology Letters, 40*(4), 689−696. Available from https://doi.org/10.1007/s10529-018-2513-8.

Jahromi, H., Fazaelipoor, M. H., Ayatollahi, S., & Niazi, A. (2014). Asphaltenes biodegradation under shaking and static conditions. *Fuel, 117*, 230−235. Available from https://doi.org/10.1016/j.fuel.2013.09.085.

Ji, G. D., Sun, T. H., Ni, J. R., & Tong, J. J. (2009). Anaerobic baffled reactor (ABR) for treating heavy oil produced water with high concentrations of salt and poor nutrient. *Bioresource Technology, 100*(3), 1108−1114. Available from https://doi.org/10.1016/j.biortech.2008.08.015.

Kasai, Y., Kishira, H., & Harayama, S. (2002). Bacteria belonging to the genus cycloclasticus play a primary role in the degradation of aromatic hydrocarbons released in a marine environment. *Applied and Environmental Microbiology, 68*(11), 5625−5633. Available from https://doi.org/10.1128/AEM.68.11.5625-5633.2002.

Ke, L., Bao, W., Chen, L., Wong, Y. S., & Tam, N. F. Y. (2009). Effects of humic acid on solubility and biodegradation of polycyclic aromatic hydrocarbons in liquid media and mangrove sediment slurries. *Chemosphere, 76*(8), 1102−1108. Available from https://doi.org/10.1016/j.chemosphere.2009.04.022.

Kulkarni, S. (2015). Biological wastewater treatment for phenol removal: A review. *International Journal of Research, 2*(2), 593−598.

Kulkarni, S., & Kherde, P. (2015). Research on advanced biological effluent treatment: a review. *International Journal of Research and Review, 2*(8), 508−512.

Kumar, S., Saha, T., & Sharma, S. (2015). Treatment of pulp and paper mill effluents using novel biodegradable polymeric flocculants based on anionic polysaccharides: a new way to treat the waste water. *International Research Journal of Engineering and Technology, 2*(4), 1415−1428.

Lettinga, G., Field, J. A., Sierra-Alvarez, R., Van Lier, J. B., & Rintala, J. (1991). Future perspectives for the anaerobic treatment of forest industry wastewaters. *Water Science and Technology, 24*(3−4), 91−102. Available from https://doi.org/10.2166/wst.1991.0466.

Lin, C. K., Tsai, T. Y., Liu, J. C., & Chen, M. C. (2001). Enhanced biodegradation of petrochemical wastewater using ozonation and bac advanced treatment system. *Water Research, 35*(3), 699−704. Available from https://doi.org/10.1016/S0043-1354(00)00254-2.

Liu, G. H., Ye, Z., Tong, K., & Zhang, Y. H. (2013). Biotreatment of heavy oil wastewater by combined upflow anaerobic sludge blanket and immobilized biological aerated filter in a pilot-scale test. *Biochemical Engineering Journal, 72*, 48−53. Available from https://doi.org/10.1016/j.bej.2012.12.017.

Llop, A., Pocurull, E., & Borrull, F. (2009). Evaluation of the removal of pollutants from petrochemical wastewater using a membrane bioreactor treatment plant. *Water, Air, and Soil Pollution, 197*(1−4), 349−359. Available from https://doi.org/10.1007/s11270-008-9816-7.

Lu, M., Gu, L. P., & Xu, W. H. (2013). Treatment of petroleum refinery wastewater using a sequential anaerobic-aerobic moving-bed biofilm reactor system based on suspended ceramsite. *Water Science and Technology, 67*(9), 1976–1983. Available from https://doi.org/10.2166/wst.2013.077.

Ma, F., Guo, J. b, Zhao, L. j, Chang, C. c, & Cui, D. (2009). Application of bioaugmentation to improve the activated sludge system into the contact oxidation system treating petrochemical wastewater. *Bioresource Technology, 100*(2), 597–602. Available from https://doi.org/10.1016/j.biortech.2008.06.066.

Ma, K. L., Li, X. K., Wang, K., Zhou, H. X., Meng, L. W., & Zhang, J. (2015). 454-Pyrosequencing reveals microbial community structure and composition in a mesophilic UAFB system treating PTA wastewater. *Current Microbiology, 71*(5), 551–558. Available from https://doi.org/10.1007/s00284-015-0884-9.

Maiti, S. (2004). *Handbook of methods in environmental studies: Water and waste water analysis.* ABD Publishers.

Marques, J. J., Souza, R. R., Souza, C. S., & Rocha, I. C. C. (2008). Attached biomass growth and substrate utilization rate in a moving bed biofilm reactor. *Brazilian Journal of Chemical Engineering, 25*(4), 665–670. Available from https://doi.org/10.1590/S0104-66322008000400004.

McKinney, R. E. (1963). *Biological Treatment of Petroleum Refinery Wastes.*

McKinney, R. E. (1967). Biological treatment systems for refinery wastes. *Journal of the Water Pollution Control Federation, 39*(3), 346–359. Available from https://www.jstor.org/stable/25035752.

Mielczarek, A. T., Kragelund, C., Eriksen, P. S., & Nielsen, P. H. (2012). Population dynamics of filamentous bacteria in Danish wastewater treatment plants with nutrient removal. *Water Research, 46*(12), 3781–3795. Available from https://doi.org/10.1016/j.watres.2012.04.009.

Mittal, A. (2011). Biological wastewater treatment. *Water Today, 32–44.* Available from https://www.watertoday.org/Article%20Archieve/Aquatech%2012.pdf.

Mohee, R., & Mudhoo, A. (2012). *Bioremediation and sustainability: Research and applications.* John Wiley and Sons. Available from https://doi.org/10.1002/9781118371220.

Mrayyan, B., & Battikhi, M. N. (2005). Biodegradation of total organic carbons (TOC) in Jordanian petroleum sludge. *Journal of Hazardous Materials, 120*(1–3), 127–134. Available from https://doi.org/10.1016/j.jhazmat.2004.12.033.

Nasirpour, N., Mousavi, S. M., & Shojaosadati, S. A. (2015). Biodegradation potential of hydrocarbons in petroleum refinery effluents using a continuous anaerobic-aerobic hybrid system. *Korean Journal of Chemical Engineering, 32*(5), 874–881. Available from https://doi.org/10.1007/s11814-014-0307-9.

Newman, L. A., & Reynolds, C. M. (2004). Phytodegradation of organic compounds. *Current Opinion in Biotechnology, 15*(3), 225–230. Available from https://doi.org/10.1016/j.copbio.2004.04.006.

Pal, S., Banat, F., Almansoori, A., & Abu Haija, M. (2016). Review of technologies for biotreatment of refinery wastewaters: Progress, challenges and future opportunities. *Environmental Technology Reviews, 5*(1), 12–38. Available from https://doi.org/10.1080/21622515.2016.1164252.

Prabu, S. L., Suriyaprakash, T. N. K., Kandasamy, R., & Rathinasabapathy, T. (2015). Effective waste water treatment and its management. *Toxicity and waste management using bioremediation, 312–334.* Available from https://doi.org/10.4018/978-1-4666-9734-8.ch016, IGI Global.

Prasad, M., & Freitas, H. (2003). Metal hyperaccumulation in plants-biodiversity prospecting for phytoremediation technology. *Electron Journal of Biotechnology, 6, 275–321.*

Rafael, L., Lorenzo, H.-D., Campelo, J. M., Clark, J. H., Hindalgo, J. M., Diego, L., ... Romero, A. A. (2008). Biofuels: A technological perspective. *Energy & Environmental Science, 542.* Available from https://doi.org/10.1039/b807094f.

Rahman, W. U., Khan, M. D., Khan, M. Z., & Halder, G. (2018). Anaerobic biodegradation of benzene-laden wastewater under mesophilic environment and simultaneous recovery of methane-rich biogas. *Journal of Environmental Chemical Engineering, 6*(2), 2957–2964. Available from https://doi.org/10.1016/j.jece.2018.04.038.

Rasheed, Q. J., Pandian, K., & Muthukumar, K. (2011). Treatment of petroleum refinery wastewater by ultrasound-dispersed nanoscale zero-valent iron particles. *Ultrasonics Sonochemistry, 18*(5), 1138–1142. Available from https://doi.org/10.1016/j.ultsonch.2011.03.015.

Renou, S., Givaudan, J. G., Poulain, S., Dirassouyan, F., & Moulin, P. (2008). Landfill leachate treatment: Review and opportunity. *Journal of Hazardous Materials, 150*(3), 468–493. Available from https://doi.org/10.1016/j.jhazmat.2007.09.077.

Rincón, N., Chacín, E., Marín, J., Fernández, N., Torrijos, M., & Moletta, R. (2003). Anaerobic biodegradability of water separated from extracted crude oil. *Environmental Technology (United Kingdom), 24*(8), 963–970. Available from https://doi.org/10.1080/09593330309385634.

Saien, J., & Nejati, H. (2007). Enhanced photocatalytic degradation of pollutants in petroleum refinery wastewater under mild conditions. *Journal of Hazardous Materials, 148*(1–2), 491–495. Available from https://doi.org/10.1016/j.jhazmat.2007.03.001.

Santo, C. E., Vilar, V. J. P., Bhatnagar, A., Kumar, E., Botelho, C. M. S., & Boaventura, R. A. R. (2013). Biological treatment by activated sludge of petroleum refinery wastewaters. *Desalination and Water Treatment, 51*(34–36), 6641–6654. Available from https://doi.org/10.1080/19443994.2013.792141.

Satyawali, Y., & Balakrishnan, M. (2008). Wastewater treatment in molasses-based alcohol distilleries for COD and color removal: A review. *Journal of Environmental Management, 86*(3), 481–497. Available from https://doi.org/10.1016/j.jenvman.2006.12.024.

Sawyer, C., McCarty, P., & Parkin, G. (2003). *Chemistry for environmental engineering and science. McGraw-Hill series in civil and environmental engineering.* McGraw-Hill.

Seif, H. (2001). Physical treatment of petrochemical wastewater. In *Sixth international water technology conference,* IWTC, Alexandria, Egypt. http://www.iwtc.info/2001_pdf/10-1.pdf.

Shokrollahzadeh, S., Azizmohseni, F., Golmohammad, F., Shokouhi, H., & Khademhaghighat, F. (2008). Biodegradation potential and bacterial diversity of a petrochemical wastewater treatment plant in Iran. *Bioresource Technology, 99*(14), 6127–6133. Available from https://doi.org/10.1016/j.biortech.2007.12.034.

Siddique, M. N. I., Munaim, M. S. A., & Ab. Wahid, Z. (2016). Role of hydraulic retention time in enhancing bioenergy generation from petrochemical wastewater. *Journal of Cleaner Production, 133,* 504–510. Available from https://doi.org/10.1016/j.jclepro.2016.05.183.

Siddique, M. N. I., Munaim, M. S. A., & Wahid, Z. B. A. (2017). The combined effect of ultrasonic and microwave pretreatment on bio-methane generation from codigestion of petrochemical wastewater. *Journal of Cleaner Production*, *145*, 303–309. Available from https://doi.org/10.1016/j.jclepro.2017.01.061.

Squillace, P. J., Zogorski, J. S., Wilber, W. G., & Price, C. V. (1996). Preliminary assessment of the occurrence and possible sources of MTBE in groundwater in the United States, 1993-1994. *Environmental Science and Technology*, *30*(5), 1721–1730. Available from https://doi.org/10.1021/es9507170.

Tajrishy, M., & Abrishamchi, A. (2005). Integrated approach to water and wastewater management for Tehran, Iran, water conservation, reuse and recycling. *Proceedings of the Iranian-American Workshop*.

Tchobanoglous, G., Burton, & Stensel. (2003). Wastewater engineering (treatment disposal reuse).

Thevenieau, F., Fardeau, M. L., Ollivier, B., Joulian, C., & Baena, S. (2007). Desulfomicrobium thermophilum sp. nov., a novel thermophilic sulphate-reducing bacterium isolated from a terrestrial hot spring in Colombia. *Extremophiles: Life Under Extreme Conditions*, *11*(2), 295–303. Available from https://doi.org/10.1007/s00792-006-0039-9.

Tong, K., Zhang, Y., Liu, G., Ye, Z., & Chu, P. K. (2013). Treatment of heavy oil wastewater by a conventional activated sludge process coupled with an immobilized biological filter. *International Biodeterioration and Biodegradation*, *84*, 65–71. Available from https://doi.org/10.1016/j.ibiod.2013.06.002.

Veil, J., Puder, M., Elcock, & Redweik, R. (2004). A White paper describing produced water from production of crude oil, natural gas and coal bed methane.

Vendramel, S., Bassin, J. P., Dezotti, M., & Sant'Anna, G. L. (2015). Treatment of petroleum refinery wastewater containing heavily polluting substances in an aerobic submerged fixed-bed reactor. *Environmental Technology (United Kingdom)*, *36*(16), 2052–2059. Available from https://doi.org/10.1080/09593330.2015.1019933.

Venkateswaran, K., Hoaki, T., Kato, M., & Maruyama, T. (1995). Microbial degradation of resins fractionated from Arabian light crude oil. *Canadian Journal of Microbiology*, *41*(4–5), 418–424. Available from https://doi.org/10.1139/m95-055.

Verma, S., Prasad, B., & Mishra, I. M. (2010). Pretreatment of petrochemical wastewater by coagulation and flocculation and the sludge characteristics. *Journal of Hazardous Materials*, *178*(1–3), 1055–1064. Available from https://doi.org/10.1016/j.jhazmat.2010.02.047.

Vieira, D. S., Sérvulo, E. F. C., & Cammarota, M. C. (2005). Degradation potential and growth of anaerobic bacteria in produced water. *Environmental Technology*, *26*(8), 915–922. Available from https://doi.org/10.1080/09593332608618499.

Wake, H. (2005). Oil refineries: A review of their ecological impacts on the aquatic environment. *Estuarine, Coastal and Shelf Science*, *62*(1–2), 131–140. Available from https://doi.org/10.1016/j.ecss.2004.08.013.

Wang, S., Ghimire, N., Xin, G., Janka, E., & Bakke, R. (2017). Efficient high strength petrochemical wastewater treatment in a hybrid vertical anaerobic biofilm (HyVAB) reactor: A pilot study. *Water Practice and Technology*, *12*(3), 501–513. Available from https://doi.org/10.2166/wpt.2017.051.

Wang, X. B., Chi, C. Q., Nie, Y., Tang, Y. Q., Tan, Y., Wu, G., & Wu, X. L. (2011). Degradation of petroleum hydrocarbons (C6-C40) and crude oil by a novel Dietzia strain. *Bioresource Technology*, *102*(17), 7755–7761. Available from https://doi.org/10.1016/j.biortech.2011.06.009.

Wu, C., Zhou, Y., Wang, P., & Guo, S. (2015). Improving hydrolysis acidification by limited aeration in the pretreatment of petrochemical wastewater. *Bioresource Technology*, *194*, 256–262. Available from https://doi.org/10.1016/j.biortech.2015.06.072.

Xiao, Y., & Roberts, D. J. (2010). A review of anaerobic treatment of saline wastewater. *Environmental Technology*, *31*(8–9), 1025–1043. Available from https://doi.org/10.1080/09593331003734202.

Xiao-ling, Z. (2015). Treatment of heavy oil wastewater by UASB–BAFs using the combination of yeast and bacteria. *Environmental Technology*, 2381–2389. Available from https://doi.org/10.1080/09593330.2015.1030346.

Yakimov, M. M., Giuliano, L., Gentile, G., Crisafi, E., Chernikova, T. N., Abraham, W. R., ... Golyshin, P. N. (2003). Oleispira antarctica gen. nov., sp. nov., a novel hydrocarbonoclastic marine bacterium isolated from Antarctic coastal sea water. *International Journal of Systematic and Evolutionary Microbiology*, *53*(3), 779–785. Available from https://doi.org/10.1099/ijs.0.02366-0.

Yuan, S. Y., Chen, S. J., & Chang, B. V. (2011). Anaerobic degradation of tetrachlorobisphenol-A in river sediment. *International Biodeterioration and Biodegradation*, *65*(1), 185–190. Available from https://doi.org/10.1016/j.ibiod.2010.11.001.

Zeppilli, M., Villano, M., Aulenta, F., Lampis, S., Vallini, G., & Majone, M. (2015). Effect of the anode feeding composition on the performance of a continuous-flow methane-producing microbial electrolysis cell. *Environmental Science and Pollution Research*, *22*(10), 7349–7360. Available from https://doi.org/10.1007/s11356-014-3158-3.

Zhao, X., Wang, Y., Ye, Z., Borthwick, A. G. L., & Ni, J. (2006b). Oil field wastewater treatment in biological aerated filter by immobilized microorganisms. *Process Biochemistry*, *41*(7), 1475–1483. Available from https://doi.org/10.1016/j.procbio.2006.02.006.

Zheng, T. (2015). A compact process for treating oilfield wastewater by combining hydrolysis acidification, moving bed biofilm, ozonation and biologically activated carbon techniques. *Environmental Technology*, *37*(9), 1171–1178. Available from https://doi.org/10.1080/09593330.2015.1105301.

Zhukov, D. V., Murygina, V. P., & Kalyuzhnyi, S. V. (2007). Kinetics of the degradation of aliphatic hydrocarbons by the bacteria Rhodococcus ruber and *Rhodococcus erythropolis*. *Applied Biochemistry and Microbiology*, *43*(6), 587–592. Available from https://doi.org/10.1134/S0003683807060038.

Zou, X. (2015). Treatment of heavy oil wastewater by UASB–BAFs using the combination of yeast and bacteria. *Environmental Technology*, *36*(18), 2381–2389. Available from https://doi.org/10.1080/09593330.2015.1030346.

Chapter 5

Prospects of green technology in the management of refinery wastewater: application of biofilms

Taghreed Al-Khalid[1], Riham Surkatti[2] and Muftah H. El-Naas[3]

[1]Department of Chemical and Petroleum Engineering, College of Engineering, UAE University, Al-Ain, United Arab Emirates, [2]Gas Processing Center, College of Engineering, Qatar University, Doha, Qatar, [3]Gas Processing Center, Qatar University, Doha, Qatar

Introduction

Pollution problem

Industrial and domestic facilities continue to contribute to the extensive generation of wastewaters worldwide with a widely diverse range of pollutants. The oil and gas industry, particularly refineries, produces through its different operations effluents containing large quantities of highly toxic organic compounds, such as phenols. The petroleum industry is also challenged by another issue that is the huge amount of produced water resulting from crude oil extraction, which also contains several inorganic and organic compounds, considered to be of high environmental impact.

The disposal of these streams into water bodies represents a serious threat as they can significantly affect the aquatic ecosystem. Aquatic life could suffer from detrimental effects, which could be further worsened by the formation of a layer of immiscible oily matter on the surface of water that acts as a barrier to sunlight and oxygen, and thus leading to anoxic conditions for aquatic life due to deficiency of dissolved oxygen. It has been shown that these conditions are highly linked to an increased mortality rate of aquatic organisms as well as neural and physiological abnormalities in the embryo stage (Al-Khalid & El-Naas, 2018; Jain, Majumder, Ghosal, & Gupta, 2020). The quality of groundwater could also deteriorate as these toxic compounds have high solubility in water, and due to their persistence, they can leak into groundwater upon long transportation (Al-Khalid & El-Naas, 2018; Jain et al., 2020). Additionally, it has been reported that prolonged exposure to these toxic hydrocarbons is highly associated with different types of anomalies (Jain et al., 2020). Aliphatic and aromatic compounds such as benzene, toluene, ethylbenzene, xylene (BTEX), phenols, and polycyclic aromatic hydrocarbons (PAHs) were reported by the US Environmental Protection Agency (USEPA) as priority pollutants confirmed to be carcinogenic, mutagenic, or teratogenic. Their adverse effect may destroy the metabolic and enzymatic activities of aquatic microorganisms even at low concentrations, resulting in a negative impact on the entire array of ecological processes. Jain et al. (Jain et al., 2020) provided an informative summary on the various environmental impacts of some main components present in wastewater effluents from petroleum refineries and petrochemical plants.

The rise in the global energy demand, with increasing urbanization and industrialization, continues to pose several challenges regarding the discharge of industrial wastewaters as a critical issue for the preservation of the environment. During the transformation of crude oil, more than 2500 species of end products are produced such as gasoline, kerosene, liquefied petroleum gas, gas oil, jet fuel, diesel, petrochemical raw materials, and lubricating oils. All of these are obtained through a train of separation processes ranging from initial desalination, distillation, conversion via catalytic and thermal cracking, catalytic and steam reforming, isomerization, alkylation, and lubricating oils production, as well as treatment processes (such as sour water treatment, diesel desulfurization, and catalyst regeneration). The production of all these products requires different configurations of process units and refineries at different levels of complexity, which leads to the discharge of effluents containing biorefractory and highly toxic compounds. In the last few decades,

Petroleum Industry Wastewater. DOI: https://doi.org/10.1016/B978-0-323-85884-7.00006-0

water pollution has been globally realized as a challenging problem owing to the increased disposal of industrial wastes into water bodies, and therefore, with a growing concern on the environment raised by social and political authorities, there has been an urgent need to implement effective policies for wastewater management and to develop eco-friendly wastewater treatment technologies, mainly aimed at the complete mineralization of petroleum refinery wastewater (PRWW) (Al-Khalid & El-Naas, 2018; Jain et al., 2020).

In light of this alarming fact, PRWW should be treated to comply with the environmental legislative norms before being discharge into aqueous ecosystems. As frequently reported (Al-Khalid & El-Naas, 2018; Diya'Uddeen, Daud, & Abdul Aziz, 2011; Jain et al., 2020), these effluents are generally characterized by high toxicity, high chemical oxygen demand (COD), considerable biochemical oxygen demand (BOD), total dissolved solids (TDS), total suspended solids (TSS), total organic carbon (TOC), turbidity, and phenolic content. However, the qualitative and quantitative analysis of PRWW effluents depend on their origin and vary considerably with the location and the type of processing unit, as well as the process configuration and complexity (Diya'Uddeen et al., 2011). Therefore, the characterization of the PRWW is a preliminary important step towards the selection of the optimum treatment method for PRWW, to be also guided by environmental and economic aspects within the local regulatory framework (Al-Khalid et al., 2020). In this context, traditional treatment methods are often incapable of meeting the standards set by the environmental agencies, which shifts research to new frontiers in the scope of proficient technologies.

Overview of treatment options

The concern about water safety and quality for human consumption and environmental requirements has strongly pushed the search for several alternative methods to treat polluted water. The different treatment methods are generally categorized into classical single-step methods, and hybrid or combined (integrated) methods (Prabakar et al., 2018). Another classification is based on the three process categories: physical, chemical, and biological. Membrane filtration and activated carbon adsorption are the most widely used physical processes. The chemical processes include the traditional chemical precipitation by the use of flocculants and coagulants, comprising electrocoagulation, electroflocculation, electrodeposition, and electroflotation. However, oxidation is the most common chemical process for the treatment of PRWW, including both electrochemical oxidation and the more advanced oxidative processes (AOPs) such as photocatalytic degradation, Fenton process, and microwave-assisted catalytic wet air oxidation, which have attracted great attention because of their ability to completely degrade a lot of organic substances, resulting in complete mineralization (Al-Khalid & El-Naas, 2018; Al-Khalid, Surkatti, & El-Naas, 2020; Malakar, Saha, Baskaran, & Rajamanickam, 2017; Al-Khalid et al., 2020; Malakar et al., 2017). Diya'Uddeen et al. (2011); Rasalingam et al. (2014) reviewed different treatment technologies for PRWW, with a focus on photocatalytic degradation as an advanced oxidation process (AOP). Biological treatment is considered as one of the most preferred and least expensive options for the decontamination of PRWW, either under aerobic or anaerobic prevailing conditions. Biological methods rely on the fundamental role of the enzymatic activity of microorganisms such as bacteria, algae, and fungi, to catalyze metabolic reactions for energy conversion using the carbon content in the organic waste, hence converting the toxic substances into less harmful intermediates, leading to complete mineralization of the toxic organic contaminants to water and CO_2. A wide variety of processes and techniques are considered for the biological treatment of PRWW, including but not limited to sequencing batch reactors (SBRs), rotating biological contactor (RBC), trickling filters, biological aerated filters, membrane bioreactor (MBR), moving bed bioreactor (MBBR), anaerobic membrane bioreactors (AMBRs), and upflow anaerobic sludge blanket (UASB) (Singh and Borthakur, 2018; Al-Ghouti et al., 2019), in addition to the conventional and most commonly used activated sludge process (ASP) that has long been adopted as the standard and prevalent biological treatment of PRWW. Jain et al. (2020) presented a summary of the various biological treatment processes used for the treatment of petrochemical waste in terms of operating conditions, hydraulic retention time (HRT), and performance level. A brief description of the main treatment processes as applied for the removal of volatile organic compounds present in petroleum refinery effluents, their salient features, main advantages, and drawbacks is provided in a comparative study by Malakar et al. (2017). The topic was also reviewed by Al-Khalid & El-Naas (2018) and Al-Khalid et al. (2020). Adham, Hussain, Minier-Matar, Janson, & Sharma (2018) discussed comprehensively different case studies covering a broad range of membrane technology applications for the remediation of wastewater in the oil and gas industry.

Although many wastewater treatment processes have attained good removal efficiencies, there is still no established physicochemical or biological treatment technique that is completely effective in removing persistent organic pollutants from all types of RWW. A plausible solution is sought after and it is strongly recommended to employ integrated systems of physical, chemical, and biological treatment methods. Mei et al. (Mei et al., 2020) developed a green membrane system for phenols separation and recovery from simulated PRWW; the polypropylene-hollow-fiber supported liquid membrane (SLM) system was designed to use vegetable oil as the liquid membrane phase so that the SLM system can efficiently recover phenols before the biotreatment. In general, the track of PRWW treatment is depicted by three main stages: pretreatment (primary), main treatment step (secondary), and final polishing treatment step (tertiary). In the primary step, the aim is the reduction or removal of the suspended solids, oil and grease, and turbidity, mainly attained by physical or mechanical techniques, such as dissolved air floatation, equalization tanks, and gravity separation, such as the American Petroleum Institute (API) separator. This step is critical for the effective and sustained performance of the secondary treatment step (Al-Khalid & El-Naas, 2018; Diya'Uddeen et al., 2011; El-Naas, Alhaija, & Al-Zuhair, 2014). The remaining oil, degradable organic substances, and a fraction of biorefractory organic contaminants are targeted in the second stage, which is mainly a bioprocess (Jain et al., 2020). Tertiary treatment follows to lower the concentration of the contaminants to meet the specific allowable discharge limits set by environmental authorities. In this context, various AOPs, such as photocatalysis, wet air oxidation, and Fenton process are employed, in preference to adsorption due to the possibility of complete mineralization as a parent compound is destroyed into its nontoxic inorganic constituents, whereas adsorption merely transfers the organic pollutants from aqueous medium to a solid medium (Jain et al., 2020). Al-Khalid et al. (2020) presented an extensive discussion on the adoption of combined or integrated schemes for the treatment of industrial wastewater.

Biotreatment as a green technology

Most of the aforementioned physicochemical methods are obstructed by noticeable drawbacks. In addition to the incapability of treating heavily loaded PRWW with high COD levels, they are restricted by inherent limitations such as technical complexity, high energy consumption as well as high capital and operating costs, which question their technical and economic feasibility. Basically, most of these processes do not render the toxic material mineralized, but rather move it to another phase (Abdelwahab, Amin, & El-Ashtoukhy, 2009; Al-Khalid & El-Naas, 2018; El-Naas, Al-Zuhair, & Alhaija, 2010). As for the conventional treatment processes, the handling of a tremendous amount of oily sludge produced during the process is highly problematic. Accordingly, there is a pressing demand for the development of eco- and cost-effective as well as energy-efficient wastewater treatment technologies (Jain et al., 2020). Research has been aimed at developing more robust and sustainable technologies for the removal of persistent pollutants from PRWW. Compared to other alternatives, biotreatment is often preferred for the decontamination of PRWW due to its many advantages. Biological treatment is inexpensive, and with the possibility of complete mineralization of toxic compounds, it is considered to be environment-friendly. It can also be an integral stage in synergic schemes devised for enhanced efficiency, where it is combined with other treatment techniques in a cost-effective treatment sequence. As a green technology, biotechnology has proved to play a key role in line with the recent trend towards sustainability, hence developing novel reactor systems and effective biocatalysts have become the topic of priority research in the area, focused on the biodegradation of major contaminants in industrial wastewater (Al-Khalid et al., 2020). In the following sections, light is shed on the biotreatment of PRWW, limited mainly to bacterial cultures, as to applications and different related aspects of acclimatization and immobilization.

Biotreatment options for refinery wastewater

Biological methods are driven by the metabolic and enzymatic capabilities of microbial species, mainly bacteria, algae, and fungi for detoxification of PRWW, where microorganisms employ enzymatic mechanisms to decompose hydrocarbon substances into smaller molecules and use them for their metabolism as carbon and energy sources. As a result, dissolved organic constituents and ammonia are mineralized into water and CO_2 and nitrates/nitrites, respectively (Al-Khalid et al., 2020).

Occasionally, some substances require additional biotreatment techniques to enhance the rate of the natural process of their biodegradation; biostimulation and bioaugmentation are the most common. In biostimulation, this is achieved by adding nutrients to the contaminated medium, which stimulates the growth of the microorganisms present and accelerates the biodegradation of the contaminants. This is exemplified by the significantly stimulated growth of degrading bacteria and fungi caused by the addition of nitrogen and phosphorus, which are naturally existing nutrients in low concentrations in the oceans. Leoncio et al. (Leoncio et al., 2020) conducted pioneering research on the evaluation of accelerated biodegradation of aggregates of oil-suspended particulate matter collected from PRWW. It was proved that the indigenous bacteria were sufficiently biostimulated by the addition of nutrients (NH_4Cl, $NaNO_3$, and KH_2PO_4) best at the lowest concentration, which enhanced the biodegradation of Total Petroleum Hydrocarbons (TPH) averaged at 98.65%. The methodology was suggested as an efficient and eco-sustainable application to a water treatment unit that is appended to an oil refinery, with further tests on the minimum required nutrient concentration to cut on biostimulation costs.

As for bioaugmentation, it is a specially designed process that depends on the selective addition of pure strains or mixed cultures to wastewater treatment reactors, to improve the enzymatic breakdown of certain pollutants, for example, recalcitrant organics, or overall COD (Herrero & Stuckey, 2015). Worldwide, the effect of bioaugmentation on biodegradation has been reported, showing that native bacterial consortium gets tolerant to increased levels of hydrocarbon toxicity (Varjani, Pandey, & Upasani, 2020). In their study, Varjani et al. (2020) pointed out the effectiveness of bioaugmentation for the degradation of oily sludge by a novel bacterial consortium, which incorporates hydrocarbon-degrading microorganisms. The treatment was recommended for remediation of soil polluted with oily sludge. Recently, the potential of this option has been emphasized in view of the advantages offered by advanced immobilization techniques and modern bioreactor design, coupled with significant advances in multiple areas like microbial ecology, and molecular biology. The feasibility of bioaugmentation is a serious technical issue that should dominantly be assessed based on field application, because the assembly of the microbial community in biological systems is a highly complicated and multifaceted process that involves different microorganisms interacting using different types of cell signaling and working synergically to treat the effluent. At the moment, the process is still inadequately understood, with the basic knowledge required on the role of various key factors. This is not limited to the selection of strain (or custom-made consortium), but also extends to the manner this complex community is formed and protected to help the selected microorganisms maintain their activities, as well as the different aspects of acclimation, immobilization, scaling up, and reactor design. Herrero and Stuckey (2015) provided a detailed review on bioaugmentation and its application for wastewater treatment, with emphasis placed on emerging areas of research such as nanomaterials and protein engineering. Zhou et al. (2020) carried out both treatments of biostimulation and bioaugmentation to study the effect of native biosurfactant-producing bacterial strain, isolated from hydraulic fracturing flowback and produced water (HF-FPW), on COD removal and degradation of petroleum hydrocarbons in HF-FPW obtained from an oil-producing area. Biosurfactants are a class of molecules that can enhance the removal of contaminants by a mechanism that relies on reducing surface tension and hence increasing the solubility and bioavailability of many hydrophobic hazardous compounds. The study was claimed the first attempt for the microbial treatment of HF-FPW by using native biosurfactant-producing cultures that shed light on the development of this treatment. The addition of the native bacteria could remarkably boost the removal of *n*-alkanes, COD, and PAHs. Moreover, an evident modification of the bacterial community was manifested by enriching the hydrocarbon-degrading consortia. Accordingly, the study recommended the incorporation of indigenous biosurfactant-producing microorganisms as a favorable in situ bioremediation technology for treating HF-FPW.

In general, a biofilm is described as a biomass collection featured by cells that are attached to a surface or interface (Abu Bakar et al., 2018). The following discussion will address three categories of biotreatment applications: first, the conventional aerated tank application known as activated sludge process (ASP) by suspended (free) bacterial cultures, second the biofilm applications modified from the conventional ASP by film attachment (surface immobilization), and third the special category of immobilized cell applications based on novel immobilization techniques of cells entrapment in a microporous matrix. The second and third categories will be discussed after a discussion on acclimatization and immobilization.

Conventional biotreatment with free or suspended bacterial culture activated sludge process (ASP)

The conventional ASP has long been widely used for the decontamination of PRWW, often combined with mechanical and physicochemical pretreatment (or posttreatment) in an integrated approach (Al-Khalid & El-Naas, 2014; Al-Khalid and El-Naas, 2018; El-Naas et al., 2014). Conventional ASP systems are based on two controlling mechanisms: biodegradation and bioflocculation. Bioflocculation is prompted by the presence of suspended sludge, which motivates the formation of flocs (activated sludge); soluble and insoluble materials will get adsorbed within the matrix of these flocs. Thus, their physical installation incorporates two separate tanks: (i) an aeration tank that involves a biological step of mineralization reactions and (ii) a settling tank that involves a flocculation step for the settlement of the activated sludge from the treated water, followed by physical or mechanical separation of these flocs. Although it has been reported that removal efficiency of 98%−99% of hydrocarbon contaminants can be achieved within an average retention time of 20 days (Al-Ghouti, Al-Kaabi, Ashfaq, & Da'na, 2019; Jiménez, Micó, Arnaldos, Medina, & Contreras, 2018), the conventional ASP system suffers from noticeable drawbacks. In fact, an inherent obstacle is the very long HRT (Jain et al., 2020). The separation of aeration and settling tanks is a technical disadvantage, in addition to excessive sludge formation. Other deficiencies include the high energy demand by aeration and recycling units as well as the requirement of large floor area and low biomass concentrations in the aeration tank (Al-Khalid et al., 2020; Nancharaiah & Kiran Kumar Reddy, 2018).

Aside from conventional ASP, few studies are documented on the use of free cells for the treatment of PRWW; most studies employing free or suspended bacterial cultures were performed under batch conditions with model solutions containing synthetic wastewater. Agarry et al. (2008) treated PRWW in a study aimed at the investigation of the phenol-biodegrading capability of two indigenous *Pseudomonas* species under batch fermentation process.

Immobilization: why needed?

Biodegradation is a favorable and attractive technology for the treatment of PRWW, however traditional processes face a lot of operational problems, mainly the inhibition effect detected at high concentrations of toxic content, in addition to the previously mentioned problems associated with ASP including long start-up intervals, and weak tolerance to shock loads (Al-Khalid & El-Naas, 2018; Al-Khalid et al., 2020) has also been reported that biological methods will not assure complete removal of recalcitrant organic content usually existent in PRWW, and although bioaugmentation has been proposed as a solution, a main issue of bioaugmentaion arises when the process is moved to a full-scale operation, that is the uncertainty of reproducibility (Al-Khalid & El-Naas, 2018). Biofilm reactors, employing immobilized cells, have been proposed for significantly improved bacterial activity and superior salient features over suspended cell reactors (Al-Khalid & El-Naas, 2018; Al-Khalid et al., 2020). The immobilization technique is a novel approach that allows the utilization of higher density biomass and provides the microbial cells with protection against the inhibition effect at high concentrations of toxic contaminants. Problems of solid waste disposal are minimized thanks to the effective treatment by the immobilized microorganism with little sludge formation, as well as the ease of separation and reutilization of the biomass that results in cost reduction (Wang et al., 2007; Liu et al., 2009). Moreover, the solid residence time could be increased while still maintaining safe operation in continuous flow systems with minimal clogging (Ismail & Khudhair, 2018) no biomass washout (Al-Khalid & El-Naas, 2012).

Various immobilization methods have been used to enhance biological activity like bead entrapment, cell coating, film attachment, carrier binding, and encapsulation (Al-Khalid & El-Naas, 2018). Generally, these methods can be categorized into two main distinct techniques: cell attachment and cell entrapment. In the first method, microorganisms adhere by self-adhesion to the surfaces of different suitable materials including polyurethane foam (PUF), synthetic foams, nylon sponges, and others. On the other hand, in the cell-entrapment process, microorganisms are physically confined and trapped within the fibrous or porous structural matrix of suitable polymeric materials like Ca-alginate (calcium-alginate gel beads), chitosan (a natural nontoxic biopolymer), polyvinyl alcohol (PVA), and cellulose derivatives. Several studies have been documented on the use of both methods for biotreatment of PRWW (Banerjee & Ghoshal, 2017). A more elaborate discussion on this topic is presented in section 3.

Although cell-entrapment immobilization is evidently advantageous, it may be problematic considering diffusion limitation imposed by the resistance of the protective structure, which in turn considerably affects the intrinsic reaction kinetics, a topic that was addressed by several studies to account for these limitations by proposing comprehensive diffusion-reaction models (Banerjee et al., 2001; Tepe and Dursun, 2008). Al-Khalid & El-Naas (2018) reported bacterial immobilization in PVA gel matrices as a novel approach in biotreatment, which is based on repeated cycles of freezing-thawing to induce crosslinking in the gel matrix, thus transforming it into an elastic rubbery structure with high porosity and enhanced mechanical strength. Fig. 5.1 shows the distribution of immobilized bacteria (*Pseudomonas putida*) in PVA (El-Naas et al., 2014). Reference was made to several studies by El-Naas and coworkers for details on employing this technique for biodegradation of different organic contaminants in PRWW. Fig. 5.2 shows biodegradation rates of 2,4-dichlorophenol (2, -DCP) by free and PVA gel-immobilized *P. putida* during the acclimatization stage (Al-Khalid, 2014). Ca-alginate has long been widely used in the field of biocatalysis, PVA was a favored option due to its inherent properties being more durable and less resistant to mass transfer. The same technique of cyclic freezing-thawing with PVA gel cubes was also used by others (Luo, Liu, Zhang, & Jin, 2009; Wang et al., 2007; Partovinia & Naeimpoor, 2013). Several studies on the treatment of different types of petroleum effluents confirmed higher biodegradation rates by immobilized cells over free cells, with reported values such as 33% higher, indicating improved tolerance towards various toxic organic compounds (Ismail & Khudhair, 2018). Higher specific reaction rates in a continuous mode operation are possibly attained due to higher biomass concentrations coupled with enhanced capabilities of the biomass towards the sorption of contaminants (Morgan-Sagastume et al., 2019).

In light of the fact that the biodegradation mechanism essentially involves a chain of enzyme-catalyzed reactions, growing attention has been recently oriented to the application of immobilized active enzymes obtained from potent degrading microorganisms for the decontamination of wastewater, which is an environmental credit due to reduced energy consumption. Improved stability and tolerance against severe changes in environmental conditions are major advantages related to the enzyme performance resulting from enhanced catalytic efficiency. Other benefits associated

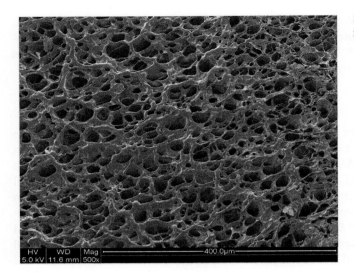

FIGURE 5.1 SEM photo for polyvinyl alcohol gel matrix for biomass immobilization (El-Naas et al., 2014).

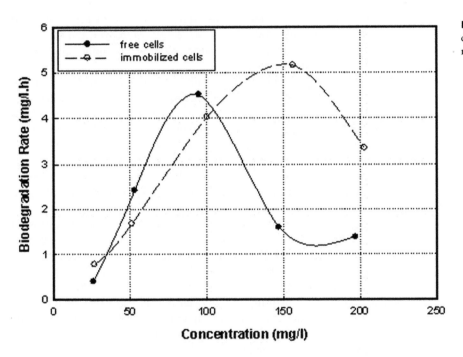

FIGURE 5.2 Biodegradation rates of 2,4-dichlorophenol by free and immobilized bacteria during acclimatization (Al-Khalid, 2014).

with enzyme immobilization include easy handling, increased reusability, and extended lifetime. Among various enzymes, peroxidases, in particular, have received a focal interest due to their potency to catalyze reactions of a wide range of toxic contaminants (Jun et al., 2019). The phenol removal efficiency of 99.9% could be achieved by immobilized peroxidase compared to 70% phenol removal by free enzyme under the same operating conditions. According to Jun et al. (2019), with the promising prospects of nanotechnology, a breakthrough is expected in wastewater treatment by the incorporation of immobilized peroxidase on nanomaterials. Notwithstanding the attractiveness of immobilized-enzyme treatment as a novel sustainable technique, it is still limited by issues of mass transfer and operational feasibility of scaled-up continuous systems, thus lending these challenges to further research. A few recent studies documented the enzymatic treatment of organic pollutants, the topic was the focus of extensive discussion in reviews by Demarche, Junghanns, Nair, and Agathos (2012) and Jun et al. (Jun et al., 2019). In recent years, considerable attention has been placed on developing SBRs utilizing aerobic granules which form through a process of self-immobilization of microorganisms resulting from cell-to-cell adhesion, in a novel sustainable biotechnique that eliminates the need for a carrier medium (Al-Khalid & El-Naas, 2018; Al-Khalid et al., 2020; Khan, Mondal, & Sabir, 2013). The concept of aerobic granulation is driven by the growth of high-density biomass (up to 15,000 mgL^{-1}) that is capable of decontaminating highly loaded wastewater (up to 15 kg COD m^{-3} day) (Liu et al., 2007), thus alleviating some of the restrictions of conventional ASP systems. However, this technology is still limited by an early development stage. The status of research of this mechanism and its applications can be found in reviews by Khan et al. (2013) and Demarche et al. (2012), and Nancharaiah & Kiran Kumar Reddy (2018).

The role of acclimatization

Acclimatization of biomass is an important step in any biological treatment system. This step is conducted to allow the microbial community to tolerate the toxic contaminants present in the wastewater. In most cases, the acclimatization of biomass is carried out using simple carbon sources such as glucose or sucrose. The application of the acclimatization process has been studied by several investigators for the removal of several pollutants from RWW. The acclimatization of mixed culture of microalgae was obtained to improve the removal of organics present in wastewater. In this study, sucrose was added to the synthetic wastewater solution that has a COD concentration 6797 mgL^{-1}. The biodegradation by acclimatized and nonacclimatized biomass was tested for the removal of organics from wastewater. Results showed that acclimated biomass including *S. obliquus, C. vulgaris and C. sorokiniana* resulted in higher biomass production than those without acclimation. The acclimatized microalgae achieved a COD reduction of 87%, which indicated the importance of the acclimatization step in the biological treatment to assist the treatment of highly polluted wastewater streams (Hu, Meneses, Stratton, & Wang, 2019). Phenols are the most common pollutants present in PRWW in high concentrations, in which they show a high inhibitory effect on the growth of many microorganisms (Léonard, Youssef, Destruhaut, Lindley, & Queinnec, 1999). Thus, several studies were conducted to improve the biodegradation of organic pollutants from wastewater. The acclimatization of immobilized-mixed culture for phenol degradation was studied to remove phenol from wastewater at initial concentrations up to 800 mgL^{-1}. In the beginning, the mixed culture was acclimatized using glucose as the sole source of carbon. Results showed that acclimatized bacteria tolerated the toxic effect of phenol and degraded phenol from wastewater with an initial concentration of up to 500 mgL^{-1}. However, the higher concentration required more biodegradation time in order to allow more biomass growth that was hindered due to the toxicity of high phenol concentrations (Saravanan, Pakshirajan, & Saha, 2008).

The application of glucose as an acclimatization source with the addition of phenols was studied by El-Naas, Al-Muhtaseb, and Makhlouf (2009). In this work, the acclimatization of *P. putida* immobilized in PVA gel was carried out using glucose. The biomass was acclimatized firstly in a solution that contains glucose as a sole source of carbon, through a gradual increase of the concentration of glucose to 1000 mgL^{-1} over the course of five days. The activated bacteria were then acclimatized in a solution containing phenol, by reducing glucose concentration gradually and increasing phenol concentration in the solution. After the acclimatization, the immobilized biomass could effectively degrade phenol at concentrations up to 300 mgL^{-1}. Results showed that the acclimatized bacteria can remove all the phenol from wastewater. The SEM analysis confirmed the bacterial growth within the immobilization matrix Fig. 5.1. The biodegradation of phenol at high concentration was investigated using glucose acclimatized bacteria, where the acclimatization was carried out in solutions at phenol: glucose ranging from 200:800 mgL^{-1} to 1000:0 mgL^{-1}. When tested for phenol degradation, the acclimatized cells achieved a high biodegradation rate at an initial concentration of up to 1000 mgL^{-1}. However, the simultaneous biodegradation of phenol with the addition of glucose in the biodegradation media enhanced the biodegradation efficiency at higher phenol concentration levels (3900 mgL^{-1}), without any inhibition effect (Mamma et al., 2004). Although most of the acclimatization steps have been carried out using simple carbon sources such as glucose, the activation and acclimatization of biomass in phenolic wastewater were investigated by several researchers (Tosu, Luepromchai, & Suttinun, 2015). It was proven that the role of glucose in biomass activation varies according to the type of organic pollutants that will be removed from wastewater. Al-Khalid and El-Naas (Al-Khalid & El-Naas, 2014) investigated the aerobic biodegradation of phenol and 2,4-DCP by acclimatized *P. putida*. The performance of glucose-acclimatized bacteria was compared with that of the biomass acclimatized in both solutions of phenol and 2,4-DCP with concentrations up to 200 mgL^{-1}. As for phenol, the nonglucose acclimatized bacteria showed better biodegradation performance compared to the glucose-acclimatized bacteria. However, the acclimatization using glucose enhanced the biodegradation rate of the 2,4-DCP. After a long adaptation period, all acclimatized bacteria reached the same biodegradation rate, thus indicating the absence of the glucose role in the enhancement of the biodegradation step, especially in the case of phenol degradation. The acclimatization role varies according to whether free or immobilized biomass is used. Non- acclimatized bacteria in the immobilized form were used for the treatment of PRWW. The immobilized bacteria were able to achieve 62% COD reduction, compared with the free cells that achieved only 28% COD removal (Ismail & Khudhair, 2018). Thus, the application of the acclimatization process has a significant role in enhancing biodegradation performance in general and has a high influence on the biodegradation efficiency of the immobilized biomass.

Immobilized-biofilm application

Immobilization carrier selection is very important in the application of biomass immobilization in the area of wastewater treatment. The choice of suitable immobilization matrices has a direct effect on the feasibility of immobilized biomass and consequently affects the efficiency of biological wastewater treatment. The immobilization carrier should have several properties including insolubility, nontoxicity, nonbiodegradability, and ease of separation and handling properties; in addition to the high cell mass loading capacity, high biological, mechanical, and chemical stability (Abdelwahab et al., 2009; Al-Khalid & El-Naas, 2018; El-Naas et al., 2010; Bouabidi et al., 2019). These biomass carriers must also be biocompatible and have an optimum nutrient diffusion. Several immobilization matrices have been applied in the area of biological wastewater treatment and can be categorized as inorganic immobilization matrices such as clay, activated charcoal, ceramic, zeolite, porous glass, and anthracite; in addition to the organic immobilization carriers classified as natural and synthetic carriers as shown in Table 5.1.

The selectivity of the immobilization matrix is one of the most important steps in the application of biofilm. Several immobilization matrices have been used in biomass immobilization, among them, alginates, PVA gel PUF are the most widely used. Table 5.2 shows the application of several immobilization matrices that have been used in the biological treatment systems for the degradation of organic pollutants from the wastewater.

The biological treatment using immobilized biomass can be obtained through several immobilization methods, such as covenant attachment, immobilization through enzyme cross-linking, immobilization by adsorption, entrapment, and encapsulation immobilization. It can be said that there is no single immobilization method or support applicable to all enzymes and their various applications due to the different properties of the substrates, the diverse applications of the products obtained, and the different physicochemical characteristics of each enzyme. Fig. 5.3 shows the most common types of immobilization methods that have been widely applied in the area of wastewater treatment. More details about these types are given in Table 5.3.

Biotreatment by film attachment (surface immobilization)

In light of the inherent limitations of ASP, the substitution of this process continues to be a favored topic of priority research to tackle the different issues of conventional biological processes. In this regard, development has been reported, represented by hybrid bioreactors, which combine a biofilm reactor with an aerated tank (suspended culture). This development allows for the maximized benefits of the two combined types of growth, wherein the features of stability and capability of handling shock loads from a biofilm system are added to the simplicity of operation and economic advantages offered by a suspended growth system (Dey & Mukherjee, 2013). A wide variety of techniques and processes like RBC and trickling filters, which are most common of a fixed film treatment, have been developed for the biological treatment of PRWW (Al-Ghouti et al., 2019; Singh & Borthakur, 2018). Additionally, biological aerated filters and SBRs can be considered. It was reported that SBR, with its characteristic simplicity, the flexibility of operation, and cost-effectiveness for small-scale treatment facilities is proposed as an attractive substitution for conventional biological wastewater treatment systems (El-Naas, Al-Zuhair, & Makhlouf, 2010; Sahinkaya & Dilek, 2006). The reactor type spectrum may extend to encompass microbial fuel cells, hollow fiber membrane contactors, rotating rope bioreactors, pulsed plate bioreactors, and two-phase partitioning bioreactors. A thorough discussion on reactor types and their design aspects can be found elsewhere (Al-Khalid & El-Naas, 2012).

Needless to say, anaerobic treatment could also be considered in the evaluation of cost-effective alternatives to aerobic biotreatment, especially in cases of PRWW with a high organic load (Jiménez et al., 2018). There has recently been a

TABLE 5.1 Advantages and disadvantages of natural and synthetic organic immobilization matrices.

Immobilization matrix	Natural	Synthetic
Example	Chitin, Agar, Alginate, Carrageenan	PVA, polypropylene ammonium, polyurethane, acrylamide
Advantages	High diffusion rates, environmentally friendly	High stability, good durability, good mechanical strength
Disadvantages	Poor mechanical strength, nontoxic	May be toxic to microorganisms

TABLE 5.2 Common immobilization matrices applied in the removal of organic pollutants in the wastewater.

Enzyme	Support materials	Results obtained	References
Alginates			
Bacillus cereus	Ca-alginate	COD removal of 99.2%, TOC removal of 95.4% and phenols removal of 99/8%	Banerjee and Ghoshal (2017)
Candida tropicalis YMEC14	Ca-alginate	69.7% COD removal 69.2% monophenols removal 55.3% polyphenols reduction after 24 h fermentation cycle	Ettayebi et al. (2003)
Chlorella vulgaris	Ca-alginate	100% NH4 $^+$-N removal. 95% PO4 3-P removal within 24 h	Tam and Wong (2000)
Ralstonia eutropha	calcium-alginate	68% phenol removal	Léonard et al. (1999)
Pseudomonas putida	Sodium alginate	More than 90% phenols removal was achieved at initial concentration up to 1000 mgL^{-1}.	Chung, Wu, and Juang (2005)
Polymers			
Pseudomonas putida	PVA gel	100% phenol removal	El-Naas et al. (2009)
Pseudomonas putida	PVA gel	100% p-cresol removal	Surkatti and El-Naas (2014)
Acinetobacter sp. Sphingomonas sp.	PVA gel	95% phenol removal	Liu, Zhang and Wang (2009)
Bacillus cereus	Polurethane (PU)	90% removal of TOC, COD and phenols	Banerjee and Ghoshal (2017)
Bacteria from recycled sludge	Polyurethane (PU)	94% phenol removal	Hsien and Lin (2005)
Composites			
Activated sludge	sodium alginate with polyvinyl alcohol	61.7% COD removal 66.6% petroleum hydrocarbons removal	Ismail and Khudhair (2018)
Bacillus amyloliquefaciens	Sodium alginate-Chitosan-Sodium alginate (ACA)	100% phenol removal	Lu et al. (2012)
Chlorophenol degrading bacteria	Carrageenan-chitosan gel	100% chlorophenol	Wang and Qian (1999)
Sphingomonas sp. GY2B	Polyvinyl alcohol—alginate—kaolin beads	99% phenol removal	Ruan et al. (2018)
Phenol degraded strains	Bentonite and Carboxymethyl Cellulose Gel	96% phenol removal	Duan, Wang, Sun, and Xie (2016)

  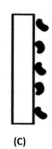 

(A)   (B)   (C)   (D)

FIGURE 5.3 Common types of immobilization methods; (A) cross-linking, (B) encapsulation, (C) adsorption, and (D) matrix entrapment.

rising interest in employing anaerobic processes for the treatment of oily wastewaters, mainly due to the merit of a complementary role to aerobic systems that results in the production of methane as a bioenergy gas, with a prospect for heat or electricity generation. While the mechanism of aerobic treatment depends on using oxygen for oxidizing the organic matter into CO_2 anaerobic processes basically result in the reduction of organic matter into methane, which serves a dual

TABLE 5.3 The definition of common immobilization methods; (A) cross-linking, (B) encapsulation, (C) adsorption, (D) matrix entrapment

Immobilization type	Description
Cross-linking	Cross-linking of enzyme is a free-standing method of biomass immobilization. It allows the application of highly stable and recyclable enzymes that have high retention activity. This method can be used for the immobilization of any type of enzyme and it has many environmental and economic benefits especially in the industrial biocatalysts applications.
Encapsulation	Biomass encapsulation is a special type of entrapment immobilization. In this method, the biomass is entrapped in a semipermeable membrane, to allow free movement of substrate and nutrients within the immobilization matrix and at the same time restricts the flow of biomass through the membrane walls (Krishnamoorthi, Banerjee, & Roychoudhury, 2015).
Adsorption (surface)	Immobilization using enzymes adsorption is the simplest biomass immobilization method. It is achieved through the interaction between the immobilization matrix and enzymes from either hydrophobic or ionic interactions, depending on the surface chemistry of the immobilization carrier and amino acid type present in the surface of the enzyme (Kilonzo, Margaritis, & Bergougnou, 2011).
Matrix entrapment	Entrapment immobilization is based on the entrapment of biomass in a polymeric network or gel network. This immobilization method requires the formation of a polymer network using a gelation reaction at low temperatures (Bayat, Hassanshahian, & Cappello, 2015).

purpose of energy recovery and COD reduction, which in turn leads to additional savings in energy caused by the reduction in needed aeration owing to the decreased organic load before going through the aerobic treatment (Morgan-Sagastume et al., 2019). A study by Sagastume et al. (Morgan-Sagastume et al., 2019) aimed at evaluating the feasible application of an anaerobic moving bed biofilm reactor (AnMBBR) for the anaerobic treatment for oil-contaminated wastewater from an oil-recovery facility. The biofilm-based process was proved feasible for COD removal and biogas production, with noticeably improved tolerance to high organic loading rates. The RBC is a good example of a fixed-film biological process, characterized by alternating operational schemes between aerobic and anaerobic modes, with the main advantage of a high interfacial area inherent in the rotating disks of the reactor, which enhances contact between the biomass and contaminant. Consequently, the organic contaminant material in the liquid will be easily adsorbed on the biofilm grown on the disks. Other benefits include an eco-friendly stable operation that is based on a simple and cheap design. The RBC treatment was proved efficient for the treatment of high-strength synthetic phenolic wastewater by activated sludge in a study by Rana, Gupta, and Rana (2018) Integrated anaerobic-aerobic wastewater treatment techniques were also discussed in detail by Chan, Chong, Law, and Hassell (2009). The anaerobic baffled reactor (ABR) is composed of multiple upflow UASB reactors separated by baffles. Its utilization was demonstrated by Mahdavianpour, Moussavi, and Farrokhi (2018) for the treatment of synthetic wastewater loaded with a high content of p-cresol where complete removal of 1000 mgL^{-1} p-cresol could be achieved within an HRT of 24 h.

Polyurethane (PU) foam is highly resistant to organic solvents and microbial attack and has several advantages including high porosity and mechanical strength; thus it has been widely applied as an immobilization matrix (Oh, Maeng, & Kim, 2000). Additional advantages have been noticed by many researchers such as the feasibility of large-scale application at a low price, easy control of the pore size, stable maintenance of the extent of biological growth. Jou and Huang (Jou & Huang, 2003) conducted a fixed biofilm application for the treatment of real PRWW where a pilot-scale bioreactor system was assembled using a highly porous PU foam that incubated biomass at high concentrations up to 8000 mgL^{-1}. The tests emphasized the superior performance of this bio-technique compared to conventional ASP, as mainly demonstrated by the less sludge formation that is one-third of the sludge generated by the conventional ASP. Furthermore, a higher removal rate was attained, and greater stability of the system was perceived. A COD removal exceeding 85% and almost 100% degradation of phenol were regularly achieved with 8-h HRT compared to 50%−60%, 99%, and 25 h, respectively, with conventional ASP. The system exhibited extreme stability under real-world conditions (Jou & Huang, 2003).

Extensive research has been directed to the development of novel membrane technologies. Recalcitrant organics represent a serious challenge to petroleum wastewater treatment, as they can drastically and adversely affect the biological floc formation and biomass settleability in traditional clarifiers. Hence, MBRs are frequently utilized either as standalone treatment units for contaminants removal or as pretreatment facilities to desalination membranes, where the biomass is continuously separated from the effluent of treated wastewater by membrane filtration thus alleviating the effects of settleability, leading to a more feasible process. Adham et al. (Adham et al., 2018) assessed the biotreatability of a mixed effluent from a gas field, consisting of produced and process water, over a range of HRTs (16−32 h), solids residence times (60−120 days), and temperatures (22°C−38°C) employing the technology of hollow fiber membrane. Starting with a COD of 1300 mgL^{-1} and TDS of 5200 mgL^{-1}, the COD and TOC removals were on the average of 60%, as confirmed by the results over 8 months testing period. Viero et al. (2008) utilized a submerged membrane bioreactor (SMBR) for the treatment of PRWW. In essence, this may be considered a combined process that incorporates activated sludge and a granulated activated carbon filter. The process is featured for robustness that allows the feasible operation under hydraulic fluctuations and shock loading rates. The results showed improved COD and TOC removal efficiencies by 17% and 20%, respectively, which emphasized the vital role of the membrane in this improvement. This is justified by the ability of the SMBR to sustain high strength PRWW during long-run operation, thus attaining high removal efficiencies of phenols (Al-Khalid & El-Naas, 2018). In another study by Sambusiti et al. (2020), a pilot-scale SMBR was tested in a continuous application for the decontamination of oil and gas wastewater. Efficient removals were recorded as to phenols (up to 100%); BTEX (up to 100%); PAHs (up to 100%); heavy metals (from 29% to 97%); and TOC (96%−98%). As such, those results are very promising for the treatment of PRWW with a characteristic composition comparable in terms of salinity, metals, and hydrocarbons content to the synthetic PRWW tested in this study. The membrane is usually made of ceramic or polymeric material; the former being often preferred with complex wastewaters owing to the advantages of higher chemical and thermal resistance compared to polymeric membranes. Furthermore, their major advantage is the ease of cleaning, being backwashable with water and chemicals like chlorine. The membrane may assume various geometries including hollow fibers, tubular, and flat sheet modules. Al-Khalid & El-Naas (2012) reported several studies on the use of hollow fibers polymeric membranes for the decontamination of high strength synthetic phenolic wastewater, where enhanced degradation efficiency was noticeable, thanks to the positive effect of a thicker biofilm layer formed on the outer surface of the fibers (Chung et al., 2005).

The search for green methods pushed towards the development of sustainable approaches and paved the path for rising technologies, which evolved based on nanoparticles or nanomaterials to take a role in biodegradation applications. With their characteristic larger surface/volume ratio, nanomaterials would significantly improve the biodegradation process by increasing the adsorption of contaminant substrate to their surfaces. Nzila (2018) tested this approach under both aerobic and methanogenic (anaerobic) operating conditions. Bavandi, Emtyazjoo, Saravi, Yazdian, and Sheikhpour (2019) evaluated this technology for the removal of trinitrophenol from synthetic wastewater by utilizing carbon-based nanostructures and metal nanoparticles along with bacterial biomass. Removal efficiencies higher than 90% for an initial trinitophenol concentration of 1301 mgL^{-1} could be achieved.

In line with the trend of green technology, the utilization of microbial desalination cell (MDC) has been reported as a promising technology for the biotreatment of petroleum refinery effluents (Salman & Ismail, 2020). The technique can allow for the production of purified water in tandem with electric power generation through the desalination process. The method was applied by Ashwaniy, Perumalsamy, and Pandian (2020) for reinforcing the synergistic interaction of microalgae and bacteria for the reduction of organic contaminant content in real PRWW. MDC is an eco-friendly, sustainable, and novel technology that involves a multifunctional single process for desalination, wastewater treatment, and bioelectricity generation. In this integrated system, algae grown in PRWW functions as an electron acceptor or a biocathode that acts as a substrate to promote bacterial growth in the anode, resulting in biomass cultivation in parallel with clean and economical energy production. It was reported that MDC is an emerging multipurpose technology that stemmed from the traditional MFC process, basically designed to achieve the dual purpose of wastewater treatment and the recovery of clean energy in the form of electricity (Al-Ghouti et al., 2019). This development of MDC originating from MFC was attained by incorporating a desalination chamber in-between the cathode and the anode chambers, in a reactor assembly. Al-Khalid et al. (2020) documented studies on the utilization of MFC for the simultaneous removal of organic pollutants from wastewater and bioelectricity generation. The method was even employed for the quick quantitative assessment of wastewater toxicity through the successful development of MFC as a biomonitoring system that allows a rapid response when needed for toxicity assessment in wastewater treatment plants (Lu, Yu, Zhou, & Xing, 2019).

Jain et al. (2020) emphasized the importance of constructed wetlands as a cost-effective, and eco-friendly treatment method, with the significant merit of low sludge production, for effective removal of recalcitrant organic contaminants from high strength refinery and petrochemical wastewater by the involvement of a variety of mechanisms such as phytodegradation, sorption, and microbial degradation, averaging a COD removal of 80%−90%. They have been recommended for easy operation as treatment techniques that can handle large amounts of petrochemical and PRWW naturally and sustainably, which makes them a highly viable option, especially for developing countries. Moreover, their advantages extend to social aspects, including habitat enrichment and promoting the esthetic value of a wastewater treatment site. Constructed wetlands use plants that can be selected based on availability nearby. Mustafa, Azim, Raza, and Kori (2018) validated the performance of free water surface wetland in a pilot-scale study aimed at the removal of

BTEX present in produced water from a gas field, with achieved removal efficiencies of 92.6% for benzene, 93.4% for toluene, 98.3% for ethylbenzene, 91.3% form and p-xylene and 87.4% for o-xylene.

Biotreatment by cell entrapment

The entrapment of biomass in an immobilization matrix is the most common technique in cell immobilization; it is an irreversible technique in which immobilized microorganisms are entrapped inside the support material. The entrapment of biomass creates a barrier that protects microorganisms and ensures their prolonged activity and biomass storage during treatment storing processes (Górecka & Jastrzębska, 2011). Generally, the entrapment in the polymeric matrices can be obtained through the polymerization or crosslinking processes. In this process immobilization of microorganisms is achieved through the entrapment of biomass in a polymeric matrix with a porous structure that is used to capture microbial cells from suspension solution. The porosity structure of the immobilization carrier increases the diffusion of several pollutants and nutrients within the matrix and thus enhances the biological treatment process (Verma, Brar, Blais, Tyagi, & Surampalli, 2006). The entrapment of biomass in alginate gel is the most common type due to the simplicity of the production and application. Several studies compared the application of the free cells and immobilized biomass in alginate for the treatment of hazardous pollutants in PRWW. The attachment of microbial cells in the crosslinked alginate is mostly obtained by the aggregation of the cells within the cavities of the immobilization matrix. This was confirmed by Abarian, Hassanshahian, and Esbah (2019) when they studied the entrapment of *P. putida* and *A. scleromae* in sodium-alginate beads. They concluded that the entrapped bacteria have more biodegradation efficiency compared to the free cells due to the distribution of microbial cells within the crosslinked gel matrix. The application of immobilization matrices increases the contact surface between microbial biomass and immobilization carrier. Banerjee & Goshal (2011) studied the immobilization of two *B. cereus* strains in Ca-alginate gel for the treatment of phenol at high concentrations. The immobilized cells were compared with the free cells for the removal of phenol in the concentration range from 100 to 2000 mgL^{-1}. It was shown that the entrapped bacteria in the alginate matrix achieved high degradation performance at high initial phenol concentration (1500−2000 mgL^{-1}), in contrast at low phenol concentration both free and immobilized bacteria showed similar biodegradation performance. The authors pursued the study with actual petroleum wastewater for the investigation of biodegradation performance of the two hyper-phenol tolerant bacterial strains both immobilized in two packed bed systems: one in Ca-alginate beads (entrapment immobilization) and one in PU foam (surface immobilization) (Banerjee & Ghoshal, 2017). The performance was evaluated by measuring the decrease in TOC, COD, total phenolic content, total phosphate, and ammonium-nitrogen ($NH4^+-N$) levels. A better performance by the Ca-alginate immobilized system was confirmed by the results of TOC, COD, and phenolic reduction. The results, in general, strongly supported the successful implication of the immobilized bacterial strains in the biodegradation PRWW, as evidenced by a reduction in the initial COD, TOC content, and concentration of phenolics exceeding 90% by the biological treatment in continuous mode. Although entrapment of biomass has often been in wide application in the field of wastewater treatment, it has some limitations including; high immobilization cost, cell leakage, diffusion limitation, and the possibility of deactivation during immobilization. In addition to the scratching of immobilization material and low biomass loading capacity where the biomass should be incorporated inside the immobilization carrier (Krekeler, Ziehr, & Klein, 1991; Seung, Suk, Park, & Yoo, 2005).

In addition to biomass entrapment, cells can be immobilized through encapsulation or hybrid entrapment immobilization technique. In the encapsulation process, the immobilized biomass is restricted with the membrane walls to form capsules. The membrane is semipermeable and allows the flow of nutrients and substrates and at the same time keeps the biocatalyst inside it (Kreuzer & Massey, n.d.). The pore size of the membrane is the dominant factor that controls this phenomenon. The membrane that is applied in this method allows limited access of pollutants to the interior microcapsules, which ensures the protection of the biocatalyst from harsh environmental and operation conditions (in the bioreactor system). It also prevents the leakage of biocatalysts and increases process efficiency (Park & Chang, 2000). The immobilization material applied in encapsulation doesn't require any chemical modification for the core material. All these advantages make the encapsulation method suitable for most microorganism immobilization and also for the sequential enzymatic reactions in multienzyme systems.

Encapsulation of biomass was studied by several investigators, in this process, the inoculum of biomass is added through the preparation of the immobilization matrices. This method allows the distribution of microbial cells within the immobilization matrix at the beginning and allows them later to grow inside the pores of the immobilization carrier. Biomass encapsulation can be obtained through several materials including polyurethane, PVA gel, and alginates, in addition to the composites that are composed of two or more materials. The encapsulation of bacterial cells in PVA gel for PRWW treatment was widely studied in pure form or composites form. It was shown that *P. putida* immobilized in PVA prepared by the iterative freezing-thawing method was applied for the removal of several organic contaminants present in PRWW. The bacterial biomass was immobilized through the encapsulation method where cells are allowed to grow inside the pores of the gel. Therefore, the biomass activity is directly affected by the porous structure of the physically crosslinked PVA gel prepared by repeated freezing-thawing cycles; the encapsulation of the cells within the immobilization matrix will also prevent the microorganisms from the high concentration of toxic pollutants (Encapsulation of biomass) was studied by several investigators; in this process, the inoculum of biomass is added through the preparation of the immobilization matrices. This method allows the distribution of microbial cells within the immobilization matrix at the beginning and allows them later to grow inside the pores of the immobilization carrier. Biomass encapsulation can be obtained through several materials including polyurethane, PVA gel, and alginates, in addition to the composites that are composed of two or more materials. The encapsulation of bacterial cells in PVA gel for PRWW treatment was widely studied in pure form or composites form. It was shown that *P. putida* immobilized in PVA prepared by the iterative freezing-thawing method was applied for the removal of several organic contaminants present in PRWW. The bacterial biomass was immobilized through the encapsulation method where cells are allowed to grow inside the pores of the gel. Therefore, the biomass activity is directly affected by the porous structure of the physically crosslinked PVA gel prepared by repeated freezing-thawing cycles; the encapsulation of the cells within the immobilization matrix will also prevent the microorganisms from the high concentration of toxic pollutants (El-Naas, Mourad, & Surkatti, 2013). The application of PVA gel as an immobilization matrix that is characterized by high porosity structure and mechanical strength allows effective removal of toxic components from wastewater. A biofilm of *P. putida* immobilized in PVA gel matrix was used in a spouted bed bioreactor (SBBR) for the removal of phenol, complete degradation could be achieved in 5 h at an initial concentration of 150 mgL^{-1} (El-Naas et al., 2009). Moreover, the immobilized *P. putida* was also examined for the biodegradation of *p*-cresol in synthetic wastewater and showed high removal efficiency. The immobilized *P. putida* achieved the removal of more than 85% *p*-cresol at an initial concentration of 200 mgL^{-1} in a continuous SBBR system (Surkatti & El-Naas, 2014). Al-Khalid (2014) used PVA gel-immobilized *P. putida* for the treatment of real PRWW in SBBR, with the immobilized bacteria already acclimatized to the real PRWW. The total phenols were significantly reduced by almost 90% and a maximum COD reduction of 59% was observed. The process was deemed promising with an obtained 100% removal efficiency for all cresols, whereas the removal efficiencies for phenol and DCP were approximately 87% and 63%, respectively. Jiang, Ruan, Li, and Li (2013) studied the performance of *Acinetobacter* sp. BS8Y isolated from activated sludge for phenol degradation when immobilized in PVA gel. Results showed that immobilized *Acinetobacter* sp. was capable of degrading phenols at high concentrations, and achieved 99.2% phenol removal at an initial concentration of 600 mgL^{-1}. The immobilized bacteria were also able to tolerate the change in operating conditions such as pH and temperature fluctuations. It was also clear that immobilization of biomass in cross-linked PVA achieved high biomass stability that was confirmed from the effective performance after 50 times reuse and the long storage period (El-Naas et al., 2013; Jiang et al., 2013).

Composites immobilization was also applied for biomass immobilization that will enhance the structure and mechanical strength and stability. Ismail and Khudhair (2018) studied the performance of immobilized activated sludge for the treatment of real-field petroleum wastewater. In this study, the activated sludge was immobilized in a composite of polysaccharide sodium alginate and PVA. The biodegradation efficiency of organic pollutants and phenol using immobilized activated sludge showed a remarkable improvement, in addition to high biomass stability that was confirmed after 35 days of the biodegradation process. Sodium alginate was used for the preparation of the immobilization carrier and reinforced by the PVA to improve its mechanical stability. The immobilization of microbial biomass was obtained by adding the inoculum during the preparation of biofilm. The immobilized bacteria were used for the treatment of real PRWW in the lab and pilot-scale SBBR. The immobilized biomass could remove 62% COD, compared to the free cells that removed 28% COD content. System analysis showed that the immobilized biomass can be used for three cycles without losing the biodegradation efficiency in the removal of organics from wastewater. Additionally, the immobilized cells can maintain storage stability at a long storage period (35 days), compared to free cells that lose their activity after a storage period of 28 days (Ismail & Khudhair, 2018).

In addition to the application of PU foams as immobilization matrices, several PU composites were applied for wastewater treatment to enhance the adsorption of the contaminants within the immobilization matrix and therefore increase the removal efficiency. PU composite was studied for the removal of ions existing in wastewater. The composite was prepared by the polymerization of diisocyanate (TDI) and polyether polyol with activated carbon fiber. Results showed that immobilized biomass in PU composite foams resulted in an excellent removal rate of Cu(II) and other organics from wastewater. PU composite showed high removal efficiency as a result of simultaneous adsorption and biodegradation processes. The system achieved a removal rate of 80 and 85% of COD and Cu(II), respectively within 4 h (Zhou et al., 2009).

A summary of various biological treatment processes employed to treat real-field petroleum effluents is given in Table 5.4. The operating conditions and performance level of these processes have been indicated.

TABLE 5.4 Performance of biological treatment methods in treating real-field petroleum effluents.

Wastewater type	Initial concentration	Biological treatment	Operating conditions	Removal efficiency	Reference
Real PRWW	Phenol = 30 mgL^{-1}	Batch fermentor (aerobic)	30°C	Phenol: 69.4%–94.5% in 72 h	Agarry et al. (2008)
Field aggregates of oil-suspeded particulate matter + artificial saline water	TPH = 492.31 mgkg^{-1}	Batch biostimulation reactor (aerobic),	28°C, salinity: 30	TPH: 98.65% in 60 days	Leoncio et al. (2020)
Hydraulic fracturing flowback and produced water (HF-FPW)	COD = 6646.7 mgL^{-1} n-alkanes: 2635.4 mgL^{-1} PAHs: 918.6 µgL^{-1}	Batch bioaugmentation and biostimulation (aerobic)	4°C–100°C, pH: 2–12, salinity: 0–100 g/L	COD: 77%, n-alkanes: 94%, PAHs: 77%, 7 days	Zhou et al. (2020)
Wastewater from an oil recovery facility	COD = 36000 mgL^{-1}	Anaerobic MBBR	37°C, HRT = 30 days Organic loading rate = 1.1 kg COD m^{-3} day	COD: 67%	Morgan-Sagastume et al. (2019)
Real PRWW	COD = 510 ± 401.9 mgL^{-1} phenol: 30 ± 6.2 mgL^{-1}	- Conventional activated sludge process (ASP) - Pilot scale fixed film bioreactor-PU foam	- HRT = 25 h - HRT = 8 h	- COD: 50%–60%, phenol: 99% - COD: > 85%, phenol: 100%	Jou and Huang (2003)
Mixture of produced and process water from a gas field	COD = 1300 mgL^{-1} TDS = 5200 mgL^{-1}	MBR	22°C–38°C HRT = 16–32 h Solids residence time = 60–120 days	COD and TOC: 54%–63%	Adham et al. (2018)
Real petroleum mixture (oily PRWW + phenolic wastewater)	COD = 1010 ± 45 mgL^{-1} Phenol = 82 ± 2 mgL^{-1}	SMBR (aerobic, acclimatized activated sludge)	25°C, HRT = 10h, Average chloride = 1.57 g/L	COD: 48%–77%	Viero et al. (2008)
Real PRWW	COD = 610 mgL^{-1} BOD = 321 mgL^{-1} TDS = 580 mgL^{-1} TSS = 340 mgL^{-1}	MDC (microalgae and bacteria)	30 ± 2°C, microalgae best grown in 50% RWW	COD: 70%, BOD: 81% Phosphorous: 67% Sulfide: 61% TDS: 67%, TSS: 62%	Ashwaniy et al. (2020)
Produced water from a gas field	Benzene = 1.57 mgL^{-1} Toluene = 0.14 mgL^{-1} Ethylbenzene = 0.29 mgL^{-1} m-and p-Xylene = 2.01 mgL^{-1}	Constructed wetland (free water surface): pilot system	Ambient temperature	Benzene: 92.6% Toluene: 93.4% Ethylbenzene: 98.3% m- and p-Xylene: 91.3%	Mustafa et al. (2018)

Wastewater	Characteristics	Treatment	Conditions	Results	Reference
Real PRWW	COD = 9200 mgL^{-1} TOC = 4548 mgL^{-1} Phenolics = 3561 mgL^{-1} PO43- P = 121.1 mgL^{-1} NH4+ N = 121.09 mgL^{-1}	Continuous PBR (aerobic): - (*Bacillus cereus* bacteria AKG1 and AKG2, immobilized in Ca-alginate beads) - (*Bacillus cereus* bacteria AKG1 and AKG2, attached to PU foam)	30°C, pH = 5.9 ± 1.5 (of original medium, no sdjustment), HRT = 2.83 h	Ca-alginate system: 241 h COD: 99.2%, TOC: 95.9% Phenolics: 99.8% PO43-: 44.4%, NH4+: 49.4% - PU system: 241 h COD: 91%, TOC: 88% Phenolics: 91.5% PO43-: 25.9%, NH4+: 42.8%	Banerjee and Ghoshal (2017)
Real PRWW	COD = 2700 mgL^{-1} TOC = 770–800 mgL^{-1} Phenols = 160–185 mgL^{-1} DCP = 30 mgL^{-1} Cresols = 100–115 mgL^{-1}	Batch SBBR (aerobic): *P. putida* immobilized in PVA gel	30°C, pH = 8.7 (of original medium, no sdjustment)	24 h treatment: COD: 59% TOC: 54% Phenols: 90% Cresols: 100% DCP: 63%	Al-Khalid (2014)
Real PRWW	COD = 1250 ± 30 mgL^{-1} TPH = 2300 ± 50 mgL^{-1} TSS = 175 ± 1 mgL^{-1} Phenol = 10 ± 00.5 mgL^{-1}	Continuous SBBR (aerobic): Non-acclimated activated sludgeimmobilized in a mixture of sodium alginate and PVA	30°C, pH = 7.2–7.5 (of original medium, no sdjustment), HRT approximately 68 min	COD: 61.7% TPH: 66.6%	Ismail and Khudhair (2018)

Conclusions

The advancement of technology and industrialization has led to severe deterioration in environmental quality. This concern has raised the management of wastewater containing high concentrations of toxic contaminants as a global challenge and directed concerted efforts and research towards innovative sustainable solutions. Biotreatment, as a green technology approach, has proven to be very effective, most economical, and most promising. Suspended growth and attached growth bioprocesses have been employed for the decontamination of petroleum refinery effluents. In this regard, acclimatization of biomass comes as an important step in the biological treatment to increase the tolerance of microorganisms towards the toxic inhibitory effect of contaminants. Immobilized biofilms have shown superior characteristics over suspended cell systems, with significant protection provided to the microbial cells against the toxicity of contaminants at high concentrations. Various immobilization techniques were successfully applied, either as surface biofilm attachment or as cell entrapment. Thus, a lot of research has been focused on developing novel immobilization techniques and innovative configurations of biofilm processes. With new frontiers still open for further research and investigation, a recent approach that employs an innovative technique for the immobilization of a potent bacterial strain (*P. putida*) in a PVA gel matrix in a specially designed spouted bed bioreactor has shown to be a very good example in this direction.

Acknowledgments

The authors would like to acknowledge the support of the Qatar National Research Fund (a member of Qatar Foundation) through Grant # NPRP-10-0129-170278.

References

Abarian, M., Hassanshahian, M., & Esbah, A. (2019). Degradation of phenol at high concentrations using immobilization of *Pseudomonas putida* P53 into sawdust entrapped in sodium-alginate beads. *Water Science and Technology, 79*(7), 1387–1396. Available from https://doi.org/10.2166/wst.2019.134.

Abdelwahab, O., Amin, N. K., & El-Ashtoukhy, E. S. Z. (2009). Electrochemical removal of phenol from oil refinery wastewater. *Journal of Hazardous Materials, 163*(2–3), 711–716. Available from https://doi.org/10.1016/j.jhazmat.2008.07.016.

Abu Bakar, S. N. H., Abu Hasan, H., Mohammad, A. W., Sheikh Abdullah, S. R., Haan, T. Y., Ngteni, R., & Yusof, K. M. M. (2018). A review of moving-bed biofilm reactor technology for palm oil mill effluent treatment. *Journal of Cleaner Production, 171*, 1532–1545. Available from https://doi.org/10.1016/j.jclepro.2017.10.100.

Adham, S., Hussain, A., Minier-Matar, J., Janson, A., & Sharma, R. (2018). Membrane applications and opportunities for water management in the oil & gas industry. *Desalination, 440*, 2–17. Available from https://doi.org/10.1016/j.desal.2018.01.030.

Agarry, S. E., Durojaiye, A. O., Yusuf, R. O., Aremu, M. O., Solomon, B. O., & Mojeed, O. (2008). Biodegradation of phenol in refinery wastewater by pure cultures of *Pseudomonas aeruginosa* NCIB 950 and Pseudomonas fluorescence NCIB 3756. *International Journal of Environment and Pollution, 32*(1), 3–11. Available from https://doi.org/10.1504/IJEP.2008.016894.

Al-Ghouti, M. A., Al-Kaabi, M. A., Ashfaq, M. Y., & Da'na, D. A. (2019). Produced water characteristics, treatment and reuse: A review. *Journal of Water Process Engineering, 28*, 222–239. Available from https://doi.org/10.1016/j.jwpe.2019.02.001.

Al-Ghouti, M. A., Li, J., Salamh, Y., Al-Laqtah, N., Walker, G., & Ahmad, M. N. M. (2010). Adsorption mechanisms of removing heavy metals and dyes from aqueous solution using date pits solid adsorbent. *Journal of Hazardous Materials, 176*(1–3), 510–520. Available from https://doi.org/10.1016/j.jhazmat.2009.11.059.

Al-Khalid, T. (2014). Aerobic Biodegradation of 2, 4 Dichlorophenol in a Spouted Bed Bio-Reactor (SBBR). PhD, UAE University.

Al-Khalid, T., & El-Naas, M. (2014). Biodegradation of phenol and 2,4-dichlorophenol: The role of glucose in biomass acclimatization. *International Journal of Engineering Research & Technology (IJERT), 3*, 1579–1586.

Al-Khalid, T., & El-Naas, M. (2018). *Organic contaminants in refinery wastewater: characterization and novel approaches for biotreatment,* 371–391. Available from https://doi.org/10.5772/57353.

Al-Khalid, T., & El-Naas, M. H. (2012). Aerobic biodegradation of phenols: A comprehensive review. *Critical Reviews in Environmental Science and Technology, 42*(16), 1631–1690. Available from https://doi.org/10.1080/10643389.2011.569872.

Al-Khalid, T., Surkatti, R., & El-Naas, M. H. (2020). Organic contaminants in industrial wastewater: Prospects of waste management by integrated approaches. *Springer Science and Business Media LLC.* Available from https://doi.org/10.1007/978-981-15-0497-6_10.

Ashwaniy, V. R. V., Perumalsamy, M., & Pandian, S. (2020). Enhancing the synergistic interaction of microalgae and bacteria for the reduction of organic compounds in petroleum refinery effluent. *Environmental Technology & Innovation, 19*, 100926. Available from https://doi.org/10.1016/j.eti.2020.100926.

Banerjee, A., & Ghoshal, A. K. (2017). Biodegradation of an actual petroleum wastewater in a packed bed reactor by an immobilized biomass of Bacillus cereus. *Journal of Environmental Chemical Engineering, 5*(2), 1696–1702. Available from https://doi.org/10.1016/j.jece.2017.03.008.

Banerjee, A., & Goshal, A. K. (2011). Phenol degradation performance by isolated *Bacillus cereus* immobilized in alginate. *International Biodeterioration & Biodegradation, 65*(7), 1052–1060. Available from https://doi.org/10.1016/j.ibiod.2011.04.011.

Banerjee, I., Modak, J. M., Bandopadhyay, K., Das, D., & Maiti, B. R. (2001). Mathematical model for evaluation of mass transfer limitations in phenol biodegradation by immobilized *Pseudomonas putida*. *Journal of Biotechnology*, *87*(3), 211–223. Available from https://doi.org/10.1016/S0168-1656(01)00235-8.

Bavandi, R., Emtyazjoo, M., Saravi, H. N., Yazdian, F., & Sheikhpour, M. (2019). Study of the capability of nanostructured zero-valent iron and graphene oxide for bioremoval of trinitrophenol from wastewater in a bubble column bioreactor. *Electronic Journal of Biotechnology*, *39*, 8–14. Available from https://doi.org/10.1016/j.ejbt.2019.02.003.

Bayat, Z., Hassanshahian, M., & Cappello, S. (2015). Immobilization of microbes for bioremediation of crude oil polluted environments: A mini review. *Open Microbiology Journal*, *9*, 48–54. Available from https://doi.org/10.2174/1874285801509010048.

Bouabidi, Z. B., El-Naas, M. H., & Zhang, Z. (2019). Immobilization of microbial cells for the biotreatment of wastewater: A review. *Environmental Chemistry Letters*, *17*(1), 241–257. Available from https://doi.org/10.1007/s10311-018-0795-7.

Chan, Y. J., Chong, M. F., Law, C. L., & Hassell, D. G. (2009). A review on anaerobic–aerobic treatment of industrial and municipal wastewater. *Chemical Engineering Journal*, *155*(1–2), 1–18. Available from https://doi.org/10.1016/j.cej.2009.06.041.

Chung, T. P., Wu, P. C., & Juang, R. S. (2005). Use of microporous hollow fibers for improved biodegradation of high-strength phenol solutions. *Journal of Membrane Science*, *258*(1–2), 55–63. Available from https://doi.org/10.1016/j.memsci.2005.02.026.

Demarche, P., Junghanns, C., Nair, R. R., & Agathos, S. N. (2012). Harnessing the power of enzymes for environmental stewardship. *Biotechnology Advances*, *30*(5), 933–953. Available from https://doi.org/10.1016/j.biotechadv.2011.05.013.

Dey, S., & Mukherjee, S. (2013). Performance study and kinetic modeling of hybrid bioreactor for treatment of bi-substrate mixture of phenol-m-cresol in wastewater: Process optimization with response surface methodology. *Journal of Environmental Sciences (China)*, *25*(4), 698–709. Available from https://doi.org/10.1016/S1001-0742(12)60096-5.

Diya'Uddeen, B. H., Daud, W. M. A. W., & Abdul Aziz, A. R. (2011). Treatment technologies for petroleum refinery effluents: A review. *Process Safety and Environmental Protection*, *89*(2), 95–105. Available from https://doi.org/10.1016/j.psep.2010.11.003.

Duan, L., Wang, H., Sun, Y., & Xie, X. (2016). Biodegradation of phenol from wastewater by microorganism immobilized in bentonite and carboxymethyl cellulose gel. *Chemical Engineering Communications*, *203*(7), 948–956. Available from https://doi.org/10.1080/00986445.2015.1074897.

El-Naas, M. H., Alhaija, M. A., & Al-Zuhair, S. (2014). Evaluation of a three-step process for the treatment of petroleum refinery wastewater. *Journal of Environmental Chemical Engineering*, *2*(1), 56–62. Available from https://doi.org/10.1016/j.jece.2013.11.024.

El-Naas, M. H., Al-Muhtaseb, S. A., & Makhlouf, S. (2009). Biodegradation of phenol by Pseudomonas putida immobilized in polyvinyl alcohol (PVA) gel. *Journal of Hazardous Materials*, *164*(2–3), 720–725. Available from https://doi.org/10.1016/j.jhazmat.2008.08.059.

El-Naas, M. H., Al-Zuhair, S., & Alhaija, M. A. (2010). Removal of phenol from petroleum refinery wastewater through adsorption on date-pit activated carbon. *Chemical Engineering Journal*, *162*(3), 997–1005. Available from https://doi.org/10.1016/j.cej.2010.07.007.

El-Naas, M. H., Al-Zuhair, S., & Makhlouf, S. (2010). Continuous biodegradation of phenol in a spouted bed bioreactor (SBBR). *Chemical Engineering Journal*, *160*(2), 565–570. Available from https://doi.org/10.1016/j.cej.2010.03.068.

El-Naas, M. H., Mourad, A. H. I., & Surkatti, R. (2013). Evaluation of the characteristics of polyvinyl alcohol (PVA) as matrices for the immobilization of Pseudomonas putida. *International Biodeterioration and Biodegradation*, *85*, 413–420. Available from https://doi.org/10.1016/j.ibiod.2013.09.006.

Ettayebi, K., Errachidi, F., Jamai, L., Tahri-Jouti, M. A., Sendide, K., & Ettayebi, M. (2003). Biodegradation of polyphenols with immobilized *Candida tropicalis* under metabolic induction. *FEMS Microbiology Letters*, *223*(2), 215–219. Available from https://doi.org/10.1016/S0378-1097(03)00380-X.

Górecka, E., & Jastrzębska, M. (2011). Review article: Immobilization techniques and biopolymer carriers. *Biotechnology and Food Science*, *75*(1), 65–86.

Herrero, M., & Stuckey, D. C. (2015). Bioaugmentation and its application in wastewater treatment: A review. *Chemosphere*, *140*, 119–128. Available from https://doi.org/10.1016/j.chemosphere.2014.10.033.

Hsien, T. Y., & Lin, Y. H. (2005). Biodegradation of phenolic wastewater in a fixed biofilm reactor. *Biochemical Engineering Journal*, *27*(2), 95–103. Available from https://doi.org/10.1016/j.bej.2005.08.023.

Hu, X., Meneses, Y. E., Stratton, J., & Wang, B. (2019). Acclimation of consortium of micro-algae help removal of organic pollutants from meat processing wastewater. *Journal of Cleaner Production*, *214*, 95–102. Available from https://doi.org/10.1016/j.jclepro.2018.12.255.

Ismail, Z. Z., & Khudhair, H. A. (2018). Biotreatment of real petroleum wastewater using nonacclimated immobilized mixed cells in spouted bed bioreactor. *Biochemical Engineering Journal*, *131*, 17–23. Available from https://doi.org/10.1016/j.bej.2017.12.005.

Jain, M., Majumder, A., Ghosal, P. S., & Gupta, A. K. (2020). A review on treatment of petroleum refinery and petrochemical plant wastewater: A special emphasis on constructed wetlands. *Journal of Environmental Management*, *272*, 111057. Available from https://doi.org/10.1016/j.jenvman.2020.111057.

Jiang, L., Ruan, Q., Li, R., & Li, T. (2013). Biodegradation of phenol by using free and immobilized cells of *Acinetobacter* sp. BS8Y. *Journal of Basic Microbiology*, *53*(3), 224–230. Available from https://doi.org/10.1002/jobm.201100460.

Jiménez, S., Micó, M. M., Arnaldos, M., Medina, F., & Contreras, S. (2018). State of the art of produced water treatment. *Chemosphere*, *192*, 186–208. Available from https://doi.org/10.1016/j.chemosphere.2017.10.139.

Jou, C. J. G., & Huang, G. C. (2003). A pilot study for oil refinery wastewater treatment using a fixed film bioreactor. *Advances in Environmental Research*, *7*(2), 463–469. Available from https://doi.org/10.1016/S1093-0191(02)00016-3.

Jun, L. Y., Yon, L. S., Mubarak, N. M., Bing, C. H., Pan, S., Danquah, M. K., … Khalid, M. (2019). An overview of immobilized enzyme technologies for dye and phenolic removal from wastewater. *Journal of Environmental Chemical Engineering*, *7*(2). Available from https://doi.org/10.1016/j.jece.2019.102961.

Khan, M. Z., Mondal, P. K., & Sabir, S. (2013). Aerobic granulation for wastewater bioremediation: A review. *Canadian Journal of Chemical Engineering, 91*(6), 1045−1058. Available from https://doi.org/10.1002/cjce.21729.

Kilonzo, P., Margaritis, A., & Bergougnou, M. (2011). Effects of surface treatment and process parameters on immobilization of recombinant yeast cells by adsorption to fibrous matrices. *Bioresource Technology, 102*, 3662−3672. Available from https://doi.org/10.1016/j.biortech.2010.11.055.

Krekeler, C., Ziehr, H., & Klein, J. (1991). Influence of physicochemical bacterial surface properties on adsorption to inorganic porous supports. *Applied Microbiology and Biotechnology, 35*(4), 484−490. Available from https://doi.org/10.1007/BF00169754.

Kreuzer, H., & Massey, A. (n.d.). *American society for microbiology.* Available from https://doi.org/10.1128/9781555817480.

Krishnamoorthi, S., Banerjee, A., & Roychoudhury, A. (2015). Immobilized enzyme technology : Potentiality and prospects review. *Enzymology & Metabolism, 1*, 1−11.

Léonard, D., Youssef, C. B., Destruhaut, C., Lindley, N. D., & Queinnec, I. (1999). Phenol degradation by Ralstonia eutropha: Colorimetric determination of 2-hydroxymuconate semialdehyde accumulation to control feed strategy in fed- batch fermentations. *Biotechnology and Bioengineering, 65*(4), 407−415, https://doi.org/10.1002/(sici)1097-0290(19991120)65:4 < 407::aid-bit5 > 3.0.co;2-%23.

Leoncio, L., de Almeida, M., Silva, M., Oliveira, O. M. C., Moreira, Í. T. A., & Lima, D. F. (2020). Evaluation of accelerated biodegradation of oil-SPM aggregates (OSAs). *Marine Pollution Bulletin, 152*, 110893. Available from https://doi.org/10.1016/j.marpolbul.2020.110893.

Liu, J., Zhang, S. M., Chen, P. P., Cheng, L., Zhou, W., Tang, W. X., ... Ke, C. M. (2007). Controlled release of insulin from PLGA nanoparticles embedded within PVA hydrogels. *In Journal of Materials Science: Materials in Medicine, 18*(11), 2205−2210. Available from https://doi.org/10.1007/s10856-007-3010-0.

Liu, Y. J., Zhang, A. N., & Wang, X. C. (2009). Biodegradation of phenol by using free and immobilized cells of Acinetobacter sp. XA05 and Sphingomonas sp. FG03. *Biochemical Engineering Journal, 44*(2−3), 187−192. Available from https://doi.org/10.1016/j.bej.2008.12.001.

Lu, D., Zhang, Y., Niu, S., Wang, L., Lin, S., Wang, C., ... Yan, C. (2012). Study of phenol biodegradation using *Bacillus amyloliquefaciens* strain WJDB-1 immobilized in alginate-chitosan-alginate (ACA) microcapsules by electrochemical method. *Biodegradation, 23*(2), 209−219. Available from https://doi.org/10.1007/s10532-011-9500-2.

Lu, H., Yu, Y., Zhou, Y., & Xing, F. (2019). A quantitative evaluation method for wastewater toxicity based on a microbial fuel cell. *Ecotoxicology and Environmental Safety, 183*, 109589. Available from https://doi.org/10.1016/j.ecoenv.2019.109589.

Luo, H., Liu, G., Zhang, R., & Jin, S. (2009). Phenol degradation in microbial fuel cells. *Chemical Engineering Journal, 147*(2−3), 259−264. Available from https://doi.org/10.1016/j.cej.2008.07.011.

Mahdavianpour, M., Moussavi, G., & Farrokhi, M. (2018). Biodegradation and COD removal of p-Cresol in a denitrification baffled reactor: Performance evaluation and microbial community. *Process Biochemistry, 69*, 153−160. Available from https://doi.org/10.1016/j.procbio.2018.03.016.

Malakar, S., Saha, P. D., Baskaran, D., & Rajamanickam, R. (2017). Comparative study of biofiltration process for treatment of VOCs emission from petroleum refinery wastewater—A review. *Environmental Technology and Innovation, 8*, 441−461. Available from https://doi.org/10.1016/j.eti.2017.09.007.

Mamma, D., Kalogeris, E., Papadopoulos, N., Hatzinikolaou, D. G., Christrakopoulos, P., & Kekos, D. (2004). Biodegradation of phenol by acclimatized pseudomonas putida cells using glucose as an added growth substrate. *Journal of Environmental Science and Health, Part A, 39*(8), 2093−2104. Available from https://doi.org/10.1081/ESE-120039377.

Mei, X., Li, J., Jing, C., Fang, C., Liu, Y., Wang, Y., ... Ding, Y. (2020). Separation and recovery of phenols from an aqueous solution by a green membrane system. *Journal of Cleaner Production, 251*. Available from https://doi.org/10.1016/j.jclepro.2019.119675.

Morgan-Sagastume, F., Jacobsson, S., Olsson, L. E., Carlsson, M., Gyllenhammar, M., & Sárvári Horváth, I. (2019). Anaerobic treatment of oil-contaminated wastewater with methane production using anaerobic moving bed biofilm reactors. *Water Research, 163*. Available from https://doi.org/10.1016/j.watres.2019.07.018.

Mustafa, A., Azim, M. K., Raza, Z., & Kori, J. A. (2018). BTEX removal in a modified free water surface wetland. *Chemical Engineering Journal, 333*, 451−455. Available from https://doi.org/10.1016/j.cej.2017.09.168.

Nancharaiah, Y. V., & Kiran Kumar Reddy, G. (2018). Aerobic granular sludge technology: Mechanisms of granulation and biotechnological applications. *Bioresource Technology, 247*, 1128−1143. Available from https://doi.org/10.1016/j.biortech.2017.09.131.

Nzila, A. (2018). Biodegradation of high-molecular-weight polycyclic aromatic hydrocarbons under anaerobic conditions: Overview of studies, proposed pathways and future perspectives. *Environmental Pollution, 239*, 788−802. Available from https://doi.org/10.1016/j.envpol.2018.04.074.

Oh, Y. S., Maeng, J., & Kim, S. J. (2000). Use of microorganism-immobilized polyurethane foams to absorb and degrade oil on water surface. *Applied Microbiology and Biotechnology, 54*(3), 418−423. Available from https://doi.org/10.1007/s002530000384.

Park, J. K., & Chang, H. N. (2000). Microencapsulation of microbial cells. *Biotechnology Advances, 18*(4), 303−319. Available from https://doi.org/10.1016/S0734-9750(00)00040-9.

Partovinia, A., & Naeimpoor, F. (2013). Phenanthrene biodegradation by immobilized microbial consortium in polyvinyl alcohol cryogel beads. *International Biodeterioration and Biodegradation, 85*, 337−344. Available from https://doi.org/10.1016/j.ibiod.2013.08.017.

Prabakar, D., Suvetha, K. , S., Manimudi, V. T., Mathimani, T., Kumar, G., Rene, E. R., & Pugazhendhi, A. (2018). Pretreatment technologies for industrial effluents: Critical review on bioenergy production and environmental concerns. *Journal of Environmental Management, 218*, 165−180. Available from https://doi.org/10.1016/j.jenvman.2018.03.136.

Rana, S., Gupta, N., & Rana, R. S. (2018). Removal of organic pollutants with the use of rotating biological contactor. *Materials today: proceedings, 5*(2), 4218−4224. Available from https://doi.org/10.1016/j.matpr.2017.11.685.

Rasalingam, S., Peng, R., & Koodali, R. T. (2014). Removal of hazardous pollutants from wastewaters: Applications of TiO_2-SiO_2 mixed oxide materials. *Journal of Nanomaterials*, 1−42. Available from https://doi.org/10.1155/2014/617405.

Ruan, B., Wu, P., Chen, M., Lai, X., Chen, L., Yu, L., ... Liu, Z. (2018). Immobilization of Sphingomonas sp. GY2B in polyvinyl alcohol—alginate—kaolin beads for efficient degradation of phenol against unfavorable environmental factors. *Ecotoxicology and Environmental Safety*, *162*, 103—111. Available from https://doi.org/10.1016/j.ecoenv.2018.06.058.

Sahinkaya, E., & Dilek, F. B. (2006). Effect of biogenic substrate concentration on the performance of sequencing batch reactor treating 4-CP and 2,4-DCP mixtures. *Journal of Hazardous Materials*, *128*(2—3), 258—264. Available from https://doi.org/10.1016/j.jhazmat.2005.08.002.

Salman, H. H., & Ismail, Z. Z. (2020). Desalination of actual wetland saline water associated with biotreatment of real sewage and bioenergy production in microbial desalination cell. *Separation and Purification Technology*, 250. Available from https://doi.org/10.1016/j.seppur.2020.117110.

Sambusiti, C., Saadouni, M., Gauchou, V., Segues, B., Ange Leca, M., Baldoni-Andrey, P., & Jacob, M. (2020). Influence of HRT reduction on pilot scale flat sheet submerged membrane bioreactor (sMBR) performances for Oil&Gas wastewater treatment. *Journal of Membrane Science*, *594*, 117459. Available from https://doi.org/10.1016/j.memsci.2019.117459.

Saravanan, P., Pakshirajan, K., & Saha, P. (2008). Growth kinetics of an indigenous mixed microbial consortium during phenol degradation in a batch reactor. *Bioresource Technology*, *99*(1), 205—209. Available from https://doi.org/10.1016/j.biortech.2006.11.045.

Seung, H. S., Suk, S. C., Park, K., & Yoo, Y. J. (2005). Novel hybrid immobilization of microorganisms and its applications to biological denitrification. *Enzyme and Microbial Technology*, *37*(6), 567—573. Available from https://doi.org/10.1016/j.enzmictec.2005.07.012.

Singh, P., & Borthakur, A. (2018). A review on biodegradation and photocatalytic degradation of organic pollutants: A bibliometric and comparative analysis. *Journal of Cleaner Production*, *196*, 1669—1680. Available from https://doi.org/10.1016/j.jclepro.2018.05.289.

Surkatti, R., & El-Naas, M. H. (2014). Biological treatment of wastewater contaminated with p-cresol using Pseudomonas putida immobilized in polyvinyl alcohol (PVA) gel. *Journal of Water Process Engineering*, *1*, 84—90. Available from https://doi.org/10.1016/j.jwpe.2014.03.008.

Tam, N. F. Y., & Wong, Y. S. (2000). Effect of immobilized microalgal bead concentrations on wastewater nutrient removal. *Environmental Pollution*, *107*(1), 145—151. Available from https://doi.org/10.1016/S0269-7491(99)00118-9.

Tepe, O., & Dursun, A. Y. (2008). Combined effects of external mass transfer and biodegradation rates on removal of phenol by immobilized *Ralstonia eutropha* in a packed bed reactor. *Journal of Hazardous Materials*, *151*(1), 9—16. Available from https://doi.org/10.1016/j.jhazmat.2007.05.049.

Tosu, P., Luepromchai, E., & Suttinun, O. (2015). Activation and immobilization of phenol-degrading bacteria on oil palm residues for enhancing phenols degradation in treated palm oil mill effluent. *Environmental Engineering Research*, *20*(2), 141—148. Available from https://doi.org/10.4491/eer.2014.039.

Varjani, S., Pandey, A., & Upasani, V. N. (2020). Oilfield waste treatment using novel hydrocarbon utilizing bacterial consortium—A microcosm approach. *Science of The Total Environment*, *745*, 141043. Available from https://doi.org/10.1016/j.scitotenv.2020.141043.

Verma, M., Brar, S. K., Blais, J. F., Tyagi, R. D., & Surampalli, R. Y. (2006). Aerobic biofiltration processes - Advances in wastewater treatment. *Practice Periodical of Hazardous, Toxic, and Radioactive Waste Management*, *10*(4), 264—276. Available from https://doi.org/10.1061/(ASCE)1090-025X(2006)10:4(264).

Viero, A. F., de Melo, T. M., Torres, A. P. R., Ferreira, N. R., Sant'Anna, G. L., Borges, C. P., & Santiago, V. M. J. (2008). The effects of long-term feeding of high organic loading in a submerged membrane bioreactor treating oil refinery wastewater. *Journal of Membrane Science*, *319*(1—2), 223—230. Available from https://doi.org/10.1016/j.memsci.2008.03.038.

Wang, J., & Qian, Y. (1999). Microbial degradation of 4-chlorophenol by microorganisms entrapped in carrageenan-chitosan gels. *Chemosphere*, *38*(13), 3109—3117. Available from https://doi.org/10.1016/S0045-6535(98)00516-5.

Wang, Y., Tian, Y., Han, B., Zhao, H. B., Bi, J. N., & Cai, B. 1 (2007). Biodegradation of phenol by free and immobilized Acinetobacter sp. strain PD12. *Journal of Environmental Sciences*, *19*(2), 222—225. Available from https://doi.org/10.1016/S1001-0742(07)60036-9.

Zhou, H., Huang, X., Liang, Y., Li, Y., Xie, Q., Zhang, C., & You, S. (2020). Enhanced bioremediation of hydraulic fracturing flowback and produced water using an indigenous biosurfactant-producing bacteria Acinetobacter sp. Y2. *Chemical Engineering Journal*, *397*, 125348. Available from https://doi.org/10.1016/j.cej.2020.125348.

Zhou, L. C., Li, Y. F., Bai, X., & Zhao, G. H. (2009). Use of microorganisms immobilized on composite polyurethane foam to remove Cu(II) from aqueous solution. *Journal of Hazardous Materials*, *167*(1—3), 1106—1113. Available from https://doi.org/10.1016/j.jhazmat.2009.01.118.

Chapter 6

Algal bioremediation versus conventional wastewater treatment

Fares Almomani[1], Abdullah Omar[1] and Ahmed M.D. Al ketife[1,2]

[1]Department of Chemical Engineering, Qatar University, Doha, Qatar, [2]Faculty of Engineering, University of Thi-Qar, Nasiriyah, Iraq

Introduction

The release of organic and inorganic pollutants generated from domestic, agricultural, and industrial water activities can have a serious impact on the environment. Conventional treatment processes (primary and secondary) have shown excellent efficiencies in removing the easily settled solids and oxidizing the organic matter (OMs) present in wastewater (Ww). However, secondary effluents such as nutrients [nitrogen (N) and phosphorus (P)] cause eutrophication and long-term environmental problems. In addition, the presence of refractory organics and heavy metals can pass the conventional treatment processes without removal and can cause serious environmental if discharged without treatment. Microalgae-based Ww treatment systems are gaining popularity in recent years and offer a simple and cost-effective tertiary biotreatment process combined with the production of valuable biomass that can be utilized for several purposes. Microalgae include eukaryotic microalgae and prokaryotic cyanobacteria, which have high potential to grow in a harsh environment and are capable of photosynthesis can utilize inorganic carbon (IC) and nutrients from Ww. As a result, microalgae can be used to remove carbon (C), N, and P from Ww and aid disinfection due to the increase in pH during photosynthesis. Oxygen produced by microalgae can support the biological treatment of Ww by providing oxygen to the heterotrophic bacteria. In addition, microalgae hold amazing potential for the removal of heavy metals, as well as some toxic organic compounds, and can be used for CO_2 biofixation from the air, without the production of secondary pollutants. The produced algae biomass can be used for the production of food, biofuels, and different chemicals. Algae can have a good capacity for the production of polymer, fatty acids, pharmaceuticals, and cosmetic products. It has been reported that microalgae are excellent at capturing CO_2 with a fixing rate of 183 tons per 100 tons of produced biomass. The biodiesel produced from microalgae is one of the very few biofuels with negative CO_2 emissions (-183 kg CO_2 M) (Chisti, 2008). Furthermore, for the production of 1 kg of algal cells up to 0.33 kg nitrogen and 0.71 kg phosphate are removed from water (Yang et al., 2011).

The water demand in developed and developing countries has increased more than six folds in the last decade generating high volumes of Ww. This huge volume of municipal Ww, if properly treated can be reused for different purposes including landscape and irrigation. The Ww treatment plants (WwTPs), in general, have excellent efficiency in removing organic matter and solids, with limited efficiency for the removal of nutrients (ammonia and phosphorus) and emerging contaminants (Almomani, 2016). Therefore, an urgent need to develop advanced treatment technologies to remove these contaminants from Ww and reduce the chance for eutrophication and enable its reuse is a priority.

Microalgae have been used to recover nutrients from municipal Ww streams. This offers the potential for prospective integration with Ww treatment (Abdel-Raouf, Al-Homaidan, & Ibraheem, 2012; Åkerström, Mortensen, Rusten, & Gislerød, 2014; Almomani et al., 2017; Arbib, Ruiz, Álvarez-Díaz, Garrido-Pérez, & Perales, 2014; Gao et al., 2018; Mohamad et al., 2017). Microalgae can uptake different constituents (N, P, and minor nutrients) from Ww and use them as a growth medium. In this way, algal cultures can solve both economic and environmental problems while simultaneously producing biofuels or other useful chemicals (Davis et al., 2014; Sivakumar et al., 2012). Deploying a sustainable green technology such as the algae solar hybrid-filter photobioreactor (PBR) for Ww treatment, carbon recovery, and reuse can eliminate and save natural gas, the nation's main source of energy (Arbib et al., 2014).

Petroleum Industry Wastewater. DOI: https://doi.org/10.1016/B978-0-323-85884-7.00011-4

However, full-scale or commercialized biological processes that convert CO_2 to biomass feedstock from a point source in a full-scale process are not established yet.

The cultivation and growth of microalgae in Ww depends on water quality parameters, which include pH and nutrients, organic matter as well as the adequate light source, O_2, and CO_2 gas concentrations. These parameters are based on the source of the Ww, as well as the algae species itself (Baral, Singh, & Sharma, 2015; Pittman, Dean, & Osundeko, 2011). Whilst, many studies have focused on the influence of Ww source (artificial, municipal, agricultural, or industrial Ww) on nutrient removal and microalgae growth rates and/or biofuel production (Arbib et al., 2014), in this chapter, we will highlight the role of microalgae in the treatment of Ww

Conventional Ww treatment technologies

Preliminary treatment (PET) is the very first step in Ww treatment, and it usually consists of screening and grit removal Fig. 6.1. The main purpose of the PET is the removal of large and settleable solids/debris before further processing of the Ww. PET reduces interference of large objects with the following unit operations that are usually sensitive to inhibitive factors, and reduces the chance for mechanical wear, and avoids further maintenance costs. Although most of the large solids are removed after the PET, the Ww effluents are not yet entirely ready to be processed through the secondary and tertiary treatment stages. This is because of the presence of readily settleable solids and floating material that are of significant size. Additionally, taste and odor-causing substances must be removed. This is where primary treatment (PT) comes in to further clarify the Ww streams before the biological treatment. The PT typically removes the biological oxygen demand (BOD), total suspended solids (TSS), and oil and grease at a rate of 25%−50%, 50%−70%, and 65%, respectively (Al-Rekabi, Qiang, & Qiang, 2007). The three most common unit operations in the PT are sedimentation, aeration, and coagulation−flocculation unit operations. The secondary treatment (ST), also known as the biological treatment, is the final step in the conventional process before the advanced treatment. In the ST residual colloidal and soluble biodegradable organics (BODs) are removed and utilized as substrate for bacterial growth (Shao et al., 2020), which in turn produces side end-products, such as CO_2, NH_3, and H_2O in the case of aerobic growth. In the ST the biological treatment tank allows the microorganisms to "clean" the water by the introduction of oxygen through aeration. The microbial growth leads to the accumulation of biomass, which has to be partially removed before recycling it back to the reactor in a process called activated sludge process (ASP) (Östman, 2018).

Advanced Ww treatment technologies

Filtration is one of the most common advanced Ww (Aww) technologies, which exists in most Ww treatment plants. In this process, the Ww passes through a porous medium that holds the suspended solids while allowing the filtrate to pass through. Although filtration is an excellent technique to remove high amounts of suspended solids, surface

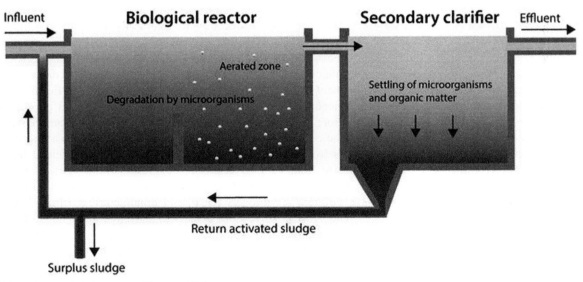

FIGURE 6.1   Activated sludge process (Östman, 2018).

limitations is a serious challenge when dealing with high concentration of dissolved contaminants. Therefore, ion exchange is often proposed (Garai & Yavuz, 2019; Guida et al., 2021; Maslova, Ivanenko, Yanicheva, & Gerasimova, 2020; Villeneuve, Perreault, Chevallier, Mikhaylin, & Bazinet, 2019; Wang, Li et al., 2020; Wang, Tian et al., 2020). The ion exchange relies on strong binding between resins and certain ions such as calcium and magnesium. While ion exchange is an inherently constituent-selective removal process, its feasibility in industrial and power generation applications is limited due to the need for ultra-pure water. Therefore, membrane technologies, including ultrafiltration (Huang & Feng, 2019) are used.

Nutrients [phosphorus (P) and nitrogen (N)] are a serious concern to Ww management and environmental engineers. P and N most notably lead to eutrophication, leading to the high oxygen uptake rate that actively harms living organisms and aquatic life in the receiving waters. Additionally, ammonia must be removed to acceptable levels to prevent its toxicity on the aquatic life and nitrogenous oxygen demand. This issue is most commonly tackled by the utilization of a sequence of anaerobic/anoxic/aerobic biological treatment. The process involves the removal of P and ammonia PAOs and the oxidation/reduction of NH_3 to N_2.

Biological treatment is not effective when dealing with recalcitrant organic contaminants that are resistant to biodegradation. Chemical oxidation, also known as the advanced oxidation process (AOP) can be used to solve this problem at a higher overall cost caused primarily by the need to continuously provide chemicals to complete the required chemical reaction (Ben, Qiang, Pan, & Chen, 2009; Zhang, Sun, & Guan, 2013).

Adsorption can also be used as an advanced treatment process. Adsorption involves the diffusion of a fluid phase to a solid surface where the adsorbate (solute) is attached to the surface of the adsorbent (solid) by either weak physical forces or chemical bonds (Seader, Henley, & Roper, 1998). This transfer phenomenon occurs until the adsorbent reaches its capacity or what is called the saturation point, after which the solid must be "regenerated" for further use. Adsorption can be used for either purification purposes (usually defined at solute wt.% $<10\%$) or bulk separation ($>10\%$). Generally, sorbents are expected to be selective, have a high capacity, chemi-thermally stable, hard, foul resistant, convenient, cost-effective, and most importantly has large specific surface areas. After going through all the treatment unit operations, the Ww effluent must pass through the disinfection stage, which is a well-established and reliable technology that mainly aims to destroy most of the pathogenic microorganisms left in the effluent while also treating some of the residual organic and inorganic residuals. Disinfection by chemical means (chlorination and ozonation) or physical ones (UV light) is very common (Azuma & Hayashi, 2021).

Algal culture for Ww treatment systems

The cultivation of microalgae inside the PBR can differ in nature and performance depending on the method of maintaining the microorganisms inside the reactor, with the most common of algae growth are *suspended* and *immobilized* microalgae cultures. Figs. 6.2 and 6.3 illustrate the general reactor setups for suspended versus immobilized microalgae cultures. Hilares et al. (2021) illustrate the treatment of poultry slaughterhouse Ww (PSW) by the use of the microalgae *Chlorella vulgaris* which is kept in suspension via a sparger in the reactor that is set for both batch and continuous processing. The study showed about 85% and 90% of COD removal after 2 and 9 days of cultivation, respectively. The "stirring" process could be conducted in many ways, including sparger, magnetic stirring bars (Su & Jacobsen, 2021), and aeration (Ayre, Mickan, Jenkins, & Moheimani, 2021). The suspension of microalgae mid-water allows for the direct contact between the microorganisms and the fresh nutrients flowing into the reactor in the continuous operation, but that also endangers the species that are unprotected from any possible damage made by impact with the water and air bubbles. This could be avoided by immobilizing the microalgae and attaching them into biocarriers in which biofilms of the species are created on the surfaces of the carriers (Moreno-Garrido, 2008). In this state of controlled microbial growth, the algae are protected and the impact damage from the turbulence caused by feeding the reactor more water and the content stirring is minimized. Additionally, attached microalgal growth can achieve higher biomass yields and growth than the suspended method (Zhuang, Li, & Hao Ngo, 2020). Some of the novel immobilization methods in the literature include attaching to fiber carriers (Zhuang et al., 2016), rotating attachment materials (Gross, Henry, Michael, & Wen, 2013), and even the use of capillary-driven PBRs (Xu et al., 2017). Furthermore, immobilization of the strain *Desmodesmus* sp. is applied in treating domestic Ww, specifically, the removal of 17β-estradiol, which is a precursor for carcinogenic compounds (Wang, Li et al., 2020; Wang, Tian et al., 2020). Up to 85%−99% of 17β-estradiol removal was achieved by the algal beads. For comparison purposes, Lin-Lan, Jing-Han, and Hong-Ying (2018) study the differences between attached and suspended algal growth and the physiological differences in properties in both applications.

The use of single microalgae strain was highly recommended by different research due to the ability to investigate the algae performance without the interference of other variables that are introduced by other species. However, in

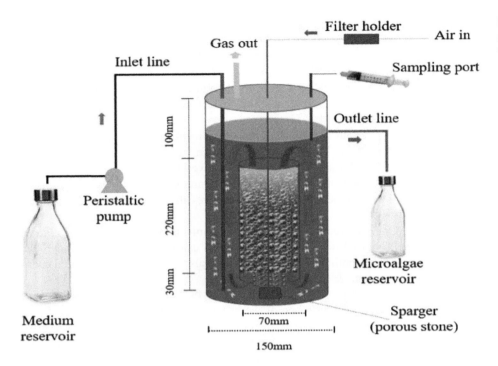

FIGURE 6.2 Suspended microalgal growth culture maintained by sparger (Hilares et al., 2021).

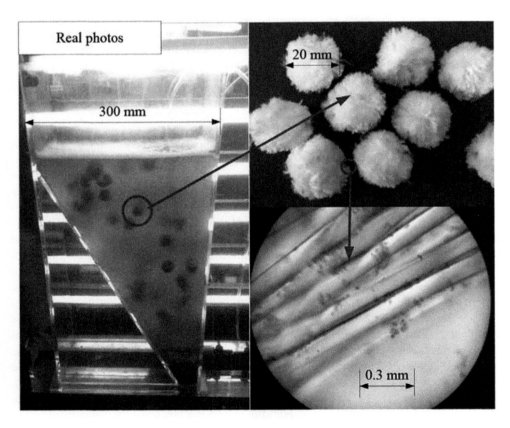

FIGURE 6.3 Immobilized/attached microalgal growth using cotton pellets (Zhuang et al., 2018).

practice, the use of a mixed culture could enhance the growth of the microorganisms and thus the nutrient/contaminant uptake rate from the Ww. This is because each species are observed to need certain types of nutrients due to their biology, so different species with varying structures would lead to a wider range of nutrients removed from the Ww. Table 6.1 summarizes research work that used mixed cultures in treating different types of Wws.

TABLE 6.1 Review of existing literature regarding the utilization of mixed algal cultures.

Paper goal	Ww	Growth type	Results	References
Investigation of the effect of HRT on removal efficiencies inside a membrane PBR	SYSWw	Immobilized growth	Higher HRT values reduced efficiency (optimum was 24 h)	Solmaz and Işık (2020)
Investigation of the potential of mixed cultures enriched from aerobic and anaerobic sludge	AEWw	Suspended growth	After 14 days of cultivation, complete removal of ammonia nitrogen was achieved	Yu, Kim, and Lee (2019)
Comparative study between monocultures and mixed cultures of microalgae in terms of biomass and treatment	SNSWw	Suspended growth	Mixed cultures generally achieved more stable and higher biomass production and nutrient removal, but could not achieve over yielding	Luo, Le-Clech, and Henderson (2020)
Effect of phosphorus supplementation on growth rate and nutrient removal of mixed microalgae culture	SARP & PFM	Suspended growth	Phosphorus removal and algal cell growth are shown to be independent, and higher amounts of internally stored phosphate led to more ammonium removal	Ruiz-Martinez et al. (2014)

Algal growth systems

Open and closed systems are used for algae growth. Open systems are normally open to the atmosphere and exposed to the sun and environments. These systems include waste stabilization pond systems (WSPSs) and high rate algal ponds (HRAPs). In general, WPs are used for Ww treatment which is considered as "green treatment". Effective Ww treatment can be accomplished through the integrated growth of microalgae and heterotrophic bacteria. The produced oxygen from the Microalgae photosynthesis process. is utilized by the heterotrophic bacteria to biooxidize the organic compounds in Ww at aerobic conditions. The CO_2 produced from the biooxidation process is consumed by microalgae in the photosynthesis process. HRAP consists of an algal reactor combined with strong oxidation ponds. The HRAP offers a much more effective Ww treatment option compared to typical oxidation ponds. The high efficiency of the HRAP is primarily caused by strong microalgae photosynthesis resulting in more oxygen as a byproduct to supply the aerobic oxidation process and consumption of nutrients by algae, which will be converted to biomass. Open PBR systems are hard to control and can be easily contaminated (Znad, 2012). Closed systems include TPBRs, mechanically stirred tanks, airlift, and bubble columns. Closed systems are easier to control and can achieve higher mass transfer rates. Table 6.2 summarizes the advantages and disadvantages of open and closed algae growth systems

A comparison between open pond (Op) and close systems (CPBR) Table 6.3 indicates open systems as being cost-effective, particularly at the large-scale process and low land costs, and simple in implementation. Against this, these systems incur a higher contamination risk, lower CO_2 fixation, and concomitantly lower biomass productivity. PBRs offer higher biomass productivity, CO_2 fixation, lower contamination risks and are fully controllable systems but are most costly and complicated in design. They are thus most suited to small-scale applications

The following sections present a description of the most common PBR used in algae-based Ww treatment technologies

Tubular photobioreactor

Tubular PBR (TPBR) consists of long helical or straight tubes configured in various geometries, aiming to maximize the use of light from the source. The growth medium can be circulated by injection of gas at an end of the tube, which contains a certain concentration of CO_2 and is allowed to exist in the system at the other end of the tube. Experimental work showed that large-scale TPBR usually fails due to oxygen accumulation (Molina Grima, Acién Fernández, García Camacho, Camacho Rubio, & Chisti, 2000) Fig. 6.4.

Mechanically stirred photobioreactor

Mechanically stirred PBR (MSPBR) uses baffles to move the growth media, to achieve the transfer of air into the growth media. The stirred growth media can have some disadvantages such as high shear stress which causes damage

TABLE 6.2 Advantages and disadvantages of open and closed algae growth systems (Sudhakar, Suresh, & Premalatha, 2011).

Parameter	Open pond	Closed PBR
Construction	Simple	More complicated-varies by design
Cost	Cheaper to construction and operation cost is cheaper	more expensive construction, operation
Water losses	High	Low
Typical biomass concentration	Low, $0.1-0.2$ g L^{-1}	High: $2-8$ g L^{-1}
Temperature control	Difficult	Easily controlled
Species control	Difficult	Simple
Contamination	High risk	Low risk
Light utilization	Poor	Very high
CO_2 losses to the atmosphere	High	Almost none
Typical Growth rate (g m^{-2} d^{-1})	Low: $10-25$	Variable: $1-500$
Area requirement	Large	Small
Depth/diameter of water	0.3 m	0.1 m
Surface: volume ratio	~6	

TABLE 6.3 Advantages and disadvantages for some common types of photobioreactors.

Configuration	Prospects	Limitations
Open ponds	less costly and more economic; easy to clean after cultivation; good for mass cultivation of algae	Difficult to control culture conditions; Not suitable for the cultivation of algal cells for long periods; poor productivity occupy a large land and use limited mass; too few strains of algae cultures are easily contaminated
Vertical-columnPBR	High mass transfer good mixing with low shear stress; low energy consumption high potentials for scalability easy to sterilize; good for immobilization of algae; reduced photo-inhibition and photo-oxidation	Small illumination surface area. Their construction requires sophisticated materials shear stress to algal cultures; decrease of illumination surface area upon scale-up
Flat-plate PBR	Large illumination surface area; suitable for outdoor cultures; good for immobilization of algae; a good light path and biomass productivities; relatively cheap easy to clean up Readily tempered low oxygen build-up	Scale-up require many compartments and support materials; difficulty in controlling culture temperature; some degree of wall growth possibility of hydrodynamic stress to some algal strains
TubularPBR	Large illumination surface area suitable for outdoor cultures; Good biomass productivities; relatively cheap	Gradients of pH, dissolved oxygen, and carbon dioxide along the tubes Fouling some degree of wall growth requires large land space

Source: Adapted from Patel, A.K., Joun, J., & Sim, S.J. (2020). A sustainable mixotrophic microalgae cultivation from dairy wastes for carbon credit, bioremediation and lucrative biofuels. Bioresource Technology, *313*, 123681; Rodrigues, L.H.R., et al. (2011). Algal density assessed by spectrophotometry: A calibration curve for the unicellular algae Pseudokirchneriella subcapitata. Journal of Environmental Chemistry and Ecotoxicology; Zkeri, E., et al. (2021). Comparing the use of a two-stage MBBR system with a methanogenic MBBR coupled with a microalgae reactor for medium-strength dairy wastewater treatment. Bioresource Technology, *323*, 124629 (Patel, Joun, & Sim, 2020; Rodrigues et al., 2011; Zkeri et al., 2021).

to the wall of the cells (Grima, 1996). On the other hand, when the growth medium is stirred slowly it will not expose all the cells to the light source and might limit the mass transfer Fig. 6.5.

Airlift photobioreactor

Airlift PBR (ALPBR) consists of a column separated into two sections, air/CO_2 is injected in one of the section sections which causes circulation of the growth medium. The injection section is called the riser and the other section is called the

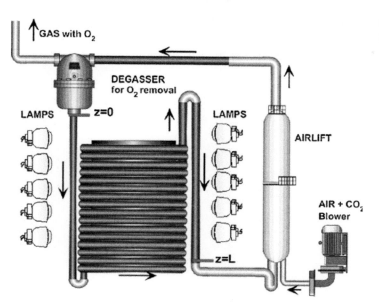

FIGURE 6.4   Tubular photobioreactor (Concas, Pisu, & Cao, 2010).

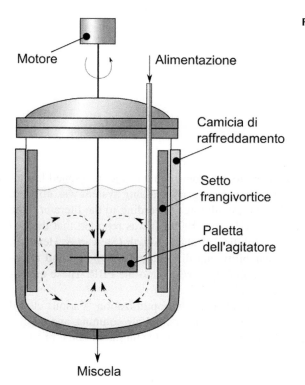

FIGURE 6.5   Mechanically stirred photobioreactor (Pugliesi, 2009).

downcomer (Miron, 2000). The ALPBR is mainly used for fermentation and Ww treatment. The main difficulty with this type of PBR is the small illumination area. Airlift PBRs showed good mass transfer, energy consumption, and mixing.

Bubble column photobioreactor

Bubble column PBR consists of the vertical cylindrical or rectangular column. The column is filled with a growth medium, CO_2/air is injected at the bottom of the column by a sparging system. It is reported that Airlift PBR can achieve a high mass transfer compared to other systems. The new version of the bubble column PBR can achieve efficient aeration and less pressure drop at high flow rates Fig. 6.6 (Poulsen & Iversen, 1998).

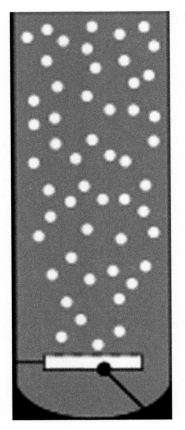

FIGURE 6.6   Bubble column photobioreactor (Chanab, 2017).

Optimum photobioreactor design

The design of PBRs underwent different modifications during the last decay to overcome some of the mentioned limitations. For example, to prevent oxygen accumulation in the system an automated oxygen degassing system was utilized. The length to flow velocity ratio was calibrated to achieve an optimal algal growth rate. According to the literature, the main failure of large-scale TPBR is the high dissolved oxygen (DO) value (Herzog, 1999). It was reported that the DO value in some of PBRs can reach as high as 20 mg L^{-1}, and can cause inhibition of algae growth (Stewart & Hessami, 2005). Researchers reported other issues with TBPR such as limited scalability and difficulties in building and maintaining the system. Other researchers tried to eliminate excess oxygen by bubbling (air/N_2) into the PBR. These techniques can lower the free oxygen in the system, but the DO remains almost unaffected.

The influence of the overall reactor design represents the starting point in algal growth optimization. Various closed PBR configurations have been considered, ranging from flat plate reactors, air-lift reactors, bubble columns, and tubular reactors. Of the system facets impacting on algae growth, however, tolerance to changes in loading is of some practical significance since

Bioaccumulation and bioassimilation

Bioaccumulation is a broad concept that describes the net positive movement of chemicals and metals into an organism for growth, and beyond growth requirements, by adsorption. Metals play an important role in the growth and operational performance of microalgae cultivation. They are an essential micronutrient for growth (Brownell & Nicholas, 1967; Takahashi, Kopriva, Giordano, Saito, & Hell, 2011), but they add to the costs of operations regarding microalgae cultivation and hydrothermal liquefaction (Jiang & Savage, 2019; Leng, Li, Wen, & Zhou, 2018). The biological uptake of metals using algae occurs primarily by the reactions between the anionic functional groups in the microalgae cell wall and the cationic metal ions (Mehta & Gaur, 2005; Siegel & Siegel, 1973), causing high adsorption capacities (q_{max}) and significant metal ions removal efficiencies (%RE). Table 6.4 summarizes the literature review of existing research in the removal of metal ions by different algae species along with the process parameters and other important variables. Table 6.5 summarizes the effects of metals (and other nutrients) on cell growth, highlighting important

TABLE 6.4 Reported algal biosorption capacities, heavy metals removal efficiencies, and micronutrients uptake.

Microalgae strain	Metal	Con. (mg L⁻¹)	Sorbent con. (g L⁻¹)	Process parameters	Functional. group, (cm⁻¹)	Model	Sorption capacity, q_{max} (mg g⁻¹)	RE (%)	References
Parachlorella sp.	Cd^{2+}	18–180	1	BP$_{TT}$: PC; 30°C and pH 7, 120 rpm, 5–6 h	3430 –OH; – NH 1653; 1540 C = O amid I and amid II; 1072 of C-O alcoholic group	IM: L; F;T	91	nr	Dirbaz and Roosta (2018)
Spirulina sp.							50		
Scenedesmus sp.							26		
Nannochloropsis sp.							60		
Mixed culture		60		BP$_{DY}$		DM: P$_{Fo}$; P$_{SO}$;	51		
Arthrospira platensis (TDB)	Cu^{2+}	100	0.5	BP$_{TT}$: 26°C–28°C	nr	IM: L	33	nr	Markou et al. (2015a)
	Ni^{2+}						57		
	Cu^{2+}			BP$_{DY}$		DM: P$_{SO}$	23		
	Ni^{2+}						12		
Spirulina platensis (Extracted bead)	Cr^{6+}	100	0.1	BP$_{TT}$: pH 5;180 rpm; 25°C; 24 h	nr	IM: BET	41	nr	Kwak et al. (2015)
Chlorella vulgaris, (RBAE)	Cr^{6+}	50	3	BP$_{TT}$: pH 4; 45°C; 30 mL; 200 rpm,	Carboxyl and Amino groups; 3421 O–H; N–H; 3010 =C–H; 2800; 3000, 1746 C–H; 1656 and 1545 amide I and amid II; 1244,1154, and 1080 P = O; 1200 and 900 C–OH and C–O–C group.	Sips Isotherm	43	nr	Xie et al. (2014)
Chlorella vulgaris Beijerinck	La^{3+}	1	(1 × 10⁴)*	BP$_{TT}$:500 mL; 25°C pH 6.0; 24 h.	nr	F	29	nr	Danièle, Jérôme, and Corinne (2016)
	Gd^{3+}						29		
	Y^{3+}						29		
Chlorella vulgaris (DGA)	Nd^{2+}	250**	0.5	BP$_{DEM}$: pH 5; 35°C; 90 min; 300 rpm	Hydroxyl; Carbonyl; sulfhydryl; sulfonate; Thioester; Amine; Secondary amine; Imine; Imidazole; Phosphonate; Phosphodiester.	DEM	160	nr	Kucuker, Nadal, and Kuchta (2016)
		2.15	2	CP$_{DEM}$: pH 5; 35°C; flow rate 20 mL min⁻¹, 47 min, 2.5 L			nr	97	
		100						50	
		50						100	
		250	5				34	nr	
Ankistrodesmus sp.	La^{3+}	10–100	0.72–0.65	BP$_{DY}$	nr	DEM	640–910	nr	Correa, Luna, and da Costa (2017)
Golenkinia sp.							700–1200		
Ankistrodesmus sp.							1300		
Golenkinia sp.							720		

(Continued)

TABLE 6.4 (Continued)

Microalgae strain	Metal	Con. (mg L⁻¹)	Sorbent con. (g L⁻¹)	Process parameters	Functional. group, (cm⁻¹)	Model	Sorption capacity, q_{max} (mg g⁻¹)	RE (%)	References
D. multivariabilis	La³⁺	≤100	0.05	BP_TT: 4 h; pH 5–7.5, 250 Ml; 350 rpm; 25°C.	nr	IM: L	100	nr	Birungi and Chirwa (2014)
C. reinhardtii							140		
S. bacillaris							51		
S. acuminutus							110		
C. saccharophilum							130		
C. vulgaris							63		
D. multivariabilis						DM: P_SO	63		
C. reinhardtii							57		
S. bacillaris							73		
S. acuminutus							64		
C. saccharophilum							53		
C. vulgaris							29		
C. minutissima (UTEX 2341)	Zn²⁺		4	BP_TT: PH 6; T 10, 28°C–37°C; 50–50 mL	nr	IM: L	120	nr	Yang et al. (2015)
	Mn²⁺						35		
	Cd²⁺						35		
	Cu²⁺						3		
Spirulina platensis	Cd²	100	0.5–5	BP_TT; pH 5–6; 25°C; 60 min; 150 rpm.	Spirulina: 340 for –OH group; 2918 for –CH; 1546–1629–1657 Amide I and II; 1029 and 1406 –S-O and –P-O respectively; 1353–1161 S = O, in 852 N-H; 545 peak of –PO and O-P-O. C. vulgaris: 3440 NH; 2070 N = C = S; 1654 Amide I and II; 1075 C-C and C-O; 580 –PO and –O-P-O.	IM: L	39[1]	93[2]	Sayadi, Rashki, and Shahri (2019)
	Pb²⁺						39[1]	94[2]	
	Cu²⁺						37[1]	81[2]	
	Cd²⁺						38[1]	88[2]	
	Pb²⁺						41[1]	90[2]	
	Cu²⁺						37[1]	85[2]	
Spirulina platensis	Cd²			5, 15, 30, 50, 75, 100 mg L⁻¹;		DM:P_SO	52	nr	
	Pb²⁺						54	nr	
	Cu²⁺						37	nr	
C. vulgaris	Cd²⁺						50	nr	
	Pb²⁺						50	nr	
	Cu²⁺						44	Nr	

Abbreviations: Con., concentration; RE, removal efficiency; refs, reference; Cd, cadmium; BP, batch process; IT, isotherm experiment; DY, dynamic experiment; PC, process conditions; IM, isotherm model; L, langmuir; F, freundlich; T, tempkin; Bolded Later predominant model; DM, dynamic model; PFO, pseudofirst order; PSO, pseudofirst order; nr, not reported; TDB, typical dry biomass; Cu, copper; Ni, nickel; Cd, cadmium; Cr, chromium; BET, Brunauer–Emmett–Teller; RBAE, residual biomass after the extraction; La, lanthanum; Gd, gadolinium; Y, yttrium; * cells concentrations cell mL-1; DGA, dried green algae; Nd, neodymium; Zn zinc; ** Nd in mixed leachate derived neodymium magnets; DEM, direct experimental mode; CP, continuous process; 1 measured at 0.5 g biomass dose; 2 measured at 5 g biomass dose, the University of Texas at Austin Culture Collection of Algae Number UTEX 2341.

TABLE 6.5 Reported macronutrient functions in algal cells metabolism pathways.

Metal species	Concn. (mg L⁻¹)	Cultivation conditions	X_{max} (g L⁻¹)	μ (d⁻¹)	Model used	% removal	Metal Y^1 (mg g biomass⁻¹)	HRT (d⁻¹)	References
Mg^{2+}	7.4	MBM; SF: N 41 mg L⁻¹; P 53 mg L⁻¹; N/P 0.77; $C_{c,g}$ 2.1%.	1.8	nr	Biok.	70	2.9	10–50	Ayed et al. (2015)
K^+	13.3	PLA; GBCPBR; N 100 mg L⁻¹; P 40 mg L⁻¹; pH 7; $C_{c,g}$ 0.03ᵃ %	1.3	nr	Ex.	13	10	nr	Markou et al. (2015b)
Cu^{2+}	19 ×10⁻³	Dse; MPBRc; N 0.78 ± 0.1 mg L⁻¹; P 14.1 ± 0.9 mg L⁻¹; pH 7.8; $C_{c,g}$ 4%.	1.72	nr	Ex.	65	1.1 × 10⁻²	2	Gao et al. (2016)
Mn^{2+}	17 × 10⁻³					100	9.9 × 10⁻³		
Fe^{3+}	153 × 10⁻³					100	8.9 × 10⁻²		
Cu^{2+} 2	100	PBRs; 30 ± 2°C, 0.1 d⁻¹, 10 L, $C_{c,g}$ 0.03ᵃ %; 360 µE m⁻².s⁻¹, 48 h	0.5	nr	Ex.	nr			Markou et al. (2015a)
Ni^{2+} 2									
Cu^{2+} 3		PBR$_{Batch}$							
Ni^{2+} 3									
La^{3+}	10–100	PBR$_{Batch}$ 23°C, 12 h, pH 7.5, cultivation medium ASM⁻¹, AND initial inoculum 0.1 g L⁻¹	0.7	nr	Biok	nr	nr	nr	Corrêa, Luna, and da Costa (2016)
		PBRC, 90 minutes, 25 mL min⁻¹, and 8 h.				80	nr		
Ca^{2+}	23	PBRBatch, P_{ww}, CPBR 350 mL, 24°C, 180 µE, 100 h	1.6	1.2	Biok	100	14.5	50	Znad et al. (2018)
S^{2-}	21					70	9		
Mg^{+2}	10					82	5		
K^{+1}	26					54	8.8		
As^{3+}	1.1 × 10⁻⁶					98	7 × 10⁻⁷		
Ni^{2+}	2.3 × 10⁻⁶					100	1.5 × 10⁻⁶		
Co^{2+}	0.4 × 10⁻⁶					100	2.5 × 10⁻⁷		
Pb^{2+}	1.1 × 10⁻⁶					100	7 × 10⁻⁷		
Zn^{2+}	26 × 10⁻⁶					100	3 × 10⁻⁷		
Cu^{2+}	38 × 10⁻⁶					100	2.5 × 10⁻⁵		
Mn^{2+}	8.2 10⁻⁶					95	5 × 10⁻⁶		
Fe^{3+}	57 × 10⁻⁶					95	3.5 × 10⁻⁵		
Cd^{2+}	0.1 × 10⁻⁶					100	6 × 10⁻⁸		
Se^{2-}	39 × 10⁻⁶					100	2.5 × 10⁻⁵		

TABLE 6.6 Reported micronutrients influence on *Chlorella vulgaris* growth.

Technique	Theory	Advantages	Disadvantage	References
Flocculation	Algal cells are grouped using a flocculants chemical (ferric sulfate, ferric chloride, and ammonium sulfate), bioagents (chitosan), or microbes (bacteria). to increase their size and settling rate	Time-saving	Required large space High chemical/ operating cost Cannot separate intact cells	Brennan and Owende (2010)
Filtration	Algal cells with cells size >70 μm can be filtered under pressure or suction. Small algal cells (<30 μm) can be separated using ultrafiltration.	Time-saving	Cannot separate intact cells. Membrane fouling and clogging High-pressure requirements	Giovannoni et al. (1990)
Flotation	Algal cells are trapped by air bubbles	Cheap compared with other methods. Time-saving	Affected by bubble distribution into the suspension.	Brennan and Owende (2010)
Sonication	Using acoustic forces to separate algal cells	Low fouling and shear problem Limited freely moving parts can be operated in continuous operation can collect intact cells. Occupy small space	High power requirements Cooling requirements High cost at large-scale	Bosma et al. (2003)
Centrifugation	Using centrifugal force to separate cells	Fast and suitable for large microalgae	High energy requirement	Schenk et al. (2008)
Precipitation	Use for large algae with self-precipitation properties	Low energy or chemical requirements. Natural processes can be operated continuous.	Used for specific species- Required optimum settling time. Not all species are self-precipitated	

parameters such as the degree of importance (DOI), which is an indicator of how much impact the ion has on the growth and metabolism, whereas there is no relationship between the concentration of a metal in the algae growth (X) and its DOI. Al Ketife, Al Momani, and Judd (2020) conducted a study in which the bioaccumulation capabilities of *C. vulgaris* are determined through the use of a mathematical model utilizing Response Surface Methodology, Box Behnken Methodology, and best-fit simulation to calculate uptake rates of the heavy metals. The results showed that the microalgae could remove up to 74%, 73% and 69% of Cd^{2+}, Cu^{2+}, and Pb^{2+} respectively. Table 6.6 reports the effects of different metal species on the cultivation of *C. vulgaris*. Considering the positive effect of metal ions on algal growth, the simultaneous treatment of Ww and microalgae cultivation for further processes such as biofuel production is a process that suggests itself, leading to decreased costs and a more environmentally friendly method of combining the two purposes. This is further illustrated by Yang, Cao, Xing, and Yuan (2015), who successfully implemented the combined lipid production (LP) and biosorption/bioaccumulation of different heavy metal ions by the microalgae strain *Chlorella minutissima*. It can be clearly seen that heavy metals have no negative effect on algal growth, but they also lead to an increase in the lipid content by 21.07% and 93.90% with the addition of Cd^{2+} and Cu^{2+}, respectively. Additionally, the maximum removal rates achieved for Zn^{2+}, Mn^{2+}, Cd^{2+} and Cu^{2+} were 62.05%, 83.68%, 74.34%, and 83.60%, respectively.

Algae harvesting

Algal cell harvesting is a key point in the microalgae cultivation process. Many techniques have been suggested for cell harvesting (Table 6.7). Regardless of the variety of the harvesting processes presented in the table, associated with

TABLE 6.7 Advantages and disadvantages of the most common harvesting techniques.

Technique	Theory	Advantages	Disadvantage	References
Flocculation	• Algal cells are grouped using a flocculants chemical (ferric sulfate, ferric chloride, and ammonium sulfate), bioagents (chitosan), or microbes (bacteria).	• Timesaving	• Required large space • High chemical/operating cost • Cannot separate intact cells	Brennan and Owende (2010)
Filtration	• Algal cells with cells size >70 µm can be filtered under pressure or suction. • Small algal cells (<30 µm) can be separated using ultrafiltration.	• Timesaving	• Cannot separate intact cells. • Membrane fouling and clogging • High-pressure requirements	Giovannoni et al. (1990)
Flotation	• Algal cells are trapped by air bubbles	• Cheap compared with other methods. • Timesaving	• Affected by bubble distribution into the suspension.	Brennan and Owende (2010)
Sonication	• Using acoustic forces to separate algal cells	• Low fouling and shear problem • Limited freely moving parts can be operated in continuous operation • can collect intact cells. • Occupy small space	• High power requirements • Cooling requirements • High cost at large-scale	Bosma et al. (2003)
Centrifugation	• Using centrifugal force to separate cells	• Fast and suitable for large microalgae	• High energy requirement	Schenk et al. (2008)
Precipitation	• Use for large algae with self-precipitation properties	• Low energy or chemical requirements. • Natural process • Can be operated in	• Used for specific species- • Required optimum settling time.	

various advantages and limitations, there is no perfect technique has been reported, concerning the large-scale process and economic feasibility. Therefore, selecting the good techniques is a function of microalgae cultivation parameters, algal strain, and the selected harvesting techniques

Conclusions

Microalgae have been widely studied as a promising biological method for Ww treatment and biomass production. The following points can be concluded:

High ability to remove various pollutants from various Ww resources, especially N and P, therefore it can be successfully used for Ww treatment.

Algal Pond PBR can be used successfully for WwT, whist the application technology in a large-scale process is still challenging and needs more investigations.

The appropriate algal strain that can be used for certain purposes should be carefully selected.

The algal harvesting process is still concerned regarding the cell's rain efficiency and process economy, whilst the most common methods are filtration and centrifugation.

To make this technology applicable for the industrial scale, in this newly emerged field of science, needs to extensive work of research to overcome the current barrier and to solve the current problems associated with this industry.

References

Abdel-Raouf, N., Al-Homaidan, A. A., & Ibraheem, I. B. M. (2012). Microalgae and wastewater treatment. *Saudi Journal of Biological Sciences, 19* (3), 257−275. Available from https://doi.org/10.1016/j.sjbs.2012.04.005.

Åkerström, A. M., Mortensen, L. M., Rusten, B., & Gislerød, H. R. (2014). Biomass production and nutrient removal by Chlorella sp. as affected by sludge liquor concentration. *Journal of Environmental Management, 144*, 118−124. Available from https://doi.org/10.1016/j.jenvman.2014.05.015.

Al Ketife, A. M. D., Al Momani, F., & Judd, S. (2020). A bioassimilation and bioaccumulation model for the removal of heavy metals from wastewater using algae: New strategy. *Process Safety and Environmental Protection, 144*, 52−64. Available from https://doi.org/10.1016/j.psep.2020.07.018.

Almomani, F. (2016). Field study comparing the effect of hydraulic mixing on septic tank performance and sludge accumulation. *Environmental Technology (United Kingdom), 37*(5), 521−534. Available from https://doi.org/10.1080/09593330.2015.1074623.

Almomani, F., Judd, S., Shurai, M., Bhosale, R., Kumar, A., & Khreisheh, M. (2017). Potential use of mixed indigenous microalgae for carbon dioxide bio-fixation and advanced wastewater treatment. In *Environmental division 2017—Core programming area at the 2017 AIChE spring meeting and 13th global congress on process safety* (pp. 58−64). AIChE.

Al-Rekabi, W. S., Qiang, H., & Qiang, W. W. (2007). Improvements in wastewater treatment technology. *Pakistan Journal of Nutrition, 6*(2), 104−110. Available from https://doi.org/10.3923/pjn.2007.104.110.

Arbib, Z., Ruiz, J., Álvarez-Díaz, P., Garrido-Pérez, C., & Perales, J. A. (2014). Capability of different microalgae species for phytoremediation processes: Wastewater tertiary treatment, CO_2 bio-fixation and low cost biofuels production. *Water Research, 49*, 465−474. Available from https://doi.org/10.1016/j.watres.2013.10.036.

Ayed, H. B. A.-B., et al. (2015). Effect of magnesium ion concentration in autotrophic cultures of *Chlorella vulgaris. Algal Research; A Journal of Science and its Applications, 9*, 291−296.

Ayre, J. M., Mickan, B. S., Jenkins, S. N., & Moheimani, N. R. (2021). Batch cultivation of microalgae in anaerobic digestate exhibits functional changes in bacterial communities impacting nitrogen removal and wastewater treatment. *Algal Research, 57*, 102338. Available from https://doi.org/10.1016/j.algal.2021.102338.

Azuma, T., & Hayashi, T. (2021). On-site chlorination responsible for effective disinfection of wastewater from hospital. *Science of The Total Environment, 776*, 145951. Available from https://doi.org/10.1016/j.scitotenv.2021.145951.

Baral, S. S., Singh, K., & Sharma, P. (2015). The potential of sustainable algal biofuel production using CO_2 from thermal power plant in India. *Renewable and Sustainable Energy Reviews, 49*, 1061−1074. Available from https://doi.org/10.1016/j.rser.2015.04.181.

Ben, W., Qiang, Z., Pan, X., & Chen, M. (2009). Removal of veterinary antibiotics from sequencing batch reactor (SBR) pretreated swine wastewater by Fenton's reagent. *Water Research, 43*(17), 4392−4402. Available from https://doi.org/10.1016/j.watres.2009.06.057.

Birungi, Z. S., & Chirwa, E. M. N. (2014). The kinetics of uptake and recovery of lanthanum using freshwater algae as biosorbents: Comparative analysis. *Bioresource Technology, 160*, 43−51.

Bosma, R., et al. (2003). Ultrasound, a new separation technique to harvest microalgae. *Journal of Applied Phycology, 15*(2), 143−153.

Brennan, L., & Owende, P. (2010). Biofuels from microalgae—A review of technologies for production, processing, and extractions of biofuels and co-products. *Renewable and Sustainable Energy Reviews, 14*(2), 557−577.

Brownell, P. F., & Nicholas, D. J. D. (1967). Some effects of sodium on nitrate assimilation and N_2 fixation in *Anabaena cylindrica. Plant Physiology, 42*(7), 915−921. Available from https://doi.org/10.1104/pp.42.7.915.

Chanab. (2017). *Multiscale simulation of mass transfer in bubbly flow.* Available from: http://chanab.com/MassTransfer/.

Chisti, Y. (2008). Biodiesel from microalgae beats bioethanol. *Trends in Biotechnology, 26*(3), 126−131. Available from https://doi.org/10.1016/j.tibtech.2007.12.002.

Concas, A., Pisu, M., & Cao, G. (2010). Novel simulation model of the solar collector of BIOCOIL photobioreactors for CO_2 sequestration with microalgae. *Chemical Engineering Journal, 157*(2−3), 297−303.

Corrêa, Fd. N., Luna, A. S., & da Costa, A. C. A. (2016). Kinetics and equilibrium of lanthanum biosorption by free and immobilized microalgal cells. *Adsorption Science & Technology, 35*(1−2), 137−152.

Correa, Fd. N., Luna, A. S., & da Costa, A. C. A. (2017). Kinetics and equilibrium of lanthanum biosorption by free and immobilized microalgal cells. *Adsorption Science & Technology, 35*(1−2), 137−152.

Danièle, P., V. Jérôme, & Corinne, R. (2016). Biotechnologies for metal extraction: Assessment of microalgae for rare earths recycling and environmental remediation. In *Proceedings of the 2nd world congress on new technologies (NewTech'16).* Budapest, Hungary World Congress on New Technologies Conference.

Davis, R. E., Fishman, D. B., Frank, E. D., Johnson, M. C., Jones, S. B., Kinchin, C. M., ... Wigmosta, M. S. (2014). Integrated evaluation of cost, emissions, and resource potential for algal biofuels at the national scale. *Environmental Science and Technology, 48*(10), 6035−6042. Available from https://doi.org/10.1021/es4055719.

Dirbaz, M., & Roosta, A. (2018). Adsorption, kinetic and thermodynamic studies for the biosorption of cadmium onto microalgae Parachlorella sp. *Journal of Environmental Chemical Engineering, 6*(2), 2302−2309.

Gao, F., et al. (2016). Removal of nutrients, organic matter, and metal from domestic secondary effluent through microalgae cultivation in a membrane photobioreactor. *Journal of Chemical Technology and Biotechnology.*

Gao, F., Peng, Y. Y., Li, C., Yang, G. J., Deng, Y. B., Xue, B., & Guo, Y. M. (2018). Simultaneous nutrient removal and biomass/lipid production by Chlorella sp. in seafood processing wastewater. *Science of the Total Environment, 640−641*, 943−953. Available from https://doi.org/10.1016/j.scitotenv.2018.05.380.

Garai, M., & Yavuz, C. T. (2019). Radioactive strontium removal from seawater by a MOF via two-step ion exchange. *Chem, 5*(4), 750−752. Available from https://doi.org/10.1016/j.chempr.2019.03.020.

Giovannoni, S., et al. (1990). Tangential flow filtration and preliminary phylogenetic analysis of marine picoplankton. *Applied and Environmental Microbiology, 56*(8), 2572.

Grima, E. (1996). A study on simultaneous photolimitation and photoinhibition in dense microalgal cultures taking into account incident and averaged irradiances. *Journal of Biotechnology,* 59−69. Available from https://doi.org/10.1016/0168-1656(95)00144-1.

Gross, M., Henry, W., Michael, C., & Wen, Z. (2013). Development of a rotating algal biofilm growth system for attached microalgae growth with in situ biomass harvest. *Bioresource Technology, 150,* 195−201. Available from https://doi.org/10.1016/j.biortech.2013.10.016.

Guida, S., Conzelmann, L., Remy, C., Vale, P., Jefferson, B., & Soares, A. (2021). Resilience and life cycle assessment of ion exchange process for ammonium removal from municipal wastewater. *Science of the Total Environment, 783,* 146834. Available from https://doi.org/10.1016/j.scitotenv.2021.146834.

Herzog, H. (1999). *An introduction to CO_2 separation and capture technologies.* Energy Laboratory Working Paper.

Hilares, R. T., et al. (2021). Acid precipitation followed by microalgae (*Chlorella vulgaris*) cultivation as a new approach for poultry slaughterhouse wastewater treatment. *Bioresource Technology,* 125284.

Huang, Y., & Feng, X. (2019). Polymer-enhanced ultrafiltration: Fundamentals, applications and recent developments. *Journal of Membrane Science, 586,* 53−83. Available from https://doi.org/10.1016/j.memsci.2019.05.037.

Jiang, J., & Savage, P. E. (2019). Using solvents to reduce the metal content in crude bio-oil from hydrothermal liquefaction of microalgae. *Industrial and Engineering Chemistry Research, 58*(50), 22488−22496. Available from https://doi.org/10.1021/acs.iecr.9b03497.

Kucuker, M., Nadal, J.-B., & Kuchta, K. (2016). *Comparison between batch and continuous reactor systems for biosorption of neodymium (ND) using microalgae* (Vol. 6).

Kwak, H. W., et al. (2015). Preparation of bead-type biosorbent from water-soluble Spirulina platensis extracts for chromium (VI) removal. *Algal Research, 7,* 92−99.

Leng, L., Li, J., Wen, Z., & Zhou, W. (2018). Use of microalgae to recycle nutrients in aqueous phase derived from hydrothermal liquefaction process. *Bioresource Technology, 256,* 529−542. Available from https://doi.org/10.1016/j.biortech.2018.01.121.

Lin-Lan, Z., Jing-Han, W., & Hong-Ying, H. (2018). Differences between attached and suspended microalgal cells in ssPBR from the perspective of physiological properties. *Journal of Photochemistry and Photobiology B: Biology, 181,* 164−169. Available from https://doi.org/10.1016/j.jphotobiol.2018.03.014.

Luo, Y., Le-Clech, P., & Henderson, R. K. (2020). Characterisation of microalgae-based monocultures and mixed cultures for biomass production and wastewater treatment. *Algal Research, 49,* 101963.

Markou, G., et al. (2015a). Biosorption of Cu^{2+} and Ni^{2+} by Arthrospira platensis with different biochemical compositions. *Chemical Engineering Journal, 259,* 806−813.

Markou, G., et al. (2015b). Exploration of using stripped ammonia and ash from poultry litter for the cultivation of the cyanobacterium *Arthrospira platensis* and the green microalga *Chlorella vulgaris. Bioresource Technology, 196,* 459−468.

Maslova, M., Ivanenko, V., Yanicheva, N., & Gerasimova, L. (2020). The effect of heavy metal ions hydration on their sorption by a mesoporous titanium phosphate ion-exchanger. *Journal of Water Process Engineering, 35.* Available from https://doi.org/10.1016/j.jwpe.2020.101233.

Mehta, S. K., & Gaur, J. P. (2005). Use of algae for removing heavy metal ions from wastewater: Progress and prospects. *Critical Reviews in Biotechnology, 25*(3), 113−152. Available from https://doi.org/10.1080/07388550500248571.

Miron, A. S. (2000). *Bubble-column and airlift photobioreactors for algal culture.*

Mohamad, S., Fares, A., Judd, S., Bhosale, R., Kumar, A., Gosh, U., & Khreisheh, M. (2017). Advanced wastewater treatment using microalgae: Effect of temperature on removal of nutrients and organic carbon. *IOP Conference Series: Earth and Environmental Science, 67*(1), 012032. Available from https://doi.org/10.1088/1755-1315/67/1/012032.

Molina Grima, E., Acién Fernández, F. G., García Camacho, F., Camacho Rubio, F., & Chisti, Y. (2000). Scale-up of tubular photobioreactors. *Journal of Applied Phycology, 12*(3−5), 355−368.

Moreno-Garrido, I. (2008). Microalgae immobilization: Current techniques and uses. *Bioresource Technology, 99*(10), 3949−3964. Available from https://doi.org/10.1016/j.biortech.2007.05.040.

Östman, M. (2018). *Antimicrobials in sewage treatment plants: Occurrence, fate and resistance.* Umeå universitet.

Patel, A. K., Joun, J., & Sim, S. J. (2020). A sustainable mixotrophic microalgae cultivation from dairy wastes for carbon credit, bioremediation and lucrative biofuels. *Bioresource Technology, 313,* 123681.

Pittman, J. K., Dean, A. P., & Osundeko, O. (2011). The potential of sustainable algal biofuel production using wastewater resources. *Bioresource Technology, 102*(1), 17−25. Available from https://doi.org/10.1016/j.biortech.2010.06.035.

Poulsen, B. R., & Iversen, J. J. L. (1998). Characterization of gas transfer and mixing in a bubble column equipped with a rubber membrane diffuser. *Biotechnology and Bioengineering, 58*(6), 633−641, https://doi.org/10.1002/(SICI)1097-0290(19980620)58:6 < 633::AID-BIT9 > 3.0.CO;2-J.

Pugliesi, D. (2009). *Agitated vessel.* [cited 2017].

Rodrigues, L. H. R., et al. (2011). Algal density assessed by spectrophotometry: A calibration curve for the unicellular algae *Pseudokirchneriella subcapitata. Journal of Environmental Chemistry and Ecotoxicology.*

Ruiz-Martinez, A., et al. (2014). Mixed microalgae culture for ammonium removal in the absence of phosphorus: Effect of phosphorus supplementation and process modeling. *Process Biochemistry, 49*(12), 2249−2257.

Sayadi, M. H., Rashki, O., & Shahri, E. (2019). Application of modified *Spirulina platensis* and *Chlorella vulgaris* powder on the adsorption of heavy metals from aqueous solutions. *Journal of Environmental Chemical Engineering, 7*(3), 103169.

Schenk, P. M., et al. (2008). Second generation biofuels: High-efficiency microalgae for biodiesel production. *Bioenergy Research, 1*(1), 20−43.

Seader, J. D., Henley, E. J., & Roper, D. K. (1998). *Separation process principles* (Vol. 25).

Shao, Y., Liu, G., Wang, Y., Zhang, Y., Wang, H., Qi, L., . . . Zhang, J. (2020). Sludge characteristics, system performance and microbial kinetics of ultra-short-SRT activated sludge processes. *Environment International, 143*, 105973. Available from https://doi.org/10.1016/j.envint.2020.105973.

Siegel, B. Z., & Siegel, S. M. (1973). The chemical composition of algal cell walls. *Critical Reviews in Microbiology, 3*(1), 1−26. Available from https://doi.org/10.3109/10408417309108743.

Sivakumar, G., Xu, J., Thompson, R. W., Yang, Y., Randol-Smith, P., & Weathers, P. J. (2012). Integrated green algal technology for bioremediation and biofuel. *Bioresource Technology, 107*, 1−9. Available from https://doi.org/10.1016/j.biortech.2011.12.091.

Solmaz, A., & Işık, M. (2020). Optimization of membrane photobioreactor; the effect of hydraulic retention time on biomass production and nutrient removal by mixed microalgae culture. *Biomass and Bioenergy, 142*, 105809.

Stewart, C., & Hessami, M. A. (2005). A study of methods of carbon dioxide capture and sequestration—The sustainability of a photosynthetic bioreactor approach. *Energy Conversion and Management, 46*(3), 403−420. Available from https://doi.org/10.1016/j.enconman.2004.03.009.

Su, Y., & Jacobsen, C. (2021). Treatment of clean in place (CIP) wastewater using microalgae: Nutrient upcycling and value-added byproducts production. *Science of the Total Environment, 785*, 147337. Available from https://doi.org/10.1016/j.scitotenv.2021.147337.

Sudhakar, K., Suresh, S., & Premalatha, M. (2011). An overview of CO_2 mitigation using algae cultivation technology. *International Journal of Chemical Research, 3*(3).

Takahashi, H., Kopriva, S., Giordano, M., Saito, K., & Hell, R. (2011). Sulfur assimilation in photosynthetic organisms: Molecular functions and regulations of transporters and assimilatory enzymes. *Annual Review of Plant Biology, 62*, 157−184. Available from https://doi.org/10.1146/annurev-arplant-042110-103921.

Villeneuve, W., Perreault, V., Chevallier, P., Mikhaylin, S., & Bazinet, L. (2019). Use of cation-coated filtration membranes for demineralization by electrodialysis. *Separation and Purification Technology, 218*, 70−80. Available from https://doi.org/10.1016/j.seppur.2019.02.032.

Wang, R., Li, F., Ruan, W., Tai, Y., Cai, H., & Yang, Y. (2020). Removal and degradation pathway analysis of 17β-estradiol from raw domestic wastewater using immobilised functional microalgae under repeated loading. *Biochemical Engineering Journal, 161*, 107700. Available from https://doi.org/10.1016/j.bej.2020.107700.

Wang, Z., Tian, S., Niu, J., Kong, W., Lin, J., Hao, X., & Guan, G. (2020). An electrochemically switched ion exchange process with self-electrical-energy recuperation for desalination. *Separation and Purification Technology, 239*, 116521. Available from https://doi.org/10.1016/j.seppur.2020.116521.

Xie, Y., et al. (2014). Kinetic simulating of Cr(VI) removal by the waste *Chlorella vulgaris* biomass. *Journal of the Taiwan Institute of Chemical Engineers, 45*(4), 1773−1782.

Xu, X. Q., Wang, J. H., Zhang, T. Y., Dao, G. H., Wu, G. X., & Hu, H. Y. (2017). Attached microalgae cultivation and nutrients removal in a novel capillary-driven photo-biofilm reactor. *Algal Research, 27*, 198−205. Available from https://doi.org/10.1016/j.algal.2017.08.028.

Yang, J., Xu, M., Zhang, X., Hu, Q., Sommerfeld, M., & Chen, Y. (2011). Life-cycle analysis on biodiesel production from microalgae: Water footprint and nutrients balance. *Bioresource Technology, 102*(1), 159−165. Available from https://doi.org/10.1016/j.biortech.2010.07.017.

Yang, J. S., Cao, J., Xing, G. L., & Yuan, H. L. (2015). Lipid production combined with biosorption and bioaccumulation of cadmium, copper, manganese and zinc by oleaginous microalgae *Chlorella minutissima* UTEX2341. *Bioresource Technology, 175*, 537−544. Available from https://doi.org/10.1016/j.biortech.2014.10.124.

Yu, H., Kim, J., & Lee, C. (2019). Potential of mixed-culture microalgae enriched from aerobic and anaerobic sludges for nutrient removal and biomass production from anaerobic effluents. *Bioresource Technology, 280*, 325−336.

Zhang, J., Sun, B., & Guan, X. (2013). Oxidative removal of bisphenol A by permanganate: Kinetics, pathways and influences of co-existing chemicals. *Separation and Purification Technology, 107*, 48−53. Available from https://doi.org/10.1016/j.seppur.2013.01.023.

Zhuang, L. L., Azimi, Y., Yu, D., Wang, W. L., Wu, Y. H., Dao, G. H., & Hu, H. Y. (2016). Enhanced attached growth of microalgae Scenedesmus. LX1 through ambient bacterial pre-coating of cotton fiber carriers. *Bioresource Technology, 218*, 643−649. Available from https://doi.org/10.1016/j.biortech.2016.07.013.

Zhuang, L. L., Li, M., & Hao Ngo, H. (2020). Non-suspended microalgae cultivation for wastewater refinery and biomass production. *Bioresource Technology, 308*. Available from https://doi.org/10.1016/j.biortech.2020.123320.

Zhuang, L.-L., et al. (2018). Effects of nitrogen and phosphorus concentrations on the growth of microalgae Scenedesmus. LX1 in suspended-solid phase photobioreactors (ssPBR). *Biomass and Bioenergy, 109*, 47−53.

Zkeri, E., et al. (2021). Comparing the use of a two-stage MBBR system with a methanogenic MBBR coupled with a microalgae reactor for medium-strength dairy wastewater treatment. *Bioresource Technology, 323*, 124629.

Znad, H. (2012). CO_2 biomitigation and biofuel production using microalgae: Photobioreactors developments and future directions. *Advances in Chemical Engineering*.

Znad, H., et al. (2018). Bioremediation and nutrient removal from wastewater by *Chlorella vulgaris*. *Ecological Engineering, 110*, 1−7.

Chapter 7

Application of microalgae in wastewater treatment: simultaneous nutrient removal and carbon dioxide bio-fixation for biofuel feedstock production

Fares Almomani[1], Abdullah Omar[1] and Ahmed M.D. Al ketife[1,2]

[1]Department of Chemical Engineering, Qatar University, Doha, Qatar, [2]Faculty of Engineering, University of Thi-Qar, Nasiriyah, Iraq

Background

Microalgae are one of the most important photosynthetic unicellular organisms that can survive individually or in chains or clusters (Richmond, 2007). The use of algae by humans dated back to 2000 years when the Chinese used it as a source of food (Spolaore et al., 2006). Recently, 35,000 algal species have been identified and described out of 200,000–800,000 species that exist on our plant (Richmond, 2007). Microalgae produce half of the atmospheric oxygen via photosynthesis activities of microalgae, consuming vast amounts of greenhouse gas specifically carbon dioxide, and contributing to the treatment of pollutants from wastewater and the production biomass (Spolaore et al., 2006).

Various microalgae strains have been widely investigated toward removing nutrients from wastewater, carbon dioxide biofixation, and coupling biomass production in the literature as shown in Fig. 7.1. Algae have a fast-growing trend approximately $1.5\ \mathrm{d}^{-1}$ and high biomass accumulation of $2\ \mathrm{g\ L}^{-1}$. Therefore their usage in wastewater treatment can

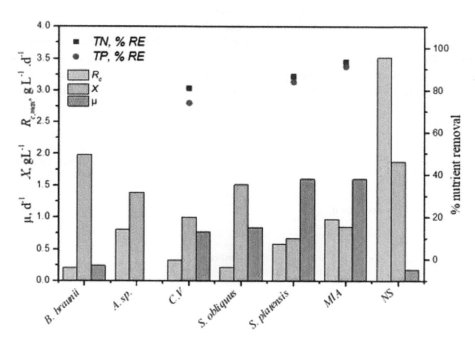

FIGURE 7.1 Different algal strains' potential ability to treat waste gas and water, along with the production of algal biomass. *B. braunii*: *Botryococcus braunii*, *A*. sp.: *Anabaena* sp., *C.V*: *Chlorella vulgaris*, *S. obliquus*: *Scenedesmus obliquus*; *MIA*: mixed indigenous microalgae, *NS*: *Nannochloris*. sp., *TN*: total nitrogen; *TP*: total phosphorus; *RE*: removal efficiency; *X*: the accumulated biomass; *RC*: carbon dioxide fixation rate.

Petroleum Industry Wastewater. DOI: https://doi.org/10.1016/B978-0-323-85884-7.00012-6

result in more than 90% of nitrogen and more than 85% phosphorus removal (Alketife, Judd, & Znad, 2017). In addition to this, algae have a significant carbon fixation rate of 3.5 g L^{-1} d^{-1} (Basu et al., 2014; Honda et al., 2012; Lam & Lee, 2013; López et al., 2009; Pires et al., 2014; Ruangsomboon, 2012; Sydney et al., 2010; Tang et al., 2011; Znad et al., 2018; Al Ketife, Judd, & Znad, 2016; Alketife, Judd, & Znad, 2017; Gonçalves, Simões, & Pires, 2014; Ho, Chen, & Chang, 2010; Ruiz-Marin, Mendoza-Espinosa, & Stephenson, 2010; Znad, Al Ketife, & Judd, 2019), compared with common expensive and low-efficiency industrial methods (Bhatnagar et al., 2011; Freed, 2007; Gong & Jiang, 2011; Rogers et al., 2014).

Green microalgae are photoautotrophic eukaryotic that have chlorophyll and other pigments in their cells. These pigments capture sunlight energy and use it in a plant-like oxygenic photosynthesis process Fig. 7.2. The mechanism involved the use of light energy (photons) by arrays of so-called light-harvesting or antenna pigments. The captured photons are then transferred to what is so-called reaction chlorophyll centers where the captured photon energy is converted, in a two-photosystem process, to chemical energy. The pigments in PSII (P680) can absorb light with a wavelength less 680 nm, producing a strong oxidant that can split water into protons (H^+), electrons (e^-), and O_2. The electrons are moved by different electron carriers and cytochrome complexes to reach the PSI. The pigments in PSI can absorb solar irradiation or light photons with a wavelength < 700 nm. The absorbed photons are stored in the cell raising the electron's energy level. Electrons with high energy react with ferredoxin (Fd) and/or nicotinamide adenine dinucleotide phosphate ($NADP^+$) reducing their oxidation level and putting them at a stable state. On other hand, the produced proton contributes to the production of adenosine triphosphate (ATP) that supports the reduction and uptake of CO_2. The ATP and NADPH reduce CO_2 via a reductive pentose phosphate pathway or Calvin cycle for cell growth. The reduced form of carbon can be uptake and stored in the cells as carbohydrates (CH_2O) and/or lipids (Amos, 2004).

Nitrogen is the second most important nutrient for algae growth following carbon (Becker, 1994). Nitrogen is associated with the primary metabolism of microalgae due to its role in building proteins and nucleic acid (Green & Durnford, 1996). Algae culture tends to have lower growth rates and productivity under low nitrogen content (Znad et al., 2012). N-starvation can lead to more lipid production at the expense of other components such as proteins (Wong, 2012). The third most important nutrient for algae growth is phosphorus. Phosphorus should be added in excess because not all phosphorus is bioavailable (Znad et al., 2012). Other metal traces and vitamins are also required for effective cultivation, such as (Cu, Mg, Zn, and B12 vitamin) (Becker, 1994; UTEX. Spir Solution 2 RecipeSpir Solution 2 Recipe, 2014).

Regarding algal growth measurement, the "amount" of microalgae can be estimated in either a concentration unit, colonies number, or inferred through relation between the water's absorbance and the microalgae concentration (water's color density). The most common methods of algal mass estimation are weight difference, square counting method, and optical density measurement. The premise of the weight difference method is simple, but some issues distinguish microorganisms from other nonliving total suspended solids. A certain volume of a wastewater sample is collected after making sure the distribution of microalgae in the source is maintained to be uniform (stirring), and the sample is dried before weighing the difference in mass of the vessel with and without the algae. This method could either give the dry or wet weight of microalgae per litter of the sample. It should be noted that because of the hygroscopic property of dried cells in many microorganisms ("Weighing Technique for Determining Bacterial Dry Mass Based on Rate of Moisture Uptake," 1983), direct measurement of cell mass in the open environment will produce measurement errors. When errors are neither tolerated nor accounted for in the calculation of mass, this can be remedied by the use of

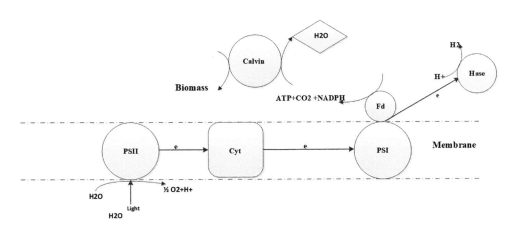

FIGURE 7.2 Schematic mechanisms of photosynthesis and biophotolysis of the photoautotrophic microbes (Richmond, 2007).

a microfiber membrane filter (Lu et al., 2015). Recently, centrifugation is commonly used to completely separate cells from the water (Amit, Nayak, & Ghosh, 2020; Santana et al., 2017). The counting method is less expensive than the former, but it takes more time as the microalgae are injected onto the surface of an agar plate (or pipetted into a sterile plate before adding the medium in the case of the pour plate method) and allowed to grow to become colonies that can be observed via a microscope. The number of colonies is counted, and the measurement unit is expected to be the number of colony-forming cells per millimeter of the original sample. Though in the case of high cell concentration, the original sample is usually diluted and that is accounted for in the calculation. A Hemocytometer is an accurate digital device that is used for these exact measurements. In this device, the plate is readily lined and split into many squares for easier measurement, where the number of cells squares is measured and the number is average for all the squares. Optical density measurements are considered a reliable, simple, and easy method of indirectly measuring algal cell concentration using a spectrophotometer. In general, the measured absorbance is plotted against time to show the algal growth pattern. The exact relation between absorbance and cell concentration differs depending on the type of algae as well as other growth conditions For example, Rodrigues et al. (2011) develop a calibration curve that can be used to calculate the algal cell density for *Pseudokirchneriella subcapitata* at a wavelength of 684.

Algae and wastewater treatment

Wastewater origin

Table 7.1 presents the characteristics of different types of wastewater (Ww). Diary wastewater (DWw) has the highest nutrient contents, including nitrogen (N), phosphorus (P), and high organic content of 383, 396.87, and 17806.36 mg L^{-1}, respectively. The pH of this Ww ranges from 6 to 8, indicating an excellent medium for microalgae cultivation. Whereas, textile wastewater (TWw) shows less affinity to support a high algal growth rate due to the low concentrations of N and P and high value of pH. It should be noted that a high concentration of dissolved organic matter (OM) does not necessarily support the growth of all kinds of algal strains. High nutrient content in petrochemical wastewater (PhWw) can successfully promote high algal cell growth, although pathogenic content in this Ww is still a concern. Other types of Ww have a relatively moderate level of nutrients (mainly; N, P, and COD) that can support a low level of algal growth, but the benefits of nutrient removal by algal cells still promising.

Industrial wastewater treatment

Wastewater from food Industry

Algal technology showed an excellent potential of treating wastewater generated from the food industry. Patel et al. (2020) investigated the treatment and biofuel production from dairy waste using mixotrophic cultivation of *Auxenochlorella protothecoides*. It was reported that 40% pretreated whey medium gave the highest biomass (X_{max}) and lipid productivity (LP) of 500 and 200 mg L^{-1} d^{-1}, respectively. Excellent treatment efficiency of 99.7%, 91%, and 100% of organic (COD) and inorganic (TN and TP) was achieved after 9 days of cultivation of 25% of the pretreated whey. The sole use of biotreatment (algae) was 92.6% and 48.5%−98.4%, respectively. Additionally, Zkeri et al. (2021) applied treated DWw by two setups: methanogenic MBBR followed by an aerobic MBBR (AnMBBR + AeMBBR) and in the second setup, the AeMBBR was replaced by a sequencing batch reactor (SBR) utilizing *Chlorella sorokiniana*. Results showed that after 7 days of microalgae cultivation in the An MBBR, the biomass concentration reached approximately 320 mg L^{-1} with 38%−44% proteins, 27.9% lipids, and 3.6% starch). The removal of COD by the microalgae exceeds 98%, while the removal efficiencies of NH$_4$-N and PO$_4$-P were 74% and 58.6%, respectively. Lu et al. (2015) compared indoor bench-scale and outdoor pilot-scale treatment of raw DWw using *Chlorella* sp. for nutrients removal and biofuel production (B$_Y$P). the maximum biomass productivity (BP) of 260 mg L^{-1} d^{-1} was achieved at indoor conditions, as opposed to outdoor productivity of 110 mg L^{-1} d^{-1}. The maximum removal rate of COD, TN, and TP are achieved at indoor conditions, and their values are found to be 88.3, 38.34, and 2.03 mg L^{-1} d^{-1}, as opposed to 41.31, 6.58 and 2.74 mg L^{-1} d^{-1} for the outdoor reactor. Utilizing the two-stage optimization method in the cultivation of *Tetraselmis* sp. in simulated DWw is studied by Swain, Tiwari, and Pandey (2020), who achieved a maximum optimized BP of 300.4 mg L^{-1} d^{-1} (51.65 DCW%), which was about two folds higher than the growth at control conditions. The reported COD removal was 95%. Gramegna et al. (2020) recycle DWw through nutrient removal using *Auxenochlorellla protothecoides* and *Chlamydomonas reinhardtii* along with other strains, namely Chlorella. The microalgae strains, *A. protothecoides* had the highest X of 3.3 g L^{-1} (18.5% DW lipids) after 10 days. *C. reinhardtii* made possible removals of COD, phosphorus, total nitrogen, and protein nitrogen of 76%,

TABLE 7.1 Typical characterization of different types of wastewater.

Ww resource	pH	TOC	BOD	COD	N	P	S	References
	(–)	mg L^{-1}	mg L^{-1}	mg L^{-1}	mg L^{-1}	mg L^{-1}	mg L^{-1}	
DWw	5–6	nr	nr	17806.36	383	396.87	nr	Patel, Joun, and Sim (2020)
	7.75 ± 0.6	nr	nr	2499 ± 812	89.1 ± 34.8	20.7 ± 9.6	nr	Zkeri et al. (2021)
	8.18 ± 0.03	nr	nr	2593 ± 15.56	283 ± 12.73	115.9 ± 7.5	nr	Lu et al. (2015)
	7.31	nr	nr	1260	30	50	nr	Makut, Das, and Goswami (2019)
	6.3 ± 0.1	160–190	nr	nr	84–89	9–10	12–91	Daneshvar et al. (2019)
	7	nr	nr	721	91	24	-	Hülsen et al. (2018)
	6.45–6.72	160–190	nr	622–722	84–89	8.6–9.6	12–89	Daneshvar et al. (2018)
TWw	12.1 ± 0.1	nr	136 ± 18	387 ± 1	nr	nr	nr	Dhaouefi et al. (2018)
	8.7 ± 0.1	nr	nr	2200 ± 150	380.5 ± 12	94 ± 3	nr	Kumar et al. (2018)
	12–14	nr	8500–9500	42000–48000	360–470	75–150	nr	Oyebamiji et al. (2019)
	10.5	nr	nr	4458	28.35	22.3	nr	Wu et al. (2020)
	11.8	nr	nr	2330	27.84	23.2	nr	Wu et al. (2017)
	9.16 ± 0.11	nr	57.4 ± 1.7	336 ± 2.08	nr	247.16 ± 2.12	462 ± 2.64	Sinha et al. (2016)
	8.45 ± 0.2	nr	nr	39,800 ± 511	nr	nr	nr	Lin, Nguyen, and Lay (2017)
	9 ± 0.35[a]	72 ± 3.61	460 ± 19.2	850 ± 27.1	205.5	11 ± 1.62	nr	Yadav, Dash, and Sen (2019)
PhWw	1.46	nr	15,160	42,112	268	361	nr	Hemalatha and Venkata (2016)
	6.8 ± 0.5	380 ± 5	nr	5750 ± 100	525 ± 10	342 ± 10	nr	Nayak and Ghosh (2019)
	10.5[b]	nr	41.4	438	nr	nr	nr	Singh, Ummalyma, and Sahoo (2020)
	nr	nr	nr	199–243[c]	79–86	3.4–4.1	39–45	García-Galán et al. (2020)
PEWw	8.1	nr	321	610	nr	1.7	15	Ashwaniy, Perumalsamy, and Pandian (2020)
	8.3 ± 0.24	nr	14.0 ± 1.36	nr	4.5	0.16	nr	Lacerda et al. (2011)
PET	8.3 ± 0.2	123 ± 8.4	630–950	285 ± 16.1	63.5 ± 2.0	17 ± 3.2	nr	Abid, Saidane, and Hamdi (2017)
POWw	8.42	nr	nr	nr	9.68	nr	1746.15	Mohammadi et al. (2018)
BGWw	7.09 ± 0.24	nr	nr	2875.26 ± 6.81	165.16 ± 4.04	60.12 ± 1.72	nr	Xie et al. (2018)
AQWw	8.3	nr	nr	160	20	11	nr	Andreotti et al. (2020)
MCWw	7.3 ± 0.4	nr	nr	714 ± 35.7	47.6 ± 2.4	3.7 ± 0.2	nr	You et al. (2021)
SPWw	7–7.5	194–209	nr	nr	290–314	33–42	nr	Chu et al. (2015)

MPWw	6.8	nr	nr	1868	nr	154.6	126.9	nr	Hu, Meneses, and Aly (2020)
PPWw	–	nr	nr	3000	nr	19	14	nr	Rasouli et al. (2018)
PoWw	7.6	nr	nr	3534	nr	405	15	nr	Hülsen et al. (2018)
PouWw	6.6	nr	nr	981	nr	26	24	nr	
RMWw	6.2	nr	nr	1777	nr	61	19.2	nr	
SUVw	6.4	nr	nr	3527	nr	121	51	nr	

Ww: wastewater; *TOC*: Total organic carbon; *BOD*: Biological oxygen demand; *COD*: Chemical oxygen demand; *N*: Nitrogen; *P*: Phosphorus; *S*: Sulfur; *DWw*: Diary Ww; *TWw*: textile Ww; *PEWw*: Petroleum Refinery Efiluent Ww; *PET*: Petrochemical Ww; *POWw*: Power plant Ww; *BGWw*: Biogas Ww; *AQWw*: Aquaculture; *MCWw*: Mariculture SPWw Starch Processing; *MPWw*: Meat Processing; *PPWw*: Meat Processing; *PPWw*: Poark Ww; *PouWw*: Poultry Ww; *RMWw*: Red meat Ww; *SUWw*: Suger Mill Ww.

[a] *Textile-based + food processing industries; PhWw, pharmaceutical Ww.*

[b] *River WW contaminated with pharmaceutical effluent.*

[c] *Domestic WW fed with pharmaceutical components.*

87%, 65%, and 65%, respectively. *A. protothecoides* also gave promising removal efficiencies of 65%, 77%, 43%, and 43%, respectively. Cultivating mixed microalgae culture (MMC) is demonstrated to be viable in the integration of DWw and bioethanol production by Hemalatha et al. (2019), with the mixed culture X reached 1.4 g L^{-1} at the sixth day of cultivation, and the composition of the biomass was 38% carbohydrates, 15% proteins, and 22% lipids. The removal efficiencies of COD, nitrates, phosphates, and sulfates were 90%, 65.5%, 73%, and 69.2%, respectively. Algal-bacterial consortium containing the two microalgae *Chlorella sorokiniana* and Chlorella sp. was utilized by Makut et al. (2019) the biomass production (X) for artificial DWw and real DWw were 284 and 287 mg L^{-1} d^{-1}, respectively. The percentage removal of nitrate and COD for the ADWW were 93.59% and 82.27%, while 84.69% and 90.49% were reported for RDWW, respectively. Freshwater microalgae *Scenedesmus quadricauda* and marine microalgae *Tetraselmis suecica* can efficiently remove nutrients from both raw and recycled DWw in 2 cycles according to Daneshvar et al. (2019). The X concentration for *S. quadricauda* and *T. suecica* were 0.43 and 0.58 g L^{-1} after 12 days in raw wastewater, and 0.36 and 0.65 g L^{-1} in recycled DWw. After two cycles of mixotrophic cultivation of microalgae, the percentage removal of TN, PO$_4$ $^{3-}$, SO$_4$ $^{2-}$ and TOC were 92.15%, 100%, 100%, and 76.77% by *S. quadricauda*, and 83.17%, 100%, 14.44%, and 28.72% by *T. suecica*. Daneshvar et al. (2018) cultivate *S. quadricauda* and *T. suecica* in DWw for nutrient removal, biomass production (X) production and tetracycline biosorption. The X $_{max}$ productivity for *S. quadricauda* and *T. suecica* were 58.75 and 50.83 mg L^{-1} d^{-1}, respectively. *S. quadricauda* removed up to 86.21%, 89.83%, and 64.47% of TN, PO$_4$ $^{3-}$ and TOC, respectively, while for *T. suecica* removed 44.92%, 42.18%, and 40.16%, respectively. Following lipid extraction, *S. quadricauda* and *T. suecica* biomasses showed adsorption capacities of 295.34 and 56.25 mg g^{-1} for tetracycline. Finally, Hülsen et al. (2018) utilize a microalgae culture containing *Chlorella vulgaris* and a Scenedesmus species in the production of single-cell proteins and treatment of wastewater from five different sources, including DWw. The produced microalgae biomass contained crude protein of 370 mg CP g^{-1} VSS and the reported percentage removal of soluble COD, NH$_4$-N and PO^4-P were up 63%, 100%, and 80%, respectively, which was significantly higher than those obtained by phototrophic purple bacteria.

The food industry also includes food processing, and on that subject, Hu et al. (2020) employ immobilized bacteria/microalgae for the treatment of meat processing Ww pretreated with sodium hypochlorite. Microalgae beads used included Scenedesmus obliquus, *Chlorella vulgaris* and *Chlorella sorokiniana*. The X_{max} production occurred after the seventh day with 0.2 mg L^{-1} NaClO conc., yielding about 2000 mg L^{-1}, while the value was only 1700 for untreated Ww. NaClO pretreatment generally had a negative effect on nutrient removal by bacteria/microalgae, with the increase in concentration even further decreasing removal efficiency. The percentage removal of COD, TN, and PO$_4$ $^{3-}$ without pretreatment were 82%, 75%, and 92%, respectively. Moreover, single-cell proteins are produced by Hülsen et al. (2018) using *Chlorella vulgaris* and a Scenedesmus species culture cultivated in pork and red meat wastewater, in which the crude protein production was 900 and 650 mg CP g^{-1} VSS. Meanwhile, the percentage removal of soluble COD, NH$_4$-N and PO$_4$-P by microalgae were 80%, 68%, and 50% from pork wastewater−grown microalgae, and 85% and 78% of soluble COD and PO$_4$-P, respectively.

Wastewater from agricultural activities

Agriculture wastewater (AWw) is a feasible wastewater medium source as is illustrated by many studies, including that of Chu et al. (2015), which summarized the outdoor treatment of starch processing-based anaerobically digested Ww by the cultivation of *Chlorella pyrenoidosa*. A relatively high hydraulic retention time (HRT) of 10 d during the summer provided the best treatment efficiency, and a biomass productivity (BP) of 176.9 ± 12.6 mg L^{-1} d^{-1}, while an HRT value of 4 days gave the highest biomass (X_{max}) and lipids productivities (LP) of 342.6 ± 12.8 and 43.37 ± 7.43 mg L^{-1} d^{-1}, respectively. The removal efficiencies (%RE) of TOC, TN and TP at an HRT of 10 days were 61.9% ± 7.6%, 78.7% ± 5.7%, and 97.2% ± 0.6%, respectively. Additionally, Rasouli et al. (2018) achieved nutrient recovery as a single cell protein fed by algal-bacterial consortium biomass cultivated in potato processing wastewater. The growth rate (μ) for the microalgae *Chlorella sorokiniana* reached up to 3.67 d^{-1} and its concentration at the end of the cycle is expected to be close to 683.25 ± 0.17 mg L^{-1} achieved by other studies. Between the microalgae and the algal−bacterial consortium, the sole use of microalgae achieved the highest %RE of COD and TP of 96% and 62%, respectively. The consortium's attained a % RE of 91% and 43%, respectively. However, it was observed that the consortium had a higher %RE of TN approximately 67% as opposed to the microalgae's 40%. The study by Hülsen et al. (2018), mentioned previously also includes poultry and sugar mill Ww. The proteins produced from the two sources were 650 and 140 mg CP g^{-1} VSS, respectively, and the %RE of soluble COD, NH$_4$-N, and PO$_4$-P were 91%, 91%, and 73% for poultry, and 90%, 50%, and 15% for sugar mill Ww, respectively.

Andreotti et al. (2020) demonstrate the feasibility of *Tetraselmis suecica* cultivation for the bioremediation of AWw, followed by simulation of the experimental setup to calculate results. The initial wastewater contains 20 and 10 mg L^{-1} of N and P. The BP at HRT of 10 days was found to be 66.55 ± 3.95 mg L^{-1} d^{-1} (9.13% carbohydrates, 20.01% lipids, and 37.27% proteins) with the maximum nitrogen and phosphorus %RE of $99.82\% \pm 0.03\%$ and $97.18\% \pm 0.01\%$, respectively. Furthermore, mariculture wastewater treatment by an integrated bacterial fermentation–algal cultivation growing *Chlorella vulgaris* is documented by You et al. (2021). At the acidogenic fermentation dilution rate of 10% and a pH of 8, A maximum biomass production (X_{max}) of 5.6 g L^{-1} was obtained after the sixth day. The %RE of NH_4^+–N, TP, COD, and VFAs were 54%, 78.6%, 82.8%, and 86%, respectively.

Wastewater from textile industry

Textile wastewater (TWw), which tends to contain environmentally harmful dyes and coloring agents among other pollutants both organic and inorganic can also be treated by algae. Oyebamiji et al. (2019) employed six different microalgae strains (*Micractinium* sp., *Chlorella sorokiniana* 1665, *Chlorella* sp. KU211a, *Chlorella* sp. KU211b, *Chlorella sorokiniana* 246, and *Chlorella* sp. CB4) for the biotreatment (BT) of TWw along with biomass generation. The biomass production (X) after two weeks at a textile wastewater concentration of 1% varied in the range 0.8–1.2 g L^{-1} depending on the strain of cultivation, and the %RE of Al^{2+} and Cu^{2+} were 44%–67%, while Pb^{2+} and Se^{2-} were removed completely. Decolorization efficiency (%D_ER) was 47.10%–70.03%, with a relatively higher concentration of TWw giving better values and C. sorokiniana 246 removing the most. Additionally, Wu et al. (2020) employ mixotrophic cultivation of immobilized *Chlorella vulgaris* and *Chlorella* sp. microalgae granules in the remediation of TWw at different concentrations. After 7 days of cultivation, the X in the immobilized granules for *C. vulgaris* and *Chlorella* sp. were 50 and 55 mg L^{-1}, respectively. Meanwhile, the NH_4^+-N, COD and color %RE were up to 90%, 75%, and 60%, respectively, depending on the dilution ratios. Sinha et al. (2016) aimed to additionally remove inorganic pollutants alongside the main targets in the treatment of textile wastewater from Direct Red-31 dye along with other pollutants using photobioreactor cultivating *Chlorella pyrenoidosa*. The %RE and algal biomass uptake were 96% and 30.53 mg g^{-1} were achieved with a dye dosage of 40 mg L^{-1} and a pH of 3. The removal efficiencies of TDS, total hardness, Chloride, Alkalinity, COD, BOD, Sulfate and phosphate were 84.18%, 22.85%, 96%, 82.5%, 82.73%, 56.44%, 54.54%, and 19.88%, respectively, when utilizing the continuous method as opposed to batch. The CO_2, a major greenhouse gas, can also be mitigated by basic algal WwT, and this is further illustrated in the study by Yadav et al. (2019) in which *Chlorella* sp. and *Chlorococcum* sp. are cultivated in Ww produced primarily from the textile industry—partly from the food industry—for nutrients removal and CO_2 sequestration. At 5% CO_2 concentration, the highest biomass productivity was 208.93 and 105.42 mg L^{-1} d^{-1} for *Chlorella* sp. and *Chlorococcum* sp., respectively. *Chlorella* sp. reduced TOC by $84.3\% \pm 3.23\%$, while *Chlorococcum* sp. reduced this by $72.9\% \pm 2.41\%$. *Chlorella* sp. removed up to 91.9%, 95.9%, 100%, and 98.8% of COD, PO_4, NH_4, and NO_3, respectively, while *Chlorococcum* sp. removed the nutrients in the range of 85.2%–100%. Meanwhile, the maximum CO_2 fixation rate (RC) for *Chlorella* sp. and *Chlorococcum* sp. were 187.65 and 94.68 mg L^{-1} d^{-1}, respectively, and the effective nutrient removal was more than 75% on the fifth day of cultivation. Lin et al. (2017) proposed a three-stage textile wastewater treatment process consisting of granular activated carbon adsorption, mesophilic anaerobic digestion, and microalgae cultivation of *Scenedesmus* sp. The peak specific growth rate (μ) of microalgae was 0.53 d^{-1} at 3.8 g COD L^{-1} digester effluent, while the %RE of color and COD from the microalgae cultivation was 90.8% and $71.1\% \pm 0.7\%$ respectively. Meanwhile, the %RE of color, COD, carbohydrates, and organic acids from the entire process were 92.4%, 89.5%, 97.4%, and 94.7%, respectively.

Utilizing consortiums combining multiple microalgal and/or algal-bacterial species is not uncommon. According to Dhaouefi et al. (2018), synthetic textile wastewater treatment (STWw) is feasible in an anoxic–aerobic photobioreactor (PBR) cultivating an algal–bacterial symbiosis and using treated water for irrigation, where the microalgae species include *Acutodesmus obliquus* (68.4%), *Scenedesmus quadricauda* (21.4%), and *Scenedesmus tenuispina* (10.2%). At a HRT of 10 days, the %RE of TOC, TN, and TP achieved are $48\% \pm 3\%$, $87\% \pm 11\%$, and $57\% \pm 5\%$, respectively. Decolorization of wastewater is achieved, and the removal rates of dispersing orange-3 and blue-1 dyes were approximately 80% and 75% respectively. Moreover, Ayed et al. (2020) investigated the possibility of treatment of textile effluent from COD and reactive diazo dye blue-40 using an algal-bacterial mixed culture consortium containing a *Chlorella* species in a batch reactor and optimizing the results using response surface methodology. The optimum results were a COD and color removal rates of 89% and 99%, respectively when the initial dye concentration is determined to be 1000 ppm, and these results were obtained at the optimum inoculum size (10%), temperature (35°C), residence time (6 days) and consortium composition (33% algae, 66% bacteria). One more example of consortium use is the cultivation of a mixed microalgae consortium dominated by *Chlorella* and *Scenedesmus* species in TWw inside a fed-batch reactor (FBR) operating in five different cycles published by Kumar et al. (2018). At the third cycle (20 days long), peak performance was witnessed, with a BP of $490 \, mg \, L^{-1} \, d^{-1}$ and a decolorization efficiency of 70.1%. Meanwhile, the %RE of COD, TN and TP were $70.7\% \pm 0.8\%$, $99.6\% \pm 0.3\%$ and 100%, respectively.

In addition of chemicals and biobased substances could demonstrably enhance the process of treatment. For example, Wu et al. (2017) focus on the optimization of environmental parameters (including the addition of dipotassium phosphate and urea) for the cultivation of *Chlorella* sp. in TWw for X production. When $4 \, mg \, L^{-1}$ of K_2HPO_4 is added, the maximum BP of $137 \, mg \, L^{-1} \, d^{-1}$ is achieved. The addition of Urea as a nitrogen source led to a higher color %RE of 50% for both with and without aeration. COD and NH_4^+-N %RE up to >75% were achieved after 7 days of cultivation. Furthermore, Lebron, Moreira, and Santos (2019) achieved high treatment efficiencies in their treatment of STWw containing the dyes rhodamine B (1), methylene blue (2) and eriochrome black T (3) through the cultivation of three microalgae species (*Fucus vesiculosus*, *Arthrospira maxima*, and *Chlorella pyrenoidosa*) with the addition of H_3PO_4 and $ZnCl_2$. The maximum dye biosorption was 1162.9 and $212.0 \, mg \, g^{-1}$ for *F. vesiculosus* (with H_3PO_4) and *C. pyrenoidosa* (with $ZnCl_2$), respectively. The respective removal efficiencies for dyes 1, 2, and 3 after the first cycle of biosorption in synthetic textile wastewater were approximate as follows: 94.47%, 95.86%, and 90.02% for *F. vesiculosus* with H_3PO_4, 65.18%, 99.70%, and 63.93% for A. maxima with $ZnCl_2$ and 88.72%, 96.08% and 76.61% for C. pyrenoidosa with $ZnCl_2$. Meanwhile, the removal rates of other pollutants such as Sodium, Calcium, Magnesium and Potassium were 65%−90%, 20%−100%, 64%−100% and 75%−100%, respectively. Biochar could also be supplemented as a nutrient source for microalgae according to Behl et al. (2019), and in their research exploring that scenario, they achieve lipid production and diazo Direct Red-31 dye decolorization by cultivating *Chlorella pyrenoidosa* in a medium supplemented with biochar. The maximum X production achieved in 12 days by the microalgae alone was $0.65 \, mg \, L^{-1}$, while the supplementation of biochar lead to a value of $0.81 \, mg \, L^{-1}$. %RE of dye from the aqueous solution after 5 days was 98.25% without adding NaN_3, which was demonstrated to have a negative impact on the performance (42.59%). When parameters are investigated individually, contact time and adsorbent (microalgae-biochar) dosages had a positive effect on the RE, while pH and initial dye dosage concentration had a negative effect.

Wastewater from pharmaceutical industries

Hemalatha and Venkata (2016) demonstrated the feasibility of microalgae cultivation in pharmaceutical wastewater (PWw) previously treated in an SBR for bioremediation and lipid production (LP), and in their study, the maximum biomass production (X) was observed at the 6th day and its concentration was $2.8 \, g \, L^{-1}$ (17.2% total lipid content and 6.2% neutral lipid content under light condition) which decreased after that. Carbon removal efficiency (%RE) at the end of the biomass growth phase was 73%, while the %RE of COD, nitrates, and phosphates from the microalgae-based tertiary unit were 74.48%, 62%, and 100%, respectively, leading to a total process (SBR + Microalgae Unit) %RE of 97.1%, 77.6%, and 100%, respectively. Additionally, Amit et al. (2020) set up cultivation of *Tetraselmis indica* in a batch photobioreactor utilizing PWw with (PW1) and without (PW2) anaerobic pretreatment in a microbial fuel cell for biodiesel and electricity production. X production and productivity after 24 days of cultivation for PW1 and PW2 were $0.78 \pm 0.05 \, g \, L^{-1}$ and $46.85 \, mg \, L^{-1} \, d^{-1}$ (35% lipids) and $1.067 \pm 0.02 \, g \, L^{-1}$ and $49.04 \, mg \, L^{-1} \, d^{-1}$ (32% lipids), respectively. Without the anaerobic pretreatment, microalgae cultivation removed COD, TOC, nitrate, and phosphate up to a value of 66.30%, 78.14%, 67.17%, and 70.03%, respectively, with CO_2 fixation rate (RC) of $0.089 \, g \, L^{-1} \, d^{-1}$. However, combining the photobioreactor (PBR) with the MFC pretreatment of wastewater resulted in an overall %RE of COD, TOC, nitrate, and phosphate and CO_2 fixation rate (RCR) of 90.29%, 97.05%, 81.60%, and 94.87%, and $0.085 \, g \, L^{-1} \, d^{-1}$, respectively. Meanwhile, micronutrients (Al^{2+}, Fe^{3+}, Ca^{2+}, Na^{2+}, Mg^{2+}, K^+, Cu^{2+}, Mn^{2+}, and Zn^{2+}) overall reduction was 80%−95%. Moreover, Nayak and Ghosh (2019) apply the treatment of raw PW in a microalgae-based PBR cultivating *Scenedesmus abundans* before further treatment in a photosynthetic MFC for biomass and biodiesel production. After 21 days of cultivation, the X production and BP by the PBR and PMFC were $0.97 \pm 0.01 \, g \, L^{-1}$ and $0.064 \, g \, L^{-1} \, d^{-1}$ and $0.94 \pm 0.01 \, g \, L^{-1}$ and $0.0633 \, g \, L^{-1} \, d^{-1}$, respectively. The %RE of COD, nitrate and phosphate from the PBR were $77\% \pm 3\%$, $82\% \pm 0.5\%$, and $65.9\% \pm 0.4\%$, and further treatment in the PMFC upped those values to $88\% \pm 3\%$, 97.12%, and 93.71%, respectively.

Another type of wastewater relevant to this section is wastewater contaminated by pharmaceuticals (CPWw), such as that investigated by Singh et al. (2020), who looked into the biotreatment of river wastewater contaminated with pharmaceutical effluent along with biomass production by the cultivation of three microalgae species (*Chlorella* sp., *Chlorococcum* sp., and *Neochloris* sp.). The X_{max} production for the three respective species were 498, 450, and $520 \, mg \, L^{-1}$ after 16 days of incubation, and the maximum lipid yield was obtained from *Chlorococcum* sp. ($129 \, mg \, L^{-1}$, 28%). Meanwhile, after 8 days of incubation, %RE of BOD for the microalgae were 84%, 83%, and 91%, respectively, while the COD %RE after 10 days of incubation was 84.5%, 88.4%, and 90%, respectively. Also, García-Galán et al. (2020) feed pharmaceutical chemicals to domestic wastewater (DWw) in the cultivation of microalgal cultures in high-rate algae ponds (HRAPs) for the biodegradation of 12 different pharmaceuticals along with 26 of their metabolites and transformation products. After 12 days of cultivation, a BP of approximately 22 g VSS m^{-2} d was observed from the HRAP that follows primary settler (control), while the other one that doesn't (test) gave productivity of 16 g VSS m^{-2} d. %RE of total COD, TN, and PO$_4$$^{3-}$P for HRAP-control were 60.8%, 71.7%, and 35.3%, respectively. Meanwhile, %RE for most of the pharmaceuticals ranged from moderate (40%−60%) to high (>60%), while O-desmethylvenlafaxine, which is known to have very low biodegradability, was removed at a range of 13%−39%. Furthermore, Escudero et al. (2020) highlight the ability of *Chlamydomonas acidophila* to remove common pharmaceutical residues found in anaerobically treated effluent contaminated with pharmaceuticals along with nutrient recovery and the microalga's resilience against the existing chemicals. The growth of microalgae was not inhibited by concentrations of the pharmaceuticals that are not above urban wastewater level, and the %RE of erythromycin and clarithromycin were 65%−93% and 50%−64%, respectively, while assimilation of 9 and 3 mg L^{-1} d^{-1}of NH$_4$ and PO$_4$ was observed, respectively.

Direct biodegradation of pharmaceuticals is also explored by many researchers, some of which will be highlighted here. According to Xiong et al. (2016), the biodegradation of anticonvulsant medication Carbamazepine (CBZ) is possible by freshwater microalgae *Chlamydomonas mexicana* and Scenedesmus obliquus. The chemical had insignificant inhibitive effects on the microalgal growth up to a concentration of 50 mg L^{-1}, but at X of 200 mg L^{-1}, the μ of *C. mexicana* and *S. obliquus* was inhibited by 11.7% and 74%, respectively. After 10 days, the total %RE of CBZ by *C. mexicana* and *S. obliquus* were 37% and 30%, respectively. Meanwhile, the biodegradation rates of the chemical by microalgae were 35% and 28%, respectively at a concentration of 1 mg L^{-1}, with higher concentration leading to lower %RE. In addition, Habibzadeh, Chaibakhsh, and Naeemi (2018) report the removal of anticancer drug Flutamide (FLU) by living and dead biomass of *Chlorella vulgaris*, and further optimization of environmental parameters such as pH, drug dosage, and biosorption time using RSM. The algal cells were more promising in its sorption capacity of 26.8 mg g^{-1}, and on the other hand, the sorption capacity by the dead biomass was 12.5 mg g^{-1}. When the pH, drug dosage, and sorption time are 7.4, 50 μM and 10 min, respectively, the maximum drug removal efficiency of 98.5% was achieved. Finally, Hena, Gutierrez, and Croué (2020) grow *Chlorella vulgaris* in an aqueous medium containing the antibiotic/antiprotozoal medication Metronidazole (MDZ) and study the effects of inoculum size and chemical dosage on the performance. MDZ had a negative effect on X production, unlike the inoculum size, and the highest yields are observed at the 12th−13th day in the range of 1.2−1.4 g L^{-1} before eventually dropping to 0.6−1 g L^{-1} after 20 days. MDZ is removed entirely after 18−20 days of cultivation at an initial concentration of 5 μM. Meanwhile, lower and higher concentrations decrease the %RE (50% at μM 1 and 60% at μM).

Algae and energy production

Microalgae cells can successfully satisfy the double benefits of treated wastewater generated from the energy industry, which is usually rich in organic pollutants, but also contains significant amounts of sulfur and phosphorus, among others while producing biomass. Mohammadi et al. (2018) report the feasibility of S^{2-} removal from power plant wastewater along with biomass production using five different microalgae strains (*Chlorella* sp., *Chlamydomonas* sp., *Oocystis* sp., *Scenedesmus* sp. and *Fischerella* sp.). The biomass productivity (PB) and S^{2-} removal of the five respective species (in their order) was 50 mg L^{-1} d^{-1} and 22.47% ± 0.76%, 20 mg L^{-1} d^{-1} and 26.05% ± 1.10%, 25.24 mg L^{-1} d^{-1} and 32.00% ± 0.51%, 26.19 mg L^{-1} d^{-1} and 24.68% ± 1.08% and 13.33 mg L^{-1} d^{-1} and 22.81% ± 1.37% after 21 days of cultivation. In addition, Xie et al. (2018) develop a process for simultaneous biodiesel production (PDP) and biotreatment (of biogas project anaerobically digested effluent by the cultivation of *Chlorella vulgaris*. At a soluble COD concentration of 272 mg L^{-1}, the maximum biomass yield after 10 days of cultivation was 1466 mg L^{-1}, and its lipid content was approximately 44%. At the end of the cultivation period (10 days) and an initial SCOD concentration of 167 mg L^{-1}, total nitrogen and total phosphorus were completely removed, while the SCOD removal efficiency was remarkably high (97.43%).

Regarding the combination of microalgae and bacteria in the treatment of wastewater, Ashwaniy et al. (2020) studied the synergistic interactions between microalgae (*Scenedesmus abundans*) and bacteria (microbial desalination cell) in the desalination of seawater, organic pollutant removal of petroleum refinery effluent (PRE), biomass production and electricity generation. After 25 days of cultivation, the maximum algae growth was observed when the species was grown in a BG11 medium that is 50% PRE. The %RE of COD, BOD, phosphorus, sulfide, TDS and TSS were 70%, 81%, 67%, 61%, 67%, and 62%, respectively. When the algae are the electron acceptor (as opposed to being a substrate for bacterial growth), the NaCl %RE is 60% on day 1 and 5% at the end of the cultivation period. In addition, CO$_2$

sequestration and pollutant removal from petroleum wastewater using an airlift bioreactor (ALPBR) that utilized hydro-carbonoclastic bacteria and the microalgae *Spongiochloris* sp. is detailed in the study by Abid et al. (2017). After the 120 days long treatment process, the X production, and PB were 8.51 g L^{-1} and 1.5 ± 0.3 g L^{-1} d^{-1}, respectively. Meanwhile, the total hydrocarbon efficiency and CO_2 biofixation (RC) rate were measured to be 99.18% and 2.9205 g L^{-1} d^{-1}, respectively. Lacerda et al. (2011) employ mathematical models in simulating the X production and CO_2 biofixation rate from refinery wastewater by the cultivation of *Aphanothece microscopica* Nageli for 1000 h. After adding a BGN medium containing 25% salts (w/v) to the Ww, the optimum X yield and CO_2 fixation rate were 1.41 and 2.61 kg L^{-1}, respectively at the end of the treatment process.

Algae and biofuel production

The utilization of algal biomass for the production of commercially competitive biofuels is increasingly common and continues to see more progress. This greener technology would lead to the production of biofuels while also maintaining significantly lower carbon emissions, alongside the sequestration of CO_2 in the cultivation of algae. Additionally, the price of algal-derived petroleum is about 20 times higher than it was 10 years ago. Other facts that highlight the attractiveness of algae as a source for biofuel production include their relatively fast growth rate (Pate, Klise, & Wu, 2011) and their ability to be grown in a lot of mediums, including sewage water or saltwater, with no need for fertile soil/crops (Ullah et al., 2015). This conversion of algal biomass to biofuels could be thermochemical, biochemical, or directly chemical.

Thermochemical conversion includes a diverse set of reactions that share one thing in common that is high heat and pressure to varying degrees. Hydrothermal liquefaction (HTL), which is considered the mildest form of thermochemical conversion, utilizes high pressures and temperatures to help initiate the hydrolysis of large components of the algae such as proteins, carbohydrates, and lipids into smaller monomers, followed by further decomposing of the monomers into very small compounds that reorganize to make the constituents of biofuels (aromatics, phenols, and fatty acids, among others) (Gu et al., 2020; Guo et al., 2015; Yang & Yang, 2019), and because this conversion process is conducted at relatively low P and T ranging between 200°C and 450°C and 4−35 MPa inside a liquid medium, the product sought after and expected is usually mostly liquid form (biooil). HTL is a reliable conversion technique that is capable of producing biooil yield (BoY) up to higher than 50 wt.% as is proven by He et al. (2020), who conducted HTL on *Nannochloropsis* sp. and *Sargassum* sp. and recorded BoYs of 54.11 wt.% and 9.49%, respectively. The liquid medium most often used is water, but other options are shown to be viable such as methanol, ethanol and isopropanol, and glycerol (Biswas et al., 2017; Cui et al., 2020; Rahman et al., 2019). Yang et al. (2017) report that the use of acid catalysts leads to a faster biooil formation and has a noticeable impact on the H/C ratio of the product. HTL units could also be fed mixed feeds as illustrated by Zhu et al. (2020), where the synergistic effects between microalgae (*Chlorella* sp. and *Tetraselmis* sp.) and wood (Pine and C&D) are studied. Liquefaction of blends evidently gave higher yields of biooil (40%−45%) than pure feeds. Gasification is the most extreme form of thermochemical conversion in which the biomass is converted to biogas at relatively high temperatures (800°C −1200°C) in the presence of a gasifying agent. Because of the extreme conditions of gasification, the desired components of such process are usually in gas form (e.g. syngas and methane) although it should be noted that gasification could be performed at lower temperatures, affecting the resulting product as well. The mechanisms of gasification are discussed by Raheem et al. (2018) and they are summarized by the set of reactions (Eqs. 7.1−7.7). Additionally, Hanchate et al. (2020) classify the different types of reactors used in gasification according to reaction conditions, gasifying agent, and mode of operation.

$$C_x H_y O_z \rightarrow Syngas + biochar + tar + steam \tag{7.1}$$

$$C + \frac{1}{2}O_2 \rightarrow CO \tag{7.2}$$

$$C + O_2 \rightarrow CO_2 \tag{7.3}$$

$$C + CO_2 \Leftrightarrow 2CO \tag{7.4}$$

$$C + 2H_2 \Leftrightarrow CH_4 \tag{7.5}$$

$$CO + H_2O \Leftrightarrow H_2 + CO_2 \tag{7.6}$$

$$CH_4 + H_2O \Leftrightarrow CO + 3H_2 \tag{7.7}$$

The gasification of mixed microalgal culture is performed by Soares, Martins, and Gonçalves (2020) in a downdraft commercial gasifier, producing syngas up to 2.8 Nm^3 kg^{-1}—dry biomass at an ER value of 0.23, and the hydrogen and carbon monoxide fractions were 11.9% and 19.5%, which is the value recommended for production. Raheem et al. (2018) worked on a central composite design (CCD) to study the effect of ZnO−Ni−CaO catalyst on the gasification of *Chlorella vulgaris* and to determine the optimum parameters including the catalyst loading for the process. At the optimum conditions of 851°C, 16.4 wt.% catalyst loading rate, and 28.8 min reaction time, the fraction and yield of hydrogen gas were 48.95% and 18.77 mol kg^{-1} biomass. Other possible catalysts shown to affect gasification include NaOH (Xu et al., 2019), waste eggshell (CaO) (Raheem et al., 2019), and zeolites (Xie et al., 2019). The third common thermochemical process is pyrolysis, which, conditions−wise, in the middle of the two previously mentioned processes. The principles of pyrolysis and the most recent advances made are discussed in detail by Yang et al. (2019) and Lee et al. (2020), and to sum it up, the pyrolysis of microalgae is the degradation of the microorganism's contents in the total absence of oxygen supply, and it can be classified according to the heating rate (slow vs intermediate vs fast pyrolysis). This process is usually conducted in the temperature range of 400°C−900°C, and because of its intermediate nature in comparison to HTL and gasification, its mechanism and products are a diverse blend between the two (solids to gases), unless extreme temperatures (either low or high) are specified, which would bring this process's outcomes closer to one of the other two. For example, fast pyrolysis at 900°C is performed by Hu, Ma, and Li (2013), yielding 91.09 wt.% biofuels and giving the highest CO and H$_2$ emissions in the temperature range of 800°C−900°C. pyrolysis can be assisted by the use of microwaves (Beneroso et al., 2013; Du et al., 2011) and it is versatile, able to apply to different types of wastewater such as sewage wastewater (Ashokkumar et al., 2019) and kitchen wastewater (Wang et al., 2020). Much like the other thermochemical processes, pyrolysis can be catalyzed, and some examples of viable catalysts are oyster shells (CaCO$_3$) (Choi et al., 2019), ZnCl$_2$, SiC, MgO, and activated carbon (Hu et al., 2014). Finally, it is worth mentioning that it is possible to combine different processes, and this is demonstrated by Xu et al. (2019), who designed a combined HTL-gasification process in which the effluent of the HTL unit goes into the gasification unit for further energy recovery.

The common biochemical conversion technique used to turn microalgae into biofuels is fermentation, which is a process that emphasizes the synergies between microalgae and bacteria. Simply put, bacteria can consume many compounds to produce biofuels (e.g. biogas, alcohols, etc.), and these nutrients could be obtained from the breaking down of microalgae constituents, primarily carbohydrates, lipids, and proteins which are hydrolyzed to produce starch and glucose, glycerol, fatty acids and alcohols, and amino acids and peptides, respectively. To illustrate, Efremenko et al. (2012) performed ABE fermentation by the immobilized bacteria *Clostridium acetobutylicum* on different microalgal strains. *Nannochloropsis* sp. led to a hydrogen production rate of up to 8.5 mmol L^{-1} medium day^{-1}, while *Arthrospira platensis* gave butanol and ethanol yields up to 117% and 35%, respectively. The biochemical conversion of *Chlorella vulgaris* into ethanol and methane is further examined by Ho et al. (2013) and Llamas et al. (2020). A product of particular interest for commercial use is Hythane; a gas blend that is made of primarily methane but also contains hydrogen that improves engine combustion and reduced CO$_2$ and NO$_x$ emissions (David et al., 2019). The production of Hythane is possible through a 2-stage anaerobic digestion process, where each component of the blend (H$_2$/CH$_4$) is fermented at a separate stage before being blended at high rates of production (Chen et al., 2020; Sun et al., 2019). The fermentation process could be further improved by the pretreatment methods, either the low-cost biological options such as the fungal pretreatment (Mushlihah et al., 2020) that led to fermentation efficiencies up to 92% from *Kappaphycus alvarezii*, chemical pretreatments (Salaeh et al., 2019) or physical pretreatment options such as ultrasound and microwave (Ha et al., 2020). Fermentation could be combined with other processes in an integrated large design to optimize production rate and cost. This includes integrating fermentation of microalgae with HTL (Rahman et al., 2019) and CO$_2$ fixation and pyrolysis (Deng et al., 2020; Serrà et al., 2020). El-Dalatony et al. (2019) and Sivaramakrishnan and Incharoensakdi (Sivaramakrishnan & Incharoensakdi, 2018) propose an integrated chemical transesterification−fermentation processes for high bioethanol and higher alcohols yields. Moreover, cultural variables such as pH, temperature, cultivation period, and the existence of supplements have a noticeable impact on the process performance as concluded by Chandra, Shukla, and Mallick (2020).

The chemical conversion of microalgae to biofuels, and more specifically, biodiesel, is done through transesterification, which is explained in detail alongside current applications, challenges, and room for improvements (Karpagam, Jawaharraj, & Gnanam, 2021; Kim et al., 2019; Zhang et al., 2013). The most important constituents of the microalgal biomass for this process are lipids, which can be turned esterified by alcohol to produce biodiesel (fatty acid alkyl esters) and other side products such as glycerol in a reversible and usually catalytic reaction. Sivaramakrishnan and Incharoensakdi (2018) report that methyl ester yields up to 92% was obtainable through the transesterification of *Scenedesmus* sp., highlighting the feasibility of the process for biodiesel production and the importance of process parameters (temperature, catalyst weight ratio, and reaction time). Methanol is the most commonly use alcohol, while some of the catalysts used for the chemical reaction include alkali catalysts (NaOH) (Karthikeyan, Periyasamy, & Prathima, 2020), enzymatic catalysts (lipase) (Sivaramakrishnan & Incharoensakdi, 2018), and acidic catalysts (sulfuric acid) (Ha et al., 2020; Shwetharani & Balakrishna, 2016) and even iron nanoparticles (Vignesh et al., 2020). Hess and Quinn (2018) compare the use of an acidic catalyst and the sole use of methanol at supercritical conditions to produce biofuels from Nannochloropsis salina by transesterification, and the results showed that the two respective methods achieved biofuel productions of 0.28 and 0.32 g g^{-1} microalgae, and FAME recovery efficiency of 89% and 100%, respectively. Because chemical reaction requires the continuous supply of alcohol, on a commercial scale, it could end up being a quite costly method, which is why it is important to be on the lookout for new ways of decreasing the process costs. Ashokkumar et al. (2019) attempt this with their design that not only aims to increase the process's cost-efficiency by the utilization of municipal sewage for the cultivation of microalgae (*Chlorella* sp. and *Sargassum* sp.), but also integrating this process in a combined transesterification−pyrolysis process, with a maximum biodiesel (FAME) yield of 93%. Other attempts at increasing the economic attractiveness of the process include the use of wastelands (Phukan et al., 2020) and landfill−generated wastewater (Dogaris et al., 2019). Transesterification of microalgae mixed with other microbial communities (bacteria and viruses, among others) was reported to yield a product with a FAME content of 485 (w/w) (Bell et al., 2019).

References

Abid, A., Saidane, F., & Hamdi, M. (2017). Feasibility of carbon dioxide sequestration by *Spongiochloris* sp. microalgae during petroleum wastewater treatment in airlift bioreactor. *Bioresource Technology, 234,* 297−302.

Al Ketife, A. M. D., Judd, S., & Znad, H. (2016). A mathematical model for carbon fixation and nutrient removal by an algal photobioreactor. *Chemical Engineering Science, 153,* 354−362.

Alketife, A. M., Judd, S., & Znad, H. (2017). Synergistic effects and optimization of nitrogen and phosphorus concentrations on the growth and nutrient uptake of a freshwater *Chlorella vulgaris. Environmental Technology, 38*(1), 94−102.

Amit, J. K., Nayak., & Ghosh, U. K. (2020). Microalgal remediation of anaerobic pretreated pharmaceutical wastewater for sustainable biodiesel production and electricity generation. *Journal of Water Process Engineering, 35,* 101192.

Amos, W. A. (2004). *Updated cost analysis of photobiological hydrogen production from Chlamydomonas reinhardtii green algae: milestone completion report.* National Renewable Energy Lab., Golden, CO.(US).

Andreotti, V., Solimeno, A., Rossi, S., Ficara, E., Marazzi, F., Mezzanotte, V., & García, J. (2020). Bioremediation of aquaculture wastewater with the microalgae *Tetraselmis suecica*: Semi-continuous experiments, simulation and photo-respirometric tests. *Science of The Total Environment, 738,* 139859.

Ashokkumar, V., et al. (2019). Cultivation of microalgae *Chlorella* sp. in municipal sewage for biofuel production and utilization of biochar derived from residue for the conversion of hematite iron ore (Fe2O3) to iron (Fe)−Integrated algal biorefinery. *Energy, 189,* 116128.

Ashwaniy, V. R. V., Perumalsamy, M., & Pandian, S. (2020). Enhancing the synergistic interaction of microalgae and bacteria for the reduction of organic compounds in petroleum refinery effluent. *Environmental Technology & Innovation, 19,* 100926.

Ayed, L., et al. (2020). Decolorization and phytotoxicity reduction of reactive blue 40 dye in real textile wastewater by active consortium: Anaerobic/aerobic algal-bacterial-probiotic bioreactor. *Journal of Microbiological Methods,* 106129.

Basu, S., et al. (2014). CO$_2$ biofixation and carbonic anhydrase activity in *Scenedesmus obliquus* SA1 cultivated in large scale open system. *Bioresource Technology, 164,* 323−330.

Becker. (1994). *Microalgae: Biotechnology and microbiology.* New York: Cambridge University.

Behl, K., et al. (2019). One-time cultivation of *Chlorella pyrenoidosa* in aqueous dye solution supplemented with biochar for microalgal growth, dye decolorization and lipid production. *Chemical Engineering Journal, 364,* 552−561.

Bell, T. A. S., et al. (2019). Contributions of the microbial community to algal biomass and biofuel productivity in a wastewater treatment *lagoon* system. *Algal Research, 39,* 101461.

Beneroso, D., et al. (2013). Microwave pyrolysis of microalgae for high syngas production. *Bioresource Technology, 144,* 240−246.

Bhatnagar, A., et al. (2011). Renewable biomass production by mixotrophic algae in the presence of various carbon sources and wastewaters. *Applied energy, 88*(10), 3425−3431.

Biswas, B., et al. (2017). Effects of temperature and solvent on hydrothermal liquefaction of *Sargassum tenerrimum* algae. *Bioresource Technology, 242,* 344−350.

Chandra, N., Shukla, P., & Mallick, N. (2020). Role of cultural variables in augmenting carbohydrate accumulation in the green microalga *Scenedesmus acuminatus* for bioethanol production. *Biocatalysis and Agricultural Biotechnology (Reading, Mass.), 26,* 101632.

Chen, C., et al. (2020). Sustainable biohythane production from algal bloom biomass through two-stage fermentation: Impacts of the physicochemical characteristics and fermentation performance. *International Journal of Hydrogen Energy, 45*(59), 34461−34472.

Choi, D., et al. (2019). Catalytic pyrolysis of brown algae using carbon dioxide and oyster shell. *Journal of CO$_2$ Utilization, 34,* 668−675.

Chu, H.-Q., et al. (2015). Continuous cultivation of *Chlorella pyrenoidosa* using anaerobic digested starch processing wastewater in the outdoors. *Bioresource Technology, 185,* 40−48.

Cui, Z., et al. (2020). Roles of co-solvents in hydrothermal liquefaction of low-lipid, high-protein algae. *Bioresource Technology, 310,* 123454.

Daneshvar, E., et al. (2019). Sequential cultivation of microalgae in raw and recycled dairy wastewater: Microalgal growth, wastewater treatment and biochemical composition. *Bioresource Technology, 273,* 556−564.

Daneshvar, E., et al. (2018). Versatile applications of freshwater and marine water microalgae in dairy wastewater treatment, lipid extraction and tetracycline biosorption. *Bioresource Technology, 268*, 523−530.

David, B., et al. (2019). *Biohythane production from food wastes, in biohydrogen* (pp. 347−368). Elsevier.

Deng, C., et al. (2020). Improving gaseous biofuel yield from seaweed through a cascading circular bioenergy system integrating anaerobic digestion and pyrolysis. *Renewable and Sustainable Energy Reviews, 128*, 109895.

Dhaouefi, Z., et al. (2018). Assessing textile wastewater treatment in an anoxic-aerobic photobioreactor and the potential of the treated water for irrigation. *Algal research, 29*, 170−178.

Dogaris, I., et al. (2019). Study of landfill leachate as a sustainable source of water and nutrients for algal biofuels and bioproducts using the microalga *Picochlorum oculatum* in a novel scalable bioreactor. *Bioresource Technology, 282*, 18−27.

Du, Z., et al. (2011). Microwave-assisted pyrolysis of microalgae for biofuel production. *Bioresource Technology, 102*(7), 4890−4896.

Efremenko, E., et al. (2012). Production of biofuels from pretreated microalgae biomass by anaerobic fermentation with immobilized *Clostridium acetobutylicum* cells. *Bioresource Technology, 114*, 342−348.

El-Dalatony, M. M., et al. (2019). Whole conversion of microalgal biomass into biofuels through successive high-throughput fermentation. *Chemical Engineering Journal, 360*, 797−805.

Escudero, A., et al. (2020). Pharmaceuticals removal and nutrient recovery from wastewaters by *Chlamydomonas acidophila*. *Biochemical Engineering Journal, 156*, 107517.

Freed, T. (2007). Wastewater industry moving toward enhanced nutrient removal standards. *WaterWorld*.

García-Galán, M. J., et al. (2020). Fate of priority pharmaceuticals and their main metabolites and transformation products in microalgae-based wastewater treatment systems. *Journal of Hazardous Materials, 390*, 121771.

Gonçalves, A., Simões, M., & Pires, J. (2014). The effect of light supply on microalgal growth, CO_2 uptake and nutrient removal from wastewater. *Energy Conversion and Management, 85*, 530−536.

Gong, Y., & Jiang, M. (2011). Biodiesel production with microalgae as feedstock: From strains to biodiesel. *Biotechnology Letters, 33*(7), 1269−1284.

Gramegna, G., et al. (2020). Exploring the potential of microalgae in the recycling of dairy wastes. *Bioresource Technology Reports, 12*, 100604.

Green, B. R., & Durnford, D. G. (1996). The chlorophyll-carotenoid proteins of oxygenic photosynthesis. *Annual Review of Plant Physiology and Plant Molecular Biology, 47*(1), 685−714.

Gu, X., et al. (2020). Recent development of hydrothermal liquefaction for algal biorefinery. *Renewable and Sustainable Energy Reviews, 121*, 109707.

Guo, Y., et al. (2015). A review of bio-oil production from hydrothermal liquefaction of algae. *Renewable and Sustainable Energy Reviews, 48*, 776−790.

Ha, G.-S., et al. (2020). Energy-efficient pretreatments for the enhanced conversion of microalgal biomass to biofuels. *Bioresource Technology, 309*, 123333.

Habibzadeh, M., Chaibakhsh, N., & Naeemi, A. S. (2018). Optimized treatment of wastewater containing cytotoxic drugs by living and dead biomass of the freshwater microalga, *Chlorella vulgaris*. *Ecological Engineering, 111*, 85−93.

Hanchate, N., et al. (2020). Biomass gasification using dual fluidized bed gasification systems: A review. *Journal of Cleaner Production*, 123148.

He, S., et al. (2020). Hydrothermal liquefaction of low-lipid algae *Nannochloropsis* sp. and *Sargassum* sp.: Effect of feedstock composition and temperature. *Science of the Total Environment, 712*, 135677.

Hemalatha, M., & Venkata, S. (2016). Mohan, microalgae cultivation as tertiary unit operation for treatment of pharmaceutical wastewater associated with lipid production. *Bioresource Technology, 215*, 117−122.

Hemalatha, M., et al. (2019). Microalgae-biorefinery with cascading resource recovery design associated to dairy wastewater treatment. *Bioresource Technology, 284*, 424−429.

Hena, S., Gutierrez, L., & Croué, J.-P. (2020). Removal of metronidazole from aqueous media by *C. vulgaris*. *Journal of Hazardous Materials, 384*, 121400.

Hess, D., & Quinn, J. C. (2018). Impact of inorganic contaminants on microalgal biofuel production through multiple conversion pathways. *Biomass and Bioenergy, 119*, 237−245.

Ho, S.-H., et al. (2013). Bioethanol production using carbohydrate-rich microalgae biomass as feedstock. *Bioresource Technology, 135*, 191−198.

Ho, S.-H., Chen, W.-M., & Chang, J.-S. (2010). Scenedesmus obliquus CNW-N as a potential candidate for CO_2 mitigation and biodiesel production. *Bioresource Technology, 101*(22), 8725−8730.

Honda, R., et al. (2012). Carbon dioxide capture and nutrients removal utilizing treated sewage by concentrated microalgae cultivation in a membrane photobioreactor. *Bioresource Technology, 125*, 59−64.

Hu, X., Meneses, Y. E., & Aly, A. (2020). Hassan, integration of sodium hypochlorite pretreatment with coimmobilized microalgae/bacteria treatment of meat processing wastewater. *Bioresource Technology, 304*, 122953.

Hu, Z., et al. (2014). The catalytic pyrolysis of microalgae to produce syngas. *Energy conversion and management, 85*, 545−550.

Hu, Z., Ma, X., & Li, L. (2013). The characteristic and evaluation method of fast pyrolysis of microalgae to produce syngas. *Bioresource Technology, 140*, 220−226.

Hülsen, T., et al. (2018). Simultaneous treatment and single cell protein production from agri-industrial wastewaters using purple phototrophic bacteria or microalgae−A comparison. *Bioresource Technology, 254*, 214−223.

Karpagam, R., Jawaharraj, K., & Gnanam, R. (2021). Review on integrated biofuel production from microalgal biomass through the outset of transesterification route: A cascade approach for sustainable bioenergy. *Science of The Total Environment, 766*, 144236.

Karthikeyan, S., Periyasamy, M., & Prathima, A. (2020). Combustion analysis of a CI engine with *Caulerpa racemosa* algae biofuel with nano additives. *Materials Today: Proceedings, 33*, 3324−3329.

Kim, B., et al. (2019). Simplifying biodiesel production from microalgae via wet in situ transesterification: A review in current research and future prospects. *Algal Research, 41*, 101557.

Kumar, G., et al. (2018). Evaluation of gradual adaptation of mixed microalgae consortia cultivation using textile wastewater via fed batch operation. *Biotechnology Reports, 20*, e00289.

Lacerda, L. M. C. F., et al. (2011). Improving refinery wastewater for microalgal biomass production and CO_2 biofixation: Predictive modeling and simulation. *Journal of petroleum science and engineering, 78*(3–4), 679–686.

Lam, M. K., & Lee, K. T. (2013). Effect of carbon source towards the growth of *Chlorella vulgaris* for CO_2 bio-mitigation and biodiesel production. *International Journal of Greenhouse Gas Control, 14*, 169–176.

Lebron, Y. A. R., Moreira, V. R., & Santos, L. V. S. (2019). Studies on dye biosorption enhancement by chemically modified Fucus vesiculosus, *Spirulina maxima* and *Chlorella pyrenoidosa* algae. *Journal of Cleaner Production, 240*, 118197.

Lee, X. J., et al. (2020). State of art review on conventional and advanced pyrolysis of macroalgae and microalgae for biochar, bio-oil and bio-syngas production. *Energy Conversion and Management, 210*, 112707.

Lin, C.-Y., Nguyen, M.-L. T., & Lay, C.-H. (2017). Starch-containing textile wastewater treatment for biogas and microalgae biomass production. *Journal of Cleaner Production, 168*, 331–337.

Llamas, M., et al. (2020). Microalgae-based anaerobic fermentation as a promising technology for producing biogas and microbial oils. *Energy, 206*, 118184.

López, C. G., et al. (2009). Utilization of the cyanobacteria *Anabaena* sp. ATCC 33047 in CO_2 removal processes. *Bioresource Technology, 100*(23), 5904–5910.

Lu, W., et al. (2015). Cultivation of *Chlorella* sp. using raw dairy wastewater for nutrient removal and biodiesel production: Characteristics comparison of indoor bench-scale and outdoor pilot-scale cultures. *Bioresource Technology, 192*, 382–388.

Makut, B. B., Das, D., & Goswami, G. (2019). Production of microbial biomass feedstock via cocultivation of microalgae-bacteria consortium coupled with effective wastewater treatment: A sustainable approach. *Algal Research, 37*, 228–239.

Mohammadi, M., et al. (2018). Cultivation of microalgae in a power plant wastewater for sulfate removal and biomass production: A batch study. *Journal of Environmental Chemical Engineering, 6*(2), 2812–2820.

Mushlihah, S., et al. (2020). Fungal pretreatment as a sustainable and low cost option for bioethanol production from marine algae. *Journal of Cleaner Production, 265*, 121763.

Nayak, J. K., & Ghosh, U. K. (2019). Post treatment of microalgae treated pharmaceutical wastewater in photosynthetic microbial fuel cell (PMFC) and biodiesel production. *Biomass and Bioenergy, 131*, 105415.

O'Toole, D. K. (1983). Weighing technique for determining bacterial dry mass based on rate of moisture uptake. *Applied and Environmental Microbiology, 46*(2), 506.

Oyebamiji, O. O., et al. (2019). Green microalgae cultured in textile wastewater for biomass generation and biodetoxification of heavy metals and chromogenic substances. *Bioresource Technology Reports, 7*, 100247.

Pate, R., Klise, G., & Wu, B. (2011). Resource demand implications for US algae biofuels production scale-up. *Applied Energy, 88*(10), 3377–3388.

Patel, A. K., Joun, J., & Sim, S. J. (2020). A sustainable mixotrophic microalgae cultivation from dairy wastes for carbon credit, bioremediation and lucrative biofuels. *Bioresource Technology, 313*, 123681.

Phukan, M. M., et al. (2020). Leveraging microalga feedstock for biofuel production and wasteland reclamation using remote sensing and ex situ experimentation. *Renewable Energy, 159*, 973–981.

Pires, J., et al. (2014). Effect of light supply on CO_2 capture from atmosphere by *Chlorella vulgaris* and *Pseudokirchneriella subcapitata*. *Mitigation and Adaptation Strategies for Global Change, 19*(7), 1109–1117.

Raheem, A., et al. (2018). Catalytic gasification of algal biomass for hydrogen-rich gas production: Parametric optimization via central composite design. *Energy Conversion and Management, 158*, 235–245.

Raheem, A., et al. (2019). Gasification of lipid-extracted microalgae biomass promoted by waste eggshell as CaO catalyst. *Algal Research; a Journal of Science and its Applications, 42*, 101601.

Rahman, Q. M., et al. (2019). A combined fermentation and ethanol-assisted liquefaction process to produce biofuel from *Nannochloropsis* sp. *Fuel, 238*, 159–165.

Rasouli, Z., et al. (2018). Nutrient recovery from industrial wastewater as single cell protein by a coculture of green microalgae and methanotrophs. *Biochemical Engineering Journal, 134*, 129–135.

Richmond, A. (2007). *Handbook of microalgal culture*.

Rodrigues, L. H. R., et al. (2011). Algal density assessed by spectrophotometry: A calibration curve for the unicellular algae *Pseudokirchneriella subcapitata*. *Journal of Environmental Chemistry and Ecotoxicology*.

Rogers, J. N., et al. (2014). A critical analysis of paddlewheel-driven raceway ponds for algal biofuel production at commercial scales. *Algal research, 4*, 76–88.

Ruangsomboon, S. (2012). Effect of light, nutrient, cultivation time and salinity on lipid production of newly isolated strain of the green microalga, *Botryococcus braunii* KMITL 2. *Bioresource Technology, 109*(0), 261–265.

Ruiz-Marin, A., Mendoza-Espinosa, L. G., & Stephenson, T. (2010). Growth and nutrient removal in free and immobilized green algae in batch and semicontinuous cultures treating real wastewater. *Bioresource Technology, 101*(1), 58–64.

Salaeh, S., et al. (2019). Feasibility of ABE fermentation from *Rhizoclonium* spp. hydrolysate with low nutrient supplementation. *Biomass and Bioenergy, 127*, 105269.

Santana, H., et al. (2017). Microalgae cultivation in sugarcane vinasse: Selection, growth and biochemical characterization. *Bioresource Technology, 228*, 133–140.

Serrà, A., et al. (2020). Circular zero-residue process using microalgae for efficient water decontamination, biofuel production, and carbon dioxide fixation. *Chemical Engineering Journal, 388*, 124278.

Shwetharani, R., & Balakrishna, R. G. (2016). Efficient algal lipid extraction via photocatalysis and its conversion to biofuel. *Applied Energy, 168,* 364–374.

Singh, A., Ummalyma, S. B., & Sahoo, D. (2020). Bioremediation and biomass production of microalgae cultivation in river water contaminated with pharmaceutical effluent. *Bioresource Technology, 307,* 123233.

Sinha, S., et al. (2016). Self-sustainable *Chlorella pyrenoidosa* strain NCIM 2738 based photobioreactor for removal of Direct Red-31 dye along with other industrial pollutants to improve the water-quality. *Journal of Hazardous Materials, 306,* 386–394.

Sivaramakrishnan, R., & Incharoensakdi, A. (2018). Utilization of microalgae feedstock for concomitant production of bioethanol and biodiesel. *Fuel, 217,* 458–466.

Soares, R. B., Martins, M. F., & Gonçalves, R. F. (2020). Experimental investigation of wastewater microalgae in a pilot-scale downdraft gasifier. *Algal Research, 51,* 102049.

Spolaore, P., et al. (2006). Commercial applications of microalgae. *Journal of Bioscience and Bioengineering, 101*(2), 87–96.

Sun, C., et al. (2019). Life-cycle assessment of biohythane production via two-stage anaerobic fermentation from microalgae and food waste. *Renewable and Sustainable Energy Reviews, 112,* 395–410.

Swain, P., Tiwari, A., & Pandey, A. (2020). Enhanced lipid production in *Tetraselmis* sp. by two stage process optimization using simulated dairy wastewater as feedstock. *Biomass and Bioenergy, 139,* 105643.

Sydney, E. B., et al. (2010). Potential carbon dioxide fixation by industrially important microalgae. *Bioresource Technology, 101*(15), 5892–5896.

Tang, D., et al. (2011). CO_2 biofixation and fatty acid composition of *Scenedesmus obliquus* and *Chlorella pyrenoidosa* in response to different CO_2 levels. *Bioresource Technology, 102*(3), 3071–3076.

Ullah, K., et al. (2015). Assessing the potential of algal biomass opportunities for bioenergy industry: A review. *Fuel, 143,* 414–423.

UTEX. (2014). *Spir solution 2 Recipe Spir solution 2 Recipe.* [cited 2016; Spirulina Medium].

Vignesh, N. S., et al. (2020). Sustainable biofuel from microalgae: Application of lignocellulosic wastes and bio-iron nanoparticle for biodiesel production. *Fuel, 278,* 118326.

Wang, S., et al. (2020). Study on the co-operative effect of kitchen wastewater for harvest and enhanced pyrolysis of microalgae. *Bioresource Technology, 317,* 123983.

Wong, N. (2012). *Microalgal biomass - Review.*

Wu, J.-Y., et al. (2020). Immobilized Chlorella species mixotrophic cultivation at various textile wastewater concentrations. *Journal of Water Process Engineering, 38,* 101609.

Wu, J.-Y., et al. (2017). Lipid accumulating microalgae cultivation in textile wastewater: Environmental parameters optimization. *Journal of the Taiwan Institute of Chemical Engineers, 79,* 1–6.

Xie, B., et al. (2018). Biodiesel production with the simultaneous removal of nitrogen, phosphorus and COD in microalgal-bacterial communities for the treatment of anaerobic digestion effluent in photobioreactors. *Chemical Engineering Journal, 350,* 1092–1102.

Xie, L.-F., et al. (2019). Hydrothermal gasification of microalgae over nickel catalysts for production of hydrogen-rich fuel gas: Effect of zeolite supports. *International Journal of Hydrogen Energy, 44*(11), 5114–5124.

Xiong, J.-Q., et al. (2016). Biodegradation of carbamazepine using freshwater microalgae *Chlamydomonas mexicana* and *Scenedesmus obliquus* and the determination of its metabolic fate. *Bioresource Technology, 205,* 183–190.

Xu, D., et al. (2019). Catalytic supercritical water gasification of aqueous phase directly derived from microalgae hydrothermal liquefaction. *International Journal of Hydrogen Energy, 44*(48), 26181–26192.

Yadav, G., Dash, S. K., & Sen, R. (2019). A biorefinery for valorization of industrial waste-water and flue gas by microalgae for waste mitigation, carbon-dioxide sequestration and algal biomass production. *Science of The Total Environment, 688,* 129–135.

Yang, C., et al. (2019). Pyrolysis of microalgae: A critical review. *Fuel Processing Technology, 186,* 53–72.

Yang, W., et al. (2017). Catalytic upgrading of bio-oil in hydrothermal liquefaction of algae major model components over liquid acids. *Energy Conversion and Management, 154,* 336–343.

Yang, J., & Yang, L. (2019). A review on hydrothermal co-liquefaction of biomass. *Applied Energy, 250,* 926–945.

You, X., et al. (2021). Integrating acidogenic fermentation and microalgae cultivation of bacterial-algal coupling system for mariculture wastewater treatment. *Bioresource Technology, 320,* 124335.

Zhang, X., et al. (2013). Biodiesel production from heterotrophic microalgae through transesterification and nanotechnology application in the production. *Renewable and Sustainable Energy Reviews, 26,* 216–223.

Zhu, Y., et al. (2020). Economic impacts of feeding microalgae/wood blends to hydrothermal liquefaction and upgrading systems. *Algal Research, 51,* 102053.

Zkeri, E., et al. (2021). Comparing the use of a two-stage MBBR system with a methanogenic MBBR coupled with a microalgae reactor for medium-strength dairy wastewater treatment. *Bioresource Technology, 323,* 124629.

Znad, H., Al Ketife, A. M. D., & Judd, S. (2019). Enhancement of CO2 biofixation and lipid production by *Chlorella vulgaris* using coloured polypropylene film. *Environmental Technology, 40*(16), 2093–2099.

Znad, H., et al. (2018). Bioremediation and nutrient removal from wastewater by *Chlorella vulgaris*. *Ecological Engineering, 110,* 1–7.

Znad, H., et al. (2012). CO_2 biomitigation and biofuel production using microalgae: photobioreactors developments and future directions. *Advances in Chemical Engineering.*

Chapter 8

Membrane-based treatment of petroleum wastewater

Abdelrahman M. Awad, Rem Jalab, Mustafa S. Nasser and Ibnelwaleed A. Hussein

Gas Processing Center, Qatar University, Doha, Qatar

Wastewater in the petroleum industry

The expansion in the petroleum industry is requiring more unconventional resources exploration along with the available explored conventional oil fields. The unconventional oil is usually found in low-permeability rocks and is sourced from heavy oil and oil shale (Silva, Morales-Torres, Castro-Silva, Figueiredo, & Silva, 2017). Extraction operations from unconventional resources are initiated by horizontal drilling to a distance of 2 km at several kilometers underground (Jackson et al., 2014). Following that, massive quantities of water estimated by 2–20 million gallons are mixed with sands and chemical additives before being injected at a high pressure of 700–1400 bar for fracturing the rocks (Silva et al., 2017). This water is either returned after drilling and termed by flowback water (FBW), or produced with hydrocarbon extraction again and named by produced water (PW). FBW composition is mostly water-based slurry fluid, while PW has an identical composition to the structure of the well from which it has been extracted. PW also accompanies the hydrocarbons produced from conventional oil reservoirs. Besides, the conventional operations require water for processing and cooling purposes in the downstream facilities after the extraction stage (Adham, Hussain, Minier-Matar, Janson, & Sharma, 2018). During the production stage in the refinery, crude oil is broken into useful products after the removal of nonhydrocarbon compounds. Therefore the main wastewater streams from the petroleum industry are the PW brought to the surface with extracted oil, as well as process water and refinery effluents from the downstream refinery operations (Jackson et al., 2014; Silva et al., 2017).

PW is described by spatially and temporally changing characteristics depending on the type of geological formation, hydrocarbon produced and additives injected during the fracturing activities (Silva et al., 2017). In fact, PW is brackish or very saline groundwater with high concentrations of total dissolved solids (TDS), organic, inorganic components, salts, bacteria, heavy metals, and naturally formed radioactive materials leached out from the earth subsurface (Warner et al., 2014; Table 8.1). Indeed, the complexity of PW nature is due to the available toxic organics and heavy metals that lead to environmental deterioration. In addition, petroleum refinery wastewater is considered an environmentally hazardous pollutant, since it contains various aromatic compounds such as benzene, toluene, and ethyl-benzene along with other polycyclic aromatics. The toxic effects of this wastewater result also from the presence of oil and grease, phenolic compounds, naphthenic acids, nitrogenous and sulfur compounds in the form of ammonia or hydrogen sulfide. Moreover, the quantity of PW varies with the location, type of injected water, and wells age. Overall, the worldwide PW yield possesses a growing trend that is projected to approach 340 billion barrels/year in 2020 compared to the produced 77 billion barrels/year in 1999 (Adham et al., 2018; Silva et al., 2017), see Table 8.1.

The global population is increasing and the lack of natural water resources limits the application of hydraulic fracturing activities requiring vast quantities of fresh water. Besides, the quality of PW is so complicated to be directly disposed into the environment without pretreatment. For instance, some organic compounds contained in the PW such as phenols, polycyclic aromatic hydrocarbons (PAHs) result in very toxic consequences on the environment if the water is disposed of without treatment (Millar, Couperthwaite, & Moodliar, 2016). Therefore management of the large volumes of PW and treatment is needed to recycle and reuse this wastewater. Earlier PW management options are classified to cover three scopes heavily governed by restricted regulations due to its complex characteristics: (1) reinjection to the same wells that was withdrawn from in the beginning, (2) discharge after proper treatment that meets the environmental

Petroleum Industry Wastewater. DOI: https://doi.org/10.1016/B978-0-323-85884-7.00003-5

TABLE 8.1 Petroleum wastewater characteristics.

Parameter	Composition range
Refinery wastewater	
pH	6.5–9.5
Chemical Oxygen Demand (COD, mg L^{-1})	550–7896
Biological Oxygen Demand (BOD, mg L^{-1})	174–3378
Total Organic Carbon (TOC, mg L^{-1})	220–265
Total Dissolved Salts (TDS, mg L^{-1})	1200–1500
Total Suspended Solids (TSS, mg L^{-1})	150
Oil and grease (mg L^{-1})	870
Ammonia (mg L^{-1})	13.5
Oily produced water	
TDS (mg L^{-1})	50,000–400,000
TSS (mg L^{-1})	1700
Total Oil (mg L^{-1})	565
TOC (mg L^{-1})	1700
Chloride (mg L^{-1})	29,000–250,000
Sodium (mg L^{-1})	9400–150,000
Bicarbonate (mg L^{-1})	400–15,000
Calcium (mg L^{-1})	1500–74,000
Phenols (mg L^{-1})	23

legislation, (3) reuse in the oil and gas (O&G) operations, irrigation, land restoration and drinking water after compliance with rules legislated to each use (Tao et al., 1993). Furthermore, the treatment of PW was limited to simple separation of oil and grease; however, the stringent environmental regulations require the development of advanced treatment technologies. PW can be treated with coagulation–flocculation, filtration, and sedimentation as pre or posttreatment processes (Silva et al., 2017).

In contrast, limitations on the disposal of PW and its reuse opportunities have pushed toward desalinating and treating this water using advanced treatment methods. Continuous advances in desalination are also required to ensure compliance with water reuse regulations demanding improved treated water quality (Lutz, Lewis, & Doyle, 2013). Membrane desalination is described as an advanced treatment stage compared with sedimentation, coagulation, and adsorption as it removes the dissolved salts, organics, and inorganic contaminants. Moreover, membrane treatment technologies are capable of producing treated water that meets the criteria for reuse and discharge in a cost-effective approach among all other water purification alternatives. The successful implementation of membrane-based processes in the treatment of O&G wastewater was examined in several studies (Fakhru'l-Razi et al., 2009).

Implementation of membrane processes as advanced treatment methods

Membrane separation processes are dependent on the thin film membranes fabricated from organic or inorganic materials that have selective permeability of fluids components. The separation process by membranes is based on the exclusion of most constituents based on the driving force and pore size of the chosen membrane. To explain this, membrane processes are classified into pressure, osmotically, thermally, and electrically driven processes (Adham et al., 2018; Fig. 8.1).

Tremendous advancements in membrane technology, allowing it to meet the environmental standards by being more capable of removing trace contaminants; have proved the possibility of its employment in petroleum wastewater

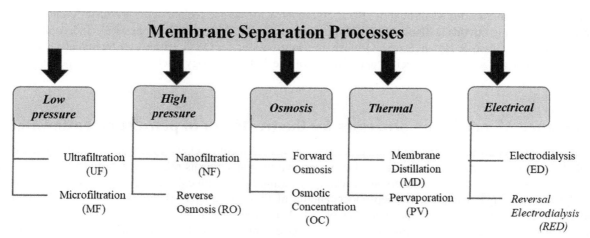

FIGURE 8.1　Classification of membrane separation processes for the treatment of oily wastewater.
Please see the online version to view the color version of the figure.

management. Membrane-based processes are featured with the high water product quality, simultaneous rejection of various pollutants, minimal chemicals requirement, low sludge generation, and reduced footprint compared to the conventional processes (Ang, Mohammad, Hilal, & Leo, 2015). These attributes render the membrane processes economically viable through the replacement of many conventional processes by a single membrane-based treatment step for achieving the same treatment duty. On the other hand, the membrane processes are impeded by two operational challenges, represented by the concentration polarization that arises from the increased concentration at the membrane surface, and the fouling due to the accumulation of contaminants on the membrane surface (Klaysom, Cath, Depuydt, & Vankelecom, 2013). Membrane fouling obviation or mitigation by the physical or chemical cleaning is preferable; since it severely deteriorates the treatment efficiency and reduces the lifespan of the membrane module. Above all, the generation of vast quantities of residual concentrate stream called brine is a drawback of certain membrane processes.

In fact, membrane treatment processes can be operated as standalone or hybrid processes. In the hybrid processes, the membrane separation step is integrated with another pretreatment process (e.g., coagulation, adsorption, or sedimentation), while the standalone process is the treatment of wastewater using only the membrane separation step. The integrated membrane processes are more eligible to produce water of better quality. Another advantage of hybrid membrane processes is facilitating an enhanced performance of the membrane, especially when the wastewater is highly loaded with natural organic matter (NOM), through the reduction of fouling risk. When wastewater of high strength is fed to the membrane directly without proper pretreatment, the membrane fouling will increase and the separation driving force will be reduced, hence poor water quality will be produced.

The membrane separation process is considered as a tertiary treatment step capable of removing the TSS, TDS, and traces of PAHs remained after passing the primary and secondary treatment steps. The removal effectiveness of the membrane processes varies according to the pore size and contaminants particle size. Indeed, the membrane removal effectiveness is described by the oil rejection according to Eq. (8.1), where higher oil removal is always preferable:

$$R_{Oil} = \frac{\text{Concentration of oil in feed} - \text{Concentration of oil in permeate}}{\text{Concentration of oil in feed}} \times 100\% \qquad (8.1)$$

In addition, the permeate flux is another key parameter indicating the process performance and relying on the membrane characteristics including porosity and hydrophilicity of fabrication materials (Changmai, Pasawan, & Purkait, 2019). The permeate flux is calculated by dividing the volume of permeate transferred across the membrane surface over the membrane area and time needed (Eq. 8.2):

$$J_p = \frac{V_p}{A \times t} \qquad (8.2)$$

Microfiltration (MF), ultrafiltration (UF), nanofiltration (NF), and reverse osmosis (RO) are effective membrane technologies applied extensively for the treatment of petroleum wastewater (Abousnina, Nghiem, & Bundschuh, 2015). Notwithstanding the successful operation of the pressure-driven membrane processes, innovative membrane treatment process such as forward osmosis (FO) utilizing the natural osmotic pressure difference is highlighted in the review of Silva et al. (2017). Moreover, membrane distillation (MD) comprises good treatment capabilities for PW from oil reservoirs. The membranes tested throughout these investigated membrane technologies are of different manufacturing materials such as polyvinylidene fluoride (PVDF), polyacrylonitrile (PAN), polysulfone (PS), and ceramic-polymeric complexes (Jamaly, Giwa, & Hasan, 2015). Osmosis-driven processes for O&G wastewater treatment include FO; pressure retarded osmosis, as well as specific applications of the FO technology such as osmotic concentration/dilution (OC/OD).

Performance of membrane processes for the treatment of petroleum wastewater

The performance of various membrane-based processes for the treatment of O&G wastewater is qualitatively compared in Table 8.2 in terms of process flexibility, product water quality, contaminants rejection, energy consumption, and pretreatment requirements. These parameters significantly determine the technical viability and economic feasibility for practical industrial implementation. A deep understanding of these parameters is vitally important to promote widespread commercial applications for the treatment of petroleum wastewater. In this section, the performance of each membrane process is discussed regarding the above-mentioned parameters to draw prior knowledge about their viability for O&G wastewater treatment.

Treatment with pressure-driven processes

In pressure-driven membranes, an externally applied pressure is utilized to transport water molecules through the membrane barrier while rejecting other constituents present in the wastewater feed. Pressure-driven processes can be classified into low-pressure processes (i.e., UF, MF) and high-pressure processes (i.e., RO, NF). In MF and UF, the wastewater is filtered through a porous structure membrane. On the other hand, NF and RO utilize dense membrane structures to desalinate wastewater and remove organic/inorganic contaminants from the feedwater (Tong, Carlson, Robbins, Zhang, & Du, 2019).

Low-pressure-driven processes

Over the past few decades, the use of MF and UF membranes has been extensively studied for the treatment of PW from O&G operations (Ahmad, Guria, & Mandal, 2020). Reported studies revealed that commercial UF and MF membranes generally generate permeate flux up to $1200 \, \text{L m}^{-2} \, \text{h}^{-1} \, \text{bar}^{-1}$ while handing feedwater with TSS and TDS up to 37,318, and 564 mg L^{-1}, respectively (Ebrahimi et al., 2009; He, Wang, Liu, Barbot, & Vidic, 2014). The performance of UF/MF is a function of the membrane properties, operating conditions, and quality of the feedwater. UF/MF demonstrated effectiveness in removing suspended solids, turbidity, oil, grease, and a few organic compounds (He et al., 2014). Compared to RO/NF, UF and MF membranes have a smaller footprint due to their compactness; hence, they are preferred in locations where space is very limited. As shown in Table 8.3 they can be applied as either a standalone process for the removal of pathogens, macromolecules, and suspended particles from wastewaters, or in a hybrid configuration as a pretreatment step before further downstream desalination to tolerate high feed salinity (He et al., 2014; Kim,

TABLE 8.2 Qualitative comparison of membrane processes for the treatment of petroleum wastewater, more stars indicate more favorable attributes.

Attribute	UF/MF	NF/RO	FO	MD	ED
Process flexibility (ability to handle high salinity feed)	Excellent	Poor	Good	Excellent	Good
Water quality	Poor	Excellent	Excellent	Good	Good
Salt rejection	Poor	Good	Excellent	Excellent	Poor
Energy efficiency	Excellent	Good	Poor	Good	Excellent
[a]Fouling propensity	Poor	Good	Excellent	Poor	Excellent
[b]Pretreatment requirements	Good	Poor	Poor	Good	Good

[a]Poor indicates that the membrane process usually has more sever fouling.
[b]Good indicates that the process requires less pretreatment.

TABLE 8.3 Application of ultrafiltration and microfiltration membranes for the treatment of oil and gas wastewater.

Feedwater characteristics	Process scheme	Performance
TDS: 23−94 mg L^{-1}, oil concentration: 32−180 mg L^{-1}, pH; 7.0−7.8	MF	Flux 1150−200 L/(m^2 h) (LMH), oil removal, 58%−82%
TDS: 200−2000 mg L^{-1}, oil concentration: 200−1000 mg L^{-1}, pH: 6.0−8.0	UF	Oil removal 99%, TOC removal 39%
TDS: 200−2000 mg L^{-1}, oil concentration: 200−1000 mg L^{-1} pH: 6.0−8.0	MF-UF-NF	Oil removal 99%, flux 400−200 Lm^{-2} h^{-1} bar^{-1}
TDS 30,000 mg L^{-1}, TOC: 0.78 mg L^{-1}; concentration of volatile organic compounds (VOCs): 2−98 μg L^{-1}	UF-MD	Flux 2.5 LMH, recovery rate 84%, VOCs rejection up to 100%, SEC 28.2 kWh m^{-3}
TDS: 23−94 mg L^{-1}, oil concentration: 32−180 mg L^{-1}, pH: 7.0−7.8	UF-NF	Oil removal 67%−80%; TOC removal 13%−27%
TDS; 6906 mg L^{-1}, turbidity; 135 NTU, COD 358 mg L^{-1}	Coagulation-UF	Permeate water with a turbidity of 0.18 NTU, COD 259.0 mg L^{-1}, and TDS 6490 mg L^{-1} COD: removal: 25%, removal of Si, Sr, and Ca: (20%−40%)

Kim, & Hong, 2018; Xiong, Zydney, & Kumar, 2016). In the former case, the UF/MF permeate can be utilized for reuse application in hydraulic fracturing and water flooding (Ebrahimi et al., 2009; Kim et al., 2018; Xiong et al., 2016)

MF membrane is mainly considered as a pretreatment stage for removing large suspended particles (>0.1 μm) from oily PW. The reduction of TSS of PW through MF significantly enhances the treatment efficiency of subsequent UF, NF, RO processes. MF can be classified as a cost-effective pretreatment method capable of removing dispersed oil droplets and large particles that remained after the primary treatment that is usually done through sedimentation or coagulation. Different polymeric and ceramic MF membranes are employed for PW treatment. Ebrahimi et al. (2009) reported that MF can attain as high as 82% oil removal from O&G wastewater. On the other hand, UF membranes have a pore size between 0.01 and 0.1 μm, which can effectively separate solids at a micro-size level. The UF process was reported to be superior in oil separation when compared to conventional physical separations (Ahmad et al., 2020). The key attributes of UF are high removal efficiency, lower energy consumption, and fewer chemicals demand (Ahmad et al., 2020; Nasiri, Jafari, & Parniankhoy, 2017). In a study by Kong et al. (2017), between 32.5% and 83.3% removal of DOM was reported for a coagulation-UF system, treating fracturing wastewater. In another work (Miller et al., 2013), the UF membrane coupled with RO achieved 99.85%−99.95% removal of hardness in PW from the Shale gas field.

Organic and colloidal fouling represents the real challenge to the widespread implementation of UF and NF for the treatment of petroleum wastewater. Many studies reported on the detrimental impact of colloidal fouling on the water flux of the membrane (Ahmad et al., 2020). A significant flux decline was observed in MF/UF membranes treating oily PW feed, highlighting this type of fouling as a constraint for the separation performance (Ahmad et al., 2020; He et al., 2014; Xiong et al., 2016). Therefore, to achieve better separation efficiency, the number of colloidal particles from raw PW must be reduced before treatment with UF or MF (Tong et al., 2019). Biological treatment has been applied as a pretreatment to membrane filtration to reduce the organic and salt concentrations in the feed stream. Riley, Oliveira, Regnery, & Cath (2016) demonstrated that the use of biologically active filtration (BAF) as a pretreatment to UF, resulted in lower membrane fouling, and the hybrid system showed over 99% and 94% removal of DOC and TDS, respectively. The performance of UF membranes can be also improved by altering the hydrophilicity and surface roughness of the membrane (Tong et al., 2019).

High-pressure-driven processes

In the RO process, water is transferred across a semipermeable membrane from a high concentration to a low concentration solution after applying hydraulic pressure to the high salinity solution side. RO membranes are widely applied for TDS removal from wastewater (Table 8.4) and they can reject dissolved and ionic components with a size of

TABLE 8.4 Application of reverse osmosis and nanofiltration membranes for desalination of oil and gas wastewater.

Feedwater characteristics	Process Scheme	Outcomes	References
TDS: 6000 mg L^{-1}, TOC: 120 mg L^{-1}, boron: 16 mg L^{-1}, hardness: 1–5; ammonia: 9.3 mg L^{-1}	Pretreatment-RO	Permeate with 143 mg L^{-1} TDS, 1–2 mg L^{-1} boron, 2–11 mg L^{-1} ammonia, and 2 mg L^{-1} TOC	–
TDS: 16,400 mg L^{-1}, TSS: mg L^{-1}, TOC: 540 mg L^{-1}, COD: 1.24 mmg L^{-1}; pH: 8.2, calcium 14.2 mg L^{-1}, magnesium 4.7 mg L^{-1}	MSBR-RO	TOC removal efficiency 92%–94%; water flux of 30 LMH, permeate with TDS 450 mg L^{-1}; oil and grasses below detection limits	–
TDS: 16,300–62,900 mg L^{-1}, TSS: 36.8–253 mg L^{-1}, TOC: 9.5–99.1 mg L^{-1}, pH: 6.6–8, calcium 454–6680 mg L^{-1}, magnesium 75.3–757 mg L^{-1}	UF-RO	Water recovery: 10%–50% at flux 2–14 LMH; salt rejection: 99.5%; inorganics rejection: 99.9%	–
TDS: 17,300 mg L^{-1}, DOC: 1360 mg L^{-1} turbidity: 110 NTU, Alkalinity 695 mg L^{-1}	Coagulation-NF	Water recovery: 50%, DOC:87%; TDS removal 75%	–
TDS: 12,331–29,618 mg L^{-1}, DOC: 35.5 mg L^{-1}	BAF-UF-NF	DOC removal between 78% and 99.6%	–
TDS: 18,900 mg L^{-1}, COD: 526.7 mg L^{-1}, turbidity: 32 NTU; Alkalinity (as CaCO3) 587 mg L^{-1}, pH: 7.8; Ca 2 + : 233 mg L^{-1}; Mg^{2+}: 26.9 mg L^{-1}	Coagulation-UF-NF	Flux 30–120 LMH; water recovery 50–85, 99.9% of turbidity, 94.2% of COD: 72.8%; removal of divalent ions: 91.7%	–
TDS: 13,462 mg L^{-1}, turbidity: 2.1 NTU, TOC: 9.1 mg L^{-1}, Conductivity 23,357 µs cm^{-1}	RO	Na$^+$ removal: 88.1%; rejection of Cl$^-$: 87.3%, rejection of multivalent ions 90%	–
TDS: 13,462 mg L^{-1}, turbidity: 2.1 NTU, TOC: 9.1 mg L^{-1}, Conductivity 23, 357 µs cm^{-1}	NF	Na$^+$ removal: 81.2%; rejection of Cl$^-$: 83.3%, rejection of multivalent ions 90%	–
TDS: 90–16000 mg L^{-1}, pH 6–8, Turbidly 0–5 NTU; Inorganic salt: 8000 mg L^{-1}; Boron 15–30 mg L^{-1}	MBR-RO-ion-exchange	Water recovery: 80%–90%; over 22 million barrels were treated and recycled. Product water had inorganic salt of 100 mg L^{-1}, boron of 0.75 mg L^{-1}; and organics below detection limits	–

0.0001 µm (Kim et al., 2018). The treatment efficiency of the RO desalination process depends on many factors such as characteristics of the wastewater feed, the target water recovery, membrane properties, and the operating conditions. Earlier researches on RO desalination of PW were unsuccessful due to poor treatment efficiency and lack of process integration (Doran, Williams, Drago, Huang, & Leong, 1999). Following that, the significant development in membrane properties and process operation has proved the potential of RO as a competitive treatment option of PW from O&G operations. Different polymeric membranes were tested by Fakhru'l-Razi et al. (2010) as posttreatment of biological process treating PW with high TOC. The RO removal efficiency of TOC was 92%–94% at a water flux of 30 L. m^{-2} h^{-1} (LMH)with product water quality was suitable for reuse applications. Although RO membranes have been applied for the treatment of feed solutions with salinity up to approximately 70,000 mg L^{-1} (Miller et al., 2013), they are only considered economically favorable for treating wastewater with TDS less than 30,000–45,000 mg L^{-1} (Henderson et al., 2011). This is because PW of high salinity requires applying hydraulic pressure that exceeds the allowable pressure of RO membranes. However, the salinity of O&G wastewater is usually of high concentration, rendering the RO process ineffective and infeasible for the treatment of high salinity feed streams (Tong et al., 2019). To withstand higher feed salinity, pretreatment is usually employed using membrane filtration (UF/NF), physicochemical, and biological processes.

NF membranes have been widely studied for water softening and multivalent ions separation (e.g., calcium, magnesium, sulfate, etc.). The operation of the NF membranes at high pressure results in a high selectivity level and makes them more susceptible to fouling. Therefore the feedwater should be pretreated to remove free suspended solids and organic compounds. Different studies have tested the NF membrane for oily PW treatment and resulted in promising outcomes (Chang, Li et al., 2019; Chang, Liu et al., 2019; Kong et al., 2018; Tong et al., 2019). Overall, NF membranes possess good salt rejection. While over 90% of Ca^{2+}, Mg^{2+}, and $SO4^{2-}$ were removed from hypersaline PW by NF membrane, only <24% removal was observed for Na^+ and Cl^- (Chang, Liu et al., 2019; Kong et al., 2018; Tong et al., 2019). The difference in the removal efficiency of these ions is assigned to the variation in the ion sizes and hydration energy (Kelewou, Lhassani, Merzouki, Drogui, & Sellamuthu, 2011). Furthermore, it has been demonstrated that higher operating pressure is favorable for higher ion removal (Riley et al., 2016). Effective removal of organics constitutes usually requires a combination of NF with other hybrid biological/or membrane filtration processes (e.g., UF) (Kong et al., 2017). Compared to RO membranes, NF usually achieves lower ion rejection rates (Chang, Liu et al., 2019; Kong et al., 2018; Tong et al., 2019). Mondal and Wickramasinghe (2008) compared the performance of commercial NF and RO membranes for TDS removal from PW stream. They reported that depending on the quality of the feed stream, and operating conditions, NF can provide a viable solution for PW treatment.

Due to their capability to reject a wide variety of pollutants, RO and NF can produce high water quality that complies with the requirement for reuse for hydraulic fracturing, direct discharge into aquifers, or agricultural irrigation (Chang, Li et al., 2019). However, both membranes (RO and NF) suffer from performance deterioration due to membrane fouling. To obtain a stable water flux, RO/NF desalination membranes usually require a pretreatment step which has been predominately based on UF/MF or biological membrane process (Chang, Li et al., 2019). The Pretreatment step is a key factor for the successful implementation of RO/NF, as it alleviates the impact of fouling and tolerating RO membranes for high feed salinity. Guo et al. (2018) applied a UF membrane as a pretreatment to RO for the treatment PW generated from gas field operations. The UF-RO system achieved over 98%, removal of TDS, COD, and Cl^-. In another study, Miller et al. (Miller et al., 2013) demonstrated a salt removal efficiency of >99% at low fouling behavior for a pilot-scale UF-RO treating petroleum wastewater generated from Texas wells.

Biological treatment using membrane bioreactor (MBR) is also widely investigated as an effective pretreatment step to RO/NF desalination (Adham et al., 2018; Riley, Ahoor, Regnery, & Cath, 2018; Shafer, 2011). This method is considered more cost-effective than other membrane processes due to its lower energy demand. Organics are effectively removed from the wastewater feed by MBRs where biomass is separated from treated effluent via a membrane barrier. Adham et al. (2018) investigated the use of MBRs for the treatment of produced and process water from Qatari processing facilities (COD 1300 mg L^{-1}, and 5200 mg L^{-1} of TDS). They reported a COD and TOC removals of 60% at membrane flux between LMH, with a minimal fouling effect on the flux. The test lasted for eight months and periodic cleaning with sodium hypochlorite was effective in removing any foulants from the membrane surface. However, tolerance to high feed salinity is the current bottleneck of MBRs and other biologically based membranes. Highly saline solutions imposed osmotic stress and plasmolysis of bacterial cells, which significantly reduces biological flocculation and biomass settlement (Kargi & Dincer, 1996).

Treatment with osmosis-driven processes

FO is classified as an osmotically driven membrane process where the natural osmotic pressure difference between two streams is exploited. In the FO process a high salinity solution *"known as draw solution (DS)"* is utilized to draw permeate (water molecules) from a low salinity feed solution across a semipermeable membrane barrier. The FO membrane forms a vigorous barrier to the organics and dissolved ions present in the wastewater feed. As more water permeates from the feed to the draw DS side, the osmotic pressure difference that drives the separation will be reduced. Therefore a further downstream process is required to recover the driving force by re-concentrating the DS, which then permits DS reuse and permeate separation (Fig. 8.2). The DS recovery step has been historically based on either, RO/ NF or a thermal membrane process (MD) for DS with very high salinity (TDS) when the operating pressure cannot be tolerated by RO/NF (Munirasu, Haija, & Banat, 2016). It is widely acknowledged that the DS recovery step is a technically challenging and economically intensive energy process.

Compared to pressure-driven desalination, the FO is featured by the lower fouling propensity, higher solute rejection, and lower energy consumption (Awad et al., 2019). These attributes primarily arise from the fact that no external hydraulic pressure is required for the FO process since it relies only on the osmotic gradient driving force. At lower applied pressure, less stress of fouling as well as lower energy is required to achieve the treatment. Furthermore, FO can tolerate a feed solution of salinity up to 200,000 mg L^{-1} TDS with low pressure requiring equipment (Tong & Elimelech, 2016). The use of FO for high saline PW and flowback water has been extensively demonstrated at laboratory and pilot scales (Awad et al., 2019). In all these cases, the technical viability of the FO technology is largely determined by the characteristics of the feedwater, the required quality of the product water, and the properties of the DS.

FO operational performance is robustly linked to the chosen DS, as an efficient DS helps in creating high osmotic pressure difference; leading to successful operation (Changmai et al., 2019). Indeed, there are around 500 inorganic

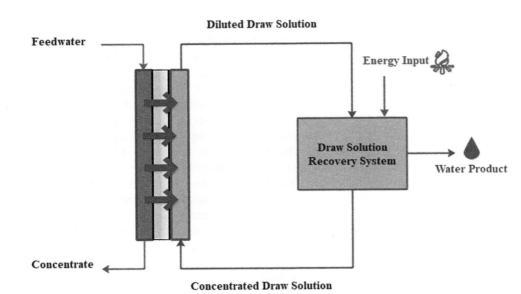

Diluted Draw Solution

Feedwater

Energy Input

Draw Solution Recovery System

Water Product

Concentrate

Concentrated Draw Solution

FIGURE 8.2 Forward osmosis process scheme.
Please see the online version to view the color version of the figure.

component options that can be utilized as DS and recovered by heating, RO, NF, MD, and precipitation by Ca(OH)$_2$ (Veil, 2015). Some organic DS such as ethanol, glucose, and fructose can be used, however, the efficiency of separation is not assured because of the low osmotic driving force generated (Ang et al., 2015). Further development of organic DS is worthy due to their advantageous biodegradability and rejection during the re-concentration stage. NaCl solutions at different concentrations were extensively used as DS for PW feeds (Awad et al., 2019).

The attainable water fluxes for osmotically based membranes relate mainly to the concentration driving force across the membrane surface. This driving force is determined by the properties of the DS (type and concentration) as well as the salinity of the feedwater. As this concentration difference increases, higher membrane fluxes can be obtained. The FO steady-state fluxes achieved at pilot-scale studies have ranged from <2.2 to approximately 8 LMH, however, larger flux values were reported at bench scale (Chakrabortty, Pal, Roy, & Pal, 2015). Water flux decline of the FO membrane due to fouling appears to be insignificant providing that proper pretreatment is used to reduce TSS and TOC of the FO feed stream. In that case, cleaning requirements for the FO process are generally benign, with no more than simple water flushing required in many studies (Awad et al., 2019). However, the presence of high TOC concentration in the feedwater can result in a significant flux decline, which necessitates the need for combined physical and chemical cleaning to restore the initial water flux.

Similar to pressure-based membranes, to obtain a high quality of the water product, pretreatment (usually with MBRs, UF, or MF) is required for FO membranes to assure successful operation by reducing suspended solids and organic before the FO stage. FO-based integrated systems usually provide a high quality of product water, that can potentially meet the requirements for different reuse applications. FO-RO systems have been demonstrated to provide a permeate product water with quality that can meet standards for drinking, irrigation, and internal reuse for hydraulic enhanced oil recovery (Awad et al., 2019). FO membranes have variable contaminants rejection as shown in Table 8.5. For instance, around 99% of all cations, anions were removed from a PW stream from the Denver-Julesburg basin, while only 92.9% rejection of boron was reported for an FO-RO system (Maltos et al., 2018). FO-distillation system treating PW from Pennsylvania State well, showed water permeate with TDS, chlorides, and barium, less than 300, 140, 0.025 mg L^{-1}, respectively which complies with requirements for surface discharge (McGinnis, Hancock, Nowosielski-Slepowron, & McGurgan, 2013).

Furthermore, FO coupled with MD has been demonstrated to be an effective hybrid system for the treatment of oily wastewater (Silva et al., 2017). FO-MD hybrid system showed 90% recovery of high saline petroleum wastewater with a complete rejection of oil and NaCl, as well as demonstrated the feasibility of water reuse (Zhang, Wang, Fu, & Chung, 2014). Another tested FO-MD system achieved a 90% recovery of shale gas flowback water at a water flux of 18 LMH after 15 h of continuous operation (Li et al., 2014). FO process has been tested at a pilot-scale level for the treatment of real PW from different O&G fields (Table 8.5) where the process scheme is illustrated in Fig. 8.3. The FO technology was examined in a membrane brine concentrator (MBC) process for the treatment of 70,000 mg L^{-1} PW using ammonium carbonate DS (NH$_3$/CO$_2$) (McGinnis et al., 2013). The MBC process comprises feed pretreatment, FO-based separation, and thermal recovery of the DS relying on a dual column distillation-condensation process.

TABLE 8.5 Application of forward osmosis-based processes for oil and gas wastewater treatment.

Feedwater	Process Scheme	Outcomes	References
PW from gas fields in Pennsylvanian. TDS 70,000 mg L^{-1},	FO-MBC[a]	62% of PW was recovered at an average flux of 2.6 LMH	–
TDS: 6490 mg L^{-1} Conductivity: 11,290 μS cm^{-1} Mg:18 mg L^{-1}; Ca: 140 mg L^{-1}; Alkalinity (as CaCO$_3$): 283 mg L^{-1}; COD: 358 mg L^{-1}, turbidity: 135 NTU	FO + VMD	FO served as pretreatment and the system achieved water recovery of 90% while producing high water quality that meets the standard for drinking water	–
PW from Haynesville shale gas field	FO-RO	85% of PW was recovered	–
TDS: 6906 mg L^{-1}; Conductivity: 11.29 mS cm^{-1}, pH: 8.12; Turbidity: 135 NTU; Calcium (mg L^{-1}) 140.2 mg L^{-1} Magnesium: 18.05 mg L^{-1}: CO: 358.5 mg L^{-1}	UF + FO	Water flux: 5–20 LMH; FO water flux ratio 80%–100%	–

[a]MBC: Membrane brine concentrator.

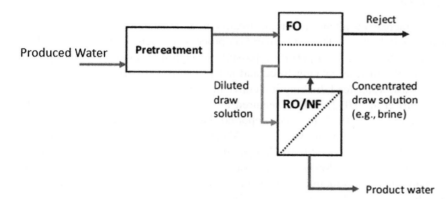

FIGURE 8.3 Forward osmosis–reverse osmosis process scheme. *Please see the online version to view the color version of the figure.*

When the quality of the final water product is not a concern, the FO can be operated as a standalone process with no need for the DS recovery step. The operation of FO without DS recovery is described by osmotic dilution/osmotic concentration where the diluted DS will be taken for another use and feed wastewater will be minimized (Cath, Hancock, Lundin, Hoppe-Jones, & Drewes, 2010). A recent pilot-scale FO system for oilfield wastewater reclamation was developed by Bear Creek Services and Hydration Technology Innovation (HTI) located in Los Angeles and Oregon in the United States, respectively (Coday & Cath, 2014). The process was operated in OD mode where drilling wastewater can be reclaimed for reuse as a completion fluid in hydraulic fracturing. The system was named the green machine and was the first energy-efficient system able to recycle and treat millions of gallons of fresh water used in the oil industry. The testing showed the process capabilities in concentrating O&G wastewater to more than 3.5 times the original salinity while achieving 75% feed recovery. Additionally, FO as the OD process was operated at a pilot scale for the treatment of reserve pit wastewater using 20 cylindrical FO membranes and 26% w/w draw solution. The OD system demonstrated the capability of recovering 70% of wastewater and rejecting all suspended solids, solutes, and heavy metals; allowing its reuse as hydraulic fracturing fluid (Coday et al., 2014).

On the other hand, FO was tested as an OC process for the volume reduction of drilling wastewater at a bench-scale level by Hickenbottom et al. (2013) in the United States. The OC trials showed promising outcomes illustrated by achieving 80% feed recovery with an initial water flux of 14 LMH which declined to 2 LMH at the end. Furthermore, the OC process mode was also implemented for the concentration of PW from Qatari's oil and gas industries at a bench-scale (Minier-Matar et al., 2015). The project demonstrated a 50% volume reduction of PW effluent and encouraged scaling up the testing to a higher level. Therefore the performance of the OC process operated as shown in Fig. 8.4 was investigated at a pilot-scale level by Jalab et al. (Jalab, Awad, Nasser, Minier-Matar, & Adham, 2020). The OC pilot-scale plant was successfully capable of recovering up to 90% of feed at an average water flux of 2.2 LMH using cellulose triacetate hollow fiber membrane type.

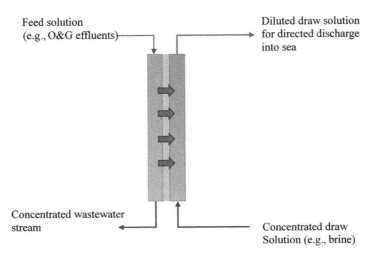

Feed solution
(e.g., O&G effluents)

Diluted draw solution
for directed discharge
into sea

Concentrated wastewater
stream

Concentrated draw
Solution (e.g., brine)

FIGURE 8.4 Osmotic concentration process scheme.
Please see the online version to view the color version of the figure.

Treatment with thermally driven processes

In recent years, the use of MD has garnered increasing attention for the treatment of high-strength and hypersaline petroleum wastewater (Alkhudhiri, Darwish, & Hilal, 2012). The MD separation process is based on the thermal differences (i.e., vapor pressure gradient) between hot feed and cold permeate passing through a hydrophobic microporous membrane barrier (Fig. 8.5). Since MD utilizes the thermal driving force and not pressure, the performance of MD is not affected by the variation in salinity of the wastewater feed, making it a convenient desalination process for PW with medium to high TDS concentrations (Alkhudhiri et al., 2012). MD is featured by the utilization of low-grade thermal energy (usually less than 100°C) (Ezugbe & Rathilal, 2020). It has been reported that MD can concentrate wastewater feed (salinity up to 350,000 mg L^{-1}) to an extent comparable to Mechanical Vapor Compression (MVC) at relatively lower temperatures (30°C−90°C) and pressure (Ahmad et al., 2020). Compared to other desalination processes (e.g., thermolytic FO, MVC), MD has a simpler configuration and lower capital costs (Rao & Li, 2016). Furthermore, due to their membrane structure (larger pore size), and operation principles (no need for external pressure), MD usually experiences less severe fouling than conventional RO. Nevertheless, considerable flux decline can be observed due to inorganic scaling when the MD is operated at relatively high water recoveries (Boo, Lee, & Elimelech, 2016).

Fig. 8.5 shows the various MD configurations adopted for PW treatment including direct contact MD (DCMD), air gap (AG), sweeping gas (SG), and vacuum MD (VMD). MD membrane fabrication has been predominantly based on polymeric materials with flat sheet DCMD being the most common module studied for wastewater treatment. Although there have been different operation modes of MD (Fig. 8.5), DCMD is the most reported process for PW treatment (Ezugbe & Rathilal, 2020). In DCMD the feed and permeate directly contact the membrane. Rao and Li (2016)demonstrated the successful operation of DCMD for treatment of PW with 133,000 mg L^{-1} TDS at a water flux of 43 LMH and feed temperature of 80°C. The investigation also revealed that increasing the TDS to 280,000 mg L^{-1} can maintain a high permeate flux of around 15 LMH. The VMD configuration was also feasible for achieving more than 99% of salt rejection from PW at a permeate flux of 5.5 LMH initially up to 2 h operation time. Thereafter and after reaching 8 h operation, the salt rejection and flux were reduced to 78% and 1.65 LMH, respectively (Heidarpour, Shi, & Chae, 2015).

The water recovery of MD can be increased by coupling MD with crystallization unit which decreases inorganic scaling and consequently reduces the impact of membrane wetting (Kim, Kwon, Lee, Lee, & Hong, 2017). Further, the development and use of anit-fouling membrane materials are crucial to enhance the performance of MD by mitigating inorganic scaling (Chang, Li et al., 2019). The technical viability of MD for generating high-quality water products from highly saline PW has been demonstrated in several studies (Table 8.6). Jang, Jeong, and Chung (2017)reported on the capability of MD to achieve >99.9% removal of all ions (including Ca^{2+}, Mg^{2+}, Na^{+}, Sr^{2+}, Ba^{2+}, Cl^{-}, and Br^{-}) from FPW. However, MD is confronted by membrane wetting phenomena which significantly undermines the quality of the product water (Kim et al., 2017). Membrane wetting occurs due to the presence of low surface energy contaminants (e.g., surfactants) in the wastewater feed. These contaminants reduce the surface tension of the feedwater, causing water to flood into the membrane pores and MD permeate (Ezugbe & Rathilal, 2020). The challenge of membrane wetting has been tackled by optimizing surface wettability through the development of highly resistant membrane materials

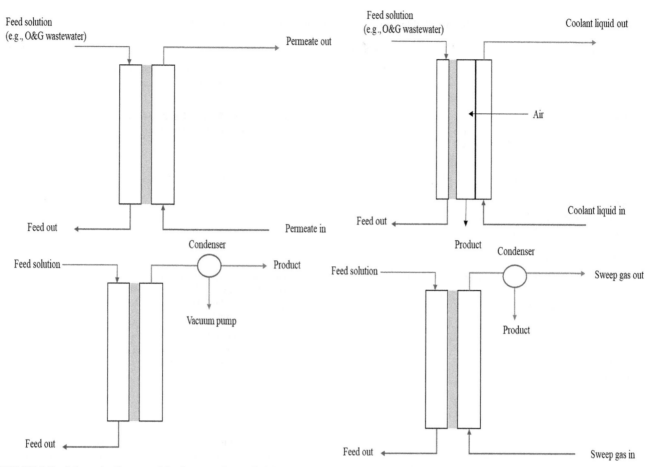

FIGURE 8.5 Schematic diagram of the four membrane distillation (MD) configurations: (A) direct contact MD, (B) air gap MD, (C) sweeping gas MD, and (E) vacuum MD (Munirasu et al., 2016).
Please see the online version to view the color version of the figure.

such as omniphobic membranes. These membranes have demonstrated performance stability while treating highly saline wastewater with high surfactant content (Boo et al., 2016).

Pervaporation (PV) is another thermal-based desalination process, which combines the concept of membrane permeation and evaporation (Subramani & Jacangelo, 2015). The separation driving force in this process is created after applying vacuum pressure to the permeate side and keeping the feed under the atmospheric pressure conditions (Fig. 8.6). The membrane forms a separation barrier between the vapor and liquid phases. Several studies have demonstrated the applicability of this process for the treatment of PW streams. PV process is capable of rejecting micro-pollutants from wastewater such as naphthalene, phenol, and 1,2-dichlorobenzene where a high rejection percentage is achieved for higher molecular sizes (Sule, Templeton, & Bond, 2016). Permeate flux of pervaporation is widely ranged from 1.7 to 15.0 mL^{-1}(m^2 h) for feed solution of 70,000 mg L^{-1} NaCl, (Sule et al., 2016). Several membrane types of polyether ester and cellulose triacetate were tested for the PV process reuse for irrigation by treating PW feed stream (Muthu & Brant, 2015). Compared to RO, pervaporation can provide salt rejection of 99.8%, while retaining more ions (Na$^+$, Cl$^-$, Ca^{2+}, As^{3+}, B^{3+}, and F$^-$). Consequently, the quality of the product water of PV can easily meet standards for drinking water and irrigation use (Muthu & Brant, 2015).

Treatment with electrically driven processes

Electrodialysis (ED) is an electrically driven process that applies the principles of a direct current electric field with the concept of ion-exchange membranes to remove TDS and other dissolved constituents from the wastewater feed. Anion exchange membrane (AEM) and cation exchange membrane (CEM) are utilized to preferentially anions and cations respectively while rejecting the diffusion of oppositely charged ions (Chang, Li, et al., 2019). Fig. 8.7 shows a schematic diagram of the ED desalination process where ions transport from the feed stream to the concentrate through the

TABLE 8.6 Application of membrane distillation for desalination of oil and gas wastewater.

Feedwater characteristics	Process Scheme	Outcomes	References
TDS: 13,3000–28,000 mg L^{-1}; oil/grease:100–260 mg L^{-1}; TSS: 1–3200 mg L^{-1}; VOCs: 35 mg L^{-1}	DCMD	Water flux: 43–15 LMH; minimal fouling effect: No pretreatment was required	–
TDS:10,000 mg L^{-1}; Conductivity: 15000 µS cm^{-1}; pH: 9; Ca^{2+}: 15 mg L^{-1}, Mg^{2+}: 38 mg L^{-1}; Na: 4148 mg L^{-1}; Cl^-: 5470 mg L^{-1}	VMD	Permeate flux: 5.5 LMH; salt rejection: 99.9%. Severe membrane fouling	–
TDS: 101,010–101,360 mg L^{-1}, pH: 5.34, Ca^{2+}: 11,820–12,675 mg L^{-1}, Mg^{2+}: 452–530 mg L^{-1}; Na: 28,060–32,030 mg L^{-1}; Sr^{2+}: 2483–2595 mg L^{-1}; Ba^{2+}: 1843–1982 mg L^{-1}; Cl^-: 66,778–73,891 mg L^{-1}	DCMD	>99% removal of all ions	–
TDS: 6490 mg L^{-1}, COD: 259.0 mg L^{-1}; Conductivity: 11,300 µS cm^{-1}; K: 393 mg L^{-1}; Na: 2,109 mg L^{-1}	UF + FO + MD	Water flux; 10–25 LMH; water recovery; 90%. Product water had conductivity of 5 µS cm^{-1}, COD 0.9 mg L^{-1}, K 0.5 mg L^{-1}, and Na 0.12 mg L^{-1}	–
TDS: 150,000 mg L^{-1}; Alkalinity (as $CaCO_3$) 200 mg L^{-1}; oil and grease: 0.018 mg L^{-1}; Na 44000 mg L^{-1}; Ca 9800 mg L^{-1}	MD + crystallization	Water recovery 62.5%, however, water recovery without pretreatment was only 20%–25% due to membrane wetting; water flux: 35 LMH	–

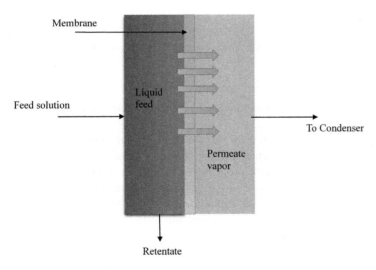

FIGURE 8.6 Pervaporation process scheme.
Please see the online version to view the color version of the figure.

alternating AEM and CEM under the electrical current field. This results in a permeate (dilute out) with lower TDS while the salinity of the concentrate stream increases. Compared with pressure-driven membrane desalination, ED usually has higher water recovery and less fouling propensity, and fewer pretreatment requirements (Chang, Li et al., 2019). However, the energy consumption of ED represents a real challenge for desalination of high salinity feed streams as the energy required for desalination is proportionally related to the concentration of ions in the feedwater. It has been claimed that desalination of PW with 40,000–90,000 TDS mg L^{-1} by the ED requires energy comparable to that of the vapor compression process (McGovern & Lienhard, 2014). The energy consumption of ED is estimated by 0.1–0.15 kWh per pound of salt removed for desalination of 30,000–72,000 TDS mg L^{-1} feedwater at stack potential of 5 V (Chang, Li et al., 2019). Therefore ED is only considered feasible for the treatment of medium to low salinity PW, or as partial desalination for high salinity stream.

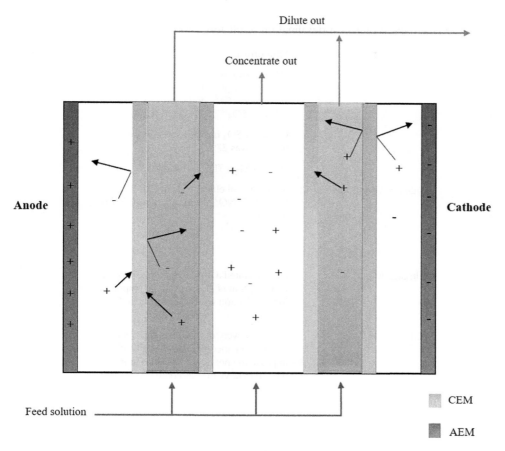

Dilute out

Concentrate out

Anode

Cathode

Feed solution

CEM

AEM

FIGURE 8.7 Schematic diagram of electrodialysis (Munirasu et al., 2016).
Please see the online version to view the color version of the figure.

Table 8.7 presents studies that applied ED for the desalination of O&G wastewater. An ED process, treating PW with a salinity of ≤ 5500 TDS mg L^{-1}, has shown the potential to generate permeate products with quality that can meet standards for drinking water (Subramani & Jacangelo, 2015). However, if the salinity of the feed is high (PW, $>60,000$ mg L^{-1}), obtaining a high-quality water product is not feasible by the ED process (Sirivedhin, McCue, & Dallbauman, 2004). The salt removal efficiency by ED desalination of PW is variable. For example, only 27% TDS removal was reported for feedwater of 65,000$-$100,000 TDS mg L^{-1}, using Na$_2$SO$_4$ or NaCl as electrolytes (concentration 90,000$-$125,000 mg L^{-1}) (Peraki, Ghazanfari, Pinder, & Harrington, 2016). However, in another study (Peraki & Ghazanfari, 2014), around 35%$-$90% of TDS removal was reported for ED systems treating flowback water containing up to 60,000 TDS mg L^{-1}. In the study by Hao, Huang, Gao, and Gao (2015) an ED system coupled with coagulation and flocculation showed that the removal of the various cation (Ca^{2+} > Na$^+$ > Mg^{2+}) and anion (Cl$^-$ > NO$_3^-$) were around 91%, while the removal of SO^{4-} was slightly lower (84.3%). In fact, the capability of ED to remove ions from the wastewater is controlled by their concentrations in feedwater at a low applied voltage (Sirivedhin et al., 2004). Furthermore, the performance of ED desalination can be improved by increasing solution pH, electrolyte conditions (temperature and concentration), and protection of the cathode membrane (Chang, Li et al., 2019).

Over the past few years, several other electrically based processes have been applied for the treatment of O&G wastewater (Chang, Li et al., 2019). These processes have included reverse electrodialysis (RED), and electrodeionization. In contrast to ED, RED does not use energy to remove salts from the feedwater. Instead, it harnesses the electricity generated by mixing two solutions with different concentrations (Hong et al., 2015). The performance of RED can be enhanced by using IX wafers in the ED cell to reduce the effect of module spacer (Hong et al., 2015). In membrane capacitive deionization (CDI), charged organic and inorganic constitutes are adsorbed by electrodes under applied electrical potential. The technical viability of CDI desalination is determined by the adsorption capacity of the electrode. The performance of membrane CDI desalination of low salinity feed (TDS < 3500 mg L^{-1}) has been evaluated at a pilot-scale where favorable energy consumption was reported for CDI compared to RO (Hong et al., 2015). Furthermore, the superior performance of membrane CDI was reported in several studies with a salt removal efficiency of membrane CDI being 32.8%$-$55.9% less than that of CDI cell (Zhao, Porada, Biesheuvel, & Van der Wal, 2013).

TABLE 8.7 Application of electrically driven membranes for desalination of oil and gas wastewater.

Feedwater characteristics	Process scheme	Outcomes	References
TDS: 5500–97,555 mg L^{-1}; pH:6.5–8.2; K$^+$: 29–130 mg L^{-1}; Na$^+$: 1,711–38,114 mg L^{-1}; Cl$^-$: 1000–58,864 mg L^{-1}; Mg^{2+}: 18 mg L^{-1}	ED	TDS removal increased linearly with applied voltage. ions removal efficiencies: Ca^{2+} and Mg^{2+} > K$^+$ >Na$^+$ > SO$_4$$^{2-}$> HCO^{3-} >Cl$^-$	–
Flowback water from Marcellus wells with TDS of 65,000–100,000 mg L^{-1}	ED	Using Na$_2$SO$_4$ or NaCl solution the TDS removal was 27%	–
TDS: 60,000 TDS mg L^{-1}; pH: 6	ED	35%–90% of TDS removal	–
Conductivity: 22,400 µS cm^{-1} TSS: 80 mg L^{-1}; pH:7; COD: 10,873 mg L^{-1}; BOD: 5477 mg L^{-1}; Turbidity: 2000 NTU; Ca^{2+}: 29–130 mg L^{-1}; Na$^+$: 055 mg L^{-1}; Cl$^-$: 6892 mg L^{-1}; Mg^{2+}: 21.7 mg L^{-1}	Coagulation–flocculation-ED	91% removal efficiencies of Ca^{2+}, Na$^+$, Mg^{2+}, Cl$^-$, NO^{3-}; removal of SO$_4$$^{2-}$ was 84.3%	–
TDS of 5800–100,000 mg L^{-1}	Multistage ICP	Demonstrated a 50% reduction of power consumption of ED by utilizing unipolar ion conduction which also enhanced salt removal	–
Conductivity: 162,000 µS cm^{-1}; pH: 6.15; Turbidity: 2200 NTU; COD: 54,100 mg L^{-1}; TSS: 21,300 mg L^{-1}; Sodium: 43,000 mg L^{-1}; Calcium: 8100 mg L^{-1}	REDI	Water recovery of 80%. Ion exchange wafers were incorporated in RED cells to improve the net power densities from 0.01 to 0.32 W m^{-2} as compared to RED	–

Ion-concentration polarization desalination (ICP) is usually applied to treat hypersaline solutions (up to 100,000 mg L^{-1} of TDS) (Subramani & Jacangelo, 2015). It has been reported that the ICP process can treat wastewater streams with TDS up to 100,000 mg L^{-1} at the salt rejection of 70% with lower energy demand as compared to MVC, and less fouling propensity than the ED process. He et al. (2014), Kim et al. (2018) and Xiong et al. (2016) demonstrated the feasibility of ICP treat feedwater of TDS approximately 30,000 mg L^{-1} with a recovery rate of 50% and salt rejection of 99% at energy demand less than 3.5 kWh m^{-3}.

Existing challenges and potentials of membrane-based processes

Although the research on membrane technologies for wastewater treatment has been active for decades (Koltuniewicz, Field, & Arnot, 1995), successful implementation of these technologies for real-world applications is still hindered by some challenges related to membrane fouling, energy consumption, and lack of experience with full-scale and pilot-scale demonstrations.

Fouling is defined as the accumulation of contaminants (colloids, organic, inorganic, or bio-foulants) on or into the surface of the membrane. This phenomenon inevitably occurs in all membrane processes. However, its severity and impact on membrane performance are different from one membrane process to the other. Membrane fouling is significantly influenced by the characteristics of the feedwater, membrane properties, and hydrodynamics or operating conditions of the membrane (Ezugbe & Rathilal, 2020). While the performance of membrane filtration (UF, MF), NF, and RO are significantly affected by fouling, osmotically and thermally driven membranes appear to be less impacted by foulant accumulation of the membrane surface. This is primarily due to the high pressure imposed by the pressure-driven membrane, which leads to a higher risk of fouling stress on the surface of the membrane. However, in all cases, fouling represents a challenge that must be addressed to accelerate the widespread implementation of membrane-based treatment of petroleum wastewater. More studies are still needed to develop a deep understanding of fouling mechanisms. It is widely acknowledged that membrane fouling can be due to either pore-blocking that causes sudden initial flux drop, or cake formation, which leads to gradual flux decline with operation time. Fouling can be mitigated by either developing new fouling resistant membrane materials or modifying existing membranes to tolerate specific conditions. Researchers also need to develop and study effective antiscalants that have the potential to alleviate the impact of inorganic scaling in MD to enhance its performance for the treatment of high-strength wastewaters. Hydrodynamics of the membrane can also control the mass transfer of foulants to the surface of the membrane, which can be achieved by playing with the module design, and the operating conditions to improve turbulence and mixing conditions. Furthermore, the development of an effective cleaning protocol and reagents can considerably control the detrimental effects of membrane fouling.

Another inherent phenomenon to all membrane processes is the concentration polarization (CP), which results from the accumulation of retained contaminants on the surface of the membrane. These contaminants can form a CP layer that emerges a concentration gradient at the membrane/solution interface. As a result, a higher backward diffusion rate of solute into the bulk solution is observed with greater resistance to permeate flow until a balance between these two components is reached. In pressure-driven membranes, CP incurs an additional osmotic pressure and considerably reduces the driving force and water flux. Furthermore, CP results in increased forwards salt diffusion through the membrane (e.g., lower solute rejection), as well as increased risk of membrane scaling or fouling. Operation at higher pressure can enhance the attainable permeate flux; however, a higher concentration of contaminants is obtained at higher pressure. Consequently, more severe CP and higher reverse solute flux of the solute diffusion backward to the bulk solution. In the case of electrically driven membranes, the concentration difference between the membrane solution interface results in a lower gradient of electric potential in the membrane, consequently, a lower separation rate and higher power consumption. Similar to fouling mitigation, CP can be generally controlled by optimizing the operating conditions, membrane material development, and module design. Higher volumetric flow rates, turbulence conditions are favorable for lower CP. These conditions lead to better solution mixing and reduce the thickness of the boundary layer, where the concentration gradient exists.

Although membrane processes are generally featured by low energy demand and costs, the applications of membrane processes for the treatment of high-strength O&G wastewater is still a challenge. This is because standalone membrane processes have a salinity limit and they need a pretreatment step, which incurs additional operating expenditure to the overall treatment costs. A system of MF coupled with IX for the treatment of wastewater was reported to cost \$18.4 m^{-3}, this is around 10−25 times greater than the cost of conventional RO desalination (Jiang, Rentschler, Perrone, & Liu, 2013). Similarly, an FO pilot unit for wastewater desalination showed an energy consumption of 275 kWh m^{-3} product water at a feed recovery rate of 64% (McGinnis et al., 2013). In fact, the energy consumption of FO-based O&G treatment has ranged from 15 to 150 kWh m^{-3}, which is significantly greater than RO and NF membranes (0.5−16 kWh m^{-3}) (Coday, Miller-Robbie, Beaudry, Munakata-Marr, & Cath, 2015). The operational energy of electrically based membranes appears to be lower than pressure-based membranes (Kim et al., 2016). Furthermore, MD can utilize low-value heat sources even though MD consumes higher specific energy compared to conventional RO desalination. These sources include industrial waste heat, solar heat, and geothermal heat (Alkhudhiri et al., 2012). A recent review on FO-based desalination processes (Awad et al., 2019)revealed that successful implementation of the FO technology is confronted by the energy requirements to recover the DS. This energy has to be reduced by 40%−50% in order for the FO to compete with conventional RO desalination. However, FO is proved to be energetically favored for specific circumstances in applications where there is no need to recover the DS, that is, the OC process.

Despite the recent surge of interest in membrane processes for the treatment of petroleum wastewater, implementation of these technologies at a large/full scale has been limited. There exist very few large installations operating at full scale for real-life applications, and there remain a limited number of commercial suppliers of dedicated membranes and systems for FO and MD (Alkhudhiri et al., 2012; Awad et al., 2019). Although many integrated systems have been proposed and tested at a laboratory scale such as an FO and MD tolerated for high salinity feedwater, performance data provided at a small scale can not be applied directly to full-scale installations. Currently, limited experience in the challenges of full-scale operation hinders the widespread applications of membrane technologies for PW treatment. Therefore researchers have to focus more on scaling up existing membrane processes and evaluating their viability under prevailing conditions to identify barriers to real-world industrial implementation. Extensive research at the pilot scale might also allow for optimization of the operating conditions at representative conditions under real wastewater matric while testing the robustness of the proposed technologies for treating a large volume of real wastewater for an extended period of operation.

Notwithstanding the above-mentioned challenges, membrane technologies have proved the capability of effective treatment of both inorganic and organic contaminants from the wastewater, while producing water quality that can potentially meet the standards for either surface discharge or reuse applications (e.g., hydraulic fracturing, and agricultural irrigation). Furthermore, membrane technologies have high process flexibility demonstrated by providing consistent product water quality regardless of influent variations. Disposal reinjection into wells has been a common practice for PW management, however, the capacity of underground aquifers is limited. Therefore, with the continual exploration and operation of unconventional wells, exploring alternative wastewater management schemes is indisputable. This includes either wastewater minimization, or reuse for internal and external applications. The relatively new OC technology can provide a suitable wastewater minimization option before rejection into the wells, while other integrated membrane systems such as FO-NF, UF-RO can be deployed effectively for wastewater reuse for agriculture purposes.

Sustainability consideration of membrane processes

Sustainable development of membrane-based treatment of O&G wastewater would promote the rapid widespread commercial applications of these technologies. For this duty, it is vitally important to consider the integration of environmental, economic, and technical aspects of the development of membrane technologies. Considering such factors can build confidence in these processes and make them appealing to major industrial players which would eventually promote their adoption in real-world applications.

In 2018 a survey of PW treatment stakeholders was conducted by People (2018) to establish priorities for wastewater management schemes or process technology by Qatari industry problem holders. Influential factors included the risk of a process failure, process flexibility, waste generation, overall cost, environmental impact, and energy consumption. Respondents included OEM/technology suppliers, Oil company employees, Consultants, Contractors, and Academics. Survey outcomes revealed the perceived importance of the low risk of process failure and cost and the perceived relative unimportance of energy consumption. Based on the outcomes of this survey the reliability of the technology is thus of paramount importance. This response is to be expected for any stakeholder directly involved with oil platform operations, where any process failure and the associated downtime incurs a very significant financial penalty. It is likely that, for such applications, safety would have scored as highly was it provided as a possible answer. Therefore, as stated earlier, the focus on scaling up existing membrane processes with pilot/full-scale testing to evaluate the process viability, flexibility and associated risk of process failure under representative conditions are vital for accelerating commercialization of the technologies.

While the operation with membrane processes has low associated health and safety risks, serious environmental implications can arise from waste generation, CO_2 emission, and the release of chemical substances. MBRs can produce large amounts of solid wastes, also considerable waste disposal is generated from RO and NF if they are operated at low water recovery. The release of chemical substances when the morphology of the membrane changes under harsh environmental conditions is another factor to be considered. The extent of ions releasing is directly related to properties of membrane materials (such as crystalline structure and particle size), and the operating conditions (such as the pH and temperature) (Muthu & Brant, 2015). Few studies have been conducted to focus on the improvement of the chemical stability of polymeric membranes (Kamali, Suhas, Costa, Capela, & Aminabhavi, 2019). Finally, measures have to be taken to address the CO_2 emission associated with the production, installation, and operation of membranes for the intended applications.

As far as the economic feasibility is concerned, operational costs arising from energy consumption is still the bottleneck of various membrane processes for wastewater treatment. While the use of pressure-driven membrane and conventional FO for treatment of petroleum wastewater has been restricted by the excessive energy use, FO-based process for applications where the DS recovery step is obviated can potentially provide an energy-efficient solution for wastewater management. In addition, the energy efficiency of the FO treatment of wastewater arises in the application of combined desalination and wastewater treatment, when a comparatively low salinity stream is available for providing osmotic dilution of a highly saline feedwater (such as PW).

Conclusions

Oil and gas exploration and production generate significant volumes of wastewater. The increasing expansion of these operations highlights the need for reliable and resilient processes for wastewater management. Membrane-based technologies have been explored for this duty for decades. These technologies have included pressure, osmosis, thermal, and electrical-based processes. Effective and efficient membrane treatment of petroleum wastewater usually requires hybrid systems. While UF/MF membranes are used as a typical pretreatment to downstream RO, FO, or MD to remove suspended solids, oil, and greases, MBRs are widely acknowledged as effective pretreatment to reduce organic content and improve the performance of posttreatment RO/NF, and FO membranes. On the other hand, RO and MD have been usually applied as posttreatment to FO-based systems for DS recovery purposes.

The use of membrane technologies for the treatment of petroleum wastewater have been extensively studied for feed source with various water quality at different operating conditions. Despite this surge of interest, currently, several challenges present real barriers to widespread industrial applications. These challenges mainly relate to membrane fouling, CP, energy requirements, and limited experiences with full/large-scale installations.

Membrane fouling and CP can result in performance deterioration, high RSF, and increased chemical consumption for cleaning which eventually impacts the economic feasibility of the process. Fouling and CP are typical challenges to all membrane process, however, their severity and impact on membrane performance are more obvious in pressure-driven membranes and MD as compared to low fouling processes such as FO and (ED.) The high fouling propensity of the pressure-driven membrane (especially RO and NF) mainly arises from the high pressure imposed on these membranes. On the other hand, successful implementation of FO is still challenged with the requirement to recover the DS which has been based on intensive energy processes (RO, MD) Nevertheless, special cases of FO where the DS recovery is obviated have demonstrated the technical and economic viability for wastewater minimization. Although MD has proved feasibility to treat O&G wastewater and can be effectively applied in combination with FO for DS recovery, successful implementation of MD is currently restricted by membrane wettability which significantly lower the treatment efficiency of the MD process.

Although several integrated membrane systems have been studied at a bench scale for O&G wastewater treatment, performance data of these hydride processes at a small laboratory scale cannot be adopted directly for full-scale installations. The application of such integrated processes at a large scale under representative operating conditions is still limited. Therefore future research studies should be directed toward scaling up existing technologies to demonstrate their technical viability and economic feasibility under conditions prevailing to those in the industry. This can allow for the identification of the barriers to real-world implementation while optimizing the operating conditions which will facilitate the adoption of membrane technologies for real-world applications.

References

Abousnina, R. M., Nghiem, L. D., & Bundschuh, J. (2015). Comparison between oily and coal seam gas produced water with respect to quantity, characteristics and treatment technologies: A review. *Desalination and Water Treatment, 54*(7), 1793–1808. Available from https://doi.org/10.1080/19443994.2014.893541.

Adham, S., Hussain, A., Minier-Matar, J., Janson, A., & Sharma, R. (2018). Membrane applications and opportunities for water management in the oil & gas industry. *Desalination, 440*, 2–17. Available from https://doi.org/10.1016/j.desal.2018.01.030.

Ahmad, T., Guria, C., & Mandal, A. (2020). A review of oily wastewater treatment using ultrafiltration membrane: A parametric study to enhance the membrane performance. *Journal of Water Process Engineering, 36*. Available from https://doi.org/10.1016/j.jwpe.2020.101289.

Alkhudhiri, A., Darwish, N., & Hilal, N. (2012). Membrane distillation: A comprehensive review. *Desalination, 287*, 2–18. Available from https://doi.org/10.1016/j.desal.2011.08.027.

Ang, W. L., Mohammad, A. W., Hilal, N., & Leo, C. P. (2015). A review on the applicability of integrated/hybrid membrane processes in water treatment and desalination plants. *Desalination, 363*, 2–18. Available from https://doi.org/10.1016/j.desal.2014.03.008.

Awad, A. M., Jalab, R., Minier-Matar, J., Adham, S., Nasser, M. S., & Judd, S. J. (2019). The status of forward osmosis technology implementation. *Desalination, 461*, 10–21. Available from https://doi.org/10.1016/j.desal.2019.03.013.

Boo, C., Lee, J., & Elimelech, M. (2016). Omniphobic polyvinylidene fluoride (PVDF) membrane for desalination of shale gas produced water by membrane distillation. *Environmental Science and Technology, 50*(22), 12275–12282. Available from https://doi.org/10.1021/acs.est.6b03882.

Cath, T. Y., Hancock, N. T., Lundin, C. D., Hoppe-Jones, C., & Drewes, J. E. (2010). A multi-barrier osmotic dilution process for simultaneous desalination and purification of impaired water. *Journal of Membrane Science, 362*(1–2), 417–426. Available from https://doi.org/10.1016/j.memsci.2010.06.056.

Chakrabortty, S., Pal, M., Roy, M., & Pal, P. (2015). Water treatment in a new flux-enhancing, continuous forward osmosis design: Transport modelling and economic evaluation towards scale up. *Desalination, 365*, 329–342. Available from https://doi.org/10.1016/j.desal.2015.03.020.

Chang, H., Li, T., Liu, B., Vidic, R. D., Elimelech, M., & Crittenden, J. C. (2019). Potential and implemented membrane-based technologies for the treatment and reuse of flowback and produced water from shale gas and oil plays: A review. *Desalination, 455*, 34–57. Available from https://doi.org/10.1016/j.desal.2019.01.001.

Chang, H., Liu, B., Yang, B., Yang, X., Guo, C., He, Q., ... Yang, P. (2019). An integrated coagulation-ultrafiltration-nanofiltration process for internal reuse of shale gas flowback and produced water. *Separation and Purification Technology, 211*, 310–321. Available from https://doi.org/10.1016/j.seppur.2018.09.081.

Changmai, M., Pasawan, M., & Purkait, M. K. (2019). Treatment of oily wastewater from drilling site using electrocoagulation followed by microfiltration. *Separation and Purification Technology, 210*, 463–472. Available from https://doi.org/10.1016/j.seppur.2018.08.007.

Coday, B. D., & Cath, T. Y. (2014). Forward osmosis: Novel desalination of produced water and fracturing flowback. *Journal - American Water Works Association, 106*(2), 37–38. Available from https://doi.org/10.5942/jawwa.2014.106.0016.

Coday, B. D., Miller-Robbie, L., Beaudry, E. G., Munakata-Marr, J., & Cath, T. Y. (2015). Life cycle and economic assessments of engineered osmosis and osmotic dilution for desalination of Haynesville shale pit water. *Desalination, 369*, 188–200. Available from https://doi.org/10.1016/j.desal.2015.04.028.

Coday, B. D., Xu, P., Beaudry, E. G., Herron, J., Lampi, K., Hancock, N. T., & Cath, T. Y. (2014). The sweet spot of forward osmosis: Treatment of produced water, drilling wastewater, and other complex and difficult liquid streams. *Desalination, 333*(1), 23–35. Available from https://doi.org/10.1016/j.desal.2013.11.014.

Doran, G. F., Williams, K. L., Drago, J. A., Huang, S. S., & Leong, L. Y. C. (1999). Pilot study results to convert oil field produced water to drinking water or reuse quality. *Proceedings - SPE International Heavy Oil Symposium*, 209–220.

Ebrahimi, M., Ashaghi, K. S., Engel, L., Willershausen, D., Mund, P., Bolduan, P., & Czermak, P. (2009). Characterization and application of different ceramic membranes for the oil-field produced water treatment. *Desalination, 245*(1–3), 533–540. Available from https://doi.org/10.1016/j.desal.2009.02.017.

Ezugbe, E. O., & Rathilal, S. (2020). Membrane technologies in wastewater treatment: A review. *Membranes, 10*(5). Available from https://doi.org/10.3390/membranes10050089.

Fakhru'l-Razi, A., Pendashteh, A., Abdullah, L. C., Biak, D. R. A., Madaeni, S. S., & Abidin, Z. Z. (2009). Review of technologies for oil and gas produced water treatment. *Journal of Hazardous Materials, 170*(2–3), 530–551. Available from https://doi.org/10.1016/j.jhazmat.2009.05.044.

Fakhru'l-Razi, A., Pendashteh, A., Abidin, Z. Z., Abdullah, L. C., Biak, D. R. A., & Madaeni, S. S. (2010). Application of membrane-coupled sequencing batch reactor for oilfield produced water recycle and beneficial re-use. *Bioresource Technology, 101*(18), 6942–6949. Available from https://doi.org/10.1016/j.biortech.2010.04.005.

Guo, C., Chang, H., Liu, B., He, Q., Xiong, B., Kumar, M., & Zydney, A. L. (2018). A combined ultrafiltration-reverse osmosis process for external reuse of Weiyuan shale gas flowback and produced water. *Environmental Science: Water Research and Technology, 4*(7), 942–955. Available from https://doi.org/10.1039/c8ew00036k.

Hao, H., Huang, X., Gao, C., & Gao, X. (2015). Application of an integrated system of coagulation and electrodialysis for treatment of wastewater produced by fracturing. *Desalination and Water Treatment, 55*(8), 2034–2043. Available from https://doi.org/10.1080/19443994.2014.930700.

He, C., Wang, X., Liu, W., Barbot, E., & Vidic, R. D. (2014). Microfiltration in recycling of Marcellus Shale flowback water: Solids removal and potential fouling of polymeric microfiltration membranes. *Journal of Membrane Science, 462*, 88–95. Available from https://doi.org/10.1016/j.memsci.2014.03.035.

Heidarpour, F., Shi, J., & Chae, S. R. (2015). Recycling of coal seam gas-associated water using vacuum membrane distillation. *Water Science and Technology, 72*(6), 908–916. Available from https://doi.org/10.2166/wst.2015.229.

Henderson, C., Acharya, H., Matis, H., Kommepalli, H., Moore, B., & Wang, H. (2011). *Cost effective recovery of low-TDS frac flowback water for re-use.*

Hickenbottom, K. L., Hancock, N. T., Hutchings, N. R., Appleton, E. W., Beaudry, E. G., Xu, P., & Cath, T. Y. (2013). Forward osmosis treatment of drilling mud and fracturing wastewater from oil and gas operations. *Desalination, 312*, 60–66. Available from https://doi.org/10.1016/j.desal.2012.05.037.

Hong, J. G., Zhang, B., Glabman, S., Uzal, N., Dou, X., Zhang, H., . . . Chen, Y. (2015). Potential ion exchange membranes and system performance in reverse electrodialysis for power generation: A review. *Journal of Membrane Science, 486*, 71–88. Available from https://doi.org/10.1016/j.memsci.2015.02.039.

Jackson, R. B., Vengosh, A., Carey, J. W., Davies, R. J., Darrah, T. H., O'Sullivan, F., & Pétron, G. (2014). The environmental costs and benefits of fracking. *Annual Review of Environment and Resources, 39*, 327–362. Available from https://doi.org/10.1146/annurev-environ-031113-144051.

Jalab, R., Awad, A. M., Nasser, M. S., Minier-Matar, J., & Adham, S. (2020). Pilot-scale investigation of flowrate and temperature influence on the performance of hollow fiber forward osmosis membrane in osmotic concentration process. *Journal of Environmental Chemical Engineering, 8*(6). Available from https://doi.org/10.1016/j.jece.2020.104494.

Jamaly, S., Giwa, A., & Hasan, S. W. (2015). Recent improvements in oily wastewater treatment: Progress, challenges, and future opportunities. *Journal of Environmental Sciences (China), 37*, 15–30. Available from https://doi.org/10.1016/j.jes.2015.04.011.

Jang, E., Jeong, S., & Chung, E. (2017). Application of three different water treatment technologies to shale gas produced water. *Geosystem Engineering, 20*(2), 104–110. Available from https://doi.org/10.1080/12269328.2016.1239553.

Jiang, Q., Rentschler, J., Perrone, R., & Liu, K. (2013). Application of ceramic membrane and ion-exchange for the treatment of the flowback water from Marcellus shale gas production. *Journal of Membrane Science, 431*, 55–61. Available from https://doi.org/10.1016/j.memsci.2012.12.030.

Kamali, M., Suhas, D. P., Costa, M. E., Capela, I., & Aminabhavi, T. M. (2019). Sustainability considerations in membrane-based technologies for industrial effluents treatment. *Chemical Engineering Journal, 368*, 474–494. Available from https://doi.org/10.1016/j.cej.2019.02.075.

Kargi, F., & Dincer, A. R. (1996). Effect of salt concentration on biological treatment of saline wastewater by fed-batch operation. *Enzyme and Microbial Technology, 19*(7), 529–537. Available from https://doi.org/10.1016/S0141-0229(96)00070-1.

Kelewou, H., Lhassani, A., Merzouki, M., Drogui, P., & Sellamuthu, B. (2011). Salts retention by nanofiltration membranes: Physicochemical and hydrodynamic approaches and modeling. *Desalination, 277*(1–3), 106–112. Available from https://doi.org/10.1016/j.desal.2011.04.010.

Kim, B., Kwak, R., Kwon, H. J., Pham, V. S., Kim, M., Al-Anzi, B., . . . Han, J. (2016). Purification of high salinity brine by multistage ion concentration polarization desalination. *Scientific Reports, 6*. Available from https://doi.org/10.1038/srep31850.

Kim, J., Kim, J., & Hong, S. (2018). Recovery of water and minerals from shale gas produced water by membrane distillation crystallization. *Water Research, 129*, 447–459. Available from https://doi.org/10.1016/j.watres.2017.11.017.

Kim, J., Kwon, H., Lee, S., Lee, S., & Hong, S. (2017). Membrane distillation (MD) integrated with crystallization (MDC) for shale gas produced water (SGPW) treatment. *Desalination, 403*, 172–178. Available from https://doi.org/10.1016/j.desal.2016.07.045.

Klaysom, C., Cath, T. Y., Depuydt, T., & Vankelecom, I. F. J. (2013). Forward and pressure retarded osmosis: Potential solutions for global challenges in energy and water supply. *Chemical Society Reviews*, 42(16), 6959−6989. Available from https://doi.org/10.1039/c3cs60051c.

Koltuniewicz, A. B., Field, R. W., & Arnot, T. C. (1995). Cross-flow and dead-end microfiltration of oily-water emulsion. Part I: Experimental study and analysis of flux decline. *Journal of Membrane Science*, 102(C), 193−207. Available from https://doi.org/10.1016/0376-7388(94)00320-X.

Kong, F. X., Chen, J. F., Wang, H. M., Liu, X. N., Wang, X. M., Wen, X., . . . Xie, Y. F. (2017). Application of coagulation-UF hybrid process for shale gas fracturing flowback water recycling: Performance and fouling analysis. *Journal of Membrane Science*, 524, 460−469. Available from https://doi.org/10.1016/j.memsci.2016.11.039.

Kong, F. X., Sun, G. D., Chen, J. F., Han, J. D., Guo, C. M., Zhang, T., . . . Xie, Y. F. (2018). Desalination and fouling of NF/low pressure RO membrane for shale gas fracturing flowback water treatment. *Separation and Purification Technology*, 195, 216−223. Available from https://doi.org/10.1016/j.seppur.2017.12.017.

Li, X. M., Zhao, B., Wang, Z., Xie, M., Song, J., Nghiem, L. D., . . . Chen, G. (2014). Water reclamation from shale gas drilling flow-back fluid using a novel forward osmosis-vacuum membrane distillation hybrid system. *Water Science and Technology*, 69(5), 1036−1044. Available from https://doi.org/10.2166/wst.2014.003.

Lutz, B. D., Lewis, A. N., & Doyle, M. W. (2013). Generation, transport, and disposal of wastewater associated with Marcellus Shale gas development. *Water Resources Research*, 49(2), 647−656. Available from https://doi.org/10.1002/wrcr.20096.

Maltos, R. A., Regnery, J., Almaraz, N., Fox, S., Schutter, M., Cath, T. J., . . . Cath, T. Y. (2018). Produced water impact on membrane integrity during extended pilot testing of forward osmosis − Reverse osmosis treatment. *Desalination*, 440, 99−110. Available from https://doi.org/10.1016/j.desal.2018.02.029.

McGinnis, R. L., Hancock, N. T., Nowosielski-Slepowron, M. S., & McGurgan, G. D. (2013). Pilot demonstration of the NH3/CO2 forward osmosis desalination process on high salinity brines. *Desalination*, 312, 67−74. Available from https://doi.org/10.1016/j.desal.2012.11.032.

McGovern, R. K., & Lienhard, V. J. H. (2014). On the potential of forward osmosis to energetically outperform reverse osmosis desalination. *Journal of Membrane Science*, 469, 245−250. Available from https://doi.org/10.1016/j.memsci.2014.05.061.

Millar, G. J., Couperthwaite, S. J., & Moodliar, C. D. (2016). Strategies for the management and treatment of coal seam gas associated water. *Renewable and Sustainable Energy Reviews*, 57, 669−691. Available from https://doi.org/10.1016/j.rser.2015.12.087.

Miller, D. J., Huang, X., Li, H., Kasemset, S., Lee, A., Agnihotri, D., . . . Freeman, B. D. (2013). Fouling-resistant membranes for the treatment of flowback water from hydraulic shale fracturing: A pilot study. *Journal of Membrane Science*, 437, 265−275. Available from https://doi.org/10.1016/j.memsci.2013.03.019.

Minier-Matar, J., Hussain, A., Janson, A., Wang, R., Fane, A. G., & Adham, S. (2015). Application of forward osmosis for reducing volume of produced/Process water from oil and gas operations. *Desalination*, 376, 1−8. Available from https://doi.org/10.1016/j.desal.2015.08.008.

Mondal, S., & Wickramasinghe, S. R. (2008). Produced water treatment by nanofiltration and reverse osmosis membranes. *Journal of Membrane Science*, 322(1), 162−170. Available from https://doi.org/10.1016/j.memsci.2008.05.039.

Munirasu, S., Haija, M. A., & Banat, F. (2016). Use of membrane technology for oil field and refinery produced water treatment - A review. *Process Safety and Environmental Protection*, 100, 183−202. Available from https://doi.org/10.1016/j.psep.2016.01.010.

Muthu, S., & Brant, J. A. (2015). Interrelationships between flux, membrane properties, and soil water transport in a subsurface pervaporation irrigation system. *Environmental Engineering Science*, 32(6), 539−550. Available from https://doi.org/10.1089/ees.2014.0519.

Nasiri, M., Jafari, I., & Parniankhoy, B. (2017). Oil and gas produced water management: A review of treatment technologies, challenges, and opportunities. *Chemical Engineering Communications*, 204(8), 990−1005. Available from https://doi.org/10.1080/00986445.2017.1330747.

People, G. (2018). Produced water treatment technology: Survey of stakeholders Dissemination method Data processing (. 1−6).

Peraki, M., & Ghazanfari, E. (2014). Electrodialysis treatment of flow-back water for environmental protection in shale gas development. In *Shale energy engineering 2014: Technical challenges, environmental issues, and public policy - proceedings of the 2014 shale energy engineering conference* (pp. 74−84). American Society of Civil Engineers (ASCE). https://doi.org/10.1061/9780784413654.008.

Peraki, M., Ghazanfari, E., Pinder, G. F., & Harrington, T. L. (2016). Electrodialysis: An application for the environmental protection in shale-gas extraction. *Separation and Purification Technology*, 161, 96−103. Available from https://doi.org/10.1016/j.seppur.2016.01.040.

Rao, G., & Li, Y. (2016). Feasibility study of flowback/produced water treatment using direct-contact membrane distillation. *Desalination and Water Treatment*, 57(45), 21314−21327. Available from https://doi.org/10.1080/19443994.2015.1119753.

Riley, S. M., Ahoor, D. C., Regnery, J., & Cath, T. Y. (2018). Tracking oil and gas wastewater-derived organic matter in a hybrid biofilter membrane treatment system: A multianalytical approach. *Science of the Total Environment*, 613−614, 208−217. Available from https://doi.org/10.1016/j.scitotenv.2017.09.031.

Riley, S. M., Oliveira, J. M. S., Regnery, J., & Cath, T. Y. (2016). Hybrid membrane bio-systems for sustainable treatment of oil and gas produced water and fracturing flowback water. *Separation and Purification Technology*, 171, 297−311. Available from https://doi.org/10.1016/j.seppur.2016.07.008.

Shafer, L. (2011). Water recycling and purification in the pinedale anticline field: Results from the anticline disposal project. In *Society of petroleum engineers - SPE americas E and P health, safety, security, and environmental conference 2011* (pp. 160−166). Society of Petroleum Engineers. https://doi.org/10.2118/141448-ms.

Silva, T. L. S., Morales-Torres, S., Castro-Silva, S., Figueiredo, J. L., & Silva, A. M. T. (2017). An overview on exploration and environmental impact of unconventional gas sources and treatment options for produced water. *Journal of Environmental Management*, 200, 511−529. Available from https://doi.org/10.1016/j.jenvman.2017.06.002.

Sirivedhin, T., McCue, J., & Dallbauman, L. (2004). Reclaiming produced water for beneficial use: Salt removal by electrodialysis. *Journal of Membrane Science*, 243(1−2), 335−343. Available from https://doi.org/10.1016/j.memsci.2004.06.038.

Subramani, A., & Jacangelo, J. G. (2015). Emerging desalination technologies for water treatment: A critical review. *Water Research, 75*, 164–187. Available from https://doi.org/10.1016/j.watres.2015.02.032.

Sule, M. N., Templeton, M. R., & Bond, T. (2016). Rejection of organic micro-pollutants from water by a tubular, hydrophilic pervaporative membrane designed for irrigation applications. *Environmental Technology (United Kingdom), 37*(11), 1382–1389. Available from https://doi.org/10.1080/09593330.2015.1116610.

Tao, F. T., Curtice, S., Hobbs, R. D., Sides, J. L., Wieser, J. D., Dyke, C. A., ... Pilger, P. F. (1993). Conversion of oilfield produced water into an irrigation/drinking quality water. In *Proceedings of the SPE/EPA exploration and production environmental conference* (pp. 571–579). Society of Petroleum Engineers (SPE).

Tong, T., & Elimelech, M. (2016). The global rise of zero liquid discharge for wastewater management: Drivers, technologies, and future directions. *Environmental Science and Technology, 50*(13), 6846–6855. Available from https://doi.org/10.1021/acs.est.6b01000.

Tong, T., Carlson, K. H., Robbins, C. A., Zhang, Z., & Du, X. (2019). Membrane-based treatment of shale oil and gas wastewater: The current state of knowledge. *Frontiers of Environmental Science and Engineering, 13*(4). Available from https://doi.org/10.1007/s11783-019-1147-y.

Veil, J. (2015). *Produced water volumes and management practices in 2012.*

Warner, N. R., Darrah, T. H., Jackson, R. B., Millot, R., Kloppmann, W., & Vengosh, A. (2014). New tracers identify hydraulic fracturing fluids and accidental releases from oil and gas operations. *Environmental Science and Technology, 48*(21), 12552–12560. Available from https://doi.org/10.1021/es5032135.

Xiong, B., Zydney, A. L., & Kumar, M. (2016). Fouling of microfiltration membranes by flowback and produced waters from the Marcellus shale gas play. *Water Research, 99*, 162–170. Available from https://doi.org/10.1016/j.watres.2016.04.049.

Zhang, S., Wang, P., Fu, X., & Chung, T. S. (2014). Sustainable water recovery from oily wastewater via forward osmosis-membrane distillation (FO-MD). *Water Research, 52*, 112–121. Available from https://doi.org/10.1016/j.watres.2013.12.044.

Zhao, R., Porada, S., Biesheuvel, P. M., & Van der Wal, A. (2013). Energy consumption in membrane capacitive deionization for different water recoveries and flow rates, and comparison with reverse osmosis. *Desalination, 330*, 35–41. Available from https://doi.org/10.1016/j.desal.2013.08.017.

Chapter 9

Management of petroleum wastewater: comparative evaluation of modern and traditional techniques

S. Joshi and S. Bhatia

Department of Chemistry, Isabella Thoburn P.G. College, Lucknow, Uttar Pradesh, India

Introduction

The use of petroleum products has significantly increased globally in the past few decades, and the adverse impact of their waste products on the environment has also become a matter of great concern (Hu, Li, & Hou, 2015; Mohammed, Mohammad, & Ali, 2018; Neff, 2005). An enormous amount of waste is generated during various processes involved and also these different processes produce waste of different nature which makes their management quite difficult. This situation is also getting aggravated due to weak environmental regulations in most countries (Danso-Boateng, Osei-Wusu, 2013; Ghazizade, Koulivand, Safari, & Heidari, 2021; Guidelines for Characterization of Offshore Drill Cuttings Piles, 2003; Onshore solid waste management in exploration & production operations, 1989). A variety of pollutants like petroleum hydrocarbons (PHCs), mercaptans, oil and grease, phenol, ammonia, sulfide, and other organic compounds are present in their complex form in the wastewater produced from the petroleum industry. Studies have shown that these pollutants, directly or indirectly have marked their harmful impact on the environment. Therefore, to sustain the use of petroleum products, various techniques like membrane technology, photocatalytic degradation, advanced oxidation process, electrochemical catalysis, resource recovery; Waste biorefinery, etc. have been developed to treat produced wastewater and to manage these in an environmentally friendly manner (Johnson & Affam, 2019). These techniques have their advantages and disadvantages. Critical analysis of these techniques will help in reducing their disadvantages and also in developing new techniques for converting waste into useful resources.

Types of waste generated in oil refineries

Oiled materials

Oily materials are of two categories as follows.

Oily sludge

The properties of crude oil get changed due to extreme changes in the surrounding temperatures. Oils also get emulsified because of the atmospheric moisture or water contents. The addition of unrequired materials during the processes also changes the basic properties of oils. Due to such changes in the properties of the oils, sludge is generated. The sludge is the remains that are generally produced in the tank bottoms and contaminated soils during the production and exploration of petroleum, The source of these remains is biotreatment sludges and desalter sludges. Generally, sludge contains silica compounds and large amounts of barium also (Jafarinejad, 2017; Johnson & Affam, 2019). Sludges are rich in Ra-226 than Ra-228. Although the concentration of radiation is lower in sludges but being more soluble sludges release more radiation to the environment (Paranhos Gazineu, De Araújo, Brandão, Hazin, & Godoy, 2005).

Petroleum Industry Wastewater. DOI: https://doi.org/10.1016/B978-0-323-85884-7.00010-2

Solid wastes

These are found in contaminated soils, soil spill debris, filter clays acids, tar rags, filter materials, packing materials, and activated carbon. Dried sludge having a low amount of oil content looks like soil and constitutes solid waste.

Produced waters

Pumped water from the oil wells leaves behind produced waters as waste after the separation of oil and gas. The ratio of produced water to oil in the conventional wells has been reported to be in the ratio of 10:1 (Liang, Ning, Liao, & Yuan, 2018). Studies have shown the production of billion barrels of waste fluids from oil and gas industries (Wei, Zhang, Sun, & Brenner, 2018). Aromatic compounds, such as benzene, toluene, ethylbenzene, xylene isomers, and phenols are present in their soluble form in the produced water. Removal of these compounds is extremely difficult due to their toxicity and complexity. Heavy metals like Cu, Pb, Zn, Cd, Fe, Mn, and Ni have been found to get extracted along during the exploration of fuel from rocks and through the oil spillage. In recent years, several new methods have been employed for producing oil and gas. In these methods known as fracking, horizontal drilling has been combined with enhanced stimulation. Due to such methods, oil and gas deposits are rich in naturally occurring radionuclides and are referred to as naturally occurring radioactive materials (NORMs). A good amount of petroleum and natural gas developed in the US was created in the earth's crust at the site of ancient seas by the decay of sea life. Radionuclides and other minerals are found to be dissolved in this shale, petroleum, and gas deposits which are present in the aquifers having seawater These elements present in the brine, get separated and settle out, thus producing wastewater at the surface. This extraction process further enhances the concentration of NORM on the surface environment and result in human contact. These wastes are classified as technologically enhanced naturally occurring radioactive material (TENORM). Studies have revealed the generation of a huge amount of TENORM sludge each year from oil production processes (API). The presence of Radium, Uranium, Thorium, and their decay products, Potassium-50, Lead 210, and Polonium-210 have been reported in the produced waters. However, their concentrations were different in different places. In general, produced waters are reinjected into deep wells or are treated for reuse. Before it's reused, produced water will have to undergo effective treatments to meet the standards of the current legislation.

Cause and effect of oil waste

Oil sludge causes corrosion of the storage tanks thus affecting their storage capacity. Accumulation of sludge in the surrounding areas causes morphological changes due to variation in the physical and chemical properties of the soil. Several studies have reported stunted growth in the vegetation of these areas because of the deficiency in nutrients of such soil. Also, the sludge being a viscous material clogs the soil pores, because of which water absorption and retention power of the soil is reduced. The adverse effect of sludge on the crop has become a global challenge. PHCs and petroleum aromatic/aliphatic hydrocarbons (PAHs), the high molecular weight components present in sludge are genotoxic and pose major health concerns. Studies have reported that PHCs infiltrate the groundwater and are harmful to the aquatic system. Resources Conservation and Recovery Act (RCRA) consider sludge as one of the hazardous wastes. These toxic wastes are a threat to the environment and their removals and proper disposal from sludge remains a big challenge before petroleum industries both in terms of economics and environment. WHO has therefore laid strict rules for Producers, refiners, and transporters of petroleum materials to take up this matter consciously. A lot of finance is involved in the proper removal, transfer, and landfilling in cleaning up the petroleum sludge. Therefore, economical and environmentally sustainable methods are developed and adopted to address this issue (Deliyanni, Kyzas, & Matis, 2017; Sebba, 1962; Varjani, Joshi, Srivastava, Ngo, & Guo, 2020).

Methods and recommendations for sustainable management of wastewater

Some recommendations for sustainable management are as follows:

(1) Petroleum industries to adopt those technologies which produce less waste. This will attain a reduction in petroleum waste generation.

(2) Use of different techniques to recover oil from the produced wastewater.

(3) Proper disposal of the wastewater through sustainable methods like reuse of waste as production material, recycling the waste into some useful material, and converting waste into energy.

Adopting these R, s can help in producing sustainable solutions to the adverse impact of produced water in accordance with environmental requirements. This chapter is a review of various techniques which are in use to recover usable materials from wastewaters in the petroleum industries and their comparative evaluation. Wastewater treatment broadly has been classified as Primary, Secondary, and Tertiary based on their methodology.

1. *Physical Methods*: Primary wastewater treatment.
2. *Chemical Methods*: Secondary wastewater treatment.
3. *Biological Methods*: Tertiary Wastewater treatment.
4. *Electrochemical*: Tertiary wastewater treatment.

Physical methods

In the primary wastewater treatment separation of oil, water, and solids occur in two stages. When concentrations of oil in the raw wastewater exceed approximately 500 mg L^{-1}, oil-water separators are used to extract free oil and remove heavy suspended solids from wastewater. In the second stage of primary treatment, the process is further repeated to remove those oils contents and suspended solids, which could not be separated in the first stage of primary treatment. The sludge generated after the second stage is pumped out of the tank and is digested. Various physical methods, like sedimentation, flotation, adsorption, and using screens, deep bed filters, barracks, etc. as barriers are used for primary treatment (Jafarinejad, 2017).

Sedimentation

The most common physical unit operation in wastewater treatment is sedimentation which uses gravitational force to separate unstable and destabilized suspended solids from wastewater. In this technique, the difference of densities between the bulk of the liquid and the solids is manipulated. Suspended particles having a density higher than that of water due to their gravity settles at the bottom of the container or tank and within a specified time are removed from the bottom of the settling tank which is known as a clarifier. These clarifiers have been designed as rectangular or circular with a radial or upward water flow pattern. These settled solids are sometimes also treated with chemical or biological reagents as in the recovery of activated sludge for recycling and thus secondary treatment is also achieved with such secondary sedimentation.

Sedimentation has four distinct types of settling as shown in the following sections.

Discrete settling

This is also known as type I settling, in which the dilute suspension of solids does not aggregate. Such settling is characteristic of the particle which is only a function of its fluid property and is not affected by the presence of other particles. When liquid is pumped from the inlet of the tank, due to the combined effect of the gravitational force the solid particles move downward and the bulk flow moves toward the outlet. Sedimented particles then effectively get removed (Walter, 1972) (Fig. 9.1).

This sedimentation is affected by various factors like:

- the depth of the sedimentation tank.
- the residence time of the particle in the liquid in the tank.
- the cross-sectional area of the tank.
- the overall volumetric flow rate through the tank.

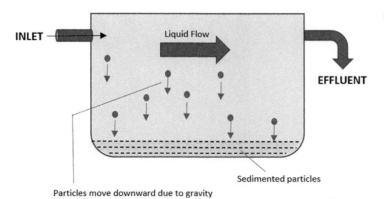

FIGURE 9.1 Diagram for discrete settling.

- size diameter of the particle.
- gravitational acceleration.
- the viscosity of the liquid.

Thus, to summarize the discrete settling is more likely associated with settling of hard particulates with high density and size without changing their basic characteristics.

Flocculent settling

In this type, the particulates at the time of settling aggregate among themselves and/or with added flocculants and form larger particulates. Due to which the mass of settling solids increases. This helps in the fast settling of the particles and also enhances their removal rate. In this mode, the sedimentation process is as in type I. Flocculent settling is used in primary clarifiers and the upper zones of secondary clarifiers. Larger particles during their settling tend to collide with other slower-settling particles, resulting in the formation of further larger particles in a latent water body. During settling, the frequency of collisions among particles increases with the depth of the tank. As the settled solids are removed expeditiously so the flocculation enhances the rate of solid removal from wastewater.

Thus, like sedimentation, flocculation also depends on various factors like:

- depth of tank,
- properties of particles,
- properties of the liquid.

Zone settling

Zone settling is also called hindered settling. In this process, particulates adhere together and form a blanket of mass at the distinct interfaces with the liquid above it. Interfaces of zone settling are determined by batch settling test. Batch settling by numerical simulation has been reported by Zheng & Bagley (1999).

Compression settling

In compression settling, the particulates accumulate at the clarifiers in the form of a compressed structure which supports the weight of the particulates that are settled in the bottom of the tank/basin. Apart from clarifiers sometimes compression also occurs in sludge thickening tanks. This type IV settling is actually a physicochemical wastewater treatment process that generates the highest concentration of suspended solids at the bottom of the tank.

Each of the above zones has different characteristics that help in further analysis.

Floatation

Floatation is a separation process using gas bubbles. Specially designed tanks are used for the process. Piccioli, Aeneson, Zhao, Dudek, and Oye (2020) in their studies have shown that various parameters like a gas bubble and oil droplet sizes, droplet-bubble attachment mechanism, interfacial properties, water composition, oil and gas properties, pressure, and temperature can affect gas flotation (Kyzas & Matis, 2018) (Fig. 9.2).

The typical laboratory arrangement is shown in Fig. 9.2. It consists of:

1. A conditioning/feed tank with a mechanical mixer;
2. A peristaltic pump;
3. Liquid rotameters;
4. The flotation column with a weir (on the top);
5. A foam collection tank;
6. An air compressor connected through a needle valve;
7. Washing trap and air flow meter;
8. Porous diaphragm at the bottom of the cell;
9. A mercury U-tube manometer and effluent tank;
10. A pH meter.

Dissolved air flotation

The dissolved air flotation (DAF) process works on Henry's law that states that the amount of dissolved gas in a liquid is proportional to its partial pressure above the liquid (Edzwald, 1995). By saturating wastewater with air under pressure

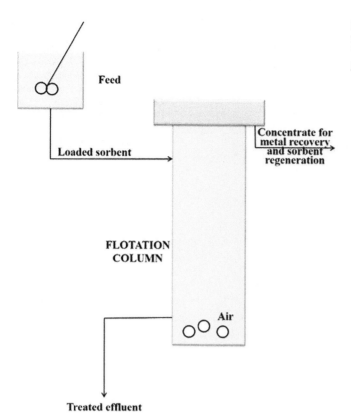

FIGURE 9.2 Schematic diagram of a counter-current dispersed-air flotation rig. *From Kyzas, G., & Matis, K. (2018). Flotation in Water and Wastewater Treatment. Processes, 6(8).* https://doi.org/10.3390/pr6080116.

and then rapidly releasing the pressure a cloud of microbubbles is produced. These bubbles attract hydrophobic particulate matter like fats; oils, grease, etc. These attached suspended matters along with bubbles then move against the gravity towards the surface of the water, forms oil foam, and are skimmed off (Huang & Li, 2008). This approach is commonly known as DAF and is quite effective in the removal of hydrocarbon compounds. The size of the particles also plays important role in DAF. Various flotation methods have been developed in the recent past and some studies have even reported complete removal of hydrocarbon pollutants (Bennett, 1988; Igunnu & Chen, 2014; Massey, Dunlap, Koblin, Luthy, & Nakles, 1976). These methods are highly efficient as well as economically favorable. Moursy et al. have used gas chromatography for the determination and identification of extracted hydrocarbon compounds using the flotation technique (Moursy & Abo El-Ela, 1982) (Fig. 9.3).

Electro flotation

Electro flotation a class of dispersed-air flotation techniques. Historically it was used in the extraction of elements from the minerals but now it is being applied in the processing of various other materials like frothers, modifiers, sorbents, and many more. During this flotation technique, through the electrolysis of water, bubbles of hydrogen and oxygen are produced. These bubbles efficiently remove suspended solids present in the produced water by the phenomena of oxygen transfer. This technique is more effective in treating oily wastewater than the DAF because of its high flotation efficiency, easy operation, low maintenance requirement, and fewer energy needs (Chen & Chen, 2010; Comninellis & Chen, 2010).

Flotation of metal ions

The flotation process exploits the difference in solubilities and mobilities of heavy metals in their separation. Ion flotation is successful in removing metal ions. In this process, surfactants or collectors with a charge opposite to that of metal ions are added to wastewater before the addition of air. These collectors have been reported to increase the rate of separation of metal ions by increasing the hydrophobicity of the surface. Hydrophobic and hydrophilic particles get separated and an increased amount of oil surfactants is accumulated at the surface before air addition. Hoseinian and his coworkers have reported that the rate of removal of metal ions depends on factors like chemical interactions and

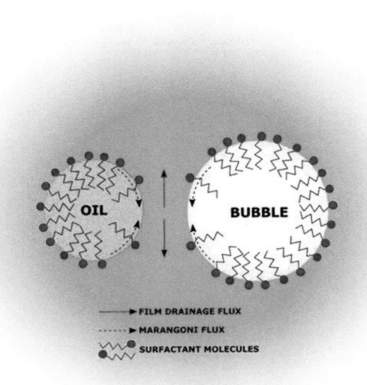

FIGURE 9.3 Oil droplet and a gas bubble in water. *From Piccioli, M., Aanesen, S. V., Zhao, H., Dudek, M., & Øye, G. (2020). Gas Flotation of Petroleum Produced Water: A Review on Status, Fundamental Aspects, and Perspectives. Energy & Fuels, 34 (12), 15579–15592. https://doi.org/ 10.1021/acs.energyfuels.0c03262.*

water entrainment in the froth phase. In this study, they observed an increased rate of removal of the Ni (II) than the removal of water suggesting the true flotation process in Ni (II) ion removal. This further indicates that the chemical conditions affect the flotation efficiency (Hoseinian, Rezai, Kowsari, & Safari, 2018).

Precipitation flotation

Precipitation when combined with the flotation technique has been reported to be effective for the removal of metal ions even if their concentration in the wastewater is low. The use of surfactant or collector increases the concentration of ions in the solution by precipitating the ion even before passing the air. The solution at this stage gets converted to dispersion form and precipitates float on the surface and can be removed easily (Polat & Erdogan, 2007). For some metals, better results have been reported with the Denver flotation cell (Harker, Backhurst, & Richardson, 2013) (Fig. 9.4).

Venbakm C. Gopalratnam et al. in their study have reported that the pH, collector dosage, and rete of air-injection rate affect the concentration of residual metal in produced water. In general, maximum removal of metals was obtained at an optimal pH of 9.18, Their studies have also reported greater removal with the nozzle air flotation system as compared to the induced air flotation system (Gopalratnam, Bennett, & Peters, 1992). Kyzas et al. have also reported the effect of pH on the biosorption of toxic metals, from an aqueous mixture of petroleum wastewater. They have reported the successive precipitation of copper hydroxide at a pH of about 5.9, zinc at pH 7.3, and nickel at pH 7.8; Calcium could not be precipitated at their range (Kyzas & Matis, 2018). Removal of calcium and magnesium was achieved by repeated precipitation using chelating ligands by Mohamed E. Mahmoud et al. They have reported a 91.75% removal value by using 30 dipping cycles, 3 min dipping time, and 76 cm^2 surface area. (Mahmoud & Obada, 2014)

Sorptive technique

Sorption is a physicochemical process that has emerged as an economically viable and highly efficient technology for wastewater management. This can be classified into absorption and adsorption based on the type of binding involved with the pollutant. Metal ions from the wastewater are removed by using ultra-fine particle size sorbents. Oil and grease

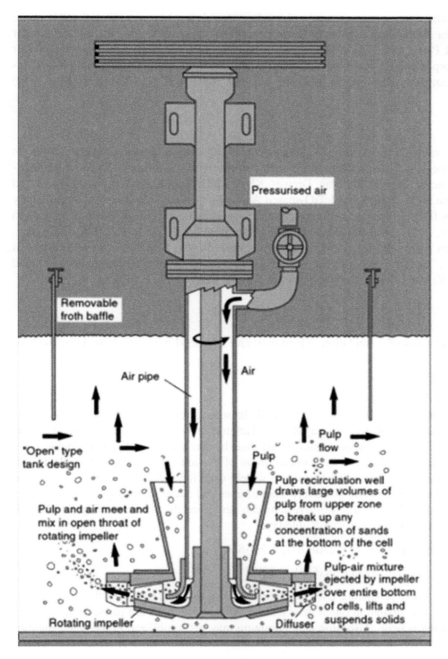

FIGURE 9.4 Denver floatation cell. *From Harker, Backhurst, and Richardson (2013). Chemical Engineering. Volume 2. Elsevier.*

classes of pollutants present in produced water have a very low affinity to water. A combination of other techniques like flotation with sorption has proved to be quite effective in managing such pollutants. Charged metal ions present in wastewater attach to the sorbents and combining this with the flotation process involves a series of independent steps and is found to be quite effective (Kyzas & Matis, 2018; Sebba, 1962; Varjani et al., 2020).

Adsorption process

Adsorption is a surface phenomenon that is being used effectively for the waste management of produced water. The process is an efficient treatment for the removal of both organic and inorganic pollutants. Due to liquid−solid intermolecular attractive forces solute present in produced water accumulates on the solid surface of the adsorbent and gets transferred from solution phase to surface of solid and can be removed easily. Different ionic, covalent, metallic, or Van der Waals forces are involved depending on the nature of adsorbent and adsorbate. Researches are being carried out to find low-cost and effective adsorbents. Different types of adsorbents like mineral, organic, biological origin, zeolites, industrial by-products, agricultural wastes, biomass, and polymeric materials have been evaluated and be quite effective and environmental friendly for the treatment of organic and inorganic effluent present in produced water (Crini, 2005; Pollard, Fowler, Sollars, & Perry, 1992). Li et al. have reported efficient separation of oil from wastewater by using coal as an adsorbent. The absorption time, coal type. coal particle size distribution, pH value, and oil concentration were reported as important parameters affecting the removal rate (Barakat, 2011; Li, Zhang, & Liu, 2010) (Table 9.1).

Absorption process

John Stevens et al. in their study have evaluated the absorption potential of 34 organic chemicals for sludge solids. In their studies, they used dry heat techniques and lyophilization along with absorption techniques and observed the highest potential for positively charged chemicals (Stevens-Garmon, Drewes, Khan, McDonald, & Dickenson, 2011). Li and Li in their studies have demonstrated that light absorption of TiO_2-based Au/Au3 + -TiO_2 photocatalyst is significantly more effective in scavenging pollutants from wastewaters. Such photocatalysts are considered to use solar energy to address environmental needs (Li and Li, 2001; Li et al., 2010; Pintor, Vilar, Botelho, Boaventura, 2016). Bari et al. reported that the application of potassium ferrate (K_2FeO_4) and ozone in combinations has high efficacy for the removal of organic matter from produced waters. The efficiency of removal was confirmed by using chemical oxygen demand (COD), fluorescence, and ultraviolet (UV) absorption techniques (Bari & Farooq, 1985).

Chemical methods

Various chemical methods involving principles of coagulation, adsorption, and advanced oxidation have been used for the treatment of petroleum effluents

Coagulation and flocculation chemicals

Sometimes the suspended solids do not settle down by gravitational pull. In such cases, settling is facilitated by the addition of certain chemicals. This process is referred to as coagulation (Tarleton & Wakeman, 2007). Coagulation is the process whereby destabilization (aggregation) of a suspension is affected by reducing the electrical double layer repulsion between particles through changes in the nature and concentration of the ions in the suspending electrolyte solution Historically, metal coagulants (alums and iron salts) have been widely used for water clarification at an amiable pH 6−7 range. When dispersed in water they generate positively charged ions and undergo hydrolysis forming sparingly soluble gelatinous hydroxides and mineral acids. The positive ions can neutralize the negative charges on the oil droplets and reduce the electrostatic repulsion of the electrical double layer

$$Al_2(SO_4)_3 + 6H_2O = 2Al(OH)_3 + 3H_2SO_4$$

$$FeCl_3 + 2H_2O = Fe(OH)_3 + 3HCl$$

The acid reduces the alkalinity of the water producing CO_2. Sometimes this CO_2 interferes with the settling as it adsorbs on the hydrous precipitate causing floc floatation rather than settling. Metal coagulants are particularly sensitive to pH. If pH is not in the range, clarification is poor. Also, iron and aluminum may get solubilized in the water increasing contamination.

Flocculation is the process whereby a long-chain polymer (or polyelectrolyte) causes particles to aggregate, often by forming bridges between them. Flocculant refers to the chemical or substance added to a suspension to accelerate the rate of flocculation or to strengthen the flocs formed during flocculation. Polyelectrolytes are large water-soluble organic polymers carrying anionic or cationic charge, or uncharged nonionic polymers. They may be synthetic or natural. Examples of natural polyelectrolytes are cationic starch, chitosan, and a polypeptide poly γ glutamic acid. While polyacrylamide is the most common synthetic cationic polyelectrolyte used. Zhao et al. have extensively reviewed the application of coagulation and flocculation techniques for petroleum effluent treatment concluding that these treatments are most economical, environmentally friendly, and easy to operate (Zhao et al., 2021).

Chemical oxidation

By chemical oxidation, organic contaminants are converted to carbon dioxide and water vapor or some small biodegradable organic molecules like alcohols, aldehyde, and carboxylic acids. Commonly used oxidants are potassium permanganate, chlorine, ozone, oxygen and hydrogen peroxide, and chlorine dioxides. The rate of oxidation depends upon the

TABLE 9.1 Suitable adsorbent for the inorganic pollutants of produced water from the petroleum industry.

S. N.	Adsorbent	Pollutant	Reference
1	Natural zeolites clinoptilolite	Pb (II), Cd (II), Zn (II), and Cu (II).	Babel (2003)
	Natural zeolites clinoptilolite	Pb (II), Cd (II), Zn (II), and Cu (II).	Bose, Aparna Bose, & Kumar (2002) and Basaldella, Va´zquez, Iucolano, & Caputo (2007)
2	Synthetic zeolite Naan zeolite	Cr (III)	Basaldella et al. (2007)
3	Synthetic zeolite used 4 A zeolite	Cu (II), Zn (II), Cr (VI), Mn (IV)	Barakat (2008)
4	Synthetic zeolite magnetically modified iron oxide	Pb (II)	Nah, Hwang, Jeon, & Choi (2006)
5	Clay–polymer composites zirconium phosphate	Heavy metals	Mouflih, Aklil, & Sebti (2005)
6	Clay–polymer composites zirconium phosphate	Heavy metals	Pan et al. (2007)
7	Clay–polymer composites calcined phosphate	Pb (II), Cu (II), and Zn (II)	Ajmal, Rao, Ahmad, and Ahmad (2000)
8	By-product from the iron foundry industry: green sands	Zn (II)	Lee, Park, & Lee (2004)
9	Iron Slag	Cu (II) and Pb (II)	Feng & Aldrich (2004)
10	Solid waste from sugar industry: bagasse fly ash,	Cd (II) and Ni (II)	Gupta et al. (2003)
11	Coal-burning	Cu (II) and Pb (II)	Aklil, Mouflih, and Sebti (2004)
12	Sawdust: 1,5-disodium hydrogen phosphate	Cr (VI)	Uysal and Ar (2007)
13	Hydrous titanium oxide	Cr (VI)	Ghosh, Dasgupta, Debnath, and Bhat (2003)
14	Hydrous titanium oxide	Cu (II)	Barakat (2005)
15	Orange peel	Ni (II)	Ajmal et al. (2000)
16	Coconut shell charcoal (CSC) modified with oxidizing agents and/or chitosan	Cr (VI)	Babel and Kurniawan (2004)
17	Pecan shells-activated carbon	Cu (II) and Zn (II)	Bansode, Losso, Marshall, Rao, and Portier (2003)
18	Potato peels charcoal	Cu (II) and Zn (II)	Aman, Kazi, Sabri, and Bano (2008)
19	Rice husk-activated carbon	Cr (VI)	Bishnoi, Bajaj, Sharma, and Gupta (2004)
20	Rice hull, containing cellulose, lignin, carbohydrate, silica and ethylenediamine.	Cr (VI)	Tang, Lee, Low, and Zainal (2003)
21	Alga wastes *Spirogyra* species	Heavy metals	Gupta, Rastogi, Saini, and Jain (2006)

(Continued)

TABLE 9.1 (Continued)

S. N.	Adsorbent	Pollutant	Reference
22	Ecklonia maxima	Heavy metals	Feng and Aldrich (2004)
23	*Ulva lactuca*	Heavy metals	El-Sikaily, Nemrel, Khaled, and Abdelwehab (2007)
24	*Oedogonium* sp. and *Nostoc* sp.	Heavy metals	Gupta, Vinod, and Rastogi (2008)
25	Brown alga *Fucus serratus*	Heavy metals	Ahmadyasbchin, Andres, Gerente, and Cloirec (2008)
26	Chitosan containing crown ether	Pb (II), Cr (III), Cd (II) and Hg (II).	Yi, Wang, and Liu (2003)
27	Crosslinked chitosan	Cd (II) Cu (II)	Crini (2005)
		As (V)	
28	crosslinked chitosan	Cu (II)	Liu, Tokura, Nishi, & Sakairi (2003)
29	Alumina/chitosan composite	Cu (II)	Liu et al. (2003)
30	Chitosan in partially converted crab shell waste	transition metal ions	Pradhan, Shukla, and Dorris (2005)
31	crosslinked chitosan beads	Cu (II)	Liu et al. (2003)
32	chitosan on the surface of nonporous glass beads.	Cu (II), Fe (III) and Cd (II).	Liu et al. (2003)
33	Crosslinked starch gel	Cu (II)	Crini (2005)
		Pb (II)	
34	Poly (ethylene glycol dimethacrylate-co-acrylamide): Hydrogel beads	Pb (II), Cd (II),	Kesenci, Say, and Denizli (2002)
		Hg (II)	
35	Poly(3-acrylamidopropyl) trimethyl ammonium chloride: Hydrogels	As(V)	Barakat and Sahiner (2008)
36	Poly (vinylpyrrolidone-co-methyl acrylate): Hydrogels	Cu (II), Ni (II), Cd (II)	Essawy and Ibrahim (2004)
37	Thiacrown ethers	Zn (II), Cd (II), Co (II), Mn (II), Cu (II), Ni (II), and Ag(I)	Saad et al. (2006)

type of oxidant used, pollutants present, contact time between oxidants and effluent, pH, and temperature (Barratt & a, 1997). Mamat et al. utilized chemical oxidation by H_2O_2 on petroleum refinery wastewater and achieved 99.83% sulfide reduction and COD concentration was reduced to 98.29% (Mamat, Bustary, & Azoddein, 2020).

Advanced oxidation

In advanced oxidation processes, highly reactive hydroxy radicals ($OH°$) are generated in sufficient quantities and have the capacity of destroying recalcitrant organic compounds which are resistant to oxidation by conventional oxidants. The mechanism of degradation is highlighted below (Olajire, 2020).

Step 1: Initiation step—involves the formation of the alkyl radical by the treatment of hydrocarbons and hydroxyl radical

$$RH + HO°H_2O + R°$$

Step 2: Propagation step—Involves the reaction of the alkyl radical with molecular oxygen to generate alkoxy radical $RO_2°$ or reaction of the alkyl radical with other neutral organic molecules

$$R^\circ + O_2 RO_2^\circ$$

$$R^\circ + R'H \ RH + R'^\circ$$

Electron transfer reactions might also occur thereby ionizing the hydroxyl radical to hydroxide ion

$$RX + HO^\circ RX^\circ + HO^-$$

The organic compounds are hydroxylated by HO° at high electron sites leading to oxidative chain reactions

$$ArH + HO^\circ ArH(HO)^\circ$$

$$ArH(HO)^\circ + O_2 ArH(HO)O_2^\circ$$

$$ArH \ (HO)O_2^\circ Ar \ (HO) + HO_{2^-}$$

Step 3: Termination step—a combination of radicals to form neutral molecules

$$R^\circ + R^\circ R - R$$

$$R^\circ + HO^\circ ROH$$

$$HO^\circ + HO^\circ H_2O_2$$

$$RO_2^\circ + y(HO^\circ/O_2) \ \times CO_2 + y/2 \ H_2O$$

The hydroxyl radicals attack organic compounds by radical addition, hydrogen abstraction, and/or electron transfer. (OH°) exhibits a faster oxidation rate as compared to H_2O_2 and $KMnO_4$ Advanced oxidation processes include cavitation (OH°) generated either by ultrasonic irradiation or using constrictions in hydraulic devices, photocatalytic oxidation using UV/near UV radiation or sunlight in presence of a semiconductor catalyst, and Fenton reagent (using a reaction between Fe ions and H_2O_2) (Ali, Al deen, Palaniandy, & Feroz, 2019).

Cavitation

Cavitation is a physical phenomenon that involves the formation, growth, and subsequent collapse of microbubbles or cavities inside a liquid resulting in very high energy densities of the order of $1-1018 \ kW \ m^{-3}$ occurring in very short intervals of time (milliseconds) and releasing large amounts of energy (Gogate, Tayal, & Pandit, 2006). Out of the various cavitation techniques Acoustic cavitation(cavities created by passing sound waves through the liquid medium of the range of 20 kHz to 1 MHz) (Rajoriya, Carpenter, Saharan, & Pandit, 2016). Hydrodynamic cavitation (involves the passage of liquid via a constriction). The released high energy creates hot spots wherein (OH°) are generated from the decomposition of water molecules (Sivakumar & Pandit, 2002). Deng et al. have effectively used hydrodynamic cavitation for the removal of oil pollutants from water achieving 98.81% efficiency (Deng et al., 2018).

Photocatalytic oxidation

Photocatalytic oxidation has garnered much interest and has been well researched for the treatment of petroleum wastewater in recent years (Twesme, Tompkins, Anderson, & Root, 2006). This method has been effectively used to oxidize recalcitrant organic compounds left unaffected after biological treatments (Vodyanitskii, Yu, Trofimov, Ya., & Shoba, 2016). TiO_2 and ZnO are the most commonly used semiconductors for the treatment of petroleum effluents (Santos, Azevedo, Sant'Anna, & Dezotti, 2006; Tetteh, Rathilal, & Naidoo, 2020). Santos et al. utilized photocatalysis as a tertiary treatment for treating petroleum refinery wastewater using TiO_2, ZnO, and TiO_2 (Santos, Azevedo et al., 2006; Santos, Goulart et al., 2006). Removal of 93% of phenols, 63% of dissolved organic carbon, and more than 50% of oil and grease were achieved using TiO_2 (P25, Degussa). Lair et al. reported degradation of recalcitrant naphthalene present in petroleum effluents using TiO_2 in presence of inorganic anions (Lair, Ferronato, Chovelon, & Herrmann, 2008). Photocatalytic reactions are initiated when a photocatalyst such as TiO_2 absorbs a photon having higher energy than its bandgap, thereby generating an electron-hole pair by the excitation of an electron from the valence band (VB) to the conduction band (CB). These electron-hole pairs migrate to the surface of TiO_2 and trigger oxidation-reduction processes. Oxygen picks up electrons from the CB to form superoxide ions (O^{2-}). The holes oxidize water to form hydroxyl radicals (OH°). The generated superoxide ions and hydroxyl radicals are responsible for the decomposition of the organic pollutants. The advantage of photolytic oxidation is fast reaction rates, no sludge production, easy operation, and ambient temperature and pressure (Fig. 9.5).

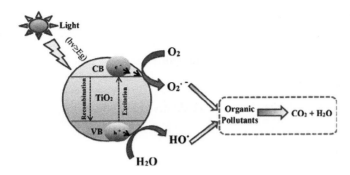

FIGURE 9.5 Mechanism of degradation of organic pollutants by TiO$_2$. *From Samsudin, E. M., Sze, N. G., Ta, Y. W., Tan, T. L., Abd. Hamid, S. B., & Joon, C. J. (2015). Evaluation on the photocatalytic degradation activity of reactive blue 4 using pure anatase nano-TiO2.* Sains Malaysiana, *44(7), 1011−1019.* https://doi.org/10.17576/jsm-2015-4407-13.

Fenton-related advanced oxygen processes

Fenton proposed that the addition of Fe^{2+} ions increased the oxidizing efficiency of H_2O_2. The mixture referred to as Fenton's reagent (mixture of Fe^{2+} and H_2O_2) when added to petroleum wastewater generates hydroxyl radicals ($OH°$). by the following equation:

$$Fe^{2+} + H_2O_2 Fe^{3+} + OH^- + OH°$$

Recently Yang et al. compared the efficiency of Fenton oxidation and activated persulphate oxidation for remediation of PHC contaminated groundwater (Yang et al., 2020). Fenton oxidation was found to have a higher reaction rate of 1.65 and 1.28 L h^{-1} for benzene and toluene respectively while with persulphate oxidation the reaction rates were 0.13 and 0.1 L h^{-1}.

The limitations of the Fenton treatment include slow regeneration of Fe^{2+}, need for separation of dissolved iron from the medium, and high volumes of iron sludge are generated which need to be treated in an acidic medium and this is an added additional expense (Olajire, 2020).

Ion exchange technology

This technology involves an interchange of toxic ions present in the petroleum effluent by nontoxic ions from the ion exchanger. Cation Ion exchangers exchange positively charged ions from the effluent while anion Ion exchanger exchange negatively charges ions. Commonly used ion exchangers are polystyrene sulfonic acids, zeolites, sodium silicates, acrylic, and *meta*acrylic resins. Ion exchange can be successfully used to remove low concentrations of both inorganic and organics (up to 250 mg L^{-1}) (Gupta, Ali, Saleh, Nayak, & Agarwal, 2012). This technology is used in combination with other pretreatment processes. Hashemi el utilized a combination of ultra-filtration, Ion exchange resin, and a multioxidant for treating oil refinery effluent effectively (Hashemi et al., 2020).

Biological treatment of petroleum effluent

These processes involve the oxidation of pollutants by microbes such as bacteria, protozoa, fungi, and algae. These microbes use pollutants as nutrients and break them down into harmless products. The domestic septic tank is the simplest example of the biological treatment process. Since the process is based on the growth of living entities, it is sensitive to temperature, pH of the medium, type of nutrients, and the presence of toxic substances. Some microbes require oxygen while others grow well in absence of oxygen. A different group of microbes grows on different nutrients. Therefore, biological process selection depends on the pollutants to be removed.

Generally, aliphatic and light aromatic fractions undergo microbial degradation, while high molecular weight aromatics, asphaltenes, and resins exhibit low rates of biodegradation (Atlas & Bragg, 2009). Microbial degradation rates increase with increasing temperature. Adapted communities of microbes, that is, those which have been previously exposed to hydrocarbons exhibit higher biodegradation rates. The mechanism of adaptation involves both selective enrichment and genetic changes increasing the hydrocarbon utilizing bacteria (Leahy & Colwell, 1990; Sarkar et al., 2020). Microbial adaptation has been observed in a variety of contaminated environments (Alidina, Li, & Drewes, 2014; Chonova et al., 2016; Pfaender, Shimp, & Larson, 1985). In the future recombinant DNA technology may be used to tailor-make bacterial and fungal strains exhibiting improved capability for biodegradation of specific pollutants, the only deterrent till now being the release of genetically engineered species in the environment.

Both Aerobic and anaerobic biological treatments are used depending on the kind of pollutants present. Conventionally, oxidation ponds (lagoons), activated sludge facilities, trickling filters, and rotating biological contractors have been used for treating petroleum effluents.

Aerobic waste treatment

The biodegradable organic matter present in petroleum effluents can be decomposed into simpler substances by microorganisms, they use organic constituents as food for their growth and clump together to form the active biomass. The common aim of all biological treatment systems for wastewater is to attain a high rate of biological oxygen demand (BOD) removal per volume and time by keeping a large amount of microbial biomass in contact with the wastewater. The biomass is then separated from wastewater by sedimentation.

$$COHNS + O_2 + microorganisms CO_2 + NH_3 + other\ products + biomass$$

In addition to carbon, several essential nutrients are required for microbial build-up. Nitrogen in the form of ammonia and sulfur in the form of sulfide and sulfate are present in process water but the absence of phosphorus might limit the growth of microbes. Thus, effluents from cooling towers and boilers that contain phosphorus are generally added. Sometimes direct phosphate salts might be added.

Oxidation ponds

Oxidation ponds are shallow lagoons that retain effluents for a few days for microbial degradation to occur. Atmospheric oxygen is available via the diffusion and photosynthetic activity of algae. Aerators may be used for augmentation. The process is efficient only in warm and sunny climates. Microbial biomass, ungraded organics, and silt deposit at the base of the pond call for regular cleaning and disposal problems. These accumulations may produce a foul odor and may even leach and contaminate the groundwater in the vicinity. Moreover, with the shortage of land, maintaining oxidation ponds is becoming difficult.

The activated sludge process

ASPIs is the most widespread method used in refineries around the world. Here the wastewater enters the aeration tank where microorganisms come in contact with it. Air is continuously injected into the system to keep the sludge aerobic. The mixture of wastewater and sludge is referred to as the "mixed liquor," and the biomass in the mixed liquor is referred to as "mixed liquor suspended solids" (MLSSs). The effluent from the aeration tank is sent to the clarifier where the biomass (SLUDGE) settles down and gets separated. A portion of the concentrated sludge, referred to as "return activated sludge" (RAS), from the clarifier is recycled back and mixed with incoming wastewater, and the remainder of the sludge is discharged as "waste activated sludge" (WAS) (IPIECA, 2010).

Supported growth biological treatment process-trickling filters

Trickling filters consists of bacteria lodged in microporous media (coke, slag, or gravel) which facilitates the growth of biological slime consisting of bacteria and infested with various worms and larvae. When the wastewater is passed down the towers, the bacterial slime decomposes the organic components in the wastewater utilizing atmospheric oxygen. With time the thickness of the slime increases cutting off the nutrient and oxygen supply to the microorganisms adhering to the solid surface leading to death. Moreover, as the bulk increases, the slime loses its ability to cling to the surface of the microporous media and is washed away. The treated effluent then enters the clarifier wherein the washed-away slime settles down as sludge. The general problem of trickling filters is their periodic clogging with microbial biomass. This requires regular renewal of the filter bed which is an expensive and disruptive process. Moreover, there is an added disadvantage of disposal of the large amount of used filter bed material generated.

Supported growth and suspended growth—rotating biological contactors

RBC is a more viable alternative to trickling filters used in some refineries as a rotating biological contactor. These are closely spaced semisubmerged rotating disks that provide a large alternately wetted and aerated surface for microbial biomass. Bacteria grow on closely spaced disks mounted on a shaft. About 40% of the disk is submerged in the effluent and the shaft rotates slowly exposing the bacteria to the atmosphere before submerging it in the effluent again. Dissolved organic compounds in the effluent are oxidized. When the microbial slime layer becomes thick it falls off due to the rotation of the disk. Thus RBC does not face clogging problems as in trickling filters. Also, microbial slime is dislodged as big chunks and thus has excellent settling characteristics.

Anaerobic waste treatment

The anaerobic process involves the decomposition of organic and inorganic matter in the absence of molecular oxygen. Organic matter is converted into CO_2 and CH_4. The wastewater is kept in a complex mix reactor system for $10-30$ days. Two groups of microbes are involved in the anaerobic digestion of waste. The first group referred to as the acid formers consist of facultative and anaerobic bacteria that bring about oxidation and fermentation of the organic waste to simple organic acids. The second group referred to as the methane formers convert these acids to methane and carbon dioxide. These bacteria have a very slow growth rate thus the main disadvantage of anaerobic decomposition is very slow and time-consuming. The advantage is that very little of the organic matter is utilized for the growth of the bacteria, of it is converted into methane which being combustible have immense usage as an energy source. Another advantage over aerobic methods is that less amount biomass is formed and the resulting solid material is well stabilized and usually suitable for disposal on the other hand the sludge obtained from aerobic treatment is large in amount with foul odor due to higher cellular organic material thus has to be digested anaerobically or dewatered and incinerated before disposal.

Anaerobic lagoons

Anaerobic lagoons are large man-made lagoons (sizes ranging from 1 to 2 acres) and up to 20 feet deep. Wastewater is piped in at the bottom of the lagoon. It settles down to form an upper liquid layer and a lower semisolid sludge layer. The liquid layer prevents oxygen to come in contact with the sludge layer resulting in anaerobic digestion of organic contaminants. This is a slow process and on average takes a few weeks.

Anaerobic sludge blanket reactors

Here the bioreactor has a free-floating layer of suspended sludge particles. As the anaerobes in the sludge blanket digest the organic matter and multiply, they flocculate and settle down. The treated water flows upwards and is collected while the evolved gases are collected by collection hoods. The anaerobic sludge blanket reactors can be of different forms.

- Up-flow anaerobic sludge blankets (UASBs): In UASB treatment, wastewater is introduced at the bottom of the bioreactor with upward flow applied. This causes the sludge blanket to float as the wastewater flows through it.
- Expanded granular sludge beds (EGSBs): EGSBs are very similar to UASB, with the key difference being that the wastewater is recirculated through the system to promote greater contact with the sludge.
- Anaerobic baffled reactors (ABRs): ABRs are constructed with semienclosed compartments that are separated by alternating baffles. The baffles compartmentalize the bioreactor and force the upward flow of effluents resulting in greater contact time with the active biomass (Fig. 9.6).

Advanced bioreactors for treating petroleum effluents

Several types of research have been carried out involving various modulations in the aerobic and anaerobic petroleum effluent treatments. These moderations involve either improvising the local environmental conditions so that the efficiency of the native microorganisms can be increased (biostimulation) or by adding more potent new microorganisms to facilitate faster and complete degradation of organic matter.

Sequential batch reactors

SBR A variation of the activated sludge process is the sequential Batch reactor. This process differs from the activated sludge method in that here both aeration and settling are carried out sequentially in the same tank. SBR is very effective for the treatment of effluents with high metal content. This method has the advantage of easy operability, low cost, and minimal sludge bulking (Malakahmad, Hasani, Eisakhani, & Isa, 2011).

Fluidized bed bioreactor

This is one of the recent methods used in effluent treatment. This method combines the advantages of the activated sludge method and trickling filtration into one method. The method involves passing petroleum effluent through a packed bed of particles at a velocity enough to fluidize the particles. Due to the upward movement of wastewater, a very dense microbial colony present on the surface of the bed can decompose the organic matter. The efficiency of

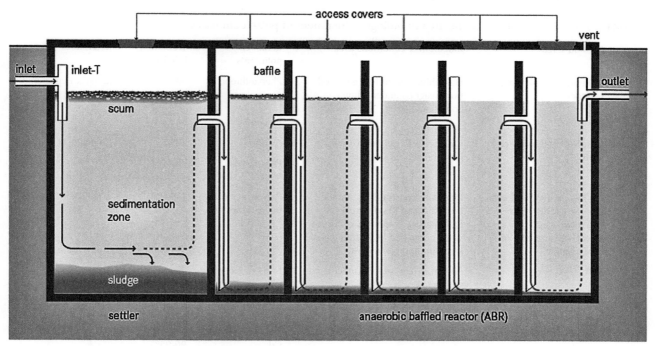

FIGURE 9.6 Schematic diagram for an anaerobic baffled reactor. *From Tilley, E., Ulrich, L., Luthi, C., Reymond, P. R. & Zurbrugg, C. (2014). Compendium of sanitation systems and technologies (2nd revised edition). Swiss Federal Institute of Aquatic Science and technology.*

FBB is 10 times the activated sludge treatment and requires 10% of the space required by the aeration tank. A common limitation observed in ASP and SBR is fairly high suspended solids in their effluents due to the low efficiency of the sedimentation process. The result is low mixed liquor volatile suspended solids (MLVSS) in the bioreactor consequently reducing the rate of biodegradation. Another limitation is that biological methods, in general, are sensitive to toxic shock loading which is very common in process water and petroleum effluents. To overcome these limitations a hybrid of biological treatment and membrane filtration technology has recently been used.

Membrane bioreactors

These are modifications on an activated sludge unit where the sedimentation tank has been replaced by a membrane filtration unit for efficient suspended solid removal. This technology has proved to be both space and time effective with the advantage that a much higher mixed liquor concentration can be used. The membranes are microporous films and are classified based on pore size as microfiltration (MF), ultrafiltration (UF), nanofiltration (NF), and reverse osmosis (RO). Recently many researchers have used MBRs to treat processed water. Membrane fouling is a common problem so these bioreactors require regular chemical cleaning thus a major limitation is the high maintenance and capital cost of membranes.

Membrane sequencing bioreactors

Recent studies have shown that alternating aerobic and anaerobic conditions in a sequential batch reactor (SBR) and replacing the settling phase by a micro or ultrafiltration membrane either submerged in the SBR or connected to it leads to better organic matter decomposition and nitrogen removal (Zhang et al., 2006). A recent study utilized MSBRs for remediation of petroleum refinery wastewater wherein air scouring and membrane relaxation has been used to minimize fouling (Pajoumshariati, Zare, & Bonakdarpour, 2017) (Table 9.2).

Electrochemical treatment of petroleum effluent

Electrochemical techniques like electrocoagulation (EC), electrodialysis (ED), and electrodialysis reversal (EDR) have great potential to replace conventional purification techniques since these methods are environmentally benign, versatile, energy-efficient, safe, can be easily automated, and are cost-efficient (Rajeshwar, Ibanez, & Swain, 1994). In

TABLE 9.2 Recent researches in the area of biological treatment of petroleum waste water.

	Method Used	Inoculum & conditions used	Removed pollutants	Treatment Efficiency	References
1.	Continuous stirred tank reactor	Halophilic consortium enriched from seagrass under saline conditions	Phenanthrene	98%	Jamal & Pugazhendi (2021)
			Fluorene	99%	
			Pyrene	84%	
2.	Fluidized-bed bioreactor (FBB)	*Rhodococcus* cultures Co-immobilized on sawdust	Alkanes	70%	Kuyukina et al. (2017)
			PAH Heavy Metal	75-96%	
3.	Upflow anaerobic sludge blanket (UASB) reactor and a two-stage biological aerated filter (BAF) system	Inoculum consisted of Candida tropicalis and Rhodotorula dairenensis). The heavy oil waste water contained large amounts of recalcitrant organic waste	Polyaromatic hydrocarbons	~90%	Zou (2015)
4.	FBB	Polypropylene support inoculated with refinery activated sludge	COD reduction	90%	Sokół (2003)
5.	SBR	Sludge from municipal AS plant. SBR operated on 8hour cycle	COD reduction	59-88%	Malakahmad et al. (2011)
			Hg	76-90%	
			Cd	96-98%	
6.	MBR (ultrafiltration)	hollow fiber membrane module. Permeate flux (L. m^{-2} h^{-1}): 15, Flow rate (L h^{-1}): 0.8 HRT (h): 27	COD reduction	88%	Capodici et al ("Treatment of Oily Wastewater with Membrane Bioreactor Systems," 2017)
			Ammonium removal & anti fouling	70%	
7.	Submerged MBR	Hollow fibre membrane	COD reduction	71%	Alsalhy et al. (2016)
			BOD reduction	60%	
			Phenol	100%	
8.	MSBR	Activate sludge inoculum	COD reduction	80%	Pajoumshariati et al. (2017)
			Oil & grease	82%	
			TPH	93.4%	

addition, instead of using chemicals and microorganisms, these technologies utilize only electrons for wastewater treatment. Thus, electrochemical treatment has garnered much interest in the area of industrial wastewater treatment.

Electrocoagulation

Electrocoagulation (EC) has been used over the past few years as a wastewater treatment technology. This method involves electrolysis of wastewater usually iron or aluminum anodes are used and inert materials like stainless steel cathodes are used. On applying electric current electrons are released from the anode which travels to the cathode.

The process involves three successive stages

In the process, anodic dissolution produces metal ions (in situ coagulants) along with hydroxyl ions and hydrogen gas at the cathode. The Fe^{2+} and Al^{3+} ions react with hydroxide ions to form insoluble hydroxides which destabilize the contaminants and break down the emulsions by reducing the electrostatic repulsion to the extent that Vander Waals attraction predominates thereby causing coagulation. The H_2 and O_2 gas bubbles adhere to the agglomerates making them rise to the surface creating a floc.

Pollutant removal efficiency is affected by electrolysis time, current strength pH, temperature, and interelectrode distance (A Review of Electrocoagulation Process for Wastewater Treatment, 2018). The efficiency of the EC cell can be

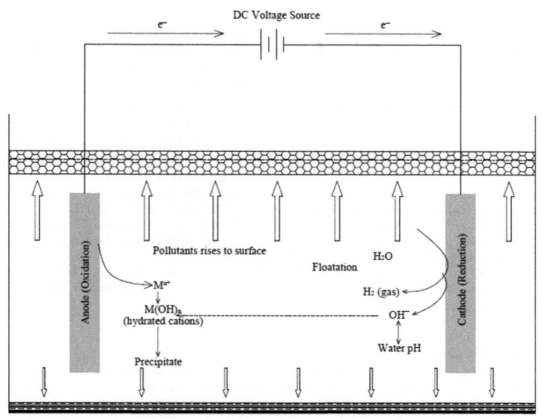

FIGURE 9.7 Bench scale two-electrode electrocoagulation reactor. *From Bharath, M., Krishana, B. M., & Kumar (2018). Review of electrocoagulation process for wastewater treatment.* International Journal of ChemTech Research, 11(3), 289–302.

increased by interchanging the polarity of the electrodes intermittently and by using EC cells with monopolar and bipolar electrodes either in parallel or series connection (Mollah et al., 2004) (Fig. 9.7).

Reactions

Anodic reactions:

$$Al \;\; Al^{+3} + 3e$$

$$Fe \;\; Fe^{+2} + 2e$$

$$2H_2O \;\; O_2 + 4\;H^+ + 4e$$

Atmospheric oxygen oxidizes Fe^{+2} to Fe^{+3}

Cathodic reactions:

$$2H_2O + 2e \;\; H_2 + 2OH^-$$

The advantage of this method is that it is a low investment, low maintenance, low sludge producing, and low space-occupying method as compared to other conventional methods of wastewater treatment. The disadvantage is that the sacrificial anode which slowly dissolves into the solution needs to be regularly replaced and over time formation of an impervious metal oxide coating on the cathode reduces efficiency.

Various researchers have used electrochemical-based technologies for the purification of petroleum effluents. In a recent study petroleum-based wastewater with initial COD of 1,6000 ppm was treated by EC utilizing Al anodes. A maximum COD removal of 88% was achieved in 5 min with 2.0 (Mohd Azli et al., 2020). A current Muftah et al. evaluated electrochemical coagulation for treating highly contaminated refinery wastewater concluded that the reactor performance was optimal at neutral pH and ambient temperature of 25°C (El-Naas, Surkatti, & Al-Zuhair, 2016).

Electrodialysis/electrodialysis reversal

After removal of oil, grease, suspended solids, and organic compounds, the dissolved solids (TDS) present in effluents need to be removed before making it usable. ED /EDR technologies involve the separation of dissolved ions from water based on selective transport of ions through ion exchange membranes. Ion exchange membranes are electrically conductive and water impermeable. The two types of membranes used in ED are:

- Anion Transfer Membranes (AEM) allow only negatively charged ions to pass through;
- Cation Transfer Membranes (CEM) allow only positively charged ions to pass through.

The ED reactor selectively removes dissolved solids based on their electrical charge by moving the wastewater through a series of alternating CEM and AEM placed between the anode and cathode. The positively charged cations migrate to the cathode by passing through the CEM and are rejected by AEM. The opposite occurs when negatively charged anions migrate to an anode. The result is an increase in ion concentration in one compartment (concentrate) and a decrease in ion concentration in the alternate compartment (dilute). EDR involves reversing the electrical charge after some time, this reversal of polarity prevents fouling and scaling of membranes (Manual of Water Supply Practices, 1995; Technical Assessment of produced water treatment technologies. An Integrated Framework for Treatment & Management of Produced Water Vols, 2009) (Fig. 9.8).

Both ED and EDR have till now only been tested at a laboratory scale. Sirivedhin et al. reported that ED could be efficiently used for desalination of low TDS-produced water (Sirivedhin, McCue, & Dallbauman, 2004). The major limitation of this method is regular membrane fouling and scaling.

Electrochemical oxidation

Electrochemical oxidation involves oxidation reaction at the anode bringing about oxidation of organic pollutants and reduction reaction at the cathode bringing about the reduction of heavy metals. The oxidation of organic pollutants can take place either by direct anodic oxidation (DAO) or Indirect or mediated electrochemical oxidation (MEO).

Direct anodic oxidation

In DAO, generally, pollutants are adsorbed on the anode surface and direct charge transfer occurs between the anode surface and the pollutant, the electrons can oxidize the organic pollutants.

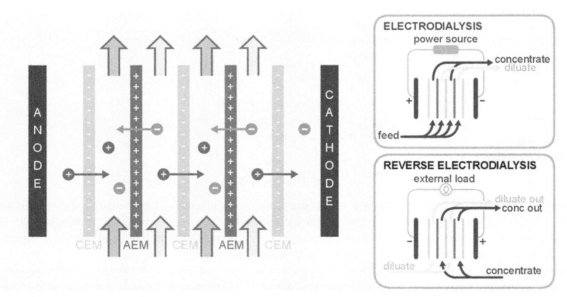

FIGURE 9.8 Principal of electrodialysis and reverse electrodialysis. *From Tedesco, M., Hamelers, H. V. M., & Biesheuvel, P. M. (2018). Nernst-Planck transport theory for (reverse) electrodialysis: III. Optimal membrane thickness for enhanced process performance.* Journal of Membrane Science, 565, 480–487. https://doi.org/10.1016/j.memsci.2018.07.090.

Mediated electrochemical oxidation

This process involves the electro generation of reactive oxygen and chlorine species at the electrode surface. Hydroxy radical ($^{\circ}OH$) is generated and adsorbed at the anode

$$M + H_2O \, M^{\circ}OH + H^+ + e^-$$

Here $M^{\circ}OH$ refers to the adsorbed hydroxyl radicals. These active anodes can bring about oxidation of organic pollutants to easily biodegradable short-chain carboxylic acids, carbon dioxide, and water. Another species that is very effective in the oxidation of recalcitrant organic pollutants is active chlorine. This is generated at the anode from the chloride ions present in the wastewater. The chlorine released at the anode undergoes hydrolysis to yield HClO, Cl^- and ClO^- these species bring about the oxidation of organic pollutants (Garcia-Segura, Ocon, & Chong, 2018).

$$2Cl - Cl_2aq + 2e$$

$$Cl_2aq + H_2O \; HClO + Cl^- + H^+$$

$$HClO \; H^+ + ClO^-$$

Active electrodes like (Pt, RuO_2, TiO_2, and IrO_2) and boron-doped diamond (BDD) electrodes are easily able to bring about oxidation of chlorine to active chlorine species leading to better remediation of organic contaminants.

Recently a lot of studies have been carried out on a laboratory scale to evaluate the effectiveness of electrochemical oxidation for the purification of petroleum wastewater. Santos et al. employed EC utilizing dimensionally stable anodes made of Ti plates coated with oxides of Ti and Ru (composition $Ti/Ru_{0.34}Ti_{0.66}O_2$). A 50% reduction in COD was observed following 70 h electrolysis at 50°C (Santos, Azevedo et al., 2006). Campos et al. evaluated electrochemical cells with Ti/Pt and BDD anode materials for treatment of petroleum effluent and concluded that BDD electrodes were most efficient removing 97% of dissolved organic matter (de Oliveira Campos, 2018). Gargouri et al. utilized electrooxidation for the purification of produced water using lead dioxide and BDD anodes. The COD removal was approximately 85% and 96% using PbO_2 and BDD after 11 and 7 h, respectively (Gargouri et al., 2014). All these studies are very encouraging, showing the effectiveness of electrochemical oxidation for treating petrochemical wastewater.

Combination of physical, chemical, biological, and electrical methods

Treatment of petroleum wastewater usually involves two stages—a mechanical pretreatment stage to remove the free oil, grease, and suspended solids followed by advanced treatments chemical, biological, electrochemical, or a combination of all to bring down pollutant levels to acceptable discharge values. Petroleum effluents are known to contain many soluble aromatic hydrocarbons, heavy metals, and even radioactive materials that are recalcitrant or might inhibit the growth of microbes thus in recent years many combinations of physicochemical, electrical technologies as pretreatments followed biological treatments have been used to reduce toxic shock from concentrated petroleum effluents to microorganisms in the bioreactors. Bahri et al. effectively used chemical oxidation via four different oxidation processes (photocatalytic, photo Fenton treatment, catalytic wet air oxidation, and wet air oxidation) as a pretreatment followed by biological treatment to reduce toxic shock for treating highly concentrated petroleum effluent (Bahri, Mahdavi, Mirzaei, Mansouri, & Haghighat, 2018). A combination of EC and the fixed-film biological process was used by Parez et al., the coupled EC-aerobic biofilter was able to remove up to 98% of the total TPH amount and over 95% of the COD load in the oil refinery wastewater (Pérez et al., 2016). Naas et al. successfully used a combination of EC, biodegradation, and adsorption to treat highly contaminated refinery wastewater and scaled up the process to pilot scale achieving 96% reduction of COD and 100% reduction of phenols (El-Naas et al., 2016).

Conclusions

Large volumes of wastewater are produced during oil and gas exploration. If this effluent is discharged directly into the environment without treatment, it may pollute the surroundings affecting the salinity, BOD, COD of the water bodies into which it is discharged. With stringent regulations on petroleum effluent discharge worldwide and shortage of water, there is an urgent need to develop advanced wastewater treatment facilities which are both operational and maintenance cost-efficient. Due to the complex nature of petroleum effluents, no single method is capable of complete removal of pollutants thus a combination of different methods has been seen to be more efficient. No one combination of treatments can cater to all oil fields worldwide as the composition of petroleum effluents are area-specific thus the treatments technologies used also vary. Though a lot of research is been carried out to develop robust advanced chemical, biological and electrical methods most of them are at the laboratory scale and are economically non-viable on large scale. Coagulation−flocculation and electrochemical treatments being environmentally benign, energy-efficient, and versatile is garnering much interest. Thus, scaling up cost-effective, environmentally benign treatment technologies is the need of the hour.

References

American Petroleum Institute (1989). Onshore solid waste management in exploration and production operations.

American Water Works Association (1995). Electrodialysis & electrodialysis reversal: Manual of water supply practices. Vol. M 38. American Water Works Association.

Ali, D., Al deen, A., Palaniandy, P., & Feroz, S. (2019). IGI Global. (pp. 99−122). Available from https://doi.org/10.4018/978-1-5225-5766-1.ch005.

Ahmadyasbchin, S., Andres, Y., Gerente, C., & Cloirec, P. (2008). Biosorption of Cu (II) from aqueous solution by Fucus serratus: Surface characterization and sorption mechanisms. *Bioresource Technology, 99*(14). Available from https://doi.org/10.1016/j.biortech.2007.12.040.

Ajmal, M., Rao, R. A. K., Ahmad, R., & Ahmad, J. (2000). Adsorption studies on Citrus reticulata (fruit peel of orange): removal and recovery of Ni (II) from electroplating wastewater. *Journal of Hazardous Materials, 79*(1−2). Available from https://doi.org/10.1016/S0304-3894(00)00234-X.

Aklil, A., Mouflih, M., & Sebti, S. (2004). Removal of heavy metal ions from water by using calcined phosphate as a new adsorbent. *Journal of Hazardous Materials, 112*(3). Available from https://doi.org/10.1016/j.jhazmat.2004.05.018.

Alidina, M., Li, D., & Drewes, J. E. (2014). Investigating the role for adaptation of the microbial community to transform trace organic chemicals during managed aquifer recharge. *Water Research, 56*, 172−180. Available from https://doi.org/10.1016/j.watres.2014.02.046.

Aman, T., Kazi, A. A., Sabri, M. U., & Bano, Q. (2008). Potato peels as solid waste for the removal of heavy metal copper (II) from waste water/industrial effluent. *Colloids and Surfaces B: Biointerfaces, 63*(1). Available from https://doi.org/10.1016/j.colsurfb.2007.11.013.

Atlas, R., & Bragg, J. (2009). Bioremediation of marine oil spills: When and when not - The Exxon Valdez experience. *Microbial Biotechnology, 2*(2), 213−221. Available from https://doi.org/10.1111/j.1751-7915.2008.00079.x.

Babel, S. (2003). Low-cost adsorbents for heavy metals uptake from contaminated water: a review. *Journal of Hazardous Materials, 97*(1−3). Available from https://doi.org/10.1016/S0304-3894(02)00263-7.

Babel, S., & Kurniawan, T. A. (2004). Cr (VI) removal from synthetic wastewater using coconut shell charcoal and commercial activated carbon modified with oxidizing agents and/or chitosan. *Chemosphere, 54*(7). Available from https://doi.org/10.1016/j.chemosphere.2003.10.001.

Bahri, M., Mahdavi, A., Mirzaei, A., Mansouri, A., & Haghighat, F. (2018). Integrated oxidation process and biological treatment for highly concentrated petrochemical effluents: A review. *Chemical Engineering and Processing - Process Intensification, 125*, 183−196. Available from https://doi.org/10.1016/j.cep.2018.02.002.

Bansode, R. R., Losso, J. N., Marshall, W. E., Rao, R. M., & Portier, R. J. (2003). Adsorption of metal ions by pecan shell-based granular activated carbons. *Bioresource Technology, 89*(2). Available from https://doi.org/10.1016/S0960-8524(03)00064-6.

Barakat, M. A. (2005). Adsorption behaviour of copper and cyanide ions at TiO2−solution interface. *Journal of Colloid and Interface Science, 291*(2). Available from https://doi.org/10.1016/j.jcis.2005.05.047.

Barakat, M. A. (2008). Removal of Cu (II), Ni (II) and Cr (III) ions from wastewater using complexation-ultrafiltration technique. *Journal of Environmental Science and Technology, 1*(3). Available from https://doi.org/10.3923/jest.2008.151.156.

Barakat, M. A. (2011). New trends in removing heavy metals from industrial wastewater. *Arabian Journal of Chemistry, 4*(4), 361−377. Available from https://doi.org/10.1016/j.arabjc.2010.07.019.

Barakat, M. A., & Sahiner, N. (2008). Cationic hydrogels for toxic arsenate removal from aqueous environment. *Journal of Environmental Management, 88*(4). Available from https://doi.org/10.1016/j.jenvman.2007.05.003.

Bari, A., & Farooq, S. (1985). Measurement of wastewater treatment efficiency by fluorescence and UV absorbance. *Environmental Monitoring and Assessment, 5*(4), 423−434. Available from https://doi.org/10.1007/BF00399469.

Barratt, A., & a, X. F. B. (1997). *Chemical Oxidation: Technologies for the Nineties, 6.*

Basaldella, E. I., Vázquez, P. G., Iucolano, F., & Caputo, D. (2007). Chromium removal from water using LTA zeolites: Effect of pH. *Journal of Colloid and Interface Science, 313*(2). Available from https://doi.org/10.1016/j.jcis.2007.04.066.

Bennett, G. F. (1988). The removal of oil from wastewater by air flotation: A review. *Critical Reviews in Environmental Control, 18*(3), 189−253. Available from https://doi.org/10.1080/10643388809388348.

Bharath, M., Krishna, B. M., & Manoj Kumar, B. (2018). A review of electrocoagulation process for wastewater treatment. *International Journal of ChemTech Research.* Available from https://doi.org/10.20902/IJCTR.2018.110333.

Bishnoi, N. R., Bajaj, M., Sharma, N., & Gupta, A. (2004). Adsorption of Cr (VI) on activated rice husk carbon and activated alumina. *Bioresource Technology, 91*(3). Available from https://doi.org/10.1016/S0960-8524(03)00204-9.

Bose, P., Aparna Bose, M., & Kumar, S. (2002). Critical evaluation of treatment strategies involving adsorption and chelation for wastewater containing copper, zinc and cyanide. *Advances in Environmental Research, 7*(1). Available from https://doi.org/10.1016/S1093-0191(01)00125-3.

Chen, X., & Chen, G. (2010). *Electroflotation. Electrochemistry for the environment* (pp. 263−277). New York: Springer. Available from https://doi.org/10.1007/978-0-387-68318-8_11.

Chonova, T., Keck, F., Labanowski, J., Montuelle, B., Rimet, F., & Bouchez, A. (2016). Separate treatment of hospital and urban wastewaters: A real scale comparison of effluents and their effect on microbial communities. *Science of the Total Environment, 542*, 965–975. Available from https://doi.org/10.1016/j.scitotenv.2015.10.161.

Comninellis, C., & Chen, G. (2010). *Electrochemistry for the environment. Electrochemistry for the environment* (pp. 1–563). New York: Springer. Available from https://doi.org/10.1007/978-0-387-68318-8.

Crini, G. (2005). Recent developments in polysaccharide-based materials used as adsorbents in wastewater treatment. *Progress in Polymer Science, 30* (1), 38–70. Available from https://doi.org/10.1016/j.progpolymsci.2004.11.002.

Danso-Boateng, E., & Osei-Wusu, A. (2013). Environmental management in the oil, gas and related energy industries in Ghana. *International Journal of Chemistry Environmental Engineeing, 4*, 116–122.

de Oliveira Campos, V. (2018). Electrochemical treatment of produced water using Ti/Pt and BDD anode. *International Journal of Electrochemical Science, 13*(8), 7894–7906. Available from https://doi.org/10.20964/2018.08.44.

Deliyanni, E. A., Kyzas, G. Z., & Matis, K. A. (2017). Various flotation techniques for metal ions removal. *Journal of Molecular Liquids, 225*, 260–264. Available from https://doi.org/10.1016/j.molliq.2016.11.069.

Deng, C., Lu, G., Zhu, M., Li, K., Ma, J., & Liu, H. (2018). Effective degradation of oil pollutants in water by hydrodynamic cavitation combined with electrocatalytic membrane. *AIP Advances, 8*(12). Available from https://doi.org/10.1063/1.5028152.

Edzwald, J. K. (1995). Principles and applications of dissolved air floatation. *Water Science and Technology, 34*, 200–207. Available from https://doi.org/10.1016/0273-1223-95.

El-Naas, M. H., Surkatti, R., & Al-Zuhair, S. (2016). Petroleum refinery wastewater treatment: A pilot scale study. *Journal of Water Process Engineering, 14*, 71–76. Available from https://doi.org/10.1016/j.jwpe.2016.10.005.

El-Sikaily, A., Nemrel, A., Khaled, A., & Abdelwehab, O. (2007). Removal of toxic chromium from wastewater using green alga Ulva lactuca and its activated carbon. *Journal of Hazardous Materials, 148*(1–2). Available from https://doi.org/10.1016/j.jhazmat.2007.01.146.

Essawy, H. A., & Ibrahim, H. S. (2004). Synthesis and characterization of poly(vinylpyrrolidone-co-methylacrylate) hydrogel for removal and recovery of heavy metal ions from wastewater. *Reactive and Functional Polymers, 61*(3). Available from https://doi.org/10.1016/j.reactfunctpolym.2004.08.003.

Feng, D., & Aldrich, C. (2004). Adsorption of heavy metals by biomaterials derived from the marine alga Ecklonia maxima. *Hydrometallurgy, 73* (1–2). Available from https://doi.org/10.1016/S0304-386X(03)00138-5.

Garcia-Segura, S., Ocon, J. D., & Chong, M. N. (2018). Electrochemical oxidation remediation of real wastewater effluents—A review. *Process Safety and Environmental Protection, 113*, 48–67. Available from https://doi.org/10.1016/j.psep.2017.09.014.

Gargouri, B., Gargouri, O. D., Gargouri, B., Trabelsi, S. K., Abdelhedi, R., & Bouaziz, M. (2014). Application of electrochemical technology for removing petroleum hydrocarbons from produced water using lead dioxide and boron-doped diamond electrodes. *Chemosphere, 117*(1), 309–315. Available from https://doi.org/10.1016/j.chemosphere.2014.07.067.

Ghazizade, M. J., Koulivand, H., Safari, E., & Heidari, L. (2021). Petrochemical waste characterization and management at pars special economic energy zone in the south of Iran. *Waste Management & Research: The Journal for a Sustainable Circular Economy, 39*(2), 199–208. Available from https://doi.org/10.1177/0734242x20922585.

Ghosh, U. C., Dasgupta, M., Debnath, S., & Bhat, S. C. (2003). Studies on management of Chromium(VI) – Contaminated industrial waste effluent using hydrous titanium oxide (HTO). *Water, Air, and Soil Pollution, 143*, 245–256. Available from https://doi.org/10.1023/A:1022814401404

Gogate, P. R., Tayal, R. K., & Pandit, A. B. (2006). Cavitation: A technology on the horizon. *Current Science, 91*.

Gopalratnam, V. C., Bennett, G. F., & Peters, R. W. (1992). Effect of collector dosage on metal removal by precipitation/flotation. *Journal of Environmental Engineering, 118*(6), 923–948. Available from https://doi.org/10.1061/(ASCE)0733-9372(1992)118:6(923).

Guidelines for characterization of offshore drill cuttings piles. (2003).

Gupta, V. K., Ali, I., Saleh, T. A., Nayak, A., & Agarwal, S. (2012). Chemical treatment technologies for waste-water recycling—An overview. *RSC Advances, 2*(16), 6380. Available from https://doi.org/10.1039/c2ra20340e.

Gupta, V. K., Rastogi, A., Saini, V. K., & Jain, N. (2006). Biosorption of copper(II) from aqueous solutions by Spirogyra species. *Journal of Colloid and Interface Science, 296*(1). Available from https://doi.org/10.1016/j.jcis.2005.08.033.

Gupta, V. K., Jain, C. K., Ali, I., Sharma, M., & Saini, V. K. (2003). Removal of cadmium and nickel from wastewater using bagasse fly ash—A sugar industry waste. *Water Research, 37*(16). Available from https://doi.org/10.1016/S0043-1354(03)00292-6.

Gupta, V. K., & Rastogi, A. (2008). Biosorption of lead(II) from aqueous solutions by non-living algal biomass Oedogonium sp. and Nostoc sp.—A comparative study. *Colloids and Surfaces B: Biointerfaces, 64*(2). Available from https://doi.org/10.1016/j.colsurfb.2008.01.019.

Harker, J. H., Backhurst, J. R., & Richardson, J. F. (2013). *Chemical engineering* (2nd ed.). Elsevier.

Hashemi, F., Hashemi, H., Shahbazi, M., Dehghani, M., Hoseini, M., & Shafeie, A. (2020). Reclamation of real oil refinery effluent as makeup water in cooling towers using ultrafiltration, ion exchange and multioxidant disinfectant. *Water Resources and Industry, 23*. Available from https://doi.org/10.1016/j.wri.2019.100123.

Hoseinian, F. S., Rezai, B., Kowsari, E., & Safari, M. (2018). Kinetic study of Ni(II) removal using ion flotation: Effect of chemical interactions. *Minerals Engineering, 119*, 212–221. Available from https://doi.org/10.1016/j.mineng.2018.01.028.

Hu, G., Li, J., & Hou, H. (2015). A combination of solvent extraction and freeze thaw for oil recovery from petroleum refinery wastewater treatment pond sludge. *Journal of Hazardous Materials, 283*, 832–840. Available from https://doi.org/10.1016/j.jhazmat.2014.10.028.

Huang, G. H., & Li, Y. P. (2008). Editorial: Emerging technologies for petroleum waste management. *Petroleum Science and Technology, 26*(7–8), 759–763. Available from https://doi.org/10.1080/10916460701824458.

Igunnu, E. T., & Chen, G. Z. (2014). Produced water treatment technologies. *International Journal of Low-Carbon Technologies*, *9*(3), 157–177. Available from https://doi.org/10.1093/ijlct/cts049.

IPIECA (2010). Petroleum refining water/waste water use & management. Operations best practice series.

Jafarinejad, S. (2017). *Solid-waste management in the petroleum industry* (pp. 269–345). Elsevier BV. Available from https://doi.org/10.1016/b978-0-12-809243-9.00007-9.

Johnson, O. A., & Affam, A. C. (2019). Petroleum sludge treatment and disposal: A review. *Environmental Engineering Research*, *24*(2), 191–201. Available from https://doi.org/10.4491/EER.2018.134.

Kesenci, K., Say, R., & Denizli, A. (2002). Removal of heavy metal ions from water by using poly (ethyleneglycol dimethacrylate-co-acrylamide) beads. *European Polymer Journal*, *38*(7). Available from https://doi.org/10.1016/S0014-3057(01)00311-1.

Kyzas, G. Z., & Matis, K. A. (2018). Flotation in water and wastewater treatment. *Processes*, *6*(8). Available from https://doi.org/10.3390/pr6080116.

Lair, A., Ferronato, C., Chovelon, J. M., & Herrmann, J. M. (2008). Naphthalene degradation in water by heterogeneous photocatalysis: An investigation of the influence of inorganic anions. *Journal of Photochemistry and Photobiology A: Chemistry*, *193*(2–3), 193–203. Available from https://doi.org/10.1016/j.jphotochem.2007.06.025.

Leahy, J. G., & Colwell, R. R. (1990). Microbial degradation of hydrocarbons in the environment. *Microbiological Reviews*, *54*(3), 305–315. Available from https://doi.org/10.1128/mmbr.54.3.305-315.1990.

Lee, T., Park, J., & Lee, J.-H. (2004). Waste green sands as reactive media for the removal of zinc from water. *Chemosphere*, *56*(6). Available from https://doi.org/10.1016/j.chemosphere.2004.04.037.

Li, X., Zhang, C., & Liu, J. (2010). Adsorption of oil from waste water by coal: Characteristics and mechanism. *Mining Science and Technology*, *20*(5), 778–781. Available from https://doi.org/10.1016/S1674-5264(09)60280-5.

Li, X. Z., & Li, F. B. (2001). Study of Au/Au3 + -TiO2 photocatalysts toward visible photooxidation for water and wastewater treatment. *Environmental Science and Technology*, *35*(11), 2381–2387. Available from https://doi.org/10.1021/es001752w.

Liang, Y., Ning, Y., Liao, L., & Yuan, B. (2018). *Special focus on produced water in oil and gas fields: Origin, management, and reinjection practice. Formation damage during improved oil recovery: Fundamentals and applications* (pp. 515–586). Elsevier. Available from https://doi.org/10.1016/B978-0-12-813782-6.00014-2.

Liu, X. D., Tokura, S., Nishi, N., & Sakairi, N. (2003). A novel method for immobilization of chitosan onto nonporous glass beads through a 1,3-thiazolidine linker. *Polymer*, *44*(4). Available from https://doi.org/10.1016/S0032-3861(02)00879-0.

Mahmoud, M. E., & Obada, M. K. (2014). Improved sea water quality by removal of the total hardness using static step-by-step deposition and extraction technique as an efficient pretreatment method. *Chemical Engineering Journal*, *252*, 355–361. Available from https://doi.org/10.1016/j.cej.2014.04.064.

Malakahmad, A., Hasani, A., Eisakhani, M., & Isa, M. H. (2011). Sequencing batch reactor (SBR) for the removal of Hg2 + and Cd2 + from synthetic petrochemical factory wastewater. *Journal of Hazardous Materials*, *191*(1–3), 118–125. Available from https://doi.org/10.1016/j.jhazmat.2011.04.045.

Mamat, M. C., Bustary, A. B., & Azoddein, A. A. M. (2020). Oxidation of sulfide removal from petroleum refinery wastewater by using hydrogen peroxide. IOP conference series: Materials science and engineering (736, 2). Institute of Physics Publishing. Available from https://doi.org/10.1088/1757-899X/736/2/022103.

Massey, M., Dunlap, R., Koblin, A., Luthy, R., & Nakles, D. (1976). Analysis of coal wastewater analytical methods: A case study of the Hygas pilot plant. Available from https://doi.org/10.2172/6415648.

Mohammed, R. S., Mohammad, R. S., & Ali, A. R. (2018). The economic impact of green marketing on petroleum residues: Theoretical and analytical study. *Journal of Advance Research in Dynamical & Control Systems*, *12*. Available from https://doi.org/10.13140/RG.2.2.20524.62087.

Mohd Azli, F. A., Mohd Azoddein, A. A., Abu Seman, M. N., Abdul Hamid, A. S., Tajuddin, T., ... Nurdin, S. (2020). *Treatment of petroleum-based industrial wastewater using electrocoagulation technology. Advances in waste processing technology* (pp. 49–59). Springer Science and Business Media LLC. Available from https://doi.org/10.1007/978-981-15-4821-5_4.

Mollah, M. Y. A., Morkovsky, P., Gomes, J. A. G., Kesmez, M., Parga, J., & Cocke, D. L. (2004). Fundamentals, present and future perspectives of electrocoagulation. *Journal of Hazardous Materials*, *114*(1–3), 199–210. Available from https://doi.org/10.1016/j.jhazmat.2004.08.009.

Mouflih, M., Aklil, A., & Sebti, S. (2005). Removal of lead from aqueous solutions by activated phosphate. *Journal of Hazardous Materials*, *119*(1–3). Available from https://doi.org/10.1016/j.jhazmat.2004.12.005.

Moursy, A. S., & Abo El-Ela, S. E. (1982). Treatment of oily refinery wastes using a dissolved air flotation process. *Environment International*, *7*(4), 267–270. Available from https://doi.org/10.1016/0160-4120(82)90116-7.

Neff, J. M. (2005). Composition, environmental, fates, and biological effects of water-based drilling muds and cuttings discharged to the marine environment.

Nah, I. W., Hwang, K.-Y., Jeon, C., & Choi, H. B. (2006). Removal of Pb ion from water by magnetically modified zeolite. *Minerals Engineering*, *19*(14). Available from https://doi.org/10.1016/j.mineng.2005.12.006.

Olajire, A. A. (2020). Recent advances on the treatment technology of oil and gas produced water for sustainable energy industry-mechanistic aspects and process chemistry perspectives. *Chemical Engineering Journal Advances*, *4*, 100049. Available from https://doi.org/10.1016/j.ceja.2020.100049.

Pajoumshariati, S., Zare, N., & Bonakdarpour, B. (2017). Considering membrane sequencing batch reactors for the biological treatment of petroleum refinery wastewaters. *Journal of Membrane Science*, *523*, 542–550. Available from https://doi.org/10.1016/j.memsci.2016.10.031.

Pan, B. C., Zhang, Q. R., Zhang, W. M., Pan, B. J., Du, W., Lv, L., ... Zhang, Q. X. (2007). Highly effective removal of heavy metals by polymer-based zirconium phosphate: A case study of lead ion. *Journal of Colloid and Interface Science*, *310*(1). Available from https://doi.org/10.1016/j.jcis.2007.01.064.

Paranhos Gazineu, M. H., De Araújo, A. A., Brandão, Y. B., Hazin, C. A., & Godoy, J. M. D. O. (2005). Radioactivity concentration in liquid and solid phases of scale and sludge generated in the petroleum industry. *Journal of Environmental Radioactivity*, *81*(1), 47−54. Available from https://doi.org/10.1016/j.jenvrad.2004.11.003.

Pérez, L. S., Rodriguez, O. M., Reyna, S., Sánchez-Salas, J. L., Lozada, J. D., Quiroz, M. A., . . . Bandala, E. R. (2016). Oil refinery wastewater treatment using coupled electrocoagulation and fixed film biological processes. *Physics and Chemistry of the Earth*, *91*, 53−60. Available from https://doi.org/10.1016/j.pce.2015.10.018.

Pfaender, F. K., Shimp, R. J., & Larson, R. J. (1985). Adaptation of estuarine ecosystems to the biodegradation of nitrilotriacetic acid: Effects of pre-exposure. *Environmental Toxicology and Chemistry*, *4*(5), 587−593. Available from https://doi.org/10.1002/etc.5620040502.

Piccioli, M., Aeneson, S. V., Zhao, H., Dudek, M., & Oye, G. (2020). Gas flotation of petroleum produced water: A review on status, fundamental aspects, and perspectives. *Energy and Fuels*, *34*(12). Available from https://doi.org/10.1021/acs.energyfuels.0c03262.

Pintor, A. M. A., Vilar, V. J. P., Botelho, C. M. S., & Boaventura, R. A. R. (2016). Oil and grease removal from wastewaters: Sorption treatment as an alternative to state-of-the-art technologies. A critical review. *Chemical Engineering Journal*, *297*, 229−255. Available from https://doi.org/10.1016/j.cej.2016.03.121.

Polat, H., & Erdogan, D. (2007). Heavy metal removal from waste waters by ion flotation. *Journal of Hazardous Materials*, *148*(1−2), 267−273. Available from https://doi.org/10.1016/j.jhazmat.2007.02.013.

Pollard, S. J. T., Fowler, G. D., Sollars, C. J., & Perry, R. (1992). Low-cost adsorbents for waste and wastewater treatment: A review. *Science of the Total Environment, The*, *116*(1−2), 31−52. Available from https://doi.org/10.1016/0048-9697(92)90363-W.

Pradhan, S., Shukla, S. S., & Dorris, K. L. (2005). Removal of nickel from aqueous solutions using crab shells. *Journal of Hazardous Materials*, *125* (1−3). Available from https://doi.org/10.1016/j.jhazmat.2005.05.029.

Rajeshwar, K., Ibanez, J. G., & Swain, G. M. (1994). Electrochemistry and the environment. *Journal of Applied Electrochemistry*, *24*(11), 1077−1091. Available from https://doi.org/10.1007/BF00241305.

Rajoriya, S., Carpenter, J., Saharan, V. K., & Pandit, A. B. (2016). Hydrodynamic cavitation: An advanced oxidation process for the degradation of bio-refractory pollutants. *Reviews in Chemical Engineering*, *32*(4), 379−411. Available from https://doi.org/10.1515/revce-2015-0075.

Saad, B., Chong, C. C., Ali, A. S. M., Bari, M. F., Rahman, I. A., Mohamad, N., & Saleh, M. I. (2006). Selective removal of heavy metal ions using sol−gel immobilized and SPE-coated thiacrown ethers. *Analytica Chimica Acta*, *555*(1). Available from https://doi.org/10.1016/j.aca.2005.08.070.

Santos, F. V., Azevedo, E. B., Sant'Anna, G. L., & Dezotti, M. (2006). Photocatalysis as a tertiary treatment for petroleum refinery wastewaters. *Brazilian Journal of Chemical Engineering*, *23*(4), 451−460. Available from https://doi.org/10.1590/S0104-66322006000400003.

Santos, M. R. G., Goulart, M. O. F., Tonholo, J., & Zanta, C. L. P. S. (2006). The application of electrochemical technology to the remediation of oily wastewater. *Chemosphere*, *64*(3), 393−399. Available from https://doi.org/10.1016/j.chemosphere.2005.12.036.

Sarkar, J., Saha, A., Roy, A., Bose, H., Pal, S., Sar, P., & Kazy, S. K. (2020). Development of nitrate stimulated hydrocarbon degrading microbial consortia from refinery sludge as potent bioaugmenting agent for enhanced bioremediation of petroleum contaminated waste. *World Journal of Microbiology and Biotechnology*, *36*(10). Available from https://doi.org/10.1007/s11274-020-02925-z.

Sebba, F. (1962). Ion flotation.

Sirivedhin, T., McCue, J., & Dallbauman, L. (2004). Reclaiming produced water for beneficial use: Salt removal by electrodialysis. *Journal of Membrane Science*, *243*(1−2), 335−343. Available from https://doi.org/10.1016/j.memsci.2004.06.038.

Sivakumar, M., & Pandit, A. B. (2002). Wastewater treatment: A novel energy efficient hydrodynamic cavitational technique. *Ultrasonics Sonochemistry*, *9*(3), 123−131. Available from https://doi.org/10.1016/S1350-4177(01)00122-5.

Stevens-Garmon, J., Drewes, J. E., Khan, S. J., McDonald, J. A., & Dickenson, E. R. V. (2011). Sorption of emerging trace organic compounds onto wastewater sludge solids. *Water Research*, *45*(11), 3417−3426. Available from https://doi.org/10.1016/j.watres.2011.03.056.

Tang, P. L., Lee, C. K., Low, K. S., & Zainal, Z. (2003). Sorption of Cr (VI) and Cu (II) in aqueous solution by ethylenediamine modified rice hull. *Environmental Technology*, *24*(10). Available from https://doi.org/10.1080/09593330309385666.

Tarleton, E. S., & Wakeman, R. J. (2007). *Pre-treatment of suspensions. Solid/liquid separation*. Elsevier. Available from https://doi.org/10.1016/B978-185617421-3/50003-1.

Technical Assessment of produced water treatment technologies. An integrated framework for treatment and management of produced water (Vols. 7122-12) (2009). Colorado School of Mines.

Tetteh, E. K., Rathilal, S., & Naidoo, D. B. (2020). Photocatalytic degradation of oily waste and phenol from a local South Africa oil refinery wastewater using response methodology. *Scientific Reports*, *10*(1). Available from https://doi.org/10.1038/s41598-020-65480-5.

Twesme, T. M., Tompkins, D. T., Anderson, M. A., & Root, T. W. (2006). Photocatalytic oxidation of low molecular weight alkanes: Observations with ZrO_2-TiO_2 supported thin films. *Applied Catalysis B: Environmental*, *64*(3−4), 153−160. Available from https://doi.org/10.1016/j.apcatb.2005.11.010.

Uysal, M., & Ar, I. (2007). Removal of Cr (VI) from industrial wastewaters by adsorption. *Journal of Hazardous Materials*, *149*(2). Available from https://doi.org/10.1016/j.jhazmat.2007.04.019.

Varjani, S., Joshi, R., Srivastava, V. K., Ngo, H. H., & Guo, W. (2020). Treatment of wastewater from petroleum industry: current practices and perspectives. *Environmental Science and Pollution Research*, *27*(22), 27172−27180. Available from https://doi.org/10.1007/s11356-019-04725-x.

Vodyanitskii, Y. N., Trofimov, S. Y., & Shoba, S. A. (2016). Promising approaches to the purification of soils and groundwater from hydrocarbons (A review). *Eurasian Soil Science*, *49*(6), 705−713. Available from https://doi.org/10.1134/S1064229316040141.

Walter, J. W. (1972). Sedimentation. Physiochemical processes for water quality control.

Wei, X., Zhang, S., Sun, Y., & Brenner, S. A. (2018). Petrochemical wastewater and produced water. *Water Environment Research*, *90*(10), 1634−1647. Available from https://doi.org/10.2175/106143018X15289915807344.

Yang, Z. H., Verpoort, F., Dong, C. D., Chen, C. W., Chen, S., & Kao, C. M. (2020). Remediation of petroleum-hydrocarbon contaminated groundwater using optimized in situ chemical oxidation system: Batch and column studies. *Process Safety and Environmental Protection, 138*, 18−26. Available from https://doi.org/10.1016/j.psep.2020.02.032.

Yi, Y., Wang, Y., & Liu, H. (2003). Preparation of new crosslinked chitosan with crown ether and their adsorption for silver ion for antibacterial activities. *Carbohydrate Polymers, 53*(4). Available from https://doi.org/10.1016/S0144-8617(03)00104-8.

Zhang, H. M., Xiao, J. N., Cheng, Y. J., Liu, L. F., Zhang, X. W., & Yang, F. L. (2006). Comparison between a sequencing batch membrane bioreactor and a conventional membrane bioreactor. *Process Biochemistry, 41*(1), 87−95. Available from https://doi.org/10.1016/j.procbio.2005.03.072.

Zhao, C., Zhou, J., Yan, Y., Yang, L., Xing, G., Li, H., ... Zheng, H. (2021). Application of coagulation/flocculation in oily wastewater treatment: A review. *Science of the Total Environment, 765*. Available from https://doi.org/10.1016/j.scitotenv.2020.142795.

Zheng, Y., & Bagley, D. M. (1999). Numerical simulation of batch settling process. *Journal of Environmental Engineering, 125*(11), 1007−1013. Available from https://doi.org/10.1061/(ASCE)0733-9372(1999)125:11(1007).

Chapter 10

Nanocomposite material-based catalyst, adsorbent, and membranes for petroleum wastewater treatment

Muneer M. Ba-Abbad[1], Ebrahim Mahmoudi[2], Abdelbaki Benamor[1] and Abdul Wahab Mohammad[2]

[1]Gas Processing Center, College of Engineering, Qatar University, Doha, Qatar, [2]Department of Chemical and Process Engineering, Faculty of Engineering and Built Environment, Universiti Kebangsaan Malaysia, Bangi, Selangor, Malaysia

Introduction

Nanotechnology has emerged as a new superior alternative approach to enhance existing techniques in various fields of science namely, food science, biomedicine, electronics, energy, agriculture, environmental remediation (Pramanik et al., 2020), and many other areas. One of the main objectives of environmental remediation is to prevent contamination from industrial wastewater onto groundwater, surface water, and soil. Among the most challenging industrial wastewater is petroleum processing wastewater (Dasgupta, Ranjan, & Lichtfouse, 2020).

Petroleum processing wastewater is one of the most complex wastewaters produced by industry to date. It usually contains a variety of free, soluble, and emulsive hydrocarbons, mixed with numerous types of heavy metals and/or solid particles. Accumulation of such harmful contaminants in the environment and specifically the human body can cause catastrophic damages. Water sources are so vital for humans, the presence of heavy metals in the environment and water has been of great concern due to its toxic nature and other adverse effects on receiving water (Huang, Yang, Zhang, & Shi, 2009). Heavy metals have a direct effect on the nervous and gastrointestinal systems of the human body. Therefore, it is necessary to minimize exposure to such contaminations by removing obnoxious heavy metals from water and the environment that are produced by various industries such as the petroleum industry (Matouq, Jildeh, Qtaishat, Hindiyeh, & Al Syouf, 2015).

Hence many researchers have focused on different methods to treat petroleum processing wastewater in various respects. The use of nanotechnologies is considered an attractive solution for the purification of such harmful wastewater. A variety of wastewater treatment methods have used nanotechnology to improve the method. Two of the most popular methods that have been reported are nanosorption (adsorption) and mixed matrixed membranes modified using nanoparticles. Both methods have been proven to be among the most effective for the treatment of petroleum processing wastewater.

Application of sorption for petroleum wastewater treatment

Adsorption is the transfer of materials from one phase (usually gas or liquid) to another phase (usually solid) at the interface between the phases. The adsorbent has a high surface area and traps organic and inorganic molecules employing intermolecular forces. The substance adsorbed or attached is called the adsorbate (Mahmoudi et al., 2020).

Adsorbents with higher porosity and surface area have proven to be more effective and efficient in the adsorption and removal of contaminants. Nanoadsorbents possess unique properties such as high surface area, porosity, and modulated surface-active functional sites, fast kinetics, and high proficiency (Mahmoudi et al., 2020). As a result, the employment of nanoadsorbents for the purification of low concentration petroleum wastewater has been under investigation by many researchers (Song et al., 2020). Since petroleum wastewater is very complex to have successful efficient adsorption, the characteristic of the wastewater should be completely investigated and the adsorbates should be

Petroleum Industry Wastewater. DOI: https://doi.org/10.1016/B978-0-323-85884-7.00014-X

identified. Subsequently, adsorbents should be chosen based on the possible mechanism suitable for the removal of the adsorbate. Adsorbates such as Hydrocarbon oils, chlorophenol (CP) (Alkindy, Naddeo, Banat, & Hasan, 2020), biphenyl, bisphenol A (BPA), Polycyclic aromatic hydrocarbons (PAHs), chlorobenzene (CB) (Li, Liao, & Zhou, 2018) and most of the organic material can be removed using the interactions of the $\pi-\pi$ (Pi−Pi) forces by the interaction between the C = C of an alkene or aromatic planes between the adsorbent and adsorbate (Angelova, Uzunov, Uzunova, Gigova, & Minchev, 2011). Most of the organic pollutants can be adsorbed through a n-π adsorption force, in this adsorption mechanism, the electron depleted sites of adsorbent interact with nonbonding electron pair of organic pollutants and remove them from the liquid phase efficiently (Angelova et al., 2011). To remove pollutants such as Surfactants, reactive dyes, chemicals, and heavy metals, adsorbents with cation-π actives sites can be employed (Ersan, Apul, Perreault, & Karanfil, 2017). In this type of adsorption π-electrons of the absorbent interacts with protonated functional groups of adsorbates resulting in the removal of pollutants from petroleum wastewater. Similarly, anionic adsorbates molecules can be adsorbed via interactions between the anion−π and the molecules. Nanoadsorbents such as activated carbon (AC) family, carbon nanotube (CNT), grapheme (G), Metal-organic frameworks (MOFs), dendrimers, nanopolymer, and nanocomposites metal oxides (Kukkar et al., 2020; Liu, Wang, Zhang, Wu, & Li, 2020; Oyewo, Elemike, Onwudiwe, & Onyango, 2020) can be an ideal adsorbent if the π-X family interactions are the primary force of adsorption. One of the major forces in adsorption for petroleum wastewater is the hydrophobic interaction between the adsorbent and the pollutants in the wastewater. It is worth mentioning that binding affinity between the adsorbent and adsorbate is highly dependent on the negative charge of the pollutants such as CP, BPA, PAHs which can be eliminated using such a mechanism, and adsorbents such as AC, CNT, graphene can be employed for such adsorption (Ji, Liu, Zhang, Zhang, & Yuan, 2020).

Electrostatic interaction can be considered as one of the main adsorption forces for the removal of heavy metals, PAHs, and other pollutants. Electrostatic interaction is highly dependent on the pH of the solution and the strength of ionic interaction between the adsorbent and the adsorbate. Materials such as nanocomposites metal oxides G, MOF, dendrimers, nanopolymer are the ideal adsorbates in this case (Wang et al., 2020).

Petroleum wastewater contains high concentrations of heavy metals and organic materials. In an attempt to remove such impurities, adsorbents possessing Lewis acid-base adsorption forces can be employed. Lewis acid-base are complex forces [namely metal ion exchange, ligand exchange, complexation, and π-electrons (base)] that have been studied by several researchers (Liu et al., 2020). Adsorbents such as AC, G, CNT, dendrimers, Nano polymers, metal oxide, and MOFs can be considered if Lewis acid-base adsorption forces are the primary forces of adsorption.

Generally, the application of nanoadsorbents in petroleum wastewater treatment has been proven efficient and cost-effective. However, the most important aspect of this process is the application of a suitable type of adsorbent based on the type of adsorbate and impurities in petroleum wastewater. Many studies have investigated and summarized the application of adsorption in petroleum wastewater treatment. A. Younis et al. have published a comprehensive study on the application of adsorption in this area (Younis, Maitlo, Lee, & Kim, 2020).

Nanocatalysis, and their composite as adsorbents

Nanoparticles can be considered as a class of inorganic materials in the nanoscale with sizes in the range of 10^{-9} m. During the 1970s and 1980s, these types of nanoparticles are defined as ultrafine particles by Granqvist and Buhrman in the United States when they first studied nanoparticles through fundamental studies (Younis et al., 2020). The main advantage of nanoparticles is mainly due to their intense properties compared to bulk particles. Nanoparticles are a link between bulk materials and atomic or molecular structures (Ba-Abbad, Kadhum, Mohamad, Takriff, & Sopian, 2012). However, the difference between the nano and bulk materials is that bulk materials have fixed physical properties which are not affected by size whereas the properties of the nanomaterials are highly dependent on the size. The interesting advantages of nanoparticles are the large surface area that leads the contributions for many applications such as adsorption and photocatalysis. One of the best removal techniques of pollutants is adsorption which has been commonly used for a wide range of wastewater treatments. This is because adsorption process is easy to design and operate, as well as cheaper in cost with high performance compared to other methods (Pal et al., 2016). Adsorption is a simple process, which consider a physical process to removing pollutants from the liquid phase as (wastewater) to the solid phase as (adsorbent) (Uddin, 2017). The adsorption process has been applied widely in industries such as petroleum wastewater, food processing, paper, textile, etc. (Uddin, 2017). In the last decade, many adsorbents have been developed using different materials to remove the organic and inorganic pollutants in the wastewater field especially, petroleum wastewater which included heavy metals, organic compounds, and hydrocarbons, and crude oil.

Methods of nanocomposite synthesis as adsorbents

A nanocomposite consists of more than one phase of solid material, by which one of these phases has 1, 2, and dimensions of less than 100 nm. However, nanocomposites can be structures having nanoscale repeated distances between the different phases that make up the material. Materials that can be included under the nanocomposite definition include many types such as porous media, colloids, gels, and copolymers (Yagub, Sen, Afroze, & Ang, 2014). There are many techniques for nanocomposite materials fabrication.

Metal nanocomposites

This class of nanocomposite materials is consisted of is relevant to a flexible metal or alloy matrix that targets the reinforcement of materials by crosslinking with nanosized materials. By combining these materials with different types of metal and ceramic, their properties such as strength, ductility, and toughness will improve and become more suitable as materials for a specific application (Camargo, Satyanarayana, & Wypych, 2009). Several different methods that have been used to synthesize metal nanocomposites are spray pyrolysis, chemical vapor deposition, colloidal method, high-energy ball milling, and rapid solidification (Camargo et al., 2009).

Ceramic nanocomposites

This class of nanocomposite materials focuses on the incorporation of energy dissipating components resulting in produced nanocomposites based on a ceramic matrix such as fibers particles and platelets. The type of composites can be synthesized by main two processes, the first one is powder process polymer precursor and the second one is sol−gel process (Allen, Tung, & Kaner, 2010).

Polymer nanocomposites

The polymer nanocomposites are commonly synthesized by loading of inorganic nanoparticles into the polymer matrix. These nanoparticles consist of nanostructured materials which are in the condensed form of bulk materials within a nanometer size range. The polymer nanocomposites have been reported to have high performance than traditionally filled polymer composites especially for specific properties such as improved thermal resistance, decreased permeability, and high chemical resistance. To prepare this type of nanocomposites, several methods are used such as In Situ Formation, Sol−Gel, Solution Mixing, and Melt Blending (Allen et al., 2010).

Carbon-based nanocomposite

Recently, carbon nanomaterials have been shown to have attractive properties thereby receiving increasing attention for applications. These carbon nanomaterials include the carbon nanotube with both types single and multiwalled, graphene, graphene oxide as well as carbon nanoparticles (Allen et al., 2010). These materials have unique structural properties including high surface area, conductivity, mechanical and thermal properties, and biocompatibility (Wu, Kuang, Zhang, & Chen, 2011). The carbon-based adsorbents have been found as a universal adsorbent for wastewater treatment applications due to their porous structure and ease of production from many kinds of raw sources (Wu et al., 2011). Besides improving adsorption performance, carbon-based adsorbents have been reported to be incorporated with other materials to improve their surface for removing a wide range of pollutants. The Carbon-based nanocomposite is divided into the AC, carbon nanotube (CNT), Graphene oxide (GO), Carbon nanofiber and AC fiber (ACF). From Table 10.1, the carbon-based adsorbents are effective for much specific applications on petroleum wastewater, which included oil, heavy metals, and organic compounds (dyes). Their performance was enhanced by hybridizing with other nanomaterials resulting in well-developed structures to improve the adsorption performance. The different types of pollutants follow different mechanisms of adsorption as described previously in Section 10.2, which clearly shows the effect of improvement of adsorbents with hybrid materials (Chua et al., 2020).

Organic and inorganic-based composite adsorbents

Over the last decades, adsorbents have been improved using several organic and inorganic materials to remove different kinds of pollutants from water and wastewater. The adsorbents are classified as organic such as carbon-based (AC), carbon materials based (GO, NT), and polymer-based (polyamid acid) and inorganic such as metallic

TABLE 10.1 Carbon-based nanocomposite adsorbent for petroleum wastewater pollutants.

Adsorbent	Type of major pollutants	Ref.
Carbon nanotubes	Oil	Kayvani Fard et al. (2016)
Carbon-based magnetic nanocomposites	Oil	Abdel-Salam et al. (2020)
hydrophobic F-rGO@wood sponge	Crude oil	Huang, Zhang, Lai, Li, and Zeng (2020)
RGO/PDA/MXene composite	Crude oil	Feng et al. (2020)
Activated carbon fiber supported TiO_2 nanotubes	Organic compounds	Wang et al. (2018)
Activated carbon composite	Organic compounds	Azad, Ghaedi, Dashtian, Hajati, and Pezeshkpour (2016)
Metal-organic polymer-activated carbon	Organic compounds	Taghipour, Karimipour, Ghaedi, and Asfaram (2018)
Activated carbon/Ag nanoparticle	Organic compounds	Naphtali Odogu et al. (2020)

based (iron oxide, zinc oxide, etc.) as well as mineral-based (bentonite and zeolite) (Awad, Jalab, Ba-Abbad, & El-Naas, 2020). Improvement of adsorbents through the development of composite materials have been getting a lot of attention by researchers because of the significant improvement and enhancement of the adsorption performance (Awad et al., 2020). Over the last few years, various works have been reported on adsorbents for wastewater treatment to remove a wide range of pollutants in petroleum wastewater. The nanocomposite adsorbents have been applied for the removal of many pollutants such as organic pollutants, heavy metals as well as color materials as dyes. Recently, conventional adsorbents have been synthesized by combining with organic materials such as polymers (functionalized groups) and doping metals to improve their properties for adsorption applications (Cheng et al., 2017). The mechanism of adsorption is more affected or controlled by the nature of the adsorbents and pollutants whereby each functionalized group and doping metal are designed to remove specific pollutants in the petroleum wastewater field including oil, heavy metals, and organic compounds. The parameters or operating conditions such as pH and ionic strength play a major role in the performance of adsorbents performance (Cheng et al., 2017). Table 10.2 listed examples of adsorbents (organic and inorganic based) that have been reported to be able to remove oil, heavy metals, and organic compounds from wastewater. These adsorbents have significant potential to be applied for petroleum wastewater treatment in the future.

Application of green adsorbents in petroleum wastewater treatment

Within the petroleum production and oil refining processing, each stage of processing can release water polluted by oil which contains organic and inorganic components and discharge of these components can result in contamination of soil and surface/groundwater. Recently, adsorption has been widely applied for wastewater treatment because the adsorption is stable with higher performance for removal of several pollutants (Bandura, Woszuk, Kolodyńska, & Franus, 2017). While the cost of adsorbents can be relatively high, various studies have tried to reduce the cost of adsorption by applying low-cost materials as adsorbents. Recently, many studies have been focused on the removal of oil, heavy metals, and dyes from industrial effluent in petroleum wastewater using green adsorbents because of their low cost, natural and eco-friendly. The green adsorbents are particularly produced from natural resources such as agricultural residues and other wastes of biomass. In general, the green adsorbents that have been reported are divided into (1) agricultural wastes and residues, (2) natural adsorbents, and (3) AC from low-cost waste materials (Crini & Lichtfouse, 2018). The removal of pollutants such as oil, heavy metal, and organic compounds has been reported to be better with higher adsorption capacities at a different range of removal for these pollutants. For example, some green adsorbents showed the ability to remove several heavy metals compared to organic compounds. Table 10.3 represents different green adsorbents applied for the removal of oil, heavy metal, and organic compounds that can be used for petroleum wastewater treatment.

TABLE 10.2 Organic and inorganic-based composite adsorbents and their application in petroleum wastewater pollutants.

Adsorbent	Pollutant	Ref.
Egg shell	Oil	Misau, El-Nafaty, Abdulsalam, and Isa (2012)
Organoclay	Oil	Younker and Walsh (2015)
Bentonite	Oil	Okiel, El-Sayed, and El-Kady (2011)
Org-bentonite	Oil	Younker and Walsh (2014)
Graphene/metal oxide composites	Organic compounds	Upadhyay, Soin, and Roy (2014)
Dendritic polymers based adsorbent	Organic compounds	Sajid, Nazal, Baig, and Osman (2018)
Organic-inorganic hybrid polymers	Heavy metals	Samiey, Cheng, and Wu (2014)
Clay minerals	Heavy metals	Sharma, McDonald, Kim, and Garg (2015)
GO and their composite	Heavy metals	Uddin (2017)
Magnetic graphene–carbon nanotube	Heavy metals	Peng, Li, Liu, and Song (2017)

TABLE 10.3 Modified and unmodified green adsorbents for petroleum wastewater field.

Adsorbent	Modification method	Pollutant	Ref.
Mango seed	Unmodified	Organic pollutants	Alencar et al. (2012)
Rice husk	Unmodified	Organic pollutants	Safa, Bhatti, Bhatti, and Asgher (2011)
Rice husk	Unmodified	Heavy metals	Sobhanardakani, Parvizimosaed, and Olyaie (2013)
Bamboo leaf powder	Unmodified	Heavy metals	Mondal, Nandi, and Purkait (2013)
Cotton fiber	Unmodified	Oil	Li, Zhu, Li, Na, and Wang (2013)
Rice husk	Unmodified	Oil	Alaa El-Din, Amer, Malsh, and Hussein (2018)
Banana peel	Unmodified	Crude oil	El Gheriany, Ahmad El Saqa, Abd El Razek Amer, and Hussein (2020)
Orange peel	Unmodified	Crude oil	Angelova et al. (2011)
Cellulose	Glycidylmethacrylate	Heavy metals	Anirudhan, Nima, and Divya (2013)
Cellulose	Methyl benzalaniline	Heavy metals	Saravanan and Ravikumar (2015)
Rice husk	pyrolysis	Crude oil	Angelova et al. (2011)
Soybean	Mercerization	Crude oil	Wong, McGowan, Bajwa and Bajwa (2016)

Cellulosic materials-based green adsorbent

Several green materials have been applied as green adsorbents because of their availability, eco-friendly, and lower cost. These materials are also commonly available as wastes. The cellulosic materials such as shells leaves, roots, seed, etc., have been used as green adsorbents in adsorption (Hokkanen, Bhatnagar, & Sillanpää, 2016). Table 10.3 shows the different types of untreated plant wastes for adsorption of numerous pollutants such as oil, heavy metal, and organic compounds. Only a few of these types showed higher adsorption capacity for adsorbing these pollutants after physical and chemical treatment. Most of the green and natural sources include cellulosic materials did not have adequate capacity to use for the removal of pollutants by adsorption (Zamparas & Tzivras, 2020). The pretreatment process can improve their adsorption performance by removing other compounds, which decrease porosity and crystallinity. The modification stage of green and natural sources through chemical and physical treatment becomes important to enhance the performance of adsorption capability (Zamparas & Tzivras, 2020).

In-depth mechanism of pollutants adsorption in petroleum wastewater field

The adsorption performance depends on the mechanism of pollutants transport on the adsorbents surface. Generally, the mechanism is described as physical adsorption, ion exchange, and surface complexation as shown in Fig. 10.1. However, the adsorption mechanism is controlled of ways interactions occur between pollutants and adsorbents where each type followed different interaction paths. The physical adsorption occurs by weak bonds hydrogen bonding, weak van der Waals forces, or hydrophobic interaction (Sandhyarani, 2019). The ion exchange focus on the exchange of the charges of the same ion on the surface of the adsorbent with adsorbate (Sandhyarani, 2019).

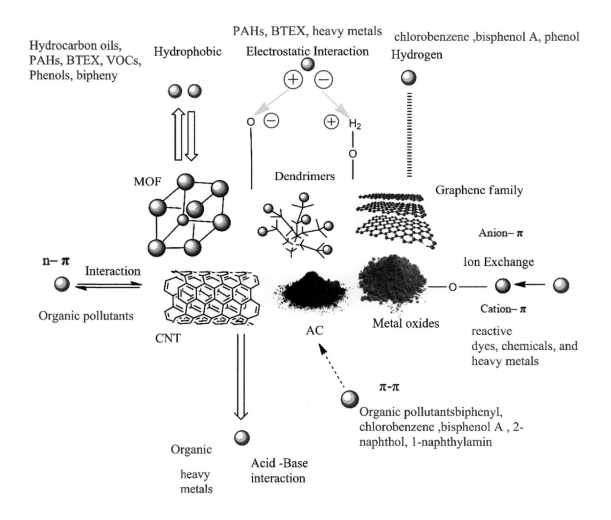

FIGURE 10.1 The suggestion adsorption mechanism types for pollutants in the petroleum wastewater field.

Additionally, surface complexation type shows as inner-sphere complex, which occurs due to a direct bond between the surface of adsorbent and adsorbate. At the same time, an outer-sphere complex in which the interaction occurs by the electrostatic attraction on the surface of a sphere (adsorbent) (Inyang et al., 2016). Other types of mechanisms have been reported such as electrostatic interaction, which occurs between different charges as a positive and negative charge on the surface of absorbent (Inyang et al., 2016). The pollutants in the petroleum wastewater field included heavy metals, organic compounds, oil and crude oil have been reported to follow different mechanisms on the surface of adsorbent as summarized in Table 10.4.

These examples showed that properties of the adsorbent surface play a major role to interact with absorbate which may control adsorption performance. The common explanation showed that hydrogen bonding as a physical mechanism is responsible for the removal of organic compounds, which included dyes and others. On the other hand, electrostatic attraction is responsible for ion exchange for the removal of heavy metal and a few cases found inner-sphere complexation as well. All mechanisms as electrostatic and hydrogen bonding interactions, as well as complexation, can be found in the case of oil removal which confirmed that mechanism involved for crude oil with slightly expected changes (Wu et al., 2018).

Application of membrane technology for petroleum wastewater treatment

Chemical engineering processes exhibit among the most significant developments compared to other disciplines, which drives new methods and strategies to fulfill chemical industry needs.

TABLE 10.4 Examples of adsorption mechanisms according to pollutants in the petroleum wastewater field.

Adsorbent	Pollutant	Mechanism	Ref.
• Metal-organic polymer-activated carbon • Reduced graphene oxide/copper oxide nanohybrids	Heavy metals	Electrostatic attraction, ion exchange	Sharma et al. (2015) and Pawar et al. (2018)
	Organic compounds	Hydrogen bonding, complexes, electrostatic interaction	Sobhanardakani et al. (2013) and Taghipour et al. (2018)
	Oil	–	
	Crude oil	–	
• Activated carbon composite	Heavy metals	Electrostatic attraction	Song et al. (2020) and Arcibar-orozco et al. (2014)
	Organic compounds	Electrostatic attraction, hydrogen bonding, ion exchange	Taghipour et al. (2018) and Azad et al. (2016)
	Oil	–	
	Crude oil	–	
• Graphene/biochar • Phosphorylated multiwalled CNT/β • Cyclodextrin/TiO$_2$-Ag	Heavy metals	Electrostatic attraction, ion exchange	Tang, Lv, Gong, and Huang (2015)
	Organic compounds	Electrostatic attraction, surface complexation	Uddin (2017) and Leudjo Taka et al. (2020)
	Oil	–	
	Crude oil	–	
• Magnetic/quaternary ammonium compound • GTP aerogel (gelatin/TiO$_2$/PEI)	Heavy metals	Inner-sphere complexation, electrostatic interaction	Upadhyay et al. (2014) and Zhang et al. (2016)
	Organic compounds	Electrostatic and hydrogen bonding	Wan Ikhsan et al. (2018) and Y. Wu et al. (2018)
	Oil	Electrostatic and hydrogen bonding interactions, complexation	
	Crude oil	–	

One of the continuous improvements highlighted in the chemical industry is separation using membrane filtration. Membrane separation has been well-adopted in most industrial operations and processes, slowly displacing conventional separation methods in certain operations such as distillation, pervaporation, mechanical filtration, wastewater reclamation, drinking water processes, brackish water treatment, water softening, and Petroleum processing wastewater treatment (Yin & Deng, 2015). The shift from conventional separation technologies has led to a rise in new separation processes, such as membrane desalination, pervaporation membrane bioreactors, catalytic membrane reactors, and enzymatic membrane reactors. Bacteria, viruses, fatty acids, and organic and inorganic particles (Jamshidi Gohari et al., 2014) are target contaminants that can be removed from the feed solutions in various processes using membranes. As a result of such rapid developments, researchers are focusing on new methods to optimize membrane filtration process costs by improving membrane characteristics, including antifouling properties, solute selectivity, solution flux, and mechanical strength (Leo, Cathie Lee, Ahmad, & Mohammad, 2012). Generally, membrane materials can be divided into two major categories: organic (polymeric) and inorganic (metal and ceramic). Inorganic membranes have outstanding mechanical and chemical stability, which is favorable for membrane applications with exposure to harsh operating conditions and that require frequent cleaning. High production cost, difficulty with scaling up the membrane module and heavyweight, however, make inorganic membranes less popular in most industries (Ratman, Kusworo, Utomo, Azizah, & Ayodyasena, 2020).

Polymeric membranes, on the other hand, appear to be a better candidate in separation processes owing to their practicality, and lower production and maintenance costs compared to inorganic membranes. Other advantages of using polymeric membranes include well-controlled pore size distribution, effective separation layer, high mechanical strength, and good resistance to chemicals in most water separation processes. However due to the complexity of petroleum wastewater, the selection of membrane material is crucial, and all the characteristics of the Wastewater should be considered before choosing the membrane material. Table 10.5 shows the properties of polymeric membrane materials and their optimum operating conditions and can be used as a reference for membrane material selection.

Polymeric membrane fabrication

Many techniques have been suggested by various researchers to produce selective and highly permeable membrane films. There are many different membrane fabrication methods and processes. However, the most widely used and applicable technique is the phase inversion process (Jamshidi Gohari et al., 2014). The phase inversion method involves split up point to an earlier homogeneous polymer solution either by exposing it to a nonsolvent atmosphere (dry process) or temperature change, through immersing the solution in a nonsolvent bath (wet process). The structures produced from these techniques are suitable to be used as membranes for various applications (Leo et al., 2012).

Depending on the parameters that induce the demixing, phase separation mechanisms can be subdivided into three main categories. Asymmetric boundaries that posed on the polymer film with a change in one of these parameters at one particular side of the film would lead to the formation of membrane structure. To reach the diffusion induced phase separation (DIPS) (Ratman et al., 2020), three different techniques have been developed: immersion of polymer solution into a nonsolvent bath, coagulation by absorption of, nonsolvent from a vapor phase, and evaporation of the solvent (Kasim et al., 2015). To obtain the desired membrane structures, combinations of various techniques are sometimes required. Once a polymer solution is exposed to a vapor comprised of a nonsolvent (a compound that is not miscible with the polymer), an asymmetric structure can then be produced. Membrane synthesized through evaporation (porous structures) requires a polymer solution composed of two solvent types as a nonsolvent which is less volatile and the second solvent as volatile (Kasim et al., 2015).

Applications of the pure unmodified membrane in petroleum wastewater treatment are not efficient due to the complexity of such wastewater. Drawbacks include fouling, lack of selectivity, being energy-intensive, and short lifespan due to harsh conditions. However, the progress of nanotechnology has allowed a significant improvement in various applications. The membrane purification process is not an exception in this development. Metal and nonmetal nanoparticles have been embedded in membranes matrix to obtain new nanohybrid membranes with enhanced performance such as higher water permeability, better hydrophilicity and, thermal and mechanical stability. Improvement of the hydrophilicity of the membrane has been reported as one of the main objectives to provide membranes with less fouling potential (Wan Ikhsan et al., 2018).

Modification polymeric membrane

Coating is a simple method of membrane modification owing to the simplicity of controlling the coating conditions. In this method, a layer of hydrophilic or hydrophobic media, depending on the target membrane application, can be

TABLE 10.5 Common polymeric membrane materials.

Polymer type	Membrane properties
Cellulose acetate (CA) and Cellulose nitrate (CN)	Cellulose membranes have some specific properties such as hydrophilic property with high crystallinity, which protect and make it lower dissolving in water. However, it has more sensitive to pH which optimum pH range is 4–6.5 (Wu et al., 2011; Ratman et al., 2020)
Nylon/polyamide (NYL/PA)	Polyamides (PA) are formed by reacting of amides with an amide group (–NHCO–), and this group is the main essential group of the polymer chain. Aliphatic polyamides, also referred to as nylon, nylon 6, nylon 4,6, and nylon 6,6 are the most frequently used nylon polymers in membrane fabrication. Nylons have great thermal and mechanical stability as well as very good solvent resistance against other solvents, such as aliphatic solvent, aromatic hydrocarbons, and some linear alcohols like N-methyl-2-pyrrolidone (NMP) and dimethylformamide (DMF). They are also very hydrophilic and have very good pH resistance, which can overcome some cellulose membrane limitations (Wu et al., 2018; Kasim et al., 2015)
Polycarbonate (PC)	PCs are used to make hydrophilic membranes and are manufactured from thin polycarbonate film with the "track-etching" method. Nonetheless, there are several reports on membrane production by phase inversion method (Yagub et al., 2014). PC membranes have good chemical and mechanical properties. They are mostly used in microfiltration membranes (Khare et al., 2007)
Polysulfone (PS)	PS may be an ideal polymeric material for membranes owing to its high mechanical properties (derived from the connecting ether oxygen), and good chemical and thermal stability. PS shows higher liquid flow than PTFE or PVDF. However, PS membranes have a high propensity to the fouling phenomenon, which has been attributed to its slightly hydrophobic surface nature (Wang et al., 2018; Jamshidi Gohari et al., 2014)
Polypropylene (PP)	PP is slightly hydrophobic. However, its resistance to a wide range of organic solvents makes it an interesting choice for industrial applications. PP is also used as pore support in RO membranes (Yagub et al., 2014; Khare et al., 2007)
Polytetrafluoroethylene (PTFE)	PTFE is used as a hydrophobic membrane and is resistant to organic solvents as well as strong acids and bases. It has low protein binding and low extractability. The main application is in the filtration of nonaqueous samples (Wu et al., 2018; Kasim et al., 2015)
Polyvinylidene difluoride (PVDF)	It is commonly used in microfiltration membranes and is hydrophobic. The chemical resistance and thermal stability of PVDF are similar to PS membranes. However, it shows lower flux than PS membranes (Yin & Deng, 2015; Abdallah et al., 2018)

coated onto the membrane surface by physical contact followed by solvent evaporation. Chemical and film formation stability provided by the grafting method is much higher than with the coating method. Membrane surfaces with highly tuneable characteristics can be easily produced by coating (either static or dynamic), which also shows no significant changes in terms of chemical and physical properties (Wan Ikhsan et al., 2018). Polyethylene glycol (PEG), zwitterionic copolymer, 2hyroxylethyl methacrylate (HEMA), methacrylic acid, acrylic acid, and sulfonyl groups have been successfully used as hydrophilic monomers in membrane modification using the grafting method. Studies have reported enhanced water permeability, surface hydrophilicity as well as reduced fouling (Wan Ikhsan et al., 2018). The coating has been applied to PSF membranes using various media, such as quaternary ammonium cellulose ether (QCMC), sodium carboxymethyl cellulose (CMCNa), and sodium p-styrene sulfonate (SSS), which show better water permeability and antifouling behavior (Jamshidi Gohari et al., 2014).

Blending is a prefabrication method that can be employed to modify membrane properties. It is one of the most important membrane modification methods as it can easily alter membrane structures and surface properties. Blending has been successfully adopted for membranes using hydrophilic polymers, such as cellulose diacetate, polyaniline (PANI) nanofibers, and sulfonated poly(ether ketone). The blending of these polymers can improve membrane behaviors like membrane permeability, flux decline, and retention capability (Kasim et al., 2015). Incorporating nanomaterials in a blending solution is another effective way that has been adopted to minimize fouling problems in oil/water separation. Nanomaterials such as MOFs, dendrimer, Al_2O_3, TiO_2, and SiO_2 has been proven effective for the purification of petroleum wastewater (Ngang, Ahmad, Low, & Ooi, 2017). These materials are reportedly effective in enhancing membrane properties, such as reducing the fouling tendency, increasing hydrophilicity, and increasing water flux

TABLE 10.6 Shows the application of polymer-based mixed matrix membranes petroleum wastewater field.

Membrane additive	Type	Enhancement in comparison to unmodified membrane				Additional functions	Ref.
		Flux	Rejection	Hydrophilicity	Fouling water recovery/foulant		
PVDF/polyethyleneimine	UF	26% average	33% average oil 0.1–0.33 g L^{-1}	25%	2.2% oil	Antimicrobial	Yin and Deng (2015) and Abdallah et al. (2018)
Cellulose/polyethyleneimine	NF	8000%	−5% negatively charged dyes	100% average	–	–	Younker and Walsh (2014) and Puspasari et al. (2018)
PES/hydrous manganese dioxide	UF	1369%	Synthetic oily wastewater 100%	76%	74% oil	–	Wang et al. (2018) and Jamshidi Gohari et al. (2014)
PES/ZnO loading, UV irradiation, and PVA coating	UF	86%	Petroleum wastewater TDS; 18.6%, COD;16.7%, NH$_3$; 87.1%	19%	85% petroleum wastewater	Quantitative fouling	Wu et al. (2011) and Ratman et al. (2020)
PSF/halloysite nanotube-hydrous ferric oxide nanoparticle	UF	1000%	99.7% 1000 ppm oily solution	34%	99.7% oil	–	Younis et al. (2020) and Wan Ikhsan et al. (2018)
PVDF/SiO$_2$-PNIPAM	UF	99%	99.47% saline oil emulsion	−3%	69.91% saline oil emulsion		Younker and Walsh (2015) and Ngang et al. (2017)
CA/polydopamine-sulfobetaine methacrylate	UF	53%	95%–99% diesel oil, food-grade oil, dodecane	11%	11.5% diesel oil, food-grade oil, dodecane	Under water oil contact angle increased by 12%	Zhang et al. (2016) and Guzman (2021)
PSF/aspartic acid functionalized graphene oxide	UF	97%	97.9% oil	18%	90% BSA		Abdalla, Wahab, and Abdala (2020)
PES/GO-SiO$_2$	UF	38%	25% oil	25%	NA		Alkindy, Naddeo, Banat, an Hasan (2020)
PSF/poly(vinyl) alcohol coated	UF	29%	Phenol 82% ammonia 92%	63%	36% petroleum refinery wastewater		Kusworo, Kumoro, and Utomo (2021)
Cellulose/GO	UF	1000 LMH gravity-driven	Hexane 100% petroleum ether 98% toluene 99%	Super hydrophilic zero contact angle	100% water recovery after 10 cycles oil/water	Super lipophobic	Ao et al. (2017)
PSF/carboxylated CNTs/GO nanohybrid	UF	1000%		25%	50% enhanced water recovery		Modi and Bellare (2019)

and solute rejection (Ngang et al., 2017). Several researchers have recently studied the possibility of using carbon-based nanomaterials to enhance membrane performance. Theoretically, carbon-based nanomaterials are required in lower amounts, which could be due to the large surface-to-volume ratio but can contribute to additional membrane structure stability (Puspasari, Huang, Sutisna, & Peinemann, 2018).

These types of mixed matrix membranes (MMMs) exhibit high pure water flux and fouling resistance and have shown outstanding separation performance of petroleum wastewater purification (Wan Ikhsan et al., 2018). The introduction of graphene and CNT nanohybrid membranes appears to be one of the ideal approaches to enhance membrane filtration performance and properties for petroleum wastewater treatment (Alkindy et al., 2020). The structure of graphene and CNT make these materials suitable for nanofillers to increase both the mechanical and chemical properties of the membranes. The incorporation of GO and CNT functionalized with metal and nonmetal nanoparticles into the membrane matrix have shown superior physicochemical properties and enhanced overall membrane performance (Alkindy et al., 2020). Generally, MMM has been established as a good option for petroleum wastewater treatment. Recently, mixed matrix polymeric membranes have emerged as a popular purification method for petroleum wastewater. This is due to their attractive features such as rigidity, heat resistance, and hard condition of the environment. Moreover, mixed matrix polymeric membranes have physical and chemical stability and can be applied in a wide range of pollutants under different characteristics of the solution. Hence, studies exploring the mixed matrix polymeric membranes are conducted using a variety of polymers and nanomaterials. Table 10.6 shows the different types of polymer-based MMMs for petroleum wastewater purification.

Conclusions

The application of nanomaterial and nanotechnology for the purification of petroleum wastewater is a developing area with huge potential. However, determining the most suitable materials for different applications is complicated and requires more effective research in this area. The fact is that most of the studies in this area are done on a lab-scale using synthetic wastewater containing one two or three pollutants, ignore real case experimental studies resulting in a lack of data on the effectiveness, cost, and scalability of these advanced nano-based purification technologies in the petroleum sector. According to the result of these shortcomings, the industrial adaptation of the nano-based purification technologies for the petroleum sector is challenging. Hence, we believe there are four main drawbacks related to the industrial implementation of nano-based purification techniques such as nanoparticle-based sorption and membrane.

1. Determination of suitable nanoadsorbent material for real case samples of petroleum bioefineries and petrochemical processing centers can be a demonstration of the nanosorption potential for purification of a complex mixture of free, soluble, and emulsive hydrocarbons.
2. MMSs (UF, NF, RO) have shown tremendous potential for this type of purification, however, large-scale production of MMMs incorporated with nanoparticles is still a challenge due to the high cost of production. Hence, more research is needed in this area.
3. Integrated nano-based processes (such as sorption, coagulation, and membrane) can be an interesting area for the application of nano-based purification in petroleum wastewater, however, lack of practical knowledge in this area is limiting the industrial implementation of the nano-based purification process.
4. Most of the published experimental work in this area ignores real case experimental studies due to the complexity of the wastewater.

References

Abdalla, O., Wahab, M. A., & Abdala, A. (2020). Mixed matrix membranes containing aspartic acid functionalized graphene oxide for enhanced oil-water emulsion separation. *Journal of Environmental Chemical Engineering*, 104269. Available from https://doi.org/10.1016/j.jece.2020.104269.

Abdallah, H., Taman, R., Elgayar, D., & Farag, H. (2018). Antibacterial blend polyvinylidene fluoride/polyethyleneimine membranes for salty oil emulsion separation. *European Polymer Journal*, *108*, 542–553. Available from https://doi.org/10.1016/j.eurpolymj.2018.09.035.

Abdel-Salam, M. O., Younis, S. A., Moustafa, Y. M., Al-Sabagha, A. M., Mostafa, M. H., & Khalild. (2020). Microwave — Assisted production of hydrophilic carbon-based magnetic nanocomposites from saw-dust for elevating oil from oil field waste water. *Journal of Cleaner Production*, 249.

Alaa El-Din, G., Amer, A. A., Malsh, G., & Hussein, M. (2018). Study on the use of banana peels for oil spill removal. *Alexandria Engineering Journal*, *57*(3), 2061–2068. Available from https://doi.org/10.1016/j.aej.2017.05.020.

Alencar, W. S., Acayanka, E., Lima, E. C., Royer, B., de Souza, F. E., Lameira, J., ... Alves, C. N. (2012). Application of *Mangifera indica* (mango) seeds as a biosorbent for removal of Victazol Orange 3R dye from aqueous solution and study of the biosorption mechanism. *Chemical Engineering Journal, 209,* 577–588. Available from https://doi.org/10.1016/j.cej.2012.08.053.

Alkindy, M. B., Naddeo, V., Banat, F., & Hasan, S. W. (2020). Synthesis of polyethersulfone (PES)/GO-SiO$_2$ mixed matrix membranes for oily wastewater treatment. *Water Science and Technology, 81*(7), 1354–1364. Available from https://doi.org/10.2166/wst.2019.347.

Allen, M. J., Tung, V. C., & Kaner, R. B. (2010). Honeycomb carbon: A review of graphene. *Chemical Reviews, 110*(1), 132–145. Available from https://doi.org/10.1021/cr900070d.

Angelova, D., Uzunov, I., Uzunova, S., Gigova, A., & Minchev, L. (2011). Kinetics of oil and oil products adsorption by carbonized rice husks. *Chemical Engineering Journal, 172*(1), 306–311. Available from https://doi.org/10.1016/j.cej.2011.05.114.

Anirudhan, T. S., Nima, J., & Divya, P. L. (2013). Adsorption of chromium (VI) from aqueous solutions by glycidylmethacrylate-grafted-densified cellulose with quaternary ammonium groups. *Applied Surface Science, 279,* 441–449. Available from https://doi.org/10.1016/j.apsusc.2013.04.134.

Ao, C., Yuan, W., Zhao, J., He, X., Zhang, X., Li, Q., ... Lu, C. (2017). Superhydrophilic graphene oxide@electrospun cellulose nanofiber hybrid membrane for high-efficiency oil/water separation. *Carbohydrate Polymers, 175,* 216–222. Available from https://doi.org/10.1016/j.carbpol.2017.07.085.

Arcibar-Orozco, J. A., Rangel-Mendez, J. R., & Diaz-Flores, P. E. (2014). Pb (Ii)-Phenol, and Cd (Ii)-Phenol by activated carbon cloth in aqueous solution. *In Simultaneous adsorption of Pb (Ii)-Cd(Ii), 226,* 2197–2206.

Awad, A., Jalab, B. N., Ba-Abbad, M., & El-Naas. (2020). Adsorption of organic pollutants by nanomaterial-based adsorbents: An overview. *Journal of Molecular Liquids,* 301.

Azad, F. N., Ghaedi, M., Dashtian, K., Hajati, S., & Pezeshkpour, V. (2016). Ultrasonically assisted hydrothermal synthesis of activated carbon-HKUST-1-MOF hybrid for efficient simultaneous ultrasound-assisted removal of ternary organic dyes and antibacterial investigation: Taguchi optimization. *Ultrasonics Sonochemistry, 31,* 383–393. Available from https://doi.org/10.1016/j.ultsonch.2016.01.024.

Ba-Abbad, M. M., Kadhum, A. A. H., Mohamad, A. B., Takriff, M. S., & Sopian, K. (2012). Synthesis and catalytic activity of TiO$_2$ nanoparticles for photochemical oxidation of concentrated chlorophenols under direct solar radiation. *International Journal of Electrochemical Science, 7*(6), 4871–4888. Available from http://www.electrochemsci.org/papers/vol7/7064871.pdf.

Bandura, L., Woszuk, A., Kolodyńska, D. K., & Franus, W. (2017). Application of mineral sorbents for removal of petroleum substances: A review. *Minerals. 7*(3).

Camargo, P. H. C., Satyanarayana, K. G., & Wypych, F. (2009). Nanocomposites: Synthesis, structure, properties and new application opportunities. *Materials Research, 12*(1), 1–39. Available from https://doi.org/10.1590/S1516-14392009000100002.

Cheng, M., Wang, Z., Lv, Q., Li, C., Sun, S., & Hu, S. (2017). Preparation of amino-functionalized Fe$_3$O$_4$@mSiO$_2$ core-shell magnetic nanoparticles and their application for aqueous Fe^{3+} removal. *Journal of Hazardous Materials,* 341.

Chua, S. F., Nouri, A., Ang, W. L., Mahmoudi, E., Mohammad, A. W., Benamor, A., & Ba-Abbad, Muneer (2020). The emergence of multifunctional adsorbents and their role in environmental remediation. *Journal of Environmental Chemical Engineering,* 9.

Crini, G. & Lichtfouse, E. (2018). *Green adsorbents for pollutant removal innovative materials.* Springer.

Dasgupta, N., Ranjan, S., & Lichtfouse, E. (2020). *Environmental Nanotechnology,* 4.

El Gheriany, I. A., Ahmad El Saqa, F., Abd El Razek Amer, A., & Hussein, M. (2020). Oil spill sorption capacity of raw and thermally modified orange peel waste. *Alexandria Engineering Journal, 59*(2), 925–932. Available from https://doi.org/10.1016/j.aej.2020.03.024.

Ersan, G., Apul, O. G., Perreault, F., & Karanfil, T. (2017). Adsorption of organic contaminants by graphene nanosheets: A review. *Water Research, 126,* 385–398. Available from https://doi.org/10.1016/j.watres.2017.08.010.

Feng, X., Yu, Z., Long, R., Sun, Y., Wang, M., Li, X., & Zeng, G. (2020). Polydopamine intimate contacted two-dimensional/two-dimensional ultrathin nylon basement membrane supported RGO/PDA/MXene composite material for oil-water separation and dye removal. *Separation and Purification Technology, 247,* 116945. Available from https://doi.org/10.1016/j.seppur.2020.116945.

Guzman, M. R. D. (2021). Increased performance and antifouling of mixed-matrix membranes of cellulose acetate with hydrophilic nanoparticles of polydopamine-sulfobetaine methacrylate for oil-water separation. *Journal of Membrane Science,* 620.

Hokkanen, S., Bhatnagar, A., & Sillanpää, M. (2016). A review on modification methods to cellulose-based adsorbents to improve adsorption capacity. *Water Research, 91,* 156–173. Available from https://doi.org/10.1016/j.watres.2016.01.008.

Huang, G., Yang, C., Zhang, K., & Shi, J. (2009). Adsorptive removal of copper ions from aqueous solution using cross-linked magnetic chitosan beads. *Chinese Journal of Chemical Engineering, 17*(6), 960–966. Available from https://doi.org/10.1016/S1004-9541(08)60303-1.

Huang, W., Zhang, L., Lai, X., Li, H., & Zeng, X. (2020). Highly hydrophobic F-rGO@wood sponge for efficient clean-up of viscous crude oil. *Chemical Engineering Journal,* 386.

Inyang, M. I., Gao, B., Yao, Y., Xue, Y., Zimmerman, A., Mosa, A., ... Cao, X. (2016). A review of biochar as a low-cost adsorbent for aqueous heavy metal removal. *Critical Reviews in Environmental Science and Technology, 46*(4), 406–433. Available from https://doi.org/10.1080/10643389.2015.1096880.

Jamshidi Gohari, R., Halakoo, E., Lau, W. J., Kassim, M. A., Matsuura, T., & Ismail, A. F. (2014). Novel polyethersulfone (PES)/hydrous manganese dioxide (HMO) mixed matrix membranes with improved antifouling properties for oily wastewater treatment process. *RSC Advance, 4*(34), 17587–17596. Available from https://doi.org/10.1039/C4RA00032C.

Ji, D., Liu, G., Zhang, X., Zhang, C., & Yuan, S. (2020). Molecular dynamics study on the adsorption of heavy oil drops on a silica surface with different hydrophobicity. *Energy and Fuels, 34*(6), 7019–7028. Available from https://doi.org/10.1021/acs.energyfuels.0c00996.

Kasim, N., Mahmoudi, E., Wahab, A., Kasima, A. W. M. N., Mahmoudi, E., & Abdullah, S. R. S. (2015). Influence of feed concentration and pH on iron and manganese rejection via nanohybrid polysulfone/Ag-GO ultrafiltration membrane. *Desalin. Water Treat, 15,* 1–13.

Kayvani Fard, A., Rhadfi, T., Mckay, G., Al-Marri, M., Abdala, A., Hilal, N., & Hussien, M. A. (2016). Enhancing oil removal from water using ferric oxide nanoparticles doped carbon nanotubes adsorbents. *Chemical Engineering Journal, 293*, 90−101. Available from https://doi.org/10.1016/j.cej.2016.02.040.

Khare, V. P., Greenberg, A. R., Kelley, S. S., Pilath, H., Roh, I. J., & Tyber, J. (2007). Synthesis and characterization of dense and porous cellulose films. *Journal of Applied Polymer Science, 105*(3), 1228−1236. Available from https://doi.org/10.1002/app.25888.

Kukkar, D., Rani, A., Kumar, V., Younis, S. A., Zhang, M., Lee, S.-S., ... Kim, K.-H. (2020). Recent advances in carbon nanotube sponge−based sorption technologies for mitigation of marine oil spills. *Journal of Colloid and Interface Science, 570*, 411−422. Available from https://doi.org/10.1016/j.jcis.2020.03.006.

Kusworo, T. D., Kumoro, A. C., & Utomo, D. P. (2021). Phenol and ammonia removal in petroleum refinery wastewater using a poly(vinyl) alcohol coated polysulfone nanohybrid membrane. *Journal of Water Process Engineering, 39*. Available from https://doi.org/10.1016/j.jwpe.2020.101718.

Leo, C. P., Cathie Lee, W. P., Ahmad, A. L., & Mohammad, A. W. (2012). Polysulfone membranes blended with ZnO nanoparticles for reducing fouling by oleic acid. *Separation and Purification Technology, 89*, 51−56. Available from https://doi.org/10.1016/j.seppur.2012.01.002.

Leudjo Taka, A., Fosso-Kankeu, E., Pillay, K., & Yangkou Mbianda, X. (2020). Metal nanoparticles decorated phosphorylated carbon nanotube/cyclodextrin nanosponge for trichloroethylene and Congo red dye adsorption from wastewater. *Journal of Environmental Chemical Engineering, 8*(3), 103602. Available from https://doi.org/10.1016/j.jece.2019.103602.

Li, D., Zhu, F. Z., Li, J. Y., Na, P., & Wang, N. (2013). Preparation and characterization of cellulose fibers from corn straw as natural oil sorbents. *Industrial and Engineering Chemistry Research, 52*(1), 516−524. Available from https://doi.org/10.1021/ie302288k.

Li, Y., Liao, M., & Zhou, J. (2018). Catechol and its derivatives adhesion on graphene: Insights from molecular dynamics simulations. *Journal of Physical Chemistry C, 122*(40), 22965−22974. Available from https://doi.org/10.1021/acs.jpcc.8b06392.

Liu, Z., Wang, Q., Zhang, B., Wu, T., & Li, Y. (2020). Efficient removal of bisphenol a using nitrogen-doped graphene-like plates from green petroleum coke. *Molecules (Basel, Switzerland), 25*(15). Available from https://doi.org/10.3390/molecules25153543.

Mahmoudi, E., Azizkhani, S., Mohammad, A. W., Ng, L. Y., Benamor, A., Ang, W. L., & Ba-Abbad, M. (2020). Simultaneous removal of Congo red and cadmium (II) from aqueous solutions using graphene oxide−silica composite as a multifunctional adsorbent. *Journal of Environmental Sciences (China), 98*, 151−160. Available from https://doi.org/10.1016/j.jes.2020.05.013.

Matouq, M., Jildeh, N., Qtaishat, M., Hindiyeh, M., & Al Syouf, M. Q. (2015). The adsorption kinetics and modeling for heavy metals removal from wastewater by Moringa pods. *Journal of Environmental Chemical Engineering, 3*(2), 775−784. Available from https://doi.org/10.1016/j.jece.2015.03.027.

Misau, U., El-Nafaty, S., Abdulsalam, Y., & Isa. (2012). *Removal of Oil from Oil Produced Water Using Eggshell, 2*, 52−63.

Modi, A., & Bellare, J. (2019). Efficiently improved oil/water separation using high flux and superior antifouling polysulfone hollow fiber membranes modified with functionalized carbon nanotubes/graphene oxide nanohybrid. *Journal of Environmental Chemical Engineering, 7*(2), 102944. Available from https://doi.org/10.1016/j.jece.2019.102944.

Mondal, D. K., Nandi, B. K., & Purkait, M. K. (2013). Removal of mercury (II) from aqueous solution using bamboo leaf powder: Equilibrium, thermodynamic and kinetic studies. *Journal of Environmental Chemical Engineering, 1*(4), 891−898. Available from https://doi.org/10.1016/j.jece.2013.07.034.

Naphtali Odogu, A., Daouda, K., Paul Keilah, L., Agbor Tabi, G., Ngouateu Rene, L., Julius Nsami, N., & Josoph Mbadcam, K. (2020). Effect of doping activated carbon based *Ricinodendron heudelotti* shells with AgNPs on the adsorption of indigo carmine and its antibacterial properties. *Arabian Journal of Chemistry, 13*(5), 5241−5253. Available from https://doi.org/10.1016/j.arabjc.2020.03.002.

Ngang, H. P., Ahmad, A. L., Low, S. C., & Ooi, B. S. (2017). Preparation of thermoresponsive PVDF/SiO$_2$-PNIPAM mixed matrix membrane for saline oil emulsion separation and its cleaning efficiency. *Desalination, 408*, 1−12. Available from https://doi.org/10.1016/j.desal.2017.01.005.

Okiel, K., El-Sayed, M., & El-Kady, M. Y. (2011). Treatment of oil−water emulsions by adsorption onto activated carbon, bentonite and deposited carbon. *Egyptian Journal of Petroleum, 20*(2), 9−15. Available from https://doi.org/10.1016/j.ejpe.2011.06.002.

Oyewo, O. A., Elemike, E. E., Onwudiwe, D. C., & Onyango, M. S. (2020). Metal oxide-cellulose nanocomposites for the removal of toxic metals and dyes from wastewater. *International Journal of Biological Macromolecules, 164*, 2477−2496. Available from https://doi.org/10.1016/j.ijbiomac.2020.08.074.

Pal, U., Sandoval, A., Madrid, S. I. U., Corro, G., Sharma, V., & Mohanty, P. (2016). Mixed titanium, silicon, and aluminum oxide nanostructures as novel adsorbent for removal of rhodamine 6G and methylene blue as cationic dyes from aqueous solution. *Chemosphere, 163*, 142−152. Available from https://doi.org/10.1016/j.chemosphere.2016.08.020.

Pawar, R. R., Lalhmunsiama., Kim, M., Kim, J. G., Hong, S. M., Sawant, S. Y., & Lee, S. M. (2018). Efficient removal of hazardous lead, cadmium, and arsenic from aqueous environment by iron oxide modified clay-activated carbon composite beads. *Applied Clay Science, 162*, 339−350. Available from https://doi.org/10.1016/j.clay.2018.06.014.

Peng, W., Li, H., Liu, Y., & Song, S. (2017). A review on heavy metal ions adsorption from water by graphene oxide and its composites. *Journal of Molecular Liquids, 230*, 496−504. Available from https://doi.org/10.1016/j.molliq.2017.01.064.

Pramanik, P., Krishnan, P., Maity, A., Mridha, N., Mukherjee, A., & Rai, V. (2020). Application of nanotechnology in agriculture. *Springer Science and Business Media LLC*. Available from https://doi.org/10.1007/978-3-030-26668-4_9.

Puspasari, T., Huang, T., Sutisna, B., & Peinemann, K. V. (2018). Cellulose-polyethyleneimine blend membranes with anomalous nanofiltration performance. *Journal of Membrane Science, 564*, 97−105. Available from https://doi.org/10.1016/j.memsci.2018.07.002.

Ratman, I., Kusworo, T. D., Utomo, D. P., Azizah, D. A., & Ayodyasena, W. A. (2020). Petroleum refinery wastewater treatment using three steps modified nanohybrid membrane coupled with ozonation as integrated pre-treatment. *Journal of Environmental Chemical Engineering, 8*(4). Available from https://doi.org/10.1016/j.jece.2020.103978.

Safa, Y., Bhatti, H. N., Bhatti, I. A., & Asgher, M. (2011). Removal of direct Red-31 and direct Orange-26 by low cost rice husk: Influence of immobilisation and pretreatments. *Canadian Journal of Chemical Engineering, 89*(6), 1554−1565. Available from https://doi.org/10.1002/cjce.20473.

Sajid, M., Nazal, M. K., Baig, N., & Osman, A. M. (2018). Removal of heavy metals and organic pollutants from water using dendritic polymers based adsorbents: A critical review. *Separation and Purification Technology*, 400–423. Available from https://doi.org/10.1016/j.seppur.2017.09.011.

Samiey, B., Cheng, C. H., & Wu, J. (2014). Organic-inorganic hybrid polymers as adsorbents for removal of heavy metal ions from solutions: A review. *Materials*, 7(2), 673–726. Available from https://doi.org/10.3390/ma7020673.

Sandhyarani, N. (2019). *Surface modification methods for electrochemical biosensors*. *Electrochemical biosensors* (pp. 45–75). Elsevier. Available from https://doi.org/10.1016/B978-0-12-816491-4.00003-6.

Saravanan, R., & Ravikumar, L. (2015). The use of new chemically modified cellulose for heavy metal ion adsorption and antimicrobial activities. *Journal of Water Resources Protection*, 7.

Sharma, V. K., McDonald, T. J., Kim, H., & Garg, V. K. (2015). Magnetic graphene-carbon nanotube iron nanocomposites as adsorbents and antibacterial agents for water purification. *Advances in Colloid and Interface Science*, 225, 229–240. Available from https://doi.org/10.1016/j.cis.2015.10.006.

Sobhanardakani, S., Parvizimosaed, H., & Olyaie, E. (2013). Heavy metals removal from wastewaters using organic solid waste-rice husk. *Environmental Science and Pollution Research*, 20(8), 5265–5271. Available from https://doi.org/10.1007/s11356-013-1516-1.

Song, Y., Song, Z., Feng, D., Qin, J., Chen, Y., Shi, Y., . . . Song, K. (2020). Phase behavior of hydrocarbon mixture in shale nanopores considering the effect of adsorption and its induced critical shifts ☆. *Industrial & Engineering Chemistry Research*, 59(17), 8374–8382. Available from https://doi.org/10.1021/acs.iecr.0c00490.

Taghipour, T., Karimipour, G., Ghaedi, M., & Asfaram, A. (2018). Mild synthesis of a Zn (II) metal organic polymer and its hybrid with activated carbon: Application as antibacterial agent and in water treatment by using sonochemistry: Optimization, kinetic and isotherm study. *Ultrasonics Sonochemistry*, 41, 389–396. Available from https://doi.org/10.1016/j.ultsonch.2017.09.056.

Tang, J., Lv, H., Gong, Y., & Huang, Y. (2015). Preparation and characterization of a novel graphene/biochar composite for aqueous phenanthrene and mercury removal. *Bioresource Technology*, 196, 355–363. Available from https://doi.org/10.1016/j.biortech.2015.07.047.

Uddin, M. K. (2017). A review on the adsorption of heavy metals by clay minerals, with special focus on the past decade. *Chemical Engineering Journal*, 308, 438–462. Available from https://doi.org/10.1016/j.cej.2016.09.029.

Upadhyay, R. K., Soin, N., & Roy, S. S. (2014). Role of graphene/metal oxide composites as photocatalysts, adsorbents and disinfectants in water treatment: A review. *RSC Advances*, 3823–3851. Available from https://doi.org/10.1039/c3ra45013a.

Wan Ikhsan, S. N., Yusof, N., Aziz, F., Misdan, N., Ismail, A. F., Lau, W. J., . . . Hayati Hairom, N. H. (2018). Efficient separation of oily wastewater using polyethersulfone mixed matrix membrane incorporated with halloysite nanotube-hydrous ferric oxide nanoparticle. *Separation and Purification Technology*, 199, 161–169. Available from https://doi.org/10.1016/j.seppur.2018.01.028.

Wang, M., Wang, M., Zhang, Z., Zhang, Z., Wang, Y., Zhao, X., . . . Yang, M. (2020). Ultrafast fabrication of metal-organic framework-functionalized superwetting membrane for multichannel oil/water separation and floating oil collection. *ACS Applied Materials and Interfaces*, 12(22), 25512–25520. Available from https://doi.org/10.1021/acsami.0c08731.

Wang, Q., Lei, X., Pan, F., Xia, D., Shang, Y., Sun, W., & Liu, W. (2018). A new type of activated carbon fibre supported titanate nanotubes for high-capacity adsorption and degradation of methylene blue. *Colloids and Surfaces A: Physicochemical and Engineering Aspects*, 555, 605–614. Available from https://doi.org/10.1016/j.colsurfa.2018.07.016.

Wong, C., McGowan, T., Bajwa, S. G., & Bajwa, D. S. (2016). Impact of fiber treatment on the oil absorption characteristics of plant fibers. *BioResources*, 11(3), 6452–6463. Available from https://doi.org/10.15376/biores.11.3.6452-6463.

Wu, B., Kuang, Y., Zhang, X., & Chen, J. (2011). Noble metal nanoparticles/carbon nanotubes nanohybrids: Synthesis and applications. *Nano Today*, 75–90. Available from https://doi.org/10.1016/j.nantod.2010.12.008.

Wu, Y., Chen, L., Long, X., Zhang, X., Pan, B., & Qian, J. (2018). Multi-functional magnetic water purifier for disinfection and removal of dyes and metal ions with superior reusability. *Journal of Hazardous Materials*, 347, 160–167. Available from https://doi.org/10.1016/j.jhazmat.2017.12.037.

Yagub, M. T., Sen, T. K., Afroze, S., & Ang, H. M. (2014). Dye and its removal from aqueous solution by adsorption: A review. *Advances in Colloid and Interface Science*, 209, 172–184. Available from https://doi.org/10.1016/j.cis.2014.04.002.

Yin, J., & Deng, B. (2015). Polymer-matrix nanocomposite membranes for water treatment. *Journal of Membrane Science*, 479, 256–275. Available from https://doi.org/10.1016/j.memsci.2014.11.019.

Younis, S. A., Maitlo, H. A., Lee, J., & Kim, K. H. (2020). Nanotechnology-based sorption and membrane technologies for the treatment of petroleum-based pollutants in natural ecosystems and wastewater streams. *Advances in Colloid and Interface Science*, 275. Available from https://doi.org/10.1016/j.cis.2019.102071.

Younker, J. M., & Walsh, M. E. (2014). Bench-scale investigation of an integrated adsorption-coagulation-dissolved air flotation process for produced water treatment. *Journal of Environmental Chemical Engineering*, 2(1), 692–697. Available from https://doi.org/10.1016/j.jece.2013.11.009.

Younker, J. M., & Walsh, M. E. (2015). Impact of salinity and dispersed oil on adsorption of dissolved aromatic hydrocarbons by activated carbon and organoclay. *Journal of Hazardous Materials*, 299, 562–569. Available from https://doi.org/10.1016/j.jhazmat.2015.07.063.

Zamparas, M., Tzivras, D., Dracopoulos, V., & Ioannides, T. (2020). Application of sorbents for oil spill cleanup focusing on natural-based modified materials: A review. *Molecules (Basel, Switzerland)*, 25, 4522–4544.

Zhang, X., Qian, J., & Pan, B. (2016). Fabrication of novel magnetic nanoparticles of multifunctionality for water decontamination. *Environmental Science and Technology*, 50(2), 881–889. Available from https://doi.org/10.1021/acs.est.5b04539.

Chapter 11

Treatment of petroleum wastewater using solar power-based photocatalysis

Farhad Qaderi and Saba Abdolalian

Civil and Environmental Engineering, Faculty of Civil Engineering, Babol Noshirvani University of Technology, Babol, Iran

Introduction

Nowadays, increasing water pollution and costs associated with healthcare and water supply have become few of the most critical global issues. As per AQUASTAT database reports published by the Food and Agriculture Organization (FAO) of United Nations, 60%−70% (around 3.5 m^3 per day) of total water consumption is used for agricultural purposes. The industry section has dedicated 18%−20%, and domestic usage is around 10%−12% of total water, 65% of the total global production supply is from fossil fuels, and this trend is expected to continue for the next 100 years (Luo, Gong, He, Zhang, & He, 2021).

Oil waste contamination affects many aspects. For example, contaminated drinking water and underground water sources threaten the health of aquatic resources and humans. It can also cause air pollution, affect crop production and destroy the natural landscape. In the oil industry, large amounts of oil waste are produced during the refining, storage, and transportation of oil and petrochemical industries. This has resulted in refine oil wastewater in a variety of ways. Much research has been performed in the field of oil waste treatment. Oil removal along with the removal of soluble organic matter, solid materials suspension, soaps, pH, sulfide, ammonia, etc. performed by conventional methods including flotation, coagulation, biological treatment, membrane separation technology, hybrid technologies, advanced oxidation process (AOP), and electrochemical decomposition (Yu & Han, 2017). Traditional methods for removal of contaminants in petroleum waste (e.g., biological treatment) are time-consuming processes, and these are unable to decompose all pollutants. Conventional methods such as adsorption, membrane filtration, and reverse osmosis, which transfer pollutants from one environment to another cause secondary pollution. Among them, advanced oxidation, including photocatalysis methods, the "Fenton" process, as well as ozonation has positive effects (Jain, Majumder, Sarathi, & Kumar, 2020). The photocatalysis is an AOP that is vastly studied and approved. This approach has the potential to destroy stable and nondegradable pollutants in petroleum wastewater, microorganisms, and inorganic pollutants (Ani, Akpan, Olutoye, & Hameed, 2018). Solar energy is the most economical form of renewable energy for large-scale usage in the future. This is why we seek the maximum use of solar energy in the industry. Photocatalysis by high-efficiency solar radiation and the least energy consumption is among the most proper and efficient technologies in the environment (Ma, Sun, Li, & Wei, 2020).

Petroleum wastewater

There are globally more than 65,000 oil and gas fields on the coast or in the sea. With growing energy demand, oil production has increased, which in turn has led to the generation of billions of gallons wastewater per year as waste in the oil and gas industry (Peng et al., 2020). In general, the oil industry has four main subsectors: (1) exploration and production, (2) hydrocarbon processing (refineries and petrochemicals), (3) storage, transportation, and distribution, and (4) sales or marketing. In each of these sections, significant oil wastewater is produced, which needs to be treated. Petroleum wastewater is generated in both upstream and downstream operations. Upstream operations include crude oil extraction, transportation, and storage processes, while downstream operations refer to crude oil refining processes. Petroleum wastewater produced in the oil industry is classified into two types: plain oil and sludge (Hu, Li, & Zeng, 2013).

Oil wastewater is generated in two ways. The first one is produced through oil processing. Because water is not an end product of the oil industry, it is assumed that about 80%−90% of the water used in this section is removed as oil

Petroleum Industry Wastewater. DOI: https://doi.org/10.1016/B978-0-323-85884-7.00009-6

effluent. The second way is oil spills during exports or leaks at oil rigs at sea, which is another form of petroleum contamination. Therefore, growth in human oil activities could accelerate the emission of various types of hazardous pollutants into the environment. Thus, there is a direct relationship between usage and surface water pollution, and oil production (Younis, Ali, Lee, & Kim, 2020). Oil refineries produce large volumes of wastewater, including water and oil generated during drilling. This wastewater generally includes recalcitrant compounds and is rich in organic pollutants, which are difficult to biologically refine. The wastewater amount produced in refineries is about 0.4—1.6 times the processed crude oil (Aljuboury, Palaniandy, Abdul aziz, & Feroz, 2017).

The tolerance of the World Bank Group (WBG) for drainage oil effluent into the environment is summarized in Table 11.1 (Jafarinejad & Jiang, 2019). Produced wastewater in the oil industry includes various organic and inorganic components that discharge to the environment case soil and underground and surface water contamination (Peng et al., 2020). Oil wastewater entering into adjacent aquatic and terrestrial ecosystems and accumulating petroleum-based hydrocarbons and other pollutants threaten the health of the environment and living and nonliving organisms. The purification of these compounds is essential because the toxic compounds in petroleum wastewater remain in the environment for a long time (Patel & Patel, 2020).

Petroleum wastewater characteristics

Wastewater sources can be categorized into two families: residential and nonresidential. Nonresidential wastewater is generally discharged by industries such as oil, steel, iron, and agricultural and commercial activities such as hospitals,

TABLE 11.1 World bank group guideline for releasing wastewater.

Parameters	WBG guidelines
Oil and grease	10 mg/L
BOD	30 mg/L
COD	125 mg/L
TSS	30 mg/L
Phenol	0.2 mg/L
pH	6.9
Benzene	0.05 mg/L
Benzo(a) Pyne	0.05 mg/L
Total cyanide	1 mg/L
Free cyanide	0.1 mg/L
Total chromium	0.5 mg/L
Hexavalent chromium	0.05 mg/L
Copper	0.5 mg/L
Iron	3 mg/L
Lead	0.1 mg/L
Nickel	0.5 mg/L
Mercury	0.03 mg/L
Arsenic	0.1 mg/L
Vanadium	1 mg/L
Total N	10 mg/L
Total P	2 mg/L
Sulfide	0.2 mg/L
Temperature	<3 at edge of mixing

restaurants, and shops. Rainfall is another kind of nonresidential wastewater. Nonresidential wastewater gradient change is based on the source. For instance, the wastewater of the textile industry includes organic dye (Falciola, Pifferi, & Testolin, 2020). Generally, wastewater pollutants include three components: inorganic matter, organic toxic microorganisms, and pathogens. Thirty percent of pollutants belong to inorganic materials, usually composed of polyatomic compounds and heavy metals, which are the product of oil, textile, steel, and agriculture industry (Bora & Dutta, 2014; Inglezakis, Loizidou, & Grigoropoulou, 2002). Sewage composition in oil wastewater depends on crude oil quality and different operating conditions during production. In an oil refinery station, crude oil is decomposed into various components into various compounds and turned into beneficial products and also, its nonhydrocarbon materials are removed (Aljuboury et al., 2017).

According to reports, oil refinery wastewater contains $500-3000$ mg L^{-1} soluble oil and grease. Also, chemical oxygen demand (COD) and biological oxygen demand (BOD) are respectively equal to $750-1600$ and $300-1000$ mg L^{-1}. Toxic organic matters, such as phenolic compounds, were measured as almost 950 mg L^{-1}, and ammonia and sulfide were reported $20-80$ and $13-17$ mg L^{-1}, respectively (Djoko, Cahyo, & Puji, 2020). Petroleum sullages are composed of grease and oil compounds, soluble gas, and hydrocarbons (Ratman, Djoko, Puji, Aulia, & Aditya, 2020). Three major groups of hydrocarbons of oil effluent include paraffin with a few numbers of C1 to C4 carbon atoms such as methane CH_4, ethane C_2H_6, and propane C_3H_8, Naphthene included cyclohexane C_6H_{12} and dimethyl cyclopentane C_7H_{14} and Aromatic substances such as Benzene C_6H_6, toluene C_7H_8, and xylene C_8H_{10}. Also, naphthenic acids are a group of compounds in the oil industry's wastewater with toxic effects and are difficult to remove. If crude oil contains significant amounts of sulfur, it is called as sour crude oil. Sour crude oil may contain various slow degradation materials and toxic compounds (Aljuboury et al., 2017).

Oil effluents are complex matrices of organic pollutants. This sewage often contains oils and greases that are very viscous and lead to obstacles and corrosion of drainage pipes and generate an unpleasant odor. Also, phenolic compounds are another dangerous materials that threaten the environment due to severe toxicity even in low dosage, and also have the ability to remain a long time (El-naas, Al-zuhair, & Alhaija, 2010). Research has shown that this sewage has nitrogen and sulfur compounds, which are respectively shown as ammonia and hydrogen sulfide, H_2S (Aljuboury et al., 2017).

Petroleum wastewater treatment technology

Wastewater treatment by modern and efficient technology is the main challenge for the oil industry (Peng et al., 2020). Petroleum wastewater treatment is categorized into three main groups, namely, physical, biological, and chemical processes. Physical methods include absorbance by active carbon, coagulation, electrodialysis, microfiltration, evaporation, freezing, and air deposition. Biological treatment includes biological degradation, activated sludge, aeration filters, and chemical−biological reactors. Chemical treatment includes chemical oxidation by photocatalytic degradation under ultraviolet or visible rays and microwave-assisted wet air oxidation methods, electrochemical process, Fenton process, ozonation, and chemical deposition. Among them, photocatalytic degradation is more effective because it uses an abundant source of sunlight for pollutant degradation (Thakur, Kumar, Singh, & Devi, 2020). Traditional treatment of wastewater in refineries is based on mechanical and physical−chemical processes such as isolation and coagulation followed by biological treatment. Treatment approaches that are studied by scholars for oil wastewater include photodegradation (Jallouli, Pastrana-martínez, Ribeiro, & Moreira, 2018), Fenton and photo-Fenton process (Pourehie & Saien, 2020), biological degradation (Ebrahimi, Kazemi, Mirbagheri, & Rockaway, 2016; Huang et al., 2020), membrane isolation (Munirasu, Haija, & Banat, 2016), coagulation, and flocculation (Singh & Kumar, 2020), and electrochemical methods (Dheeravath, Poodari, Shankar, Vurimindi, & Vidyavathi, 2014; Yavuz, Koparal, & Bak, 2010).

So far, several processes have been assigned assigned for oil wastewater treatment, including membrane bioreactor (MBR), moving bed bioreactor (MBBR) activated sludge process (ASP), upflow anaerobic sludge blanket (UASB), anaerobic membrane bioreactors (AMBR), hybrid anaerobic reactor (HAR), upflow anaerobic fixed bed (UAFB) reactor, anaerobic−aerobic-biofilm reactor (A/O-BR), microaerobic hydrolysis acidification (MHA), and membrane sequencing batch reactors (MSBR). In this kind of biological method, large amount of sludge is produced and requires skilled needed, skilled labor, high mechanical equipment, more energy consumption, and more time for setting up and treating oil wastewater (Jain et al., 2020). The gravity method could just remove floating oil and the main technical problem of the membrane separation method is the need to descale the membrane. Coagulation and flocculation are economical and environmentally friendly but time-consuming procedures (Zhao et al., 2020). The photocatalytic oxidation process (PHCOP) has many advantages over other purification processes including high productivity, low costs, long life of catalysts, brief operating conditions, and the ability to break down different types of complex pollutants into simpler compounds (Almomani, Bhosale, & Shawaqfah, 2020). Water pollutant degradation using sun-absorbing materials is an economical and highly efficient way for wastewater treatment. Using photocatalytic decomposition, it is easy to get rid of harmful organic matter in wastewater using sunlight.

Photocatalysis

AOP includes using ways such as ultrasonic, photocatalysis, hydrogen peroxide (H_2O_2), ozone, Fenton, or UV radiation, for the production of hydroxyl free radical (OH) as a vigorous oxidant to eliminate some compounds in the solution. The high reactivity and the redox potential of this free radical cause it to react with organic compounds in water and degraded pollutants (Rostam & Taghizadeh, 2020). There are different forms of advanced oxidation (see Fig. 11.1), and they are divided into homogeneous and heterogeneous processes, including chemical and photochemical processes such as ultrasonic, photocatalysis, photo-Fenton, H_2O_2 catalysis, persulfate catalyst, UV, zonation, and their combination of them (Hitam & Jalil, 2020). Among these processes, photocatalytic degradation with the help of different types of semiconductors in the presence of light is the most attractive method that destroys organic pollutants in a sol−gel or doped state (Kaur et al., 2020). The word photocatalysis is driven from two Greek words as photo means light and *katalyo* means degradation. Photocatalysis refers to a process where optical photons act as catalyst exciters and increase the rate of chemical oxidation (Bahnemann, 2004). Catalysts are welcomed in the industrial process not only due to accelerating reactions but also because of reasons such as economic efficiency, recovery from the solution, nontoxicity, chemical stability, environmentally friendly, and cost-effectiveness characteristics (Kohansal, Haghighi, & Zarrabi, 2020).

Photocatalysts are used for the inactivation of bacterial and cancer cells, antimicrobial performance, nitrogen fixation in agriculture, and removal of air pollutants such as NO_x and SO_x (Ismael, 2020). Also, the photocatalyst effect has been studied in agriculture, air pollution (Rodriguez-gonzalez, Terashima, & Fujishima, 2019; Trapalis et al., 2016), water and environment (Fujishima, Zhang, & Tryk, 2007), treatment of textile wastewater (Moghadam & Qaderi, 2019), removing BTEX (Sheikholeslami, Kebria, & Qaderi, 2018), and oil refinery wastewater treatment (Ani et al., 2018). The first evaluation on photocatalysis was performed about five decades later on TiO_2 electrodes by Fukushima and Honda in (1969) where sun radiation was used for producing hydrogen from water. Photocatalytic water splitting has been extensively used in recent decades in Japan (Fujishima et al., 2007). Seeking to discover the photocatalytic properties of TiO_2 semiconductors with the help of ultraviolet rays in water splitting reaction, much interest has been expressed by researchers for active semiconductor synthesis by visible light irradiation in pure, doped, and composite

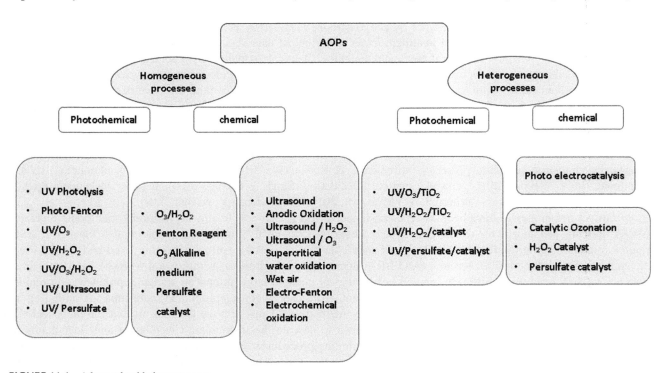

FIGURE 11.1 Advanced oxidation processes.

forms (Baig et al., 2020). Several kinds of studies have been performed in the field of using solar energy for water pollutant treatment using photocatalysts (Oturan & Aaron, 2014). Through diverse kinds of AOPs, photocatalysis has been raised as a suitable way to treat refinery wastewater (Demir-duz, Ayyildiz, Sinan, Alvarez, & Contreras, 2019; Fujishima et al., 2007).

Considering literature review, using photocatalysis for real oil wastewater treatment needs a special catalyzer with high reactivity and more time due to the presence of intricated compounds. Keramati and Ayati (2019) integrated electrocoagulation and photocatalysis methods and reached oil wastewater treatment with lower energy consumption and higher efficiency (Keramati & Ayati, 2019). Another study used the combination of ZnO solar catalyst, TiO_2, and photo-Fenton reaction for oil wastewater treatment. Results showed that the maximum removal rate for TOC and COD was 89% and 61%, respectively (Aljuboury et al., 2017). Evaluation of photocatalyst function of ZnO and TiO_2 was investigated for degradation of oil wastewater pollutants and synthesis of a suitable photocatalyst under visible light and concluded that these compounds were able to remove real oil wastewater pollutants under ultraviolet light (Ani et al., 2018).

Fundamentals and treatment mechanisms

In general, the performance of a photocatalytic reaction is based on three interdependent steps. The first stage involves light absorbance and excitation; and then in the next stage electron separation and transfer occurs from the charge layer to conduction, and in the final stage, surface photocatalytic chemical reactions take place (Li & Wang, 2016). Upon irradiating light with an appropriate wavelength on the catalyst the energy of photons is absorbed by the photocatalyst. When this energy is greater than the energy of the catalyst bandgap, enough energy is supplied to excite the electron from the valance band to the conduction. If there is no proper absorbance, the stored energy disappears in a few nanoseconds (Bahnemann, 2004). Therefore, by transferring electrons due to the absorption of a certain energy, an electron–hole is created in the valence band where is the suitable place for pollutant oxidation (Thakur et al., 2020). As mentioned before, the redox reaction is initiated by light radiation created electrons and holes. The next oxidation reaction generates the most reactive radicals such as hydroxyl and superoxide radicals. Reacting these radicals with liquid or gaseous phase pollutant leads to their destruction through the oxidation or reduction process (Trapalis et al., 2016). Produced CO_2 during the reaction could be trapped for other users to prevent further contamination of the environment (Ani et al., 2018). Theoretical issues of photocatalysis have been studied in many scientific works (Ohtani, 2011).

After the oxidation process, the produced OH^- and O^{2-} radicals in petrochemical wastewater, react with phenolic compound and petrochemical pollutants and convert them to polyphenols, aldehydes, and carboxylic acids. As a result of these reactions, the wastewater is detoxified and eventually, the pollutants are converted to carbon dioxide and mineral water. Chemical photocatalysis method has been widely applied to investigate the decomposition of numerous pollutants such as phenols, pesticides, alkanes, alkenes, etc (Thakur et al., 2020).

Best photocatalysts

The most commonly used catalysts in the industry can be classified into three categories: homogeneous, heterogeneous, and biocatalyst. Homogeneous catalysts are the catalysts that exist in the same phase of their bed, while heterogeneous catalysts exist in different phases of the substrate. Macromolecular substances derived from living organisms such as enzymes could be used as a biocatalyst. However, using heterogeneous catalysts has been considered for large-scale industrial applications due to more advantages such as the ease of use and better performance (Khalil, Kadja, & Muallifu, 2020). Numerous metal oxides have been used since 1972 as photocatalysts of treatment such as titanium dioxide (TiO_2), zinc oxide (ZnO), Fe (III) oxide (Fe_2O_3), vanadium oxide (V_2O_5), tungsten trioxide (WO_3) (Raizada, Sudhaik, & Singh, 2019). Finding the best catalyst in this process remains a challenging issue (Baig et al., 2020).

An ideal catalyst needs special properties to be photocatalytically capable under visible light. Such properties include high absorbance capability in the visible light spectrum, having a suitable surface, shape, and size to create active sites for reaction (Baig et al., 2020). Large amounts of photocatalysts are produced as oxides, nitrides, oxi nitrides, and sulfides. Especially, in the last five years, much progress has been made in providing efficient photocatalysts for water division. Different kinds of metal oxides have been evaluated as photocatalyst powder. These include transition metal oxides contained Ti^{4+}, Zr^{4+}, Nb^{5+}, or Ta^{5+} metal ions and the ordinary metal oxide possessing Ga^{3+}, In^{3+}, Ge^{4+}, Sn^{4+}, Sb^{5+} metal ions. Metal oxide, and nitride have beneficial advantages for water decomposition owing to high stability and oxygen evolution activity (Fujishima et al., 2007).

Nanotechnology and photocatalysis

With technological advancement, nano-photocatalysts have been produced at the nanoscale, which have developed their domain and efficiency (Radhika, Selvin, Kakkar, & Umar, 2016). Nanotechnology uses materials with at least one dimension less than 100 nm, which means they are active at atomic or molecular scale (Trivedi & Bergi, 2021). Nanomaterials are widely considered due to high reactivity, dimensions, electrostatics bond, hydrophilicity, and hydrophobicity properties. Nanotechnology-based methods are very effective and flexible and could be used successfully for different processes such as absorbance, analysis, membrane processes, electronic sensors, etc. Nanoparticles are very durable and have a high specific surface area and their high surface-to-volume rate cause better reaction with wastewater pollutions (Yadav, Singh, & Kumar, 2020). As mentioned before, photocatalysis is a reaction-accelerating process due to light in the presence of a suitable catalyzer. Semiconductive nanoparticles are adopted as the desired catalyzer in the photodegradation process due to having an appropriate bandgap. During photocatalysis, nanocatalysts are present in the reaction medium and are subjected to ultraviolet or visible rays with equivalent or more energy than their bandgap and then excited. Electrons in valence band are excited and transferred to conductance band. Holes created in the valence band and generated electrons are responsible for oxidation and reduction reactions, respectively (Baruah, Chaudhary, & Malik, 2019). These electron holes destruct every reactive oxidative or reductive radicals such as hydroxyl and destroy organic/mineral pollutants introduced in contaminated water by secondary reactions. These radicals photocatalytically destroy pollutant molecules through direct transfer of electrons or produced holes (Bora & Dutta, 2014). Using nano environmental technology is a good choice to solve problems related to oil products wastewater (Younis et al., 2020).

TiO_2 is widely used due to several advantages such as chemical nontoxicity, relatively low cost, high abundance, and photocatalytic activity. The noncatalytic effect of TiO_2 from the past to current and their effect on wastewater treatment has been thoroughly reviewed in the research (Horikoshi & Serpone, 2018). TiO_2 nanoparticles are used as a photocatalyst in different industries such as cosmetics, textiles, and pharmaceuticals (Thakur et al., 2020). Research has demonstrated TiO_2 efficiency to remove PhAC from the real sewage solutions under solar radiation. Immobilization and stabilization of photocatalysts in the form of thin films on substrates could significantly simplify their isolation method and increase photocatalyst progress; therefore, in the current study, the catalyst was immobilized on the surface as a thin layer film and the effect of film thickness was investigated on the photocatalytic activity of the prepared coatings (Palacios-villarreal et al., 2020). This catalyst adsorbs a UV photon and creates a pair of electrons/holes, which generates reactive oxygen species. Nano TiO_2 reaction could be improved by optimizing particle size and shape and maximizing responsive aspects (Qu, Alvarez, & Li, 2013). The major problem with TiO_2 is its bandgap, with 3.23 eV bandwidth for the anatase phase, and lower than 387.5 nm stimulation wavelength, which limited its application under visible light. To overcome this issue, it is combined with metal nanoparticles or graphene carbon nanostructures, CNT, or doped with metal and nonmetal ions. The main advantage of this is the reduction of bandgap energy. The initial observation of increased TiO_2 absorbance from the UV related to the 1960s concentrated on doping with noble metals like platinum, gold, and silver (Benalioua, Mansour, Bentouami, Boury, & Elandaloussi, 2015). Kohansal et al. (2020) synthesized $Bi_7O_9I_3$-ZnO nanocomposites using ultrasonic assisted synthesis techniques and applied them for the photocatalysis of methyl blue under solar irradiation.

Carbon nitride graphite (g-C_3N_4) is a semiconductive polymer whose properties include nontoxicity, complete chemical and thermal stability, as well as visible light absorption in the blue spectral region. Therefore, it is used as a catalyzer in various applications (Baig et al., 2020). Cadmium sulfide (CdS) has narrower bandgap energy than (g-C_3N_4), and therefore, it can absorb more visible light. Cadmium sulfide also has sufficient negative potential and could represent more electrons for photocatalytic reactions. It is therefore used as a photocatalyst in hydrogen production, carbon dioxide reduction process, and degradation of organic and inorganic pollutants in water (Fermin, Ponomarev, & Peter, 1999; Meissner, Memming, & Kastening, 1988). Also, CdS composites have shown efficient and effective photocatalytic impact with other semiconductive oxides such as TiO_2, ZnO, semiconductive sulfides like MoS_2, ZnS, and graphene (Baig et al., 2020).

Silver phosphate (Ag_3PO_4) is a visible light catalyzer that has a broad application in water oxidation and light decomposition of organic pollution. The performance of this material can be improved by combining it with other materials. For example, its combination with TiO_2 and CeO_2 or compounds like Co_3O_4, C_3N_4, and $SrTiO_3$ has been found to be very effective. Such compounds are used for the treatment of wastewater pollutions. Ag_3PO_4 causes photocatalytic degradation in long light radiation and is effective for water contaminant destruction due to stability and ease to use. Cerium dioxide (CeO_2) is used as an active photocatalyst because of oxidation performance and high strength and stability in the treatment of environmental pollutants under ultraviolet light. However, CeO_2 has low photocatalytic efficiency, because the wide gap band only responds to ultraviolet light (Borjigin, Ding, Li, & Wang, 2020). BiOI is also a photocatalyst with a narrow bandgap of about 1.8 eV and becomes active in subject to sunlight (Kohansal et al., 2020).

ZnO is a nano-photocatalyst with a 3.37 eV bandgap. This is considered because of the ease of preparation method, significant thermal stability, and low manufacturing cost. Another unique feature of ZnO is the ability to adsorb a wide spectrum of UV radiation. Besides, studies under consideration observed that the effect of ZnO for photocatalytic degradation of organic pollutants is more than TiO_2. Several techniques such as precipitation, sol−gel, microwave, hydrothermal, and sonochemical are used for ZnO nanoparticle synthesis. In synthesis with the help of ultrasonic waves, holes were created with very high temperature and pressure which change the form, dimensions, and morphology of nanoparticles. This improves photocatalytic properties and reduces preparation time significantly (Kohansal et al., 2020).

Zinc oxide has a different morphology like nanobows, nanobars, nanocombs, tower-shaped and mud-shaped nanowires, nanorings, nanosprings. This variation in the unique structure of the zinc oxide makes this nanoparticle the richest one in terms of properties (Tummapudi, Modem, Kiran, Choudary, & Rao, 2020). For comparison, TiO_2 adsorbs better UV spectrum and is activated under this radiation. Therefore, using sun radiation especially in the visible light area is not sufficient enough for TiO_2 photocatalyst reaction. According to studies, ZnO adsorbs a wide range of sun spectrum (Rajeev, Yesodharan, & Yesodharan, 2015). Sunlight can improve the pollutant photocatalysis processes by increasing their degradation rates. Therefore, treatment with sunlight could be combined with semiconductive photocatalyst like TiO_2 to create an AOP, which could increase treatment rate and use for simultaneous destruction of a wide range of multiple organisms and chemical contaminants (Gora, Zhou, Sokolowski, Hatat-fraile, & Liang, 2018).

Factors influencing photocatalysis

The efficiency of photocatalytic degradation processes is dependent on several factors. The main factors are photocatalyst, the type of material, proper transfer of solar energy to chemical energy and pollutants removal (Long, Li, Wei, Zhang, & Ren, 2020). Factors like wastewater concentration, catalyst amount, pH and inorganic nature of pollutants in wastewater can also influence the photocatalysis process. pH is a critical factor because different effluents require different pH content for their treatment. The pH also control photocatalyst surface properties. Inorganic pollutant's nature and concentration are important in photocatalyst degradation (Yadav et al., 2020). The photocatalyst treatment process strongly depends on the settings and parameters of the reactor. Reactors are usually used in two ways. The first way is using a catalyst in the form of slurry, and the second way is stabilized form like a thin film. Extensive research has been conducted on operational parameters at the laboratory scale (Qu et al., 2013). The study of Nan et al. summarizes the effects of water quality and a range of operational parameters such as TiO_2 concentration, pH, temperature, dissolved oxygen, type and concentration of pollutants, wavelength, and light intensity (Nan, Jin, Chow, & Saint, 2010). The rate of electron transfer from the capacitance band to the conduction band is affected by the light intensity and this factor affects the photocatalytic reaction rate. Light intensity is another important influencing factor for photocatalysis. Typically, the rate of the reactions increases with the increase in light intensities. However, excessive light intensities might have an adverse effect on photocatalysis, hence, this factor needs to be optimized before applying in photocatalysis processes. Photocatalyst particle size is also a very important factor. The smaller the particles, the more levels of the catalyst are exposed to light, the more electrons and holes are created, and the activity levels of the catalyst increase. As a result, the photocatalytic reaction occurs more intensely. The fine size of the particles also has negative effects, such as increasing the recombination of electrons and holes in the surface and using fewer photons. Therefore, the optimal particle size should be used (Chen et al., 2020).

Limitations

Photocatalytic treatment has disadvantages in addition to several disadvantages, including the high cost of using UV lamps. This problem can be solved by using visible sunlight. For example, photocatalytic oxidation with materials that activate under visible light is inexpensive and environmentally friendly (Zhang, Wang, & Zhang, 2014). Recovery, management, and disposal of worn nanoparticles for reuse or separation of treated effluent are few of the major problems in using these materials to treat various types of wastewater. In practice, powdered nanoparticles are unable to recover with regenerative capacity because it is difficult to separate nanoparticles in real refineries (Kazemi & Salem, 2020). Lack of nanoparticle management and disposal strategies can lead to severe secondary contamination and therefore needs further evaluation (Yadav et al., 2020). Nanoparticle synthesis methods are not dangerous, but the reagents used during nanoparticle reproduction are purely chemical. The nanoparticles must not be hazardous and should be free of secondary contamination. Low concentrations of metal nanoparticles, such as iron, gold, and silver, are toxic. It causes the growth of bacteria and damage to native plants, as well as liver failure, damage to the brain and stem cells, and causes human skin diseases (Hussain et al., 2006).

Another limitation of this method is that research studies must be conducted under realistic conditions to properly evaluate the efficiency of this technology. Another research need is to measure the long-term performance of existing technologies on a laboratory scale. Also, because these materials are nanoscale, risk assessment and management of this technology is a challenging issue and researchers should consider the potential risks of using these materials in water and wastewater treatment (Yadav et al., 2020).

Conclusions

Petroleum wastewater contains harmful compounds that need to be treated efficiently. In this chapter, the basic characteristics of photocatalytic degradation with the help of solar energy, for petroleum wastewater pollutants, were studied, and the main details and how photocatalytic degradation methods work were discussed. Also, general studies on the characteristics of petroleum wastewater as well as types of petroleum wastewater treatment were mentioned and many research sources were reviewed and categorized focusing on photocatalytic treatment and solar energy as green energy. According to reports, photocatalysis is effective under sunlight to remove oil pollutants and is anticipated to have a bright future.

References

Aljuboury, D. A. D. A., Palaniandy, P., Abdul aziz, H. B., & Feroz, S. (2017). Treatment of petroleum wastewater by conventional and new technologies — A review. *Global NEST Journal, 19*(3), 439−452.

Almomani, F., Bhosale, R., & Shawaqfah, M. (2020). Chemosphere solar oxidation of toluene over Co doped nano-catalyst. *Chemosphere, 255.* Available from https://doi.org/10.1016/j.chemosphere.2020.126878.

Ani, I. J., Akpan, U. G., Olutoye, M. A., & Hameed, B. H. (2018). Photocatalytic degradation of pollutants in petroleum refinery wastewater by TiO_2 and ZnO-based photocatalysts: Recent development. *Journal of Cleaner Production.* Available from https://doi.org/10.1016/j.jclepro.2018.08.189.

Bahnemann, D. (2004). Photocatalytic water treatment: Solar energy applications. *Solar Energy, 77,* 445−459. Available from https://doi.org/10.1016/j.solener.2004.03.031.

Baig, U., Hawsawi, A., Ansari, M. A., Gondal, M. A., Dastageer, M. A., & Falath, W. S. (2020). Characterization and evaluation of visible light active cadmium sulfide-graphitic carbon nitride nanocomposite: A prospective solar light harvesting photo-catalyst for the deactivation of waterborne pathogen. *Journal of Photochemistry & Photobiology, B: Biology, 204.* Available from https://doi.org/10.1016/j.jphotobiol.2020.111783.

Baruah, A., Chaudhary, V., & Malik, R. (2019). Nanotechnology based solutions for wastewater treatment. *Nanotechnology in Water and Wastewater Treatment.* Available from https://doi.org/10.1016/B978-0-12-813902-8.00017-4.

Benalioua, B., Mansour, M., Bentouami, A., Boury, B., & Elandaloussi, E. (2015). The layered double hydroxide route to Bi-Zn codoped TiO_2 with high photocatalytic activity under visible light. *Journal of Hazardous Materials.* Available from https://doi.org/10.1016/j.jhazmat.2015.02.013.

Bora, T., & Dutta, J. (2014). Applications of nanotechnology in wastewater treatment-A review. *Journal of Nanoscience and Nanotechnology, 14*(1), 613−626. Available from https://doi.org/10.1166/jnn.2014.8898.

Borjigin, B., Ding, L., Li, H., & Wang, X. (2020). A solar light-induced photo-thermal catalytic decontamination of gaseous benzene by using Ag/Ag_3PO_4/CeO_2 heterojunction. *Chemical Engineering Journal, 402*(6). Available from https://doi.org/10.1016/j.cej.2020.126070.

Chen, D., Cheng, Y., Zhou, N., Chen, P., Wang, Y., Li, K., ... Ruan, R. (2020). Photocatalytic degradation of organic pollutants using TiO_2-based photocatalysts: A review. *Journal of Cleaner Production.* Available from https://doi.org/10.1016/j.jclepro.2020.121725.

Demir-duz, H., Ayyildiz, O., Sinan, A., Alvarez, M. G., & Contreras, S. (2019). Approaching zero discharge concept in re fi neries by solar − Assisted photo- Fenton and photo-catalysis processes. *Applied Catalysis B: Environmental, 248,* 341−348. Available from https://doi.org/10.1016/j.apcatb.2019.02.026.

Dheeravath, B., Poodari, S., Shankar, G., Vurimindi, H., & Vidyavathi, S. (2014). Treatment of the petroleum refinery wastewater using combined electrochemical methods treatment of the petroleum refinery wastewater using combined electrochemical methods. *Desalination and water treatment, 57*(8), 1−8. Available from https://doi.org/10.1080/19443994.2014.987175.

Djoko, T., Cahyo, A., & Puji, D. (2020). Phenol and ammonia removal in petroleum refinery wastewater using a poly (vinyl) alcohol coated polysulfone nanohybrid membrane. *Journal of Water Process Engineering.* Available from https://doi.org/10.1016/j.jwpe.2020.101718.

Ebrahimi, M., Kazemi, H., Mirbagheri, S. A., & Rockaway, T. D. (2016). An optimized biological approach for treatment of petroleum re fi nery wastewater. *Journal of Environmental Chemical Engineering, 4,* 3401−3403.

El-naas, M. H., Al-zuhair, S., & Alhaija, M. A. (2010). Removal of phenol from petroleum refinery wastewater through adsorption on date-pit activated carbon. *Chemical Engineering Journal, 162*(3), 997−1005. Available from https://doi.org/10.1016/j.cej.2010.07.007.

Falciola, L., Pifferi, V., & Testolin, A. (2020). Detection methods of wastewater contaminants: State of the art and role of nanotechnology. https://doi.org/10.1016/B978-0-12-818489-9.00003-7.

Fermin, Dj, Ponomarev, E. A., & Peter, L. M. (1999). A kinetic study of CdS photocorrosion by intensity modulated photocurrent and photoelectrochemical impedance spectroscopy. *Journal of Electroanalytical Chemistry, 473,* 192−203.

Fujishima, A., Zhang, X., & Tryk, D. A. (2007). Heterogeneous photocatalysis: From water photolysis to applications in environmental cleanup. *International Journal of Hydrogen Energy, 32*, 2664−2672. Available from https://doi.org/10.1016/j.ijhydene.2006.09.009.

Gora, S., Zhou, Y. N., Sokolowski, A., Hatat-fraile, M., & Liang, R. (2018). Solar photocatalysts with modified TiO_2: Effects on NOM and disinfection byproduct formation potential. *Environmental Science Water Research & Technology*, 1361−1376. Available from https://doi.org/10.1039/c8ew00161h.

Gutierrez-Mata, A. G.S. Velazquez-Martínez, Alberto Álvarez-Gallegos, M. Ahmadi,José Alfredo Hernández-Pérez, F. Ghanbari, S. Silva-Martínez. (2017). Recent Overview of Solar Photocatalysis and Solar Photo-Fenton Processes for Wastewater Treatment Processes for Wastewater Treatment. *International Journal of Photoenergy, 2017*. Available from https://doi.org/10.1155/2017/8528063.

Hitam, C. N. C., & Jalil, A. A. (2020). A review on exploration of Fe_2O_3 photocatalyst towards degradation of dyes and organic contaminants. *Journal of Environmental Management, 258*(December 2019), 110050. Available from https://doi.org/10.1016/j.jenvman.2019.110050.

Horikoshi, S., & Serpone, N. (2018). Can the photocatalyst TiO_2 be incorporated into a wastewater treatment method? background and prospects. *Catalysis Today*. Available from https://doi.org/10.1016/j.cattod.2018.10.020.

Hu, G., Li, J., & Zeng, G. (2013). Recent development in the treatment of oily sludge from petroleum industry: A review. *Journal of Hazardous Materials, 261*, 470−490. Available from https://doi.org/10.1016/j.jhazmat.2013.07.069.

Huang, Z., He, X., Nye, C., Bagley, D., Urynowicz, M., & Fan, M. (2020). Effective anaerobic treatment of produced water from petroleum production using an anaerobic digestion inoculum from a brewery wastewater treatment facility. *Journal of Hazardous Materials, 407*, 124348. Available from https://doi.org/10.1016/j.jhazmat.2020.124348.

Hussain, S. M., Javorina, A. K., Schrand, A. M., Duhart, H. M., Ali, S. F., & Schlager, J. J. (2006). The interaction of manganese nanoparticles with PC-12 cells induces dopamine depletion. *Toxicological Sciences, 92*(2), 456−463. Available from https://doi.org/10.1093/toxsci/kfl020.

Inglezakis, V. J., Loizidou, M. D., & Grigoropoulou, H. P. (2002). Equilibrium and kinetic ion exchange studies of Pb^{2+}, Cr^{3+}, Fe^{3+} and Cu^{2+} on natural clinoptilolite. *Water Research, 36*(11), 2784−2792. Available from https://doi.org/10.1016/S0043-1354(01)00504-8.

Ismael, M. (2020). Ferrites as solar photocatalytic materials and their activities in solar energy conversion and environmental protection: A review. *Solar Energy Materials and Solar Cells, 219*. Available from https://doi.org/10.1016/j.solmat.2020.110786.

Jafarinejad, S., & Jiang, S. C. (2019). Current technologies and future directions for treating petroleum refineries and petrochemical plants (PRPP) wastewaters. *Journal of Environmental Chemical Engineering, 7*(5), 103326. Available from https://doi.org/10.1016/j.jece.2019.103326.

Jain, M., Majumder, A., Sarathi, P., & Kumar, A. (2020). A review on treatment of petroleum refinery and petrochemical plant wastewater: A special emphasis on constructed wetlands. *Journal of Environmental Management, 272*(July), 111057. Available from https://doi.org/10.1016/j.jenvman.2020.111057.

Jallouli, N., Pastrana-martínez, L. M., Ribeiro, A. R., & Moreira, N. F. F. (2018). Heterogeneous photocatalytic degradation of ibuprofen in ultrapure water, municipal and pharmaceutical industry wastewaters using a TiO_2/UV-LED system. *Chemical Engineering Journal, 334*(June 2017), 976−984. Available from https://doi.org/10.1016/j.cej.2017.10.045.

Kaur, N., Kaur, S., Shahi, J. S., Sandhu, S., Sharma, R., & Singh, V. (2020). Comprehensive review and future perspectives of efficient N-doped, Fe-doped and (N, Fe)-co-doped titania as visible light active photocatalysts. *Vacuum, 178*(April), 109429. Available from https://doi.org/10.1016/j.vacuum.2020.109429.

Kazemi, P., & Salem, S. (2020). Evaluation of packed bed photoreactor performance in photocatalytic treatment of wastewater by solar irradiation: A novel strategy for immobilization of nano-sized anatase-graphene oxide composite on extruded rod like metakaolin support. *Solar Energy, 205* (February), 12−21. Available from https://doi.org/10.1016/j.solener.2020.03.116.

Keramati, M., & Ayati, B. (2019). Petroleum wastewater treatment using a combination of electrocoagulation and photocatalytic process with immobilized ZnO nanoparticles on concrete surface. *Process Safety and Environmental Protection, 126*, 356−365. Available from https://doi.org/10.1016/j.psep.2019.04.019.

Khalil, M., Kadja, G. T. M., & Mualliful, M. (2020). Advanced nanomaterials for catalysis: Current progress in fi ne chemical synthesis, hydrocarbon processing, and renewable energy. *Journal of Industrial and Engineering Chemistry, 93*, 78−100. Available from https://doi.org/10.1016/j.jiec.2020.09.028.

Kohansal, S., Haghighi, M., & Zarrabi, M. (2020). Intensification of $Bi_7O_9I_3$ nanoparticles distribution on ZnO via ultrasound induction approach used in photocatalytic water treatment under solar light irradiation. *Chemical Engineering Science*, 116086. Available from https://doi.org/10.1016/j.ces.2020.116086.

Li L., & Wang, M. (2016). Advanced nanomatericals for solar photocatalysis. Available from http://doi.org/10.5772/62206.

Long, Z., Li, Q., Wei, T., Zhang, G., & Ren, Z. (2020). Historical development and prospects of photocatalysts for pollutant removal in water. *Journal of Hazardous Materials*, 122599. Available from https://doi.org/10.1016/j.jhazmat.2020.122599.

Lou, X.H. Gong, Z. He, P. Zhang, L. Hea. (2021). Recent advances in applications of power ultrasound for petroleum industry. *Ultrasonics - Sonochemistry, 70*. Available from https://doi.org/10.1016/j.ultsonch.2020.105337.

Ma, R., Sun, J., Li, D. H., & Wei, J. J. (2020). Review of synergistic photo-thermo-catalysis: Mechanisms, materials and applications. *International Journal of Hydrogen Energy, 45*(55), 30288−30324. Available from https://doi.org/10.1016/j.ijhydene.2020.08.127.

Meissner, D., Memming, R., & Kastening, B. (1988). Photoelectrochemistry of cadmium sulfide. 1. Reanalysis of photocorrosion and flat-band potential. *The Journal of Physical Chemistry, 92*(12), 3476−3483. Available from https://doi.org/10.1021/j100323a032.

Moghadam, M. T., & Qaderi, F. (2019). Results in physics modeling of petroleum wastewater treatment by Fe / Zn nanoparticles using the response surface methodology and enhancing the e ffi ciency by scavenger. *Results in Physics, 15*(August), 102566. Available from https://doi.org/10.1016/j.rinp.2019.102566.

Munirasu, S., Haija, M. A., & Banat, F. (2016). Use of membrane technology for oil field and refinery produced water treatment−A review. *Process Safety and Environmental Protection, 100*, 183−202. Available from https://doi.org/10.1016/j.psep.2016.01.010.

Nan, M., Jin, B., Chow, C. W. K., & Saint, C. (2010). Recent developments in photocatalytic water treatment technology: A review. *Water Research*, *44*(10), 2997–3027. Available from https://doi.org/10.1016/j.watres.2010.02.039.

Ohtani, B. (2011). Photocatalysis A to Z—What we know and what we do not know in a scientific sense. *Journal of Photochemistry & Photobiology, C: Photochemistry Reviews*, *11*(4), 157–178. Available from https://doi.org/10.1016/j.jphotochemrev.2011.02.001.

Oturan, M. A., & Aaron, J.-J. (2014). Advanced oxidation processes in water/wastewater treatment: Principles and applications. A review. *Critical Reviews in Environmental Science and Technology*, *44*(23), 2577–2641. Available from https://doi.org/10.1080/10643389.2013.829765.

Palacios-villarreal, C., Manzano, M., Jos, J., Blanco, E., Ramírez, M., & Levchuk, I. (2020). Photocatalytic degradation of pharmaceutically active compounds (PhACs) in urban wastewater treatment plants effluents under controlled and natural solar irradiation using immobilized TiO$_2$. *e Rueda-M a*, *208*(August), 480–492. Available from https://doi.org/10.1016/j.solener.2020.08.028.

Patel, K., & Patel, M. (2020). Biosurfactants producing Stenotrophomonas sp. S1VKR-26 and assessment of phyto-toxicity. *Bioresource Technology*. Available from https://doi.org/10.1016/j.biortech.2020.123861.

Peng, B., Yao, Z., Wang, X., Crombeen, M., Sweeney, D. G., & Chiu, K. (2020). Cellulose-based materials in wastewater treatment of petroleum industry. *Green Energy and Environment*, *5*(1), 37–49. Available from https://doi.org/10.1016/j.gee.2019.09.003.

Pourehie, O., & Saien, J. (2020). Homogeneous solar fenton and alternative processes in a pilot-scale rotatable reactor for the treatment of petroleum refinery wastewater. *Process Safety and Environmental Protection*. Available from https://doi.org/10.1016/j.psep.2020.01.006.

Qu, X., Alvarez, P. J. J., & Li, Q. (2013). Applications of nanotechnology in water and wastewater treatment. *Water Research*, *47*(12), 3931–3946. Available from https://doi.org/10.1016/j.watres.2012.09.058.

Radhika, N. P., Selvin, R., Kakkar, R., & Umar, A. (2016). Recent advances in nano-photocatalysts for organic synthesis. *Arabian Journal of Chemistry*. Available from https://doi.org/10.1016/j.arabjc.2016.07.007.

Raizada, P., Sudhaik, A., & Singh, P. (2019). Photocatalytic water decontamination using graphene and ZnO coupled photocatalysts: A review. *Materials Science for Energy Technologies*. Available from https://doi.org/10.1016/j.mset.2019.04.007.

Rajeev, B., Yesodharan, S., & Yesodharan, E. P. (2015). Application of solar energy in wastewater treatment: Photocatalytic degradation of αmethylstyrene in water in presence of ZnO. *Journal of Water Process Engineering*, *8*, 108–118. Available from https://doi.org/10.1016/j.jwpe.2015.09.005.

Ratman, I., Djoko, T., Puji, D., Aulia, D., & Aditya, W. (2020). Petroleum Re finery wastewater treatment using three steps modified nanohybrid membrane coupled with ozonation as integrated pre-treatment. *Journal of Environmental Chemical Engineering*, *8*(4), 103978. Available from https://doi.org/10.1016/j.jece.2020.103978.

Rodriguez-gonzalez, V., Terashima, C., & Fujishima, A. (2019). Applications of photocatalytic titanium dioxide-based nanomaterials in sustainable agriculture. *Journal of Photochemistry and Photobiology C: Photochemistry Reviews*, *40*, 49–67. Available from https://doi.org/10.1016/j.jphotochemrev.2019.06.001.

Rostam, A. B., & Taghizadeh, M. (2020). Advanced oxidation processes integrated by membrane reactors and bioreactors for various wastewater treatments: A critical review. *Biochemical Pharmacology*, 104566. Available from https://doi.org/10.1016/j.jece.2020.104566.

Sheikholeslami, Z., Kebria, D. Y., & Qaderi, F. (2018). Nanoparticle for degradation of BTEX in produced water: An experimental procedure. *Journal of Molecular Liquids*, *264*, 476–482. Available from https://doi.org/10.1016/j.molliq.2018.05.096.

Singh, B., & Kumar, P. (2020). Pre-treatment of petroleum refinery wastewater by coagulation and flocculation using mixed coagulant: Optimization of process parameters using response surface methodology (RSM). *Journal of Water Process Engineering*, *36*(April), 101317. Available from https://doi.org/10.1016/j.jwpe.2020.101317.

Thakur, A., Kumar, P., Singh, P., & Devi, P. (2020). Chapter 7 - Photocatalytic degradation of petrochemical pollutants. In *Nano-materials as photocatalysts for degradation of environmental pollutants*. Elsevier Inc, Available from https://doi.org/10.1016/B978-0-12-818598-8.00007-9.

Trapalis, A., Todorova, N., Giannakopoulou, T., Boukos, N., Speliotis, T., Dimotikali, D., ... Yu, J. (2016). TiO$_2$-graphene composite photocatalysts for NOx removal: A comparison of surfactant-stabilized graphene and reduced graphene oxide. *Applied Catalysis B: Environmental*, *180*, 637–647. Available from https://doi.org/10.1016/j.apcatb.2015.07.009.

Trivedi, R., & Bergi, J. (2021). Advanced oxidation processes for effluent treatment plants, chapter 10 - application of bionanoparticles in wastewater treatment. Advance oxidation processes for effluent treatment plants. Elsevier Inc. Available from https://doi.org/10.1016/B978-0-12-821011-6.00010-4.

Tummapudi, N., Modem, S., Kiran, N., Choudary, G., & Rao, S. (2020). Structural, morphological, optical and mechanical studies of annealed ZnO nano particles. *Physica B: Physics of Condensed Matter*, *597*(April), 412401. Available from https://doi.org/10.1016/j.physb.2020.412401.

Yadav, D., Singh, P., & Kumar, P. (2020). Inorganic pollutants in water – Chapter 12. In *Application of nanoparticles for inorganic water purification*. https://doi.org/10.1016/B978-0-12-818965-8.00012-3.

Yavuz, Y., Koparal, A. S., & Bak, U. (2010). Treatment of petroleum refinery wastewater by electrochemical methods. *Desalination*, *258*, 201–205. Available from https://doi.org/10.1016/j.desal.2010.03.013.

Younis, S. A., Ali, H., Lee, J., & Kim, K. (2020). Nanotechnology-based sorption and membrane technologies for the treatment of petroleum-based pollutants in natural ecosystems and wastewater streams. *Advances in Colloid and Interface Science*, *275*, 102071. Available from https://doi.org/10.1016/j.cis.2019.102071.

Yu, L., & Han, M. (2017). A review of treating oily wastewater. *Arabian Journal of Chemistry*, *10*, 1913–1922. Available from https://doi.org/10.1016/j.arabjc.2013.07.020.

Zhang, T., Wang, X., & Zhang, X. (2014). Recent progress in TiO$_2$-mediated solar photocatalysis for industrial wastewater treatment. *International Journal of Photoenergy*, 2014. Available from https://doi.org/10.1155/2014/607954.

Zhao, C., Zhou, J., Yan, Y., Yang, L., Xing, G., Li, H., ... Zheng, H. (2020). Application of coagulation/flocculation in oily wastewater treatment: A review. *Science of the Total Environment*, 142795. Available from https://doi.org/10.1016/j.scitotenv.2020.142795.

Chapter 12

Electrochemical treatment of petroleum wastewater: standalone and integrated processes

Omar Khalifa, Fawzi Banat and Shadi W. Hasan

Center for Membranes and Advanced Water Technology (CMAT), Department of Chemical Engineering, Khalifa University of Science and Technology, SAN Campus, Abu Dhabi, United Arab Emirates

Introduction

The steady increase of the global population places many constraints on the available resources, especially the scarce ones. It also pushed the development of new technologies that were accompanied by industrial revolutions, stifling the resources even more. Water scarcity is a pressing issue that is affecting all spheres of life. Obtaining clean water for drinking or other purposes has been the quest for scientists and engineers. Wastewater treatment is a pathway to increase the amounts of potable water. Industrial practices consume and produce huge amounts of water. The global market for industrial wastewater treatment is projected to reach $17.74 billion by 2026 (Research & Markets, 2019). Particularly, oily wastewater from the oil and gas industry is of great potential. From the upstream to the downstream activities, huge amounts of produced water and refinery wastewater are produced. For each barrel of oil produced, nine barrels of water are extracted, and it could be more with aging wells (Fakhru'l-Razi et al., 2009). Refinery wastewater production can range from 0.4 to 1.6 times the processed crude oil (Coelho, Castro, Dezotti, & Sant'Anna, 2006).

Oily wastewater is recalcitrant in nature, in which the oil emulsions are chemically emulsified ($<20\,\mu m$) or dissolved oil ($<5\,\mu m$) and would not separate readily (Bande, Prasad, Mishra, & Wasewar, 2008). The industry's conventional treatment methods include hydrocyclones, API separators, and air flotation, which fail to clarify the wastewater efficiently (Jafarinejad, 2017). Hence, advanced treatment methods are needed to further polish the wastewater and remove/degrade the stable oil emulsions. Various methods have been employed as tertiary or polishing treatments that can be classified as chemical, physical, thermal, and biological treatments (Jiménez, Micó, Arnaldos, Medina, & Contreras, 2018). Advanced treatments such as membrane filtrations, advanced oxidation processes (AOPs), and electrochemical methods have been widely investigated in the literature. Such processes can bring down the contaminants' concentration and meet environmental regulations or limits set for reutilization purposes. Yet, when operated alone, most of these technologies suffer from high economic or technical issues. Thus, it is desired to integrate processes to mitigate such drawbacks while not compromising efficiency.

Electrochemical methods are blooming, and they provide versatile solutions for wastewater treatment and organic compounds mineralization (Radjenovic & Sedlak, 2015). Considered as green technologies, no chemical addition is needed, and little waste is generated. Electrochemical methods are easy to employ, operate, and combine with other technologies (Anglada, Urtiaga, & Ortiz, 2009). This chapter addresses prominent electrochemical processes and their integration with other advanced treatment methods, namely AOPs, membrane filtration, and biological treatment for petrochemical wastewater.

Electrochemical processes fundamentals

Starting from a broader aspect, there is a close relationship between a chemical and electrochemical reaction. A chemical reaction requires the reactants to come into contact and collide to allow electrons transfer and chemical

Petroleum Industry Wastewater. DOI: https://doi.org/10.1016/B978-0-323-85884-7.00001-1

change. Such collision is governed by the probability of collisions and their duration, which is a rapid action. As such, chemical reactions are characterized by the generated heat of the reaction. To achieve the reaction while abating the collision and elongating the transfer of electrons, the separation of the reactants is entailed, and that can be done through the use of wires connecting the electrodes and a conductive solution to facilitate the movement of electron carriers (Sillanpää & Shestakova, 2017). Namely, this setting is an electrochemical cell, which provides the opportunity to conduct a chemical reaction whilst obtaining an electric current. The use of electricity in water treatment was first proposed in the late 19th century (Chen, 2004). The progress was sluggish until the last decade of the 20th century as more research and experimentation have been conducted (Mollah, Schennach, Parga, & Cocke, 2001). The gist of employing electrochemical processes in water treatment is to make use of the effects imposed by the applied electric field resulting in separating or oxidizing contaminants in the water matrix, and that is achieved by maintaining an adequate potential from an external energy source. A simple electrochemical cell, or electrolyzer, is composed of two electrodes connected to a power source and immersed in a conductive solution (or electrolyte). The governing relationship between the different factors associated with an electrolyzer was found by Michel Faraday. Two laws of Faraday has led to the famous formula: (1) the amount of mass involved in the chemical reaction is proportional to the amount of electricity that circulates the system, and (2) the liberated masses of a substance by a given amount of electricity are proportional to their equivalent weights (Bagotsky, 2005). Faraday's law is represented by the following formula:

$$m = \frac{I t M_W}{n F} \tag{12.1}$$

where m is the liberated electrode's mass (kg), I is the electric current (A), t is the electrolysis time (s), M_W is the molecular weight of the substance (g mol^{-1}), F is Faraday's constant (96,487 C mol^{-1}), and n is the valency of the ion (or the number of electrons involved in the reaction). The development of the electrode potential is a crucial step toward establishing a current flow. The interface of the electrode's surface and the solution is the instigation point, in which an active metal would dissolve into a solution that contains its salt. Given the presence of water, that is, an aqueous solution, the polarity of the water molecules plays a role in pulling positively charged metal ions into the solution. Simultaneously, electrons are freed on the surface of the electrode corresponding to the valence electrons of the produced metal cations. As such, the electrode is considered as negatively charged, and that charge would increase with more cations escaping to the solution. The formed cations in the aqueous solution become surrounded by water molecules (i.e., undergo solvation). Concurrently, there would be some cations returning to the electrode's surface by absorbing the free electrons. These two opposing reactions proceed until equilibrium is achieved (Sillanpää & Shestakova, 2017). Conversely, if the electrode is made of less reactive metals, the opposite trends are observed, in which the electrode would be positively charged, and the solution negatively charged as of the increased anions.

Electrochemical processes in water treatment

A wide range of physical/chemical phenomena are encompassed within the term electrochemical processes, which can roughly be divided into three main categories: conversion methods, separation methods, and combined methods (Sillanpää & Shestakova, 2017). Conversion methods impose changes in the neutralization of the contaminants or impurities in the water. These methods include electrooxidation (EO) and electroreduction. Separation methods are related to removing impurities without necessarily changing their surface chemistry, such as electroflotation (EF), electrodialysis, electrodeposition, and electrodeionization. Combined methods comprise two or more conversion and separation methods, including electrocoagulation (EC), electrocatalysis, and electrodisinfection. Electrocatalysis is a big group that involves the use of light, sound, and catalysts to enhance electrodegradation of contaminants during water treatment, which is in many instanced called electrochemical advanced oxidation process (EAOP). Some phenomena are more pronounced than others, reflected by the effect they impose on the solution. In a given electrochemical cell, multiple phenomena have a propensity to cooccur. With an appropriate design of the cell and a wise choice of operational variables, a phenomenon can be made dominant for a specific operation. For instance, the use of sacrificial anodes coupled with cathodes submerged in an electrolyte would exhibit EC and EF and other phenomena depending on the type of constituents present in the solution (Khalifa, Banat, Srinivasakannan, Radjenovic, & Hasan, 2020). Fig. 12.1 depicts a simple

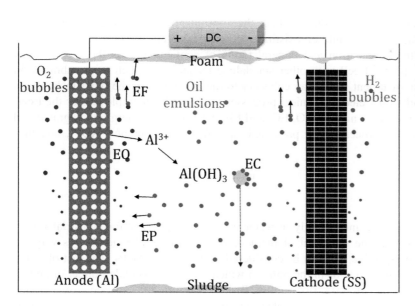

FIGURE 12.1 Electrochemical cell. Schematic of a simple electrochemical cell with Al and SS as anode and cathode, respectively. Different electrochemical phenomena are depicted. *DC*, direct current; *EF*, electroflotation; *EO*, electrooxidation; *EC*, electrocoagulation; *EP*, electrophoresis; *Al*, aluminum; *SS*, stainless steel.

electrochemical cell schematic with EC, EF, EP, and EO processes. A detailed description of each process can be found below in this section.

Several factors ought to be optimized for electrochemical processes (Garcia-Segura, Eiband, de Melo, & Martínez-Huitle, 2017; Sillanpää & Shestakova, 2017), including:

- Current density (CD);
- Overpotential or applied voltage;
- Treatment/electrolysis time;
- Type of electrolyte;
- pH of the solution;
- Reactor design (arrangement and dimensions);
- Flow type (batch or continuous);
- Type of electrodes (choice of anode and cathode metals).

Other factors include the mode of power supply (continuous/intermittent), hydraulic retention time (HRT) in the case of continuous-flow reactors, mode of electric supply (constant voltage or constant current), and type of current [direct current (DC) or alternating current (AC)]. Some factors are more effective than others for a targeted phenomenon. With novel designs and combinations with other technologies, more degrees of freedom is attained, and a larger operation realm is created. For example, adding aeration to the electrochemical cell imposes the airflow rate as an additional variable. In the context of wastewater treatment, several responses are of utmost importance in the application of electrochemical-based treatments, including:

- Current efficiency (ratio between actual to the theoretical amount of electrons consumed);
- Energy consumption (by volume of water treated, the mass of contaminant removed, or order of removal using the logarithmic ratio of the initial and final concentration of contaminants);
- Removal efficiency (specific to the targeted contaminant);
- Associated costs (includes capital and operating expenses—a techno-economic analysis is needed for scalability and applicability for scaled-up processes);
- The evaluation of the effectiveness of electrochemical-based treatments ought to be assessed over these prime indicators, in addition to other outputs related to complexity, environmental impacts, safety, and risk assessments.

Electrochemical methods are now promising for the treatment of industrial oily wastewater, in which EC and EF are the most employed (Jamaly, Giwa, & Hasan, 2015; Mikhak, Torabi, & Fouladitajar, 2019). No added chemicals are needed for most electrochemical processes, so they are labeled green and environmentally benign technologies (Jiménez et al., 2018; Mickova, 2015). Several electrochemical processes and phenomena are being employed for oily wastewater treatment, including EO, EC, and EF (Martínez-Huitle, Rodrigo, Sirés, & Scialdone, 2015). Although proved to be efficient in small-scale demonstrations, electrochemical processes can incur high energy consumption when employed at a larger scale (Anglada et al., 2009), which necessitates their optimization and integration with other technologies to make use of any positive interactions and build on existing synergies. Such combinations can be as simple as adding an energy source to facilitate the degradation and mineralization of contaminants as the case in most electro-based AOPs; or integrating an electrochemical cell with other standalone technologies such as membranes, in which the electrochemical process acts as a pretreatment step or an enhancer to the efficiency of the other technology. This section showcases the most prominent individual electrochemical processes that have been employed for the treatment of oily wastewater from the oil and gas industries, namely EO, EC, and EF. Additionally, emerging integrated and hybrid technologies incorporating electrochemical processes are presented, specifically membrane-, AOP-, and bio-based electrochemical treatments.

Electrooxidation

Electrochemical oxidation, or electrooxidation, is used increasingly in water treatment, especially for industrial wastewater (Särkkä, Bhatnagar, & Sillanpää, 2015). It is the opponent of incineration and chemical oxidation as ways of degrading organic components. From a broad perspective, EO can be considered as part of the AOPs. The gist of EO is an in situ oxidation that occurs on the surface of the anode or in the bulk, in which the former is called direct or anodic oxidation, while the latter is called indirect oxidation (Moreira, Boaventura, Brillas, & Vilar, 2017). Anodic oxidation is the most known EAOP (Martínez-Huitle et al., 2015). The process of electrochemical oxidation occurs upon water activation on the anode's surface through different mechanisms, including dissociative adsorption and electrolytic discharge (mediates oxygen evolution) (Kapałka, Fóti, & Comninellis, 2010). During the dissociative adsorption, hydroxyl radicals are formed and chemically adsorbed to the anode's surface, while in the electrolytic discharge, the formed radicals are loosely attached (physisorption). When the adsorption of hydroxyl radicals is strong on the surface, the "direct" oxidation occurs, consuming the oxidants before escaping to the bulk. When the links between the anode metal and formed hydroxyl radicals is weak, they escape to the bulk and form a variety of products, including water, oxygen, hydrogen peroxide, and ozone through the following equations (Eq. (12.2)−(12.4)) (Sáez, Rodrigo, Fajardo, & Martínez-Huitle, 2018):

$$2^\bullet OH \rightarrow H_2O + 1/2O_2 \tag{12.2}$$

$$2^\bullet OH \rightarrow H_2O_2 \tag{12.3}$$

$$2^\bullet OH \rightarrow O_3 + H_2O \tag{12.4}$$

More oxidants are formed with the rapid reaction with the hydroxyl radicals as per the availability of anions such as chloride, which ultimately forms ClO^-, and the oxidation extends to the bulk rather than being restricted to the surface of the anode. As such, the indirect oxidation efficiency is considered higher compared to the indirect oxidation as of the area of action. Electrodes' choice principally controls the EO process because the governing mechanisms depend on the electrodes' relative potential. For an efficient process, the electrode's material ought to have (Anglada et al., 2009):

- Resistance to corrosion and erosion (high stability);
- High conductivity;
- High selectivity;
- Cheap and resilient.

In the context of EO, Electrodes can be classified into active and inactive electrodes, in which the former is mostly associated with direct oxidation. The inactive electrodes are known to produce more oxidants in bulk, and hence more oxidation occurs (Sáez et al., 2018). Cathodic reactions can have an impact on the treatment of wastewater. The main hydrogen production reaction does not affect the remediation. The reductive reactions at the cathode are shown to remove halogens from water (Radjenović, Farré, Mu, Gernjak, & Keller, 2012). The production of hydrogen peroxide formed after the reduction of oxygen at nonbasic media contributes to the oxidants formed in the cell.

Produced water from an offshore oilfield was electrochemically degraded using a plug flow reactor (Yang, Jing, Chen, Li, & Yin, 2017). Using inactive electrodes, the oil was easily degraded, scoring COD removal efficiency above 75%. Petroleum refinery wastewater (PRW) was treated using an SS-based electrochemical cell and optimized using factorial designs (Thakur, Srivastava, Mall, & Hiwarkar, 2018). The treatment achieved 62 and 52% removal efficiency for COD and TOC, respectively, which aligns with the optimal conditions of initial pH 6, 182 A m^{-2} CD, an interelectrode gap of 1.5 cm, and electrolysis time lasting 145 min. Another study investigated phenol's abatement from synthetic wastewater using a ruthenium-based reactor (Yavuz & Koparal, 2006). The phenol was removed to a great extent reaching 99.7% efficiency, while the COD removal efficiency was 88.9%.

Electrocoagulation

EC is one of the most prominent electrochemical processes employed for the treatment of oily wastewater. It is efficient with the removal of immiscible pollutants that exist in the form of colloids ($<$10 µm) (Mikhak et al., 2019), which makes EC a viable option for removing recalcitrant oil droplets in petrochemical wastewaters. The working principle of electrocoagulation is essentially the same as the chemical coagulation process, except that the source of coagulants is different; in the former, coagulants are added from the external source to the reactor, whereas in electrocoagulation, the coagulants are produced in situ, hence the name. The theory behind the coagulation process lies in the interface of colloids, where the dissolved particles in the water exhibit a circumferential electric double layer (EDL) (Mollah et al., 2001). The EDL forms when a given particle's surface is negatively charged, it attracts positive counterions around its surface and repels negative ions (Fig. 12.2). A colloid is stable when the settling velocity is so small that it is virtually suspended in the solution (Barrera-Díaz, Balderas-Hernández, & Bilyeu, 2018). Thus, the aim of adding the coagulants is to destabilize the colloids by disturbing the repulsive forces acting on the surface of the pollutant. This can occur via various mechanisms, including compression of EDL, adsorption destabilization, interparticle bridging, and precipitation/enmeshment mechanism (Vepsäläinen & Sillanpää, 2020).

Aluminum and iron multivalent ions are the most efficient in destabilizing colloids or oil emulsions in the context of petrochemical wastewater (Genc & Bakirci, 2016; Khalifa et al., 2020; Thakur et al., 2018). The metal cations undergo complex equilibrium reactions forming metal hydroxides, the main precursor of neutralizing the surface charge and forming flocs with high adsorption capacity (Lee & Gagnon, 2016). In electrocoagulation, sacrificial anodes are used to release the metals into the electrolyte and hydrolyze instead of adding metal salts in chemical coagulation. As the flocs start to form, the oil emulsions agglomerate and get adsorbed/trapped within the amorphous gel-like structure. With time, the loaded flocs settle as of gravity and form a sludge layer at the bottom of the reactor, while the supernatant (treated water) is separated easily afterward. The following reactions describe the anodic and cathodic reactions using a sacrificial anode:

At the anode (Mollah et al., 2001):

$$M_{(s)} \rightarrow M^{n+}_{(aq)} + ne^- \tag{12.5}$$

$$2H_2O_{(l)} \rightarrow 4H^+_{(aq)} + O_{2(g)} + 4e^- \tag{12.6}$$

At the cathode (Mollah et al., 2004):

$$M^{n+}_{(aq)} + ne^- \rightarrow M_{(s)} \tag{12.7}$$

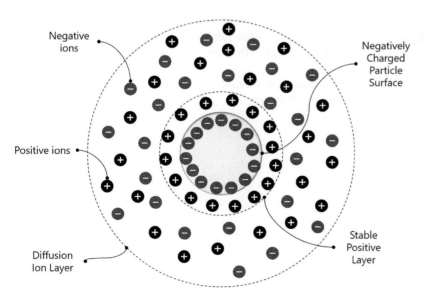

FIGURE 12.2 Electric double layer. A schematic of the electric double layer surrounding a negatively charged particle's surface.

$$2H_2O_{(l)} + 2e^- \rightarrow H_{2(g)} + 2OH^-_{(aq)} \tag{12.8}$$

In the solution (after complex reactions) (Tir & Moulai-Mostefa, 2008):

$$M^{3+}_{(aq)} + 3H_2O_{(l)} \rightarrow M(OH)_{3(s)} + 3H^+_{(aq)} \tag{12.9}$$

The EC process's efficacy in treating petroleum wastewater has been experimentally trialed and evaluated against various oil removal measures. A diesel-based synthetic PRW was successfully treated by EC using Al and SS electrodes (Safari, Azadi Aghdam, & Kariminia, 2016). The EC cell achieved a high removal of COD and diesel by 99.1% and 98.8%, respectively. Another study treated a petrochemical wastewater sample using Al electrodes, in which the EC process resulted in 97.4% turbidity removal (Giwa, Ertunç, Hapoglu, Ertunc, & Alpbaz, 2012). Using Al-based EC with polyaluminum chloride coagulant aid, a gas refinery wastewater sample was treated, resulting in 97% COD removal under optimal condition (Saeedi & Khalvati-Fahlyani, 2011). Petroleum-contaminated groundwater was treated via EC using Al, Fe, and SS (Moussavi, Khosravi, & Farzadkia, 2011). The SS-Al (anode−cathode) was the optimal electrode combination that achieved 93.4% total petroleum hydrocarbon (TPH) removal. Treatment of synthetic oily wastewater was assessed with and without aeration showing an advantage of aeration with 93.3% maximum COD removal (Khalifa et al., 2020). The volume-specific energy consumption of EC-treated produced water was found to be 4.3 times of chemical coagulation treatment at 90% removal of turbidity and COD (Khor et al., 2020). Yet, EC could prove efficient if other variables were considered, such as the transportation of the wastewater and the unit's compactness.

Electroflotation

EF is the counterpart and advanced form of conventional air flotation. The concept lies in carrying the pollutants by rising bubbles, and by buoyancy, floating them to the surface. In dissolved air flotation (DAF), the air is compressed and pumped into the treatment tank, in which the efficacy is mainly dependent on the solubility of oxygen and nitrogen in the wastewater (Russell, 2019). With recalcitrant and small-sized oil emulsions, DAF might fail to remove such traces as in the big-sized bubbles. EF comes in place to enhance the flotation process by using the gases produced as a result of the electrolysis of water, as seen in Eqs. 6 and 8. Forming small-sized bubbles in-situ adds to the efficiency of EF over DAF (Kyzas & Matis, 2016). Essentially, an EC cell with resistant electrodes (corrosion-protected or inert) is an EF cell. The oxygen and hydrogen gas bubbles form at the surface of the anode and cathode, respectively, detach from the surfaces and attach to the similarly sized pollutants. The bubble formation can be divided into three zones as seen in Fig. 12.3. The bubble-pollutant grouping then rises to the top of the reactor as of buoyancy and forms a foam layer that can then be removed by a decanter. Several factors are suggested to control the bubble size, including CD and pH of the medium (Alam & Shang, 2016; Alam, Shang, & Khan, 2017), which was estimated to vary between 20 and 70 μm (Santiago Santos, Salles Pupo, Vasconcelos, Barrios Eguiluz, & Salazar Banda, 2018). The reactor design and the choice of electrode material are of utmost importance for a successful implementation of EF, in which Ti-based inert anodes in the form of dimensional stable anodes (DAS) are the most prominent (Mohtashami & Shang, 2019). In the case of operating with sacrificial anodes such as Al and Fe, EC, and EF processes would occur concurrently.

Using EF to treat oily wastewater has shown excellent efficiency. Wastewater synthesized from crude oil was treated using EF and claimed to separate the oil content almost completely (Ibrahim, Mostafa, Fahmy, & Hafez, 2001).

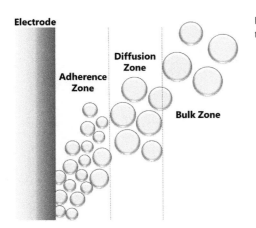

FIGURE 12.3 Electroflotation. A schematic of the zones associated with the bubbles' formation during the electroflotation process.

Another synthetic wastewater prepared from untreated crude oil was treated using Al electrodes (Bande et al., 2008). The separation achieved 90% oil removal efficiency, scoring 0.67 kWh m^{-3}. In another study, Al-based EF cells were used to treat oily water effluents from different fields in India (Tiwari & Patel, 2011). Oil removal reached 97.9% at 9.5 pH, 40 min flotation time, and 145 mg L^{-1} initial oil concentration. In another study, stainless steel sponges were used as electrodes forming an EF reactor bed to treat bilge oily wastewater (Ayten Genc & Bakirci, 2016; Khalifa et al., 2020; Thakur et al., 2018). The COD removal reached 85% with an initial concentration of 57,150 mg L^{-1} and 15 V constant voltage. Another successful implementation of EF for oily wastewater treatment achieved 94%−95% removal efficiency (Maksimov & Ostsemin, 2015).

Electrochemical-based hybrid and integrated processes

Hybridization and integration of technologies in water treatment are usually driven to mitigate individual technologies' deficiencies and extend the removal efficiency of the wastewater treatment system. Oily wastewater from the oil and gas industries is recalcitrant in nature and requires a combined effort to reach a target removal of contaminants. As such, positive synergies are sought between individual technologies if the concerned system is a one-pot solution, in which there is a direct interaction between the system constituents. This would also decrease the overall load on the system and possibly provide economic and scalable solutions.

The design of hybrid/integrated systems can take multiple stages, in which technologies are singly tested, and then combinations can emerge gradually. The gist is to use the least number of resources to meet the required treatment limits. The hybrid treatment of oily wastewater, especially the petrochemical-based effluents, can be challenging and erroneous in the design phase. Such systems' design approach would follow a top-down approach, in which discharge limits or reuse-related constraints are imposed. Hence, the hunt for the optimal design is steered toward the removal efficiency, which is then complemented by techno-economic analysis to assess how realistic the proposed solution is. Still, the bottom-up approach should not be ignored as it is suitable for innovations and novel technologies that are not yet mature; it is mostly related to gaining more understanding and seeking mechanistic justifications of what is happening. Both approaches are important and are not mutually exclusive; one would switch from a methodology to another to suit the situation.

Designs of hybrid and integrated systems are versatile given the possible combinations between different technologies and the pathways of merging them. The terms hybrid and integrated are used interchangeably in the literature to describe a combination of methods. Generally, to put the words to their respective meanings, an integrated system describes a sequence of treatments or a network of interconnected units without direct interaction. This includes systems with pre and posttreatment units, in which the effluent of a given unit is fed as the influent of another unit. Conversely, hybrid reactors resembled one-pot solutions, namely, where an explicit contact exists between different technologies (Fig. 12.4). Yet, the terms are used interchangeably in the literature. Presumably, hybrid reactors require less space than integrated systems as per the compactness. This would be ideal for places with a limited free area, such as offshore platforms. The choice between the different schemes pertains to the associated technologies and synergies (or incompatibilities). As such, a hybrid system can be a segment of an integrated system. Treatment of oily wastewater usually takes multiple stages, namely primary, secondary, and tertiary or polishing treatments (Jafarinejad & Jiang, 2019). Pre and posttreatment units are employed for multiple reasons according to the used technologies. The method that achieves efficient bulk removals would be placed first, followed by polishing technologies that reduce the contaminants concentrations to meet the constraints. Another justification for the configuration of pre and posttreatment units is the sensitivity of the technologies. For instance, a membrane, especially the pressure-driven type, ought to be accompanied by a pretreatment to mitigate the membrane's fouling and extend its lifetime. Some pretreatments aim to "prepare" the

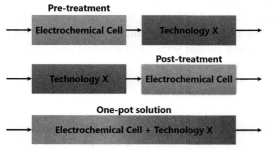

FIGURE 12.4 Configurations. Different configurations of integrated and hybrid treatment systems using electrochemical processes.

wastewater or increase the applicability of the following technology. A wastewater sample might exhibit low biodegradability, and employing an AOP as a pretreatment would increase biological treatment applicability (Pérez et al., 2016). Eventually, the design and choice of technologies are bounded by the wastewater matrix (the constituents and compositions) and the effluents' quality constraints.

Conjugating multiple electrochemical processes in one reactor can be readily applied, and sometimes it inevitably occurs. For example, using sacrificial anodes, mostly Al and Fe, would exhibit both EC and EF processes (Khalifa et al., 2020). EO can also be present to some extent due to water hydrolysis reactions. Yet, many variables can be controlled to tune the dominance/existence of certain electrochemical processes in the reactor. The mechanisms associated with EC/EF combined process include coagulation, adsorption/entrapment, precipitation, and flotation (Santiago Santos et al., 2018). Depending on the size of contaminants and the surface charge, the EC/EF process's efficacy can vary widely. Such a process's efficiency would be similar to EC cells unless designed in an integrated fashion where each process, EC and EF, occurs separately. Prime technologies have been integrated with electrochemical processes, including AOPs, membrane filtration, and biological treatments. This section showcases hybrid and integrated processes that combine electrochemical processes with these technologies.

Electrochemical-based advanced oxidation processes

AOPs are prominent methods to abate contaminants from wastewaters. Their strength lies in the formation of hydroxyl radicals that exhibit high oxidation potential (Sievers, 2011). The variety of AOP schemes makes it an attractive option to remediate recalcitrant oily wastewater, especially produced water from oil and gas fields (Jiménez et al., 2018). Fenton processes, photocatalytic reactions, and ozonation are primary technologies under the umbrella of AOPs. As discussed in Section 2.1, EO is essentially an AOP. Yet, advanced schemes of electrochemical cells that are integrated with other AOPs can be grouped separately, namely EAOP. Combining both technologies targets the reduction of the individual processes' high energy consumption, promotes the radicals' formation, and boosts the combined system's oxidative power. EAOPs are destructive compared to other electrochemical processes such as EC and EF, where the contaminants are separated into other phases and removed. As such, many advantages are sought with the use of EAOPs, including (Medel, Lugo, & Meas, 2018):

- Versatility (choice of catalysts, energy sources, design options, and persistence to higher organic load compared to counterpart technologies);
- Automation (by controlling the electrochemical-related variables);
- Environmental compatibility (with no to little waste);
- Absence of sludge/muds (as of the destructive behavior of the combined process);
- Increased biodegradability of the wastewater.

The choice of electrodes is vital for EAOPs (Medel et al., 2018), many of which proved efficient in the abatement of organics, including the boron-doped diamond (BDD), doped-SnO_2, PbO_2, and doped-TiO_2 (Chaplin, 2014). Remediation of various wastewaters is suggested using several schemes of EAOPs, including anodic oxidation, anodic oxidation/electro-generated hydrogen peroxide, and electro-Fenton, photoelectro-Fenton, among others (Moreira et al., 2017).

Other than electrochemically instigated AOP, combinations of electrochemical cells and AOPs include integrating EC with ozonation or peroxidation (Malik, Ghosh, Vaidya, & Mudliar, 2020; Tahreen, Jami, & Ali, 2020). The combined process proves to be efficient than individual performances, with 94.7% removal efficiency at optimal ozone generation rate and time (Das et al., 2021). An electrochemical-ozone hybrid system was employed to treat industrial wastewater using iron electrodes (Bernal-Martínez, Barrera-Díaz, Solís-Morelos, & Natividad, 2010). The COD was removed by 84%, surpassing the electrochemical cell and ozone's individual efficiencies, which were 43% and 60%, respectively. Another study attempted to treat biodiesel wastewater using Fe-based EC, H_2O_2, and polyaluminum chloride (PAC) (Ahmadi, Sardari, Javadian, Katal, & Sefti, 2013). O&G was removed at a maximum rate of 86%. PWR was treated using EC with the addition of Fe and aeration separately, in which around 90% COD removal was achieved by the electro-Fenton reactor (Yan et al., 2014).

Electrochemical-assisted membrane processes

Technologies accompanying membranes in hybrid processes are usually designed to mitigate technical problems associated with the membranes, such as fouling, besides increasing the concerned contaminants' removal. The same applies to electro-membrane reactors, where the electrochemical processes help remove or degrade contaminants before the membrane to reduce the susceptibility of fouling or scaling. Membrane filtration is increasingly used in treating oily wastewater, especially produced water (Dickhout et al., 2017). Depending on the pore sizes, there are multiple types of membranes, namely microfiltration, ultrafiltration, nanofiltration, and reverse osmosis, ordered from largest to smallest pores. As pressure-driven processes, the smaller the pore size, the higher the pumping power required. The membranes are also made of different materials such as polymers and ceramic; the latter is said to have better suitability with oily wastewater (Padaki et al., 2015).

EC process was coupled with ceramic MF membrane as a pretreatment to remediate synthetic oily wastewater (Yang et al., 2017). Oil removal reached 65% after the EC tank, while it reached 98.5% after the membrane, in which fouling was also reduced by the EC pretreatment. A hybrid system comprised of Ti-based EC and UF membrane was employed to treat humic acid-based wastewater (Chen, 2004). Applying higher CD resulted in higher EC activity, and hence more removal reaching 60%, while the UF was protected from cake forming on its surface. Another study implemented an integrated system of EC and forward osmosis (FO) to treat synthetic produced water (Al Hawli, Benamor, & Hawari, 2019). The EC was used to pretreat the FO membrane's draw solution, in which the whole system achieved 97.4% O&G removal efficiency. Another implementation of an Al-based EC as a pretreatment for an FO-MD integrated system was investigated to treat hydraulic fracturing produced water (Sardari, Fyfe, & Ranil Wickramasinghe, 2019). Apart from achieving high TOC removal (78%), the combination of EC with FO and MD resulted in maintaining the membranes' integrity, ensuring stable operation, and retaining high water flux.

A prominent integrated solution for the remediation of oily wastewater is the combination of EAOP and membrane filtration (Pan et al., 2019). Such integrations ought to further clarify the wastewater in a compact design and achieve high energetic efficiencies. Different configurations are suggested to integrate the EAOP with the membranes, which can act as pretreatment of the membrane's feed, posttreatment of the concentrate, or permeate. Other configurations are in the form of a hybrid design, in which the membrane is submerged along with the electrodes in the reactor. The membrane can be conductive and nonconductive; the former can be part of the electrochemical cell (acts as an electrode). The effects imposed by the electrochemical cell on the membrane are sought to mitigate fouling (B. Yang, Geng, & Chen, 2015). A membrane-EAOP integrated system was used to treat synthetic oily wastewater made from fuel oil (Li et al., 2018). The membrane was carbon-based, while the EAOP was electrochemical anodic oxidation, with the membrane acting as an anode. An EC process was employed as a pretreatment to the combined membrane/EAOP reactor. The whole system achieved higher removal of COD and TOC by 96.1% and 94.7%, respectively. Another integrated set-up investigated the synergy between ozonation and electrochemical-based filtration (Souza-Chaves, Dezotti, & Vecitis, 2020). The electro-filtration unit was made of carbon nanotubes that served as an anode as well. TOC was removed to a great extent reaching 85%, and phenol was degraded by ozonation. The kinetics of the combined process was also higher than its constituents when operated individually. Another novel design combined ozonation with an electro-membrane reactor to treat real oily wastewater from the oil and gas industry (Khalifa, Banat, Srinivasakannan, Radjenovic, et al., 2020). The electro-membrane reactor was composed of the Al-SS EC process and microfiltration ceramic membrane submerged in the reactor. With aeration, the electro-membrane removed a maximum of 18% of COD. Adding ozonation to the reactor enhanced the EC process and promoted the greater performance of the membrane, which resulted in a threefold enhancement in the removal efficiency.

Electro-biological treatment

The microbial activity can be a great mine of energy and a tool to treat wastewaters. Whether indigenous or external, the microorganism mineralizes the organic content of the wastewater and can adsorb some particles into the flocs and sludge. The applicability of biological treatment is prone to its tolerance toward the toxicity of the wastewater. Oily wastewater can be toxic to microorganisms, particularly produced water and refinery wastewater. High biodegradability is resembled by a high BOD/COD ratio and 100:5:1 ratio of C:N:P (Ishak, Malakahmad, & Isa, 2012). As such, the integration of biological treatments with other methods is desirable to increase the wastewater's biodegradability. EC using Al electrodes was integrated with a fixed film aerobic bioreactor to treat refinery wastewater (Pérez et al., 2016). The integrated system's removal efficiencies reached 98% for TPH and 95% for COD, which is higher than singly operating with EC. The biodegradability of the aerobic reactor was enhanced significantly by the EC process. Another study combined the electro-Fenton process and persulfate and biological treatment for mixed industrial wastewater (Popat, Nidheesh, Anantha Singh, & Suresh Kumar, 2019). The integrated system's COD removal efficiency reached 94%, surpassing 60% of the electrolysis cell's removal before the biological treatment. Another integrated treatment system was composed of EC, a spouted bed bioreactor (SBBR), and an activated-carbon-based adsorption bed (Fig. 12.5) (El-Naas,

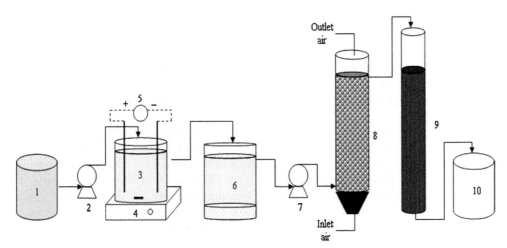

FIGURE 12.5 EC-SBBR-AC. A schematic for an integrated treatment system consisting of EC (#3), a spouted bed bioreactor (SBBR) (#8), and an activated-carbon-based adsorption bed (#9). *From El-Naas, M. H., Alhaija, M. A., & Al-Zuhair, S. (2014). Evaluation of a three-step process for the treatment of petroleum refinery wastewater.* Journal of Environmental Chemical Engineering, 2(1), 56–62. doi:10.1016/j.jece.2013.11. 024.

Alhaija, & Al-Zuhair, 2014). The system was used to treat PRW, in which EC acted as a pretreatment to the biological treatment and the adsorption-packed bed. The integrated system achieved 97 and 100% removal of COD and phenol, respectively. A pilot-scale operation of the same system achieved comparable results (El-Naas, Surkatti, & Al-Zuhair, 2016). Electrically enhanced membrane bioreactors are prominent examples of electrochemical-based biological integrated systems (Giwa et al., 2012), in which electrochemical processes are believed to enhance biodegradation and bioflocculation processes.

Microbial fuel cells (MFCs) are promising candidates that ought to combine wastewater treatment and electric power generation (Neto, Reginatto, & De Andrade, 2018). The concept is to mimic the chemical fuel cells but biologically, which is through bioelectrocatalysis. A specific type of bacteria on the anode side is used as a catalyst to transfer electrons directly or indirectly. A proton-permeable membrane separates the anode and cathode. Another alternative to the electrochemical biological method is the microbial electrolysis cell, which is similar in configuration to the microbial fuel cell (Neto et al., 2018). An MFC was to treat oil field-produced water with multiple configurations (Roustazadeh Sheikhyousefi et al., 2017). The degradation of hydrocarbons reached $96.6 \pm 1.94\%$ with 3 mW m^{-2} energy output. Another study implemented a bio-electro-Fenton microbial fuel cell (Bio-E-Fenton MFCs) yielding 99.3% COD removal and 52.5 mW m^{-2} electrical power output (Wu et al., 2018).

Challenges and outlook

Using traditional techniques, industrial oily wastewater is difficult to treat. Advanced and polishing treatments are necessary to meet environmental regulations and discharge limits. Electrochemical methods are emerging technologies that can be used as standalone and hybrid/integrated processes. Prominent candidates include EO, EC, and EF, which can effectively remove/degrade organic matter in complex wastewater matrixes. EO can readily be upgraded into EAOP by a wise choice of electrodes and catalysts, which can be integrated with other technologies such as membranes and biological treatments. EC and EF can also be efficient pretreatments of membrane systems, mitigating fouling's susceptibility and increasing the process's overall performance.

Several shortcomings of electrochemical processes ought to be investigated for efficient and practical implementation. The energy consumption of electrochemical processes might not be justifiable upon scalability, which necessitates the use of novel designs of the reactor and electrodes. The electrodes, especially the sacrificial, are prone to a continuous replacement, which entails an increase in the process's operating expenses. The cost of electrodes constitutes a big hurdle on the process's economics; low-cost electrodes would be a plausible remedy (Radjenovic & Sedlak, 2015). A major drawback of electrochemical processes is the mass transfer limitation of species in the solution. This can hinder the extent of reactions/separations, which incur efficiency losses. As such, the reactor's hydrodynamics should be carefully assessed to maximize instantaneous current efficiency and electrochemical reactivity. The use of EAOP might pose hazardous issues by forming toxic and carcinogenic byproducts depending on the composition of the wastewater. The acquaintance of the formation of such products is necessary to diminish the possibility of environmental impacts. Combinations involving EC and sludge-forming processes ought to employ sludge management systems, especially for continuous-mode operations (Khalifa, Banat, Srinivasakannan, AlMarzooqi, & Hasan, 2020). Membrane-EAOP is a potential alternative for organic wastewater degradation, and a variety of options are yet to be studied (Pan et al., 2019).

The development of electrochemical-based processes is growing, and their merits can be further exploited. As such, evolving environmental regulations can be readily abided by. The benefits of other technologies can be unlocked by synergies with electrochemical processes. The field of hybrid and integrated systems is vast, and systematic combination analysis will decrease the quest for optimal designs and positive interactions. Conducting pilot-scale experimentations would bring more confidence and momentum for such systems. The achievement of efficient electrochemical-based petroleum wastewater treatment applications, especially one-point solutions, can bring new potential for in-situ reuse in oilfields and refineries and, in turn, minimize discharge into the environment and increase wastewater recyclability.

References

Ahmadi, S., Sardari, E., Javadian, H. R., Katal, R., & Sefti, M. V. (2013). Removal of oil from biodiesel wastewater by electrocoagulation method. *Korean Journal of Chemical Engineering, 30*(3), 634−641. Available from https://doi.org/10.1007/s11814-012-0162-5.

Al Hawli, B., Benamor, A., & Hawari, A. A. (2019). A hybrid electro-coagulation/forward osmosis system for treatment of produced water. *Chemical Engineering and Processing - Process Intensification, 143*, 107621. Available from https://doi.org/10.1016/j.cep.2019.107621.

Alam, R., & Shang, J. Q. (2016). Electrochemical model of electro-flotation. *Journal of Water Process Engineering, 12*, 78−88. Available from https://doi.org/10.1016/j.jwpe.2016.06.009.

Alam, R., Shang, J. Q., & Khan, A. H. (2017). Bubble size distribution in a laboratory-scale electroflotation study. *Environmental Monitoring and Assessment, 189*(4). Available from https://doi.org/10.1007/s10661-017-5888-4.

Anglada, Á., Urtiaga, A., & Ortiz, I. (2009). Contributions of electrochemical oxidation to waste-water treatment: Fundamentals and review of applications. *Journal of Chemical Technology and Biotechnology, 84*(12), 1747−1755. Available from https://doi.org/10.1002/jctb.2214.

Bagotsky, V. S. (2005). Fundamentals of electrochemistry.

Bande, R., Prasad, B., Mishra, I., & Wasewar, K. (2008). Oil field effluent water treatment for safe disposal by electroflotation. *Chemical Engineering Journal, 137*(3), 503−509. Available from https://doi.org/10.1016/j.cej.2007.05.003.

Barrera-Díaz, C. E., Balderas-Hernández, P., & Bilyeu, B. (2018). *Electrocoagulation: Fundamentals and prospectives. Electrochemical water and wastewater treatment* (pp. 61−76). Available from https://doi.org/10.1016/B978-0-12-813160-2.00003-1.

Bernal-Martínez, L. A., Barrera-Díaz, C., Solís-Morelos, C., & Natividad, R. (2010). Synergy of electrochemical and ozonation processes in industrial wastewater treatment. *Chemical Engineering Journal, 165*(1), 71−77. Available from https://doi.org/10.1016/j.cej.2010.08.062.

Chaplin, B. P. (2014). Critical review of electrochemical advanced oxidation processes for water treatment applications. *Environmental Science: Processes Impacts, 16*(6), 1182−1203. Available from https://doi.org/10.1039/C3EM00679D.

Chen, G. (2004). Electrochemical technologies in wastewater treatment. *Separation and Purification Technology, 38*(1), 11−41. Available from https://doi.org/10.1016/j.seppur.2003.10.006.

Coelho, A., Castro, V. A., Dezotti, M., & Sant'Anna, G. L. (2006). Treatment of petroleum refinery sourwater by advanced oxidation processes. *Journal of Hazardous Materials, 137*(1), 178−184. Available from https://doi.org/10.1016/J.JHAZMAT.2006.01.051.

Das, P. P., Anweshan, A., Mondal, P., Sinha, A., Biswas, P., Sarkar, S., ... Purkait, M. K. (2021). Integrated ozonation assisted electrocoagulation process for the removal of cyanide from steel industry wastewater. *Chemosphere, 263*, 128370. Available from https://doi.org/10.1016/j.chemosphere.2020.128370.

Dickhout, J. M., Moreno, J., Biesheuvel, P. M., Boels, L., Lammertink, R. G. H., & de Vos, W. M. (2017). Produced water treatment by membranes: A review from a colloidal perspective. *Journal of Colloid and Interface Science, 487*, 523−534. Available from https://doi.org/10.1016/J.JCIS.2016.10.013.

El-Naas, M. H., Alhaija, M. A., & Al-Zuhair, S. (2014). Evaluation of a three-step process for the treatment of petroleum refinery wastewater. *Journal of Environmental Chemical Engineering, 2*(1), 56−62. Available from https://doi.org/10.1016/j.jece.2013.11.024.

El-Naas, M. H., Surkatti, R., & Al-Zuhair, S. (2016). Petroleum refinery wastewater treatment: A pilot scale study. *Journal of Water Process Engineering, 14*, 71−76. Available from https://doi.org/10.1016/j.jwpe.2016.10.005.

Fakhru'l-Razi, A., Pendashteh, A., Abdullah, L. C., Biak, D. R. A., Madaeni, S. S., & Abidin, Z. Z. (2009). Review of technologies for oil and gas produced water treatment. *Journal of Hazardous Materials, 170*(2−3), 530−551. Available from https://doi.org/10.1016/J.JHAZMAT.2009.05.044.

Garcia-Segura, S., Eiband, M. M. S. G., de Melo, J. V., & Martínez-Huitle, C. A. (2017). Electrocoagulation and advanced electrocoagulation processes: A general review about the fundamentals, emerging applications and its association with other technologies. *Journal of Electroanalytical Chemistry, 801*, 267−299. Available from https://doi.org/10.1016/j.jelechem.2017.07.047.

Genc, A., & Bakirci, B. (2016). Destabilization and treatment of emulsified oils in wastewaters by electrocoagulation. *Water Environment Research, 88*(11), 2008−2014. Available from https://doi.org/10.2175/106143016X14733681695203.

Giwa, S. O., Ertunç, S., Hapoglu, H., Ertunc, S., & Alpbaz, M. (2012). Electrocoagulation treatment of turbid petrochemical wastewater. *International Journal of Advances in Science and Technology, 5*(5), 23−32.

Ibrahim, M. Y., Mostafa, S. R., Fahmy, M. F. M., & Hafez, A. I. (2001). Utilization of electroflotation in remediation of oily wastewater. *Separation Science and Technology, 36*(16), 3749−3762. Available from https://doi.org/10.1081/SS-100108360.

Ishak, S., Malakahmad, A., & Isa, M. H. (2012). Refinery wastewater biological treatment: A short review. *Journal of Scientific & Industrial Research, 71*, 251−256.

Jafarinejad, S. (2017). *Treatment of oily wastewater. Petroleum waste treatment and pollution control* (pp. 185−267). Available from https://doi.org/10.1016/b978-0-12-809243-9.00006-7.

Jafarinejad, S., & Jiang, S. C. (2019). Current technologies and future directions for treating petroleum refineries and petrochemical plants (PRPP) wastewaters. *Journal of Environmental Chemical Engineering, 7*(5), 103326. Available from https://doi.org/10.1016/j.jece.2019.103326.

Jamaly, S., Giwa, A., & Hasan, S. W. (2015). Recent improvements in oily wastewater treatment: Progress, challenges, and future opportunities. *Journal of Environmental Sciences, 37*, 15−30. Available from https://doi.org/10.1016/J.JES.2015.04.011.

Jiménez, S., Micó, M. M., Arnaldos, M., Medina, F., & Contreras, S. (2018). State of the art of produced water treatment. *Chemosphere, 192*, 186−208. Available from https://doi.org/10.1016/J.CHEMOSPHERE.2017.10.139.

Kapałka, A., Fóti, G., & Comninellis, C. (2010). *Basic principles of the electrochemical mineralization of organic pollutants for wastewater treatment. Electrochemistry for the environment* (pp. 1−23). Available from https://doi.org/10.1007/978-0-387-68318-8_1.

Khalifa, O., Banat, F., Srinivasakannan, C., AlMarzooqi, F., & Hasan, S. W. (2020). Ozonation-assisted electro-membrane hybrid reactor for oily wastewater treatment: A methodological approach and synergy effects. *Journal of Cleaner Production*, 125764. Available from https://doi.org/10.1016/j.jclepro.2020.125764.

Khalifa, O., Banat, F., Srinivasakannan, C., Radjenovic, J., & Hasan, S. W. (2020). Performance tests and removal mechanisms of aerated electrocoagulation in the treatment of oily wastewater. *Journal of Water Process Engineering, 36*, 101290. Available from https://doi.org/10.1016/j.jwpe.2020.101290.

Khor, C. M., Wang, J., Li, M., Oettel, B. A., Kaner, R. B., Jassby, D., ... Hoek, V. E. M. (2020). Performance, energy and cost of produced water treatment by chemical and electrochemical coagulation. *Water, 12*(12), 3426. Available from https://doi.org/10.3390/w12123426.

Kyzas, G. Z., & Matis, K. A. (2016). Electroflotation process: A review. *Journal of Molecular Liquids, 220*, 657−664. Available from https://doi.org/10.1016/j.molliq.2016.04.128.

Lee, S. Y., & Gagnon, G. A. (2016). Growth and structure of flocs following electrocoagulation. *Separation and Purification Technology, 163*, 162−168. Available from https://doi.org/10.1016/j.seppur.2016.02.049.

Li, C., Feng, G., Song, C., Zhong, G., Tao, P., Wang, T., ... Shao, M. (2018). Improved oil removal ability by the integrated electrocoagulation (EC)-carbon membrane coupling with electrochemical anodic oxidation (CM/EAO) system. *Colloids and Surfaces A: Physicochemical and Engineering Aspects, 559*, 305−313. Available from https://doi.org/10.1016/j.colsurfa.2018.09.043.

Maksimov, E. A., & Ostsemin, A. A. (2015). Intensifying the cleaning of emulsion- and oil-bearing waste water from rolled-product manufacturing by electroflotation. *Metallurgist, 58*(11−12), 945−949. Available from https://doi.org/10.1007/s11015-015-0022-8.

Malik, S. N., Ghosh, P. C., Vaidya, A. N., & Mudliar, S. N. (2020). Hybrid ozonation process for industrial wastewater treatment: Principles and applications: A review. *Journal of Water Process Engineering, 35*, 101193. Available from https://doi.org/10.1016/j.jwpe.2020.101193.

Martínez-Huitle, C. A., Rodrigo, M. A., Sirés, I., & Scialdone, O. (2015). Single and coupled electrochemical processes and reactors for the abatement of organic water pollutants: A critical review. *Chemical Reviews, 115*(24), 13362−13407. Available from https://doi.org/10.1021/acs.chemrev.5b00361.

Medel, A., Lugo, F., & Meas, Y. (2018). *Application of electrochemical processes for treating effluents from hydrocarbon industries. Electrochemical water and wastewater treatment* (pp. 365−392). Available from https://doi.org/10.1016/b978-0-12-813160-2.00014-6.

Mickova, I. (2015). Advanced electrochemical technologies in wastewater treatment part I: Electrocoagulation. *American Scientific Research Journal for Engineering, Technology, and Sciences (ASRJETS), 14*, 233−257.

Mikhak, Y., Torabi, M. M. A., & Fouladitajar, A. (2019). *Refinery and petrochemical wastewater treatment. Sustainable water and wastewater processing* (pp. 55−91). Available from https://doi.org/10.1016/b978-0-12-816170-8.00003-x.

Mohtashami, R., & Shang, J. Q. (2019). Electroflotation for treatment of industrial wastewaters: A focused review. *Environmental Processes, 6*(2), 325−353. Available from https://doi.org/10.1007/s40710-019-00348-z.

Mollah, M. Y. A., Schennach, R., Parga, J. R., & Cocke, D. L. (2001). Electrocoagulation (EC)- Science and applications. *Journal of Hazardous Materials, 84*(1), 29−41. Available from https://doi.org/10.1016/S0304-3894(01)00176-5.

Mollah, M., Morkovsky, P., Gomes, J., Kesmez, M., Parga, J., & Cocke, D. (2004). Fundamentals, present and future perspectives of electrocoagulation. *Journal of Hazardous Materials, 114*(1−3), 199−210. Available from https://doi.org/10.1016/j.jhazmat.2004.08.009.

Moreira, F. C., Boaventura, R. A. R., Brillas, E., & Vilar, V. J. P. (2017). Electrochemical advanced oxidation processes: A review on their application to synthetic and real wastewaters. *Applied Catalysis B: Environmental, 202*, 217−261. Available from https://doi.org/10.1016/j.apcatb.2016.08.037.

Moussavi, G., Khosravi, R., & Farzadkia, M. (2011). Removal of petroleum hydrocarbons from contaminated groundwater using an electrocoagulation process: Batch and continuous experiments. *Desalination, 278*(1−3), 288−294. Available from https://doi.org/10.1016/j.desal.2011.05.039.

Neto, S. A., Reginatto, V., & De Andrade, A. R. (2018). *Microbial fuel cells and wastewater treatment. Electrochemical water and wastewater treatment* (pp. 305−331). Available from https://doi.org/10.1016/B978-0-12-813160-2.00012-2.

Padaki, M., Surya Murali, R., Abdullah, M. S., Misdan, N., Moslehyani, A., Kassim, M. A., ... Ismail, A. F. (2015). Membrane technology enhancement in oil-water separation. A review. *Desalination, 357*, 197−207. Available from https://doi.org/10.1016/j.desal.2014.11.023.

Pan, Z., Song, C., Li, L., Wang, H., Pan, Y., Wang, C., ... Feng, X. (2019). Membrane technology coupled with electrochemical advanced oxidation processes for organic wastewater treatment: Recent advances and future prospects. *Chemical Engineering Journal, 376*, 120909. Available from https://doi.org/10.1016/j.cej.2019.01.188.

Pérez, L. S., Rodriguez, O. M., Reyna, S., Sánchez-Salas, J. L., Lozada, J. D., Quiroz, M. A., ... Bandala, E. R. (2016). Oil refinery wastewater treatment using coupled electrocoagulation and fixed film biological processes. *Physics and Chemistry of the Earth, 91*, 53−60. Available from https://doi.org/10.1016/j.pce.2015.10.018.

Popat, A., Nidheesh, V. P., Anantha Singh, T. S., & Suresh Kumar, M. (2019). Mixed industrial wastewater treatment by combined electrochemical advanced oxidation and biological processes. *Chemosphere, 237*, 124419. Available from https://doi.org/10.1016/j.chemosphere.2019.124419.

Radjenovic, J., & Sedlak, D. L. (2015). Challenges and opportunities for electrochemical processes as next-generation technologies for the treatment of contaminated water. *Environmental Science and Technology, 49*(19), 11292−11302. Available from https://doi.org/10.1021/acs.est.5b02414.

Radjenović, J., Farré, M. J., Mu, Y., Gernjak, W., & Keller, J. (2012). Reductive electrochemical remediation of emerging and regulated disinfection byproducts. *Water Research, 46*(6), 1705−1714. Available from https://doi.org/10.1016/j.watres.2011.12.042.

Research and Markets. (2019). Global industrial wastewater treatment service market analysis 2019.

Roustazadeh Sheikhyousefi, P., Nasr Esfahany, M., Colombo, A., Franzetti, A., Trasatti, S. P., & Cristiani, P. (2017). Investigation of different configurations of microbial fuel cells for the treatment of oilfield produced water. *Applied Energy, 192*, 457−465. Available from https://doi.org/10.1016/J.APENERGY.2016.10.057.

Russell, D. (2019). *Dissolved air flotation and techniques. Practical waste water treatment* (pp. 409−417). Available from https://doi.org/10.1002/9781119527114.ch19.

Saeedi, M., & Khalvati-Fahlyani, A. (2011). Treatment of oily wastewater of a gas refinery by electrocoagulation using aluminum electrodes. *Water Environment Research, 83*(3), 256−264. Available from https://doi.org/10.2175/106143010X12780288628499.

Sáez, C., Rodrigo, M. A., Fajardo, A. S., & Martínez-Huitle, C. A. (2018). *Indirect electrochemical oxidation by using ozone, hydrogen peroxide, and ferrate.* Electrochemical *water and wastewater treatment* (pp. 165−192). Available from https://doi.org/10.1016/B978-0-12-813160-2.00007-9.

Safari, S., Azadi Aghdam, M., & Kariminia, H.-R. (2016). Electrocoagulation for COD and diesel removal from oily wastewater. *International Journal of Environmental Science and Technology, 13*(1), 231−242. Available from https://doi.org/10.1007/s13762-015-0863-5.

Santiago Santos, G. O., Salles Pupo, M. M., Vasconcelos, V. M., Barrios Eguiluz, K. I., & Salazar Banda, G. R. (2018). *Electroflotation.* Electrochemical water and wastewater treatment (pp. 77−118). Available from https://doi.org/10.1016/B978-0-12-813160-2.00004-3.

Sardari, K., Fyfe, P., & Ranil Wickramasinghe, S. (2019). Integrated electrocoagulation − Forward osmosis − membrane distillation for sustainable water recovery from hydraulic fracturing produced water. *Journal of Membrane Science, 574*, 325−337. Available from https://doi.org/10.1016/J.MEMSCI.2018.12.075.

Särkkä, H., Bhatnagar, A., & Sillanpää, M. (2015). Recent developments of electro-oxidation in water treatment - A review. *Journal of Electroanalytical Chemistry, 754*, 46−56. Available from https://doi.org/10.1016/j.jelechem.2015.06.016.

Sievers, M. (2011). *Advanced oxidation processes.* Treatise on *water science* (pp. 377−408). Available from https://doi.org/10.1016/B978-0-444-53199-5.00093-2.

Sillanpää, M., & Shestakova, M. (2017). *Introduction.* Electrochemical *water treatment methods: Fundamentals, methods and full scale applications* (pp. 1−46). Available from https://doi.org/10.1016/B978-0-12-811462-9.00001-3.

Souza-Chaves, B. M., Dezotti, M., & Vecitis, C. D. (2020). Synergism of ozonation and electrochemical filtration during advanced organic oxidation. *Journal of Hazardous Materials, 382*, 121085. Available from https://doi.org/10.1016/j.jhazmat.2019.121085.

Tahreen, A., Jami, M. S., & Ali, F. (2020). Role of electrocoagulation in wastewater treatment: A developmental review. *Journal of Water Process Engineering, 37*, 101440. Available from https://doi.org/10.1016/j.jwpe.2020.101440.

Thakur, C., Srivastava, V. C., Mall, I. D., & Hiwarkar, A. D. (2018). Mechanistic study and multi-response optimization of the electrochemical treatment of petroleum refinery wastewater. *Clean - Soil, Air, Water, 46*(3), 1700624. Available from https://doi.org/10.1002/clen.201700624.

Tir, M., & Moulai-Mostefa, N. (2008). Optimization of oil removal from oily wastewater by electrocoagulation using response surface method. *Journal of Hazardous Materials, 158*(1), 107−115. Available from https://doi.org/10.1016/j.jhazmat.2008.01.051.

Tiwari, S. P., & Patel, D. R. (2011). Removal of oil from oily effluents of North Gujarat fields (India) by electroflotation method. *Indian Journal of Science and Technology, 4*(12), 1721−1725. Available from https://doi.org/10.17485/ijst/2011/v4i12/30318.

Vepsäläinen, M., & Sillanpää, M. (2020). *Electrocoagulation in the treatment of industrial waters and wastewaters. Advanced water treatment: Electrochemical methods* (pp. 1−78). Available from https://doi.org/10.1016/B978-0-12-819227-6.00001-2.

Wu, J.-C., Yan, W.-M., Wang, C.-T., Wang, C.-H., Pai, Y.-H., Wang, K.-C., ... Thangavel, S. (2018). Treatment of oily wastewater by the optimization of Fe_2O_3 calcination temperatures in innovative bio-electron-fenton microbial fuel cells. *Energies, 11*(3), 565. Available from https://doi.org/10.3390/en11030565.

Yan, L., Wang, Y., Li, J., Ma, H., Liu, H., Li, T., ... Zhang, Y. (2014). Comparative study of different electrochemical methods for petroleum refinery wastewater treatment. *Desalination, 341*, 87−93. Available from https://doi.org/10.1016/J.DESAL.2014.02.037.

Yang, B., Geng, P., & Chen, G. (2015). One-dimensional structured IrO_2 nanorods modified membrane for electrochemical anti-fouling in filtration of oily wastewater. *Separation and Purification Technology, 156*, 931−941. Available from https://doi.org/10.1016/j.seppur.2015.10.040.

Yang, M., Jing, B., Chen, W., Li, Q., & Yin, X. (2017). Experimental study on COD composition and electrochemical degradation of wastewater in offshore oilfields. *Journal of the Chinese Chemical Society, 64*(1), 73−79. Available from https://doi.org/10.1002/jccs.201600244.

Yavuz, Y., & Koparal, A. S. (2006). Electrochemical oxidation of phenol in a parallel plate reactor using ruthenium mixed metal oxide electrode. *Journal of Hazardous Materials, 136*(2), 296−302. Available from https://doi.org/10.1016/j.jhazmat.2005.12.018.

Chapter 13

The viable role of activated carbon for the effective remediation of refinery and petrochemical wastewaters

K.L. Tan[1] and Keng Yuen Foo[2]

[1]River Engineering and Urban Drainage Research Center (REDAC), Universiti Sains Malaysia, Nibong Tebal, Malaysia, [2]River Engineering and Urban Drainage Research Center (REDAC), Engineering Campus, Universiti Sains Malaysia, Penang, Malaysia

Introduction

Activated carbon (AC), also known as activated charcoal, is broadly defined as a carbon-rich solid with a well-developed porosity and high adsorptive reactivity. Commercial AC (CAC) usually has a BET surface area between 500 and 1500 $m^2\,g^{-1}$ (Rashidi & Yusup, 2017).

AC is a multipurpose adsorbent, with a high affinity for a wide range of organic and inorganic contaminants, making it a prolific technique for environmental pollution control. The historical development of AC has been outlined in Table 13.1 (Cecen, 2011). Since its development, AC is the most established adsorbent for the treatment of industrial flue gas and wastewater, recording the world production at 12.8 million tons, with municipal water treatment taking up the largest market share (Market, 2020). Until today, AC remains to be a central strategy for air and water purification throughout the course of civilization and industrialization.

Within the refineries and petrochemical industries, AC has been an indispensable part of the wastewater treatment facilities and well-integrated with the biological activated sludge, membrane separation, and advanced oxidation processes (AOPs). Driven by the highly porous structure, wide adsorptive affinity, high chemical stability, processability of the macroscopic shape, good mechanical strength, tailorable surface functionality, and affordable cost, three distinct roles of AC are an adsorbent, catalyst support, or a growth medium for microorganisms. The main objective of this chapter is to highlight the advances thus far achieved with AC, and AC-aided technologies for the treatment of petroleum refinery and petrochemical wastewater. Specific focus will be devoted to the treatment of real industrial effluents, instead of the single component of synthetic wastewater. Challenges, compelling inadequacies, and opportunities for future research will be highlighted in the Perspective and Outlook section. This chapter is expected to provide alarming inspirations and spur further valorization of AC, and AC processes for the innovative treatment of complex wastewaters, especially effluents emitted from the petroleum refineries and petrochemical plants.

Classification, unique characteristics, and applications

In general, the CAC could be classified into four major groups, namely, (1) granular AC (GAC, 0.2−5.0 mm), (2) powder activated carbon (PAC, <0.18 mm), (3) extruded (pelletized) AC, and (4) AC cloth (Mosbah, Mechi, Khiari, & Moussaoui, 2020). Among all, exceeding 80% of them is produced in the granular or powder form, while the remaining is in the pellet form. AC contains predominantly microporous structure (Pore diameter, $D_p < 2$ nm), but a significant fraction of mesopores (2 nm $< D_p < 50$ nm), or macropores ($D_p > 50$ nm) could be formed with the changing preparation conditions and initial raw precursors.

The representative pore structure of AC is depicted in Fig. 13.1A. Microporous ACs are ideal for gas phase adsorption, while mesoporous ACs are highly desirable for liquid phase applications to accommodate the bulkier adsorbate species. As an amorphous solid (up to 90% carbon) forms a very disordered microstructure, different models have been proposed to describe the chemical structure of AC (Heidarinejad et al., 2020). It is generally accepted that AC bears

TABLE 13.1 Timeline of historical developments of activated carbon.

Year	Event
3750 BCE	Wood char (charcoal) was used by the Egyptians and Sumerians for ores processing and medicinal purposes.
450 BCE	Charcoal filter was used by the Hindus for the purification of drinking water.
1400s	Charred wooden barrels were used to store freshwater aboard ships.
1773	Scientific studies on the adsorption of gases onto the charcoal.
1786	Lowitz demonstrated the adsorptive power of charcoal in the aqueous medium.
1794	Amid the industrial revolution, wood charcoal was applied commercially for the decolorization of raw sugar syrups in England.
1808	Wood charcoal was used widely in European sugar refineries for the decolorization process.
1822	Bussy reported the thermal and chemical processing of AC production.
1854	Wood char filters were installed in sewer ventilation systems for odor removal in London.
1862	Lipscombe prepared a carbon material for the purification of potable water.
1865	Hunter reported the excellent gas adsorption properties of coconut shells-derived AC.
1872	Carbon face masks were used to remove mercury vapors in chemical industries.
1881	Kayser coined the word "adsorption" to describe the uptake of gases onto carbons.
1900	Ostreijko filed two patents on the physical and chemical activation of carbon, using CO_2, steam, or metal chlorides.
1909	Powder activated carbon (PAC) was manufactured from wood at the industrial scale in Europe, mainly used for decolorizing solutions in the chemical and food industries.
1913	First commercial production of PAC in the United States.
1914–1918	Large-scale production of granular AC (GAC) during World War I, for poisonous gas adsorption in face masks. GAC started to venture into water treatment, air purification, and chemical industries.
1929	In Germany and the United States, PAC was applied for public potable water purification. GAC filter columns were also adopted.
1935–1940	AC was used for VOC adsorption in industries and municipalities.
1970s	10,000 water and wastewater treatment plants have adopted AC in their operations.
1995	Norman and Cha pioneered the microwave-assisted production of AC from coal for gas adsorption (Norman & Cha, 1995).
2011	Foo and Hameed pioneered the microwave-assisted chemical activation of AC from agricultural biomass for aqueous dye adsorption (Foo & Hameed, 2011).

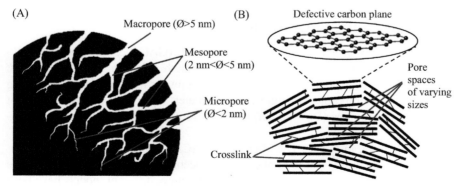

(A)

Macropore (Ø>5 nm)

Mesopore (2 nm<Ø<5 nm)

Micropore (Ø<2 nm)

(B) Defective carbon plane

Pore spaces of varying sizes

Crosslink

FIGURE 13.1 (A) Pore structure and (B) illustrative chemical structure of activated carbon.

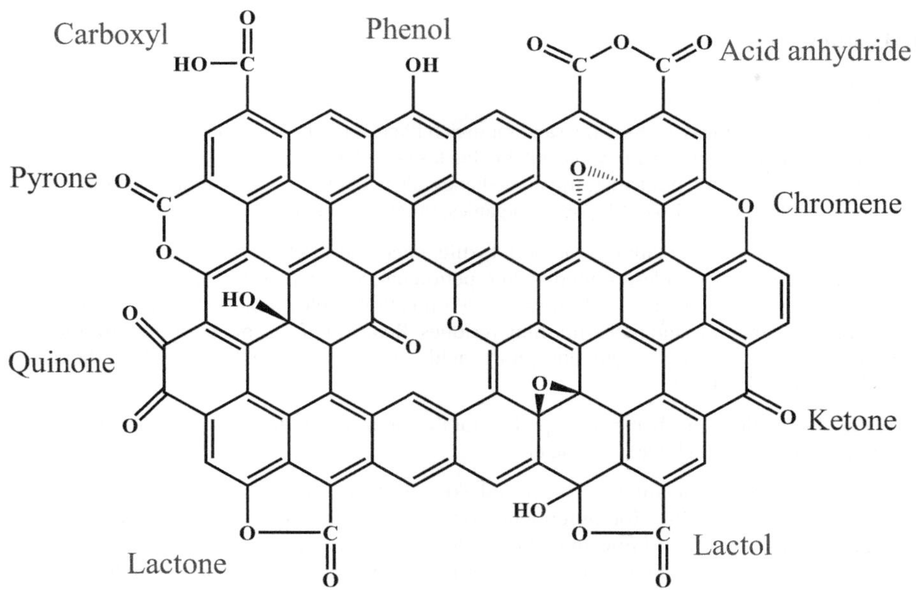

FIGURE 13.2 O-containing surface functional groups of activated carbon (Red: acidic group, Blue: basic group).

structural resemblance to graphite, but with significant disorientations of the hexagonal carbon layer planes, which are crosslinked by the aliphatic bridging groups. The sp^2-hybridized carbon atoms form a hexagonal layer or stacks of 2—4 hexagonal layers, with the interlayer spacings between 0.34 and 0.80 nm. The interlayer spacings and the spaces between the stacks are irregularly slit-shaped and interconnected to form the micropores of AC (Barton et al., 1999). The illustrative chemical structure of AC is displayed in Fig. 13.1B. From the presented model, the carbon planes are defective at the outer surface and also at the edges, in which heteroatoms are incorporated at these defect sites. These heteroatoms are mostly O and H, occasionally N, S, and P. The ability to manipulate the incorporation of heteroatoms into the carbon network via synthetic means serves as a viable tool to adjust the properties of the AC.

AC owes its adsorptive reactivity to the ample oxygenated moieties on its surface (Fig. 13.2). These O-bearing moieties which formed the pores are developed by oxidation of the carbon matrix during the activation stage. The acidity of the carbon surface is related to the O-containing functional groups like carboxyl, carboxylic anhydride, lactone, and phenolic hydroxyl (Shafeeyan, Daud, Houshmand, & Shamiri, 2010). The surface basicity is contributed mainly by the delocalized π-electrons and N-containing moieties (amide, imide, lactam, pyrrole, pyridine, nitrile), and partially by the O-bearing functional groups, notably chromene, ketone, pyrone, and quinone. The formation of these O-bearing, N-bearing, and S-bearing functional groups are driven by oxidation, nitrogenation, and sulfurization reactions, with the treatment of different activation reagents (Yang et al., 2019). The common oxidation reagents are nitric acid, sulfuric acid, hydrogen peroxide, and ozone, while the typical nitrogenation reagents are ammonia, urea, and amines. Sulfurization is usually conducted via sulfur, sulfur dioxide, hydrogen sulfide, sodium sulfide, and potassium sulfide. The introduction of O, N, and S atoms may enhance the surface polarity and hydrophilicity of AC for the adsorption of polar species; and the nonpolar, hydrophobic nature of the aromatic layers could be promoted by the reduction of the acidic O-bearing functionalities via sodium hydroxide neutralization, thermal degradation (700°C−1000°C) or hydrogen reduction. The N-rich basic AC and S-rich AC are well-known for their enhanced performance in the adsorption of acidic carbon dioxide gas and aqueous Hg^{II}. However, these chemical modifications would inevitably alter the textural characteristics of the AC samples, either favorably or adversely.

AC has found numerous applications in diverse areas of paramount importance, as summarized in Table 13.2. Notably, AC is a superior electrode material for supercapacitors in commercial electronic devices, owing to its good electronic conductivity, cyclic stability, low cost, nontoxicity, and processability. The high surface area and porous nature of AC facilitate the dense electro-adsorption of electrolyte ions on its surface.

Preparation of activated carbon

The production of AC consists of two basic procedures: carbonization and activation (Ayinla et al., 2019). Carbonization refers to the pyrolysis of the carbon precursor at 400°C−900°C under an inert atmosphere (usually

TABLE 13.2 Applications of activated carbon.

Industry	Application
Municipal water and wastewater treatment	Applied in rapid gravity filters, slow sand sandwich filters, or column adsorbers. Removes soluble impurities and contaminants, notably the taste-, color-, and odor-forming compounds, natural organic matter, pesticides, trihalomethanes, algal toxins, VOCs, pharmaceuticals, chlorine, chloramine, dissolved gases, inorganics, other micropollutants.
Industrial wastewater treatment	Treatment of processing effluents of food, textile, chemicals, petrochemicals, pharmaceuticals, pesticides, printed matter, petroleum, synthetic dyes, mineral ores, plastics, rubbers, detergents, and explosives. Purification of electroplating solutions from organic contaminants and metal finishing residues. Removal of organic impurities for the production of hydrochloric acid, phosphoric acid, alminim chloride, liquid hydrocarbons, and specialty chemicals.
Residential/domestic water purification	Water filter medium in a home water purifier for the removal of taste, odor, and contaminants transported over through the water pipelines.
Food and beverage processing	Decolorization of cane and beet sugar molasses, dextrose, and fructose. Decaffeination of tea and coffee. Purification of fermentation mixture containing amino acids and organic acids. Removal of unwanted taste/odor/color-causing compounds from edible oils, fruit juices, alcoholic beverages, glycerin. Removal of chlorine and ozone from the feedwater and gaseous impurities from CO_2 for making beer, soft drinks, and other beverages.
Natural gas processing	Regeneration of amine solutions via the adsorption of dissolved CO_2, H_2S, and hydrocarbon contaminants. Removal of mercury and sulfur from natural gas. Removal of organics impurities from H_2 stream generated by the steam methane reforming (SMR) process.
Air pollution control	Removal of VOCs, mercury, dioxins, and cadmium from combustion exhaust gases. Removal of H_2S, mercaptans, siloxane from the sewage and landfill leachate treatment facilities.
Metals recovery	Adsorbent in carbon-in-pulp (CIP), carbon-in-leach (CIL), and carbon-in-column (CIC) technologies for gold extraction. Recovery of silver, rare earth metals, uranium, copper, nickel, and zinc recovery from leach, waste, or recycled streams.
Medicine and surgery	Treatment of poisoning and overdoses by oral ingestion of AC tablets or capsules. Extracorporeal filtrative purification of human blood from toxins.
Groundwater and soil decontamination	Remediation of groundwater pollution. Soil additive to immobilize the harmful pollutants in the soil medium.
Electronic products and mechatronic systems	Electrode materials to store electrochemical energy or supercapacitors in portable devices, memory backup systems, power sources of smart grid systems, industrial equipment, and hybrid electric vehicles.
Personal protective equipment (PPE) and military gear	Adsorptive medium in face masks, shields, and safety apparel in the chemical, biological, radiological and nuclear (CBRN) industries or warfare.
Commercial premises	Adsorption of lubricating oil in commercial air compressors. Control of CO_2 and ethylene levels in the ultralow oxygen (ULO) storage of fresh fruits and vegetables.

nitrogen or argon) to remove the volatile contents and produce char, a solid residue with high fixed carbon content, and rudimentary porosity. The activation procedure is subsequently carried out to further develop the pores of the char material, and simultaneously modify the surface functionalities of the char structure. Presently, there are three main strategies of activation: (1) physical activation, (2) chemical activation, and (3) physiochemical activation. Details of these activation approaches have been provided in Table 13.3. For physical activation, carbon dioxide and steam are the most common oxidizers for char activation, reacting with the carbon matrix to release carbon monoxide and water, respectively. For chemical activation, carbonization and activation are performed in a single step (Gao, Yue, Gao, & Li, 2020). The carbon precursor is impregnated with activating chemicals and subjected to heating at $400°C-900°C$ under an inert atmosphere. Physiochemical activation is a combination of physical and chemical activation. Different types of activation agents may react with the carbon matrix under different mechanisms or pathways for the development of the pore network.

TABLE 13.3 Different methods for the preparation of activated carbon and their comparisons.

Approach	Method		Activation agent	Main process parameter	Advantage	Disadvantage
	1st step	2nd step				
Physical activation	Carbonization (400°C–900°C, inert atmosphere, 1–2 h)	Activation (800°C–1100°C, oxidizing gas, 1–3 h)	Oxidizing gas, e.g. CO_2, steam, O_2, air	Carbon precursor, precursor particle size, gas flow, heating rate, carbonization time & temperature	Clean production without secondary waste generation	High temperature, long production time, energy intensive, low carbon yield (<30%), and relatively low specific surface area
Chemical activation	One-step carbonization-cum-activation (chemicals impregnation, 400°C–900°C, inert atmosphere, 1–3 h)	—	Acidic: H_3PO_4, HNO_3, H_2SO_4 Alkaline: KOH, NaOH, K_2CO_3, Na_2CO_3 Neutral: $ZnCl_2$, $FeCl_3$, $AlCl_3$	Carbon precursor, precursor particle size, activating agent, mixing method, mass ratio of activating agent to carbon precursor, activation time & temperature	Lower temperature, shorter production time, higher carbon yield (up to over 50%), higher specific surface area and well-developed porosity	Extensive washing required for the removal of the residual activators, generation of hazardous liquid waste
Physiochemical activation	One-step carbonization-cum-activation (chemicals impregnation, 400°C–900°C, oxidizing gas, 1–3 h)	—	Activation agents for both physical and chemical activation as above	Parameters for both physical and chemical activation as above	Short process time, highly developed porosity, thorough the removal of residual chemicals from the carbon pores	Low carbon yield, laborious, materials-consuming, energy intensive

Activated carbon from renewable feedstocks and industrial waste

At present, approximately 97% of the AC production is derived from coal, while the balance is supported by coconut shells and wood (Market, 2020). With the rising demand for coal as energy fuel and its depleting reserve, there has been driving interest in the exploration of carbon-rich biomass-derived AC. The conversion of biomass waste could not only overcome the waste management problem, but also represent a renewable utilization strategy, due to the high fixed carbon content (5−70 wt.%), low ash content (0.2−10 wt.%), abundant availability, and high suitability of the textural and surface properties of the biowaste (González-García, 2018). The cellulose, hemicellulose, and lignin biopolymers of the plant-based biomass are strongly entangled to provide structural rigidity, while the relative contents of these three fractions of lignocellulosic biomass may influence the quality of the produced AC. During the pyrolysis process, a host of chemical reactions would take place (Danish & Ahmad, 2018); and concurrent with the release of gaseous products (CO, CO_2, H_2, CH_4, H_2O), the lignocellulosic macromolecules are depolymerized, reorganized, and crosslinked to form a disordered carbonaceous structure of graphite-like microcrystallites. The char product from the pyrolysis of biomass, is specifically known as "biochar," and sometimes the processing of the biomass requires significant pretreatment, including washing, drying, crushing, grinding, and sieving to desired size fraction before the carbonization process.

Similarly, industrial and municipal wastes, not limited to sewage sludge, tannery sludge, paper mill sludge, waste tire, organic food waste, animal manure, and poultry feathers have been adopted as the AC precursor (Lin & Wang, 2017). The notion of using industrial and municipal solid waste may sound feasible from a waste management perspective, but the requirement of significant pretreatment might, however, detract from the overall process economics.

Potential roles of activated carbon for the treatment of refinery and petrochemical wastewaters

Adsorbent

Agricultural bioresidues have been explored as the potential precursors for the preparation of highly porous ACs, and a typical example is the preparation date pit-derived AC by physical CO_2 activation at 900°C (El-Naas, Al-Zuhair, & Alhaija, 2010a). It was found that the adsorptive uptake of phenol onto the date-pit derived AC reduced significantly from 262.3 to 56.9 mg g^{-1} when the treatment process was conducted in the refinery wastewater instead of an aqueous phenolic solution, most possibly attributed to the competitive adsorption effect from the aromatic compounds and salinity behavior (Younker & Walsh, 2015). The adsorption mechanism was driven by the $\pi - \pi$ dispersion interaction between the benzene rings of phenols and delocalized π electrons of the graphenic layers. This hydrophobic interaction is supported by the greater performance of the NaOH-treated AC (Kurniawan, Lo, & Sillanpää, 2011). Similarly, this hydrophobic interaction played a viable effect for the adsorptive treatment of aniline and pyridine, with the uptake rate in the descending order of aniline > phenol > pyridine, indicating a stronger interaction as the aqueous solubility decreases (Li, Lei, Zhang, & Huang, 2010).

Oily sludge is the solid residue generated by crude oil processing and petroleum wastewater treatment. To circumvent the disposal of this hazardous waste (Mojoudi et al., 2019) have prepared the oil sludge-derived AC by chemical KOH activation at 800°C. The resultant AC showed a high BET surface area and total pore volume of 2263 and 1.37 cm^3 g^{-1}, respectively, and registered an adsorptive uptake of 434 mg g^{-1} for phenol at the laboratory condition, and 90.11 mg g^{-1} in the textile effluent. Conversely, the removal of phenol from the refinery effluent has been examined in a packed bed continuous operation mode, a maximum removal of 88.4% was recorded in the oily saline wastewater (El-Naas, Alhaija, & Al-Zuhair, 2017). A similar adsorptive treatment has been tested by using the coconut shell-derived AC in a fluidized bed (Kulkarni, Tapre, Patil, & Sawarkar, 2013). The fluidized bed was claimed to have demonstrated the advantages of (1) uniform temperature distribution, (2) smaller pressure drop, (3) less particle diffusion resistance, and (4) ready additions or withdrawals of solids.

Oil spills are potential accidents associated with the refinery and petroleum industries. The coconut shell (Raj & Joy, 2015), banana stem (Nazifa, Hadibarata, Yuniarto, Elshikh, & Syafiuddin, 2019), and water hyacinth derived ACs (Shokry, Elkady, & Salama, 2020), subsequent by magnetized loading of iron oxides have been successfully carried out. The hydrophobic graphitic layers could facilitate the oil uptake, while magnetic composite introduced the ease of separation after the oil adsorption treatment process. Alkaline-activated AC showed better performance, due to the larger surface area and pore size of the AC (Raj & Joy, 2015). Independent of the textural features, an ultrahigh adsorption capacity for used gas oil of 30.2 g g^{-1} was recorded on the magnetic AC (Shokry et al., 2020). The adsorption of crude oil could be enhanced by the commercial oil dispersants commonly applied in the oil spill incidents, mainly driven by the van der Waals forces between the oil adsorbate and hydrophobic AC surface (Ji, Xie, Liu, Liu, & Zhao, 2020).

Commonly present in oil refinery wastewater, benzene, toluene, ethylbenzene, and xylene (BTEX) are priority pollutants, that are toxic to the central nervous system and vital bodily organs. Accordingly, three commercial GACs were examined as a viable tool for BTEX treatment from the diethanolamine solution (Aleghafouri et al., 2015), which is commonly used in the natural gas sweetening column. The predominant removal mechanism is governed by the $\pi - \pi$ dispersive interaction between the soluble aromatics and the graphitic planes. The hydrogen-reduced Filtrasorb 400 performed better than the nitric acid-oxidized Filtrasorb 400, and the adsorptive uptake was independent of the changing solution pH, and susceptible to the "salting out" effect (Flores-Chaparro, Ruiz, Alfaro-De la Torre, & Rangel-Mendez, 2016). The uptake rate was boosted by 12%−19% in the presence of NaCl (Abdel-Aziz, Younis, Moustafa, & Khalil, 2019).

The reduction of the chemical oxygen demand (COD) of real refinery effluents has been examined using a date pit AC (El-Naas, Al-Zuhair, & Alhaija, 2010b). The removal increased with the solution pH and temperature, within the studied range of pH 2−10 and 25°C−60°C. Similar experiments have been carried out by (Oghenejoboh, Otuagoma, & Ohimor, 2016) and (Seyed Hosseini & Fatemi, 2013) in a packed bed column. Greater performance of cassava peel AC over CAC was found, with a removal rate of 73% for cassava peel AC, which was 18% higher than CAC. The steam stripping of the crude in the atmospheric and vacuum tower processing units generates a huge volume of sulfur-laden wastewater, sour water. The removal of hydrogen sulfide, thiols, and sulfides by caustic scrubbing in natural gas purification results in sulfur-laden wastewater, known as refinery spent caustic. Similarly, a two-step physiochemical activation by KOH and CO_2 treatment of three biomass precursors was implemented (Habeeb, Kanthasamy, Saber, & Olalere, 2020). Palm kernel shell AC was found to be the best for $H_2S_{(aq)}$ adsorption, with a removal efficiency of 94%. The removal of sulfide (S^{2-}) from the refinery spent caustic required a hydrophilic oxygenated carbon surface. According to Hariz, Ayni, and Monser (2014), the commercial PAC after HNO_3 or H_2O_2 oxidation treatment illustrated a double of its original capacity for sulfide treatment.

Catalyst and catalyst support

AC is the ideal carrier material for the catalytic active phase, and the synthesis of carbon supports is a relatively simple, low cost, and better scaled-up step. High surface area, tunability, chemical stability, and mechanical strength of AC provide excellent dispersion of the deposited metal active phase. The oxygen-containing groups of AC endow higher surface hydrophilicity for better interaction with the metal precursors during the impregnation stage, to serve as a greater platform for stable metal loading.

Catalyst-assisted ozonation process can accelerate the formation of hydroxyl radicals for the degradation of refractory organics. Two solid ozonation catalysts are denoted as Fe/AC (Chen, Wei, Guo, Guo, & Yan, 2014) and MnOx/ GAC (Chen, Chen, Guo, Guo, & Yan, 2014) have been developed, by loading iron and manganese oxides onto AC. The COD reduction efficiency of heavy oil refinery wastewater was highest for the (O3 + Fe/AC) and (O3 + MnOx/ GAC) systems, as compared to the uncatalyzed systems. The degradation of refractory organics proceeded via a synergistic adsorption-oxidation mechanism. AC itself is capable of decomposing O_3 into hydroxyl free radicals, and this is accelerated by the deposited catalytic metal, which takes advantage of the large dispersion area for the ozone interaction. The adsorption of organic compounds to neighboring AC surface sites promotes instant oxidative degradation, with the best COD removal recorded at 55% in 30 min for Fe/AC-catalyzed ozonation, and 75% at 80 min for MnOx/ GAC catalytic ozonation. A similar COD degrading performance of 75% COD removal in 80 min was reported by Xiong et al. (2019) for Mn@GAC catalytic ozonation. A COD degradation of 85% was reported using the PAC@Fe_3O_4 catalyst for the high-salinity petrochemical wastewater from polyvinyl chloride (PVC) manufacturing plant (Ahmadi, Kakavandi, Jaafarzadeh, & Akbar Babaei, 2017). Using H_2O_2 as the oxidant, Kakavandi and coworkers (Kakavandi & Ahmadi, 2019; Kakavandi & Babaei, 2016) have adopted the magnetic Fe_3O_4 nanoparticles impregnated AC for the heterogeneous Fenton oxidation of petrochemical wastewater, with an influent COD of 680−1000 mg L^{-1}. Deposition of AC provided strong anchorage that prevented the natural agglomeration of nanoparticles in the solution. The composite catalysts, MnP@C and MPAC could be easily recovered from the treated solution by an external magnet due to the magnetic nature of the iron-carbon composite. For MnP@C catalyst, the highest COD removal at 84% was delivered in 120 min. It was found that ultraviolet light irradiation and ultrasound could boost the catalytic degradation by the magnetic iron-carbon composite. The COD removal was increased to 88% within 80 min. The removal dropped to 80.5% in the fifth cycle, indicating the good reusability of the magnetic catalyst. Lately, a similar Fe/GAC composite catalyst has been reported to ozonize high-COD (1900−2800 mg L^{-1}) petrochemical wastewater. A longer contact time of 120 min was required for the COD removal at 88% (Jothinathan, Cai, Ong, & Hu, 2021).

The power of microwaves for handling petroleum wastewater is somewhat overlooked. The COD removal of refinery wastewater in such a microwave-irradiated Fe^0/GAC micro-electrolysis system could be accelerated and doubled to 80% (Qin & Gong, 2014). Moreover, AC is an excellent microwave absorber, and the generation of local hot spots above 1200°C on the carbon surface could contribute to the effective degradation of organic pollutants. Capitalizing on the excellent microwave absorbance of AC (Sun, Zhang, & Quan, 2008) performed catalytic wet air oxidation of petroleum wastewater by using GAC catalyst under microwave irradiation. A 90% COD removal was achieved with a concomitant 12-fold increase in the BOD_5/COD ratio to 0.47.

Growth medium

The AC-supported biofilm process has shown better advantages over the conventional suspended-growth process, notably due to the higher bacterial counts, easier population control, stronger toxicity tolerance, reduced aeration costs, better shock load tolerance, and ready retrofittability. An AC-enhanced moving bed biofilm reactor (MBBR) has been developed for the treatment of petroleum refinery wastewater (Sayyahzadeh, Ganjidoust, & Ayati, 2016). Compared with the AC-free MBBR, the COD removal or total petroleum hydrocarbon (TPH) removal was increased by 10%–40% after 22 h of hydraulic retention time (HRT). It should be noted that the organic pollutants were continuously adsorbed onto the AC, as they were degraded by the biofilm attached to the Bee-Cell media and AC. This continual biological regeneration of AC particles has significantly extended the lifespan of the AC adsorbent.

Meanwhile, the biodegradation of benzene, toluene, and xylene (BTX) was examined by GAC-supported biofilm in a continuous flow column (Muneron Mello et al., 2019). The feed concentrations of 166 mg L^{-1} BTX in the influent sample could be essentially reduced to zero at a flowrate of 3 mg L^{-1} min, and a bed height of 50 cm. The GAC-assisted biological treatment of petrochemical wastewater was conducted after primary treatment (Ahmed, Al-Dhafeeri, & Mydlarczyk, 2018). By integrating both GAC and biofilm, the wastewater stream and air were pumped counter currently through the column in a continuous flow, with suspended growth and attached growth in a single column. This innovative system, also known as the integrated fixed-film activated sludge (IFAS) process, could successfully reduce the COD/BOD_5 levels from 150/90 to 60/38 mg L^{-1} in 5 h.

Biochar—a derivative of activated carbon for the treatment of refinery and petrochemical wastewaters

The carbon-rich residue during the pyrolysis process, also known as biochar, is an alternative to AC, mainly due to the high reactivity, alterable chemical properties, and lower preparation cost. Although biochar has an underdeveloped texture compared with AC, it could outperform AC for the targeted applications. Biochar is amenable to chemical modifications, and acid treatment has been reported to significantly improve the pore development of biochar. The preparation of petroleum waste sludge-derived biochar has been conducted for the catalysis of ozonation of the reverse osmosis (RO) concentrate from a petrochemical refinery wastewater treatment plant (RWTP) (Chen et al., 2019). The HCl-treated biochar sample exhibited total organic carbon (TOC) removal between 49% and 54%. Adsorption, direct ozonation, and hydroxyl free radicals oxidation were the three dominating mechanisms to support TOC removal. The •OH generation via ozone decomposition was accelerated by the biochar, and the catalytic activity was stemmed from the organic oxygen-containing moieties and the inorganic oxides of Si, Al, Zn, Fe, and Mg of the biochar surface. The utilization of petroleum sludge-derived biochar for refinery wastewater treatment indicates a huge saving of the disposal cost and potential pollution from the incineration process.

Similarly, a rice straw-based biochar has been prepared for the adsorptive removal of naphthenic acid, and postsynthetic treatment with HCl or HNO_3 has significantly improved the surface chemistry and textural characteristics (Singh, Naik, Dutta, & Kanaujia, 2020). Chemically, the mineral acids reacted as oxidants and introduced organic O-bearing functionalities (e.g., −OH, −C = O) to facilitate the removal of metal ions from the biochar. Texturally, the mineral acids enhanced the porosity by destructing the loose humic and tarry substances blocking the porous surface. The obtained findings verified the prospects of biochar for the remediation of petroleum refinery wastewater, with a lower cost and pollution footprint.

Using the wet impregnation method, a lanthanum-loaded *Tamarix hispida* wood-derived biochar has been prepared for the treatment of petrochemical wastewater via an AOP using persulfate ($S_2O_8^{2-}$) as the oxidant in the presence of ultrasonic wave (Razmi, Ramavandi, Ardjmand, & Heydarinasab, 2019). In cooperation with the biochar catalyst, ultrasound speeds up the decomposition of persulfate into the highly reactive sulfate radicals (SO_4^-•) to oxidize complex organic compounds into simpler and biodegradable substances. The petrochemical wastewater was successfully treated by the biochar catalyst, with 88% of COD removal and total phenol removal. The functional role of lanthanum was validated, as the leaching effect from the biochar matrix showed a removal reduction to 48% at the fifth cycle of the repetitive use in the treatment process. The different properties of the biochars from different precursors could be manipulated for specific applications (Tafreshi, Sharifnia, & Moradi Dehaghi, 2019). Showed that the low-density of oak biochar made good support for floatable photocatalyst, which was desired for addressing the UV light penetration problem in the photocatalytic treatment of high-turbidity petrochemical wastewater. Consequently, with a developed ZnO/oak biochar supported photocatalyst, a high ammonia removal of 86.7% could be achieved in the 4-h treatment of turbid petrochemical wastewater.

Recent advances in the activated carbon-assisted hybrid treatment technology

Besides functioning on its own in an AC adsorption column, AC has been integrated into other treatment units to form an AC-integrated hybrid process. These hybrid processes have been operated with other treatment units in a treatment train manner to achieve the overall treatment impact. The treatment trains in the primary, secondary or tertiary processes are listed in Table 13.4. These listed processes will be further elaborated from the perspective of AC in Sections 7.1−7.3.

AC for primary treatment

GAC may improve the performance of primary treatment and produce a lower-strength feed for biological treatment. The addition of PAC into a pilot-scale dissolved air flotation (DAF) unit, placed after the flocculation unit, could greatly enhance the overall COD and BOD removal efficiency to above 70% (Hami et al., 2007). For the sequential treatment of oily wastewater, it was found that the treatment sequence played an important role. The PAC adsorption-coagulation sequence yielded 85%−89% of COD and TOC removal, which was 13%−15% higher than the coagulation-PAC adsorption sequence (Wang et al., 2017). In the treatment process, PAC served as an adsorbent for the dissolved organics as well as a seed for floc formation, in reducing the excessive usage of coagulant. Moreover, the PAC-containing flocs were easier to be separated than the PAC alone. Aromatic carboxylic acids (ACAs) could be detrimental to microbial biodegradation activity. For the effectiveness enhancement of biological treatment, the waste effluent from a Purified Terephthalic Acid (PTA) production plant has been treated by a two-step process of precipitation-adsorption (Verma et al., 2014). Sulfuric acid-assisted precipitation has successfully removed 60%−90% of ACAs, and the subsequent adsorption over GAC showed complete removal of ACAs from the pretreated effluent.

Activated carbon for secondary treatment

As membrane separation becomes increasingly popular as part of the tertiary treatment, the issue of membrane abrasion by carbon fines, along with the associated decreasing flux and shortened membrane lifespan, turns more imminent. The addition of GAC, instead of PAC, into the membrane bioreactor (MBR) remedies the undesirable side effects of PAC. The GAC-assisted MBR system offers the advantage of synergistic adsorption-biodegradation, and the EcoRightTM system was such a system jointly developed by Siemens Water Technology and Saudi Aramco Oil Refinery (Cunningham et al., 2013). The pilot study of EcoRightTM revealed various benefits of GAC addition, including shock loads attenuation, consistent effluent quality, lower membrane abrasion and fouling, reduced downtime, and extension of membrane lifespan. The refractory organics were adsorbed onto the GAC, and rendered more amenable to microbial degradation, to eventually provide a cleaner effluent. The carbon dose needed for the GAC-integrated biological treatment was 37 kg, 10 times lesser than the required amount for the conventional GAC packed column. Moreover, this post-biodegradation carbon column is prone to bacterial infestation that may block the void spaces, to the frequent requirement of backwashing. When membrane operation is not a downstream unit, PAC appears to be an attractive option for the strengthening of the activated sludge treatment process. The PAC was added to a sequencing batch reactor (SBR) system for aromatic VOCs removal from petroleum refinery wastewater (Cheng & Hsieh, 2013). The AC-assisted secondary treatments have been compiled in Table 13.5.

The biological AC (BAC) process is a treatment method that degrades the organic matter biologically over a biofilm loaded on the GAC surfaces in a packed-bed column. The integration of adsorption and biodegradation in a single column allows the hard-to-treat biodegradable matters to be retained on the carbon support and enables a longer duration for microbial breakdown reactions to be completed. The continual breakdown and release of the adsorbed organics indicates a regeneration effect on the carbon bed, for cost-saving and extending the service life of the carbon bed.

TABLE 13.4 Selected activated carbon integrated processes for the treatment of refinery wastewater. The 20 listed treatment trains consist of primary, secondary and/or tertiary treatment steps in sequence.

No.	Category	Treatment train	Role	Reference
1.	Primary	Flocculation + PAC@DAF	Adsorbent	Hami, Al-Hashimi, and Al-Doori (2007)
2.	Primary	Acid precipitation + GAC batch adsorption	Adsorbent	Verma, Prasad, and Mishra (2014)
3.	Primary	PAC batch adsorption + coagulation	Adsorbent	Wang, Shui, Ren, and He (2017)
4.	Secondary	PAC@SBR	Adsorbent & growth medium	Cheng and Hsieh (2013)
5.	Secondary	GAC@aeration basin + membrane bioreactor (MBR)	Adsorbent & growth medium	Cunningham, Felch, Smith, and Vollstedt (2013)
6.	Secondary	MBBR + biological activated carbon (BAC) filter	Adsorbent & growth medium	Schneider, Cerqueira, and Dezotti (2011)
7.	Secondary	GAC@bioelectrochemical cell	Adsorbent & growth medium	Fazli, Mutamim, Shem, and Rahim (2019)
8.	Secondary	GAC@Microbial fuel cell	Adsorbent & growth medium	Hejazi, Ghoreyshi, and Rahimnejad (2019)
9.	Secondary	PAC@Activated sludge bioreactor	Adsorbent & growth medium	Jorfi et al. (2019)
10.	Tertiary	MMF + UF + RO + MBR + GAC	Adsorbent	Widiasa et al. (2014)
11.	Tertiary	Chemical oxidation (NaOCl) + sand filter + GAC + UF + UV + RO + GAC	Adsorbent	Wong (2012)
12.	Tertiary	PAC@UF	Adsorbent	Zhang et al. (2005)
13.	Tertiary	MMF + UF + RO I + PAC@CTA-MF + RO II	Adsorbent	Zhao, Gu, and Zhang (2013)
14.	Primary	H_2O_2/UV	-	Souza, Cerqueira, Sant'Anna, and Dezotti (2011)
	Secondary	BAC filter	Adsorbent & growth medium	
15.	Primary	Ozonation	-	Lin et al. (2001)
	Secondary	BAC moving bed	Adsorbent & growth medium	
16.	Primary	Ozonation + ceramic membrane	-	Hu et al. (2020)
	Secondary	BAC filter	Adsorbent & growth medium	

(*Continued*)

TABLE 13.4 (Continued)

No.	Category	Treatment train	Role	Reference
17.	Primary	$H_2O_2/KMnO_4$ oxidation + clarifier + AC-catalyzed Fenton oxidation (fluidized bed) + AC-catalyzed Fenton oxidation (packed bed)	Catalyst	Natarajan, Jayavel, Somasundaram, and Ganesan (2020)
	Secondary	Fluidized BAC	Adsorbent & growth medium	
18.	Secondary	SBR	–	Al Hashemi, Maraqa, Rao, and Hossain Md (2014)
	Tertiary	GAC packed column	Adsorbent	
19.	Primary	Electrocoagulation (EC)	–	El-Naas, Alhaija, and Al-Zuhair (2014), El-Naas et al. (2016)
	Secondary	Spouted bed bioreactor (SBBR)	–	
	Tertiary	GAC packed column	Adsorbent	
20.	Primary	Dissolved air flotation (DAF)	–	Aliyar Zanjani, Vafaie, Mirbagheri, and Hadipour (2015)
	Secondary	Activated sludge bioreactor (ASBR)	–	
	Tertiary	GAC packed column	Adsorbent	

To support the water reclamation in the petroleum and petrochemical industries, BAC is an excellent additional secondary treatment after the MBR or MBBR process, as reported by Schneider et al. (2011) for primary-treated refinery wastewater. A BAC column was installed after the MBBR to enhance the biological degradation, while an AOP was proposed in the MBR and MBBR permeate streams before BAC treatment. A similar setup connecting H_2O_2/UV and O_3/UV before BAC has been carried out (Souza et al., 2011). The BAC bed was operated for 84 days without cleaning, while the GAC bed turned saturated only after 28 days. A treatment line of O_3 + Ceramic membrane + BAC for the polishing of MBR permeate has been suggested (Hu et al., 2020). The alumina membrane modules, immersed in the ozonation tank, may catalyze the ozonation of the MBR permeate inside the membrane nanopores. H_2O_2/UV and membrane-catalyzed ozonation were required to enhance the biodegradability of the MBR or MBBR effluents and the subsequent BAC treatment may play the purification role. According to a deeper techno-economic analysis on a capacity of 4000 m^3 d, the proposed O_3 + Ceramic membrane + BAC integrated process showed a low treatment cost of \$0.30 m^3, which was much cheaper than the currently applied Fenton-ultrafiltration (UF) step, which costs around \$0.88 m^3.

Lately, a five-step treatment train connecting three AOP units, a clarifier, and a fluidized BAC column for refinery spent caustic wastewater has been proposed (Natarajan et al., 2020). During the first step, sulfide was oxidized by $KMnO_4$ and H_2O_2 in a fluidized column, and the H_2S emissions were captured by a caustic solution. In the clarifier, suspended solids were removed by $CaCl_2$ $2H_2O$, while the third step involved the catalytic Fenton process by adopting the $FeSO_4/H_2O_2$ mixture as the oxidation reagent. The fourth step is the same catalytic Fenton process in a packed AC bed instead, and the last step was the fluidized BAC treatment. From the three AOP steps, the BOD_5/COD ratio had been raised from 0.21 to 0.32, indicating higher biodegradability, while the fluidized BAC could further lower the peak ratio to 0.03, signifying a complete decomposition of the biodegradable matters.

Activated carbon for tertiary treatment

AC adsorption is an ideal process in the tertiary treatment for refining the effluent from the biological activated sludge process. In this perspective, a three-step treatment train consists of electrocoagulation (EC)-spouted bed bioreactor (SBBR)-GAC adsorption column, corresponding to the primary, secondary and tertiary treatment has been applied for

TABLE 13.5 Activated carbon-assisted secondary treatment steps on the refinery wastewater.

Combined process	Feedwater	AC properties	AC process conditions	Overall performance	Reference
PAC@SBR	Petroleum refinery wastewater (aromatic VOCs = 52.3 mg L^{-1}, COD = 830−1160 mg L^{-1})	BET surface area = 650 m^2 g, Size = 16 μm	T = 19°C−23°C, pH = 8.0−9.5, SRT = 18−33 days, Dose = 35 kg/5800 m^3, Cycle time = 8 h	Aromatic VOCs removal = 99.996%	Cheng and Hsieh (2013)
GAC@aeration basin + MBR	Effluent of API separator of a petroleum refinery (COD = 522 mg L^{-1}, BOD5 = 195 mg L^{-1})	Granular	T = 25°C−35°C, HRT = 8.7 h, SRT = 20 d, Dose = 10 g L^{-1}, Flow = 10.9 m^3/d	COD removal = 82%, BOD$_5$ removal = 97%	Cunningham et al. (2013)
MBBR + BAC filter	Oil refinery wastewater pretreated by API separator and DAF (DOC = 29−159 mg L^{-1}, COD = 125−1095 mg L^{-1}, pH = 6.9−10)	Granular	T = 30°C, Dose = 5.5 g, Column (IDxL) = 3 × 2 cm, Flow = 0.3−0.8 mg L^{-1} min	DOC reduced to 2−4 mg L^{-1}	Schneider et al. (2011)
GAC@Bioelectrochemical cell (Two-chamber reactor)	Refinery spent caustic (pretreated) (COD = 700 mg L^{-1}, BOD$_5$ = 70 mg L^{-1}, Sulfide = 0.1 mg L^{-1})	GAC (Coconut shell, 1150 m^2 g, 6 × 12 mesh)	Cell size (cm) = 26.5 (L) × 14.5 (H) × 12(W), Flow (O$_2$) = 1.5 L/min, Dose = 5−25 g/4 L, SRT = 20 d, Anode & cathode electrode = Graphite	COD reduced to 24 mg L^{-1} BOD$_5$ reduced to 48 mg L^{-1}, Sulfide reduced to 0.02 mg L^{-1}, Voltage generated = 583 mV, Power density = 3.8 W/m^2, Coulombic efficiency = 6.15%	Fazli et al. (2019)
GAC@Microbial fuel cell (Double-pipe tubular reactor)	Petrochemical wastewater (COD = 3718 mg L^{-1}, Phenolics = 600 mg L^{-1})	Granular, BET surface area = 800 m^2 g, Pore volume = 0.47 cm^3 g	pH = 8.6, T = 25°C−45°C, Time = 0−12 h, Dose = 10−20 g/L, Recycle flow = 60 mL min, Reactor size = 0.5 m (L)x 0.075 m (ID), Anode & cathode electrode = carbon cloth	COD removal = 90%, Phenol removal = 95%, Power density = 110 mW/m^2, Coulombic efficiency = 45.8%	Hejazi et al. (2019)
PAC@Activated sludge bioreactor	Petrochemical wastewater (COD = 1268 mg L^{-1}, TDS = 20−40 g/L)	BET surface area = 745 m^2 g, Pore volume = 0.462 cm^3 g	pH = 9, T = 25°C, HRT = 12−72 h, Flow = 4−24 L d, Aeration = 6 L/min, PAC slurry feed = 1 g L^{-1}, Organic loading rate = 0.4−2.4 kg COD/m^3 · d	COD removal = 68.3%	Jorfi et al. (2019)
H$_2$O$_2$/UV + BAC filter	MBR permeate from refinery wastewater (TOC = 15−50 mg L^{-1}, COD = 60−80 mg L^{-1}, pH = 5.5−7.5)	Filtrasorb-400 (Calgon), granular	Column (IDxL) = 2 × 10 cm, Dose = 5.5 g, Flow = 0.5−1.0 mL/min	TOC dropped to 4−8.5 mg L^{-1}	Souza et al. (2011)

Process	Wastewater	Material	Operating parameters	Results	Reference
Ozonation + BAC moving bed	Petrochemical wastewater (COD = 3.2–6.3 kg COD m^3 d)	GAC, F400 Calgon, 16 × 20 US mesh	HRT = 6 h, Flow = 1.6 L h, Aeration = 120–150 nL/min	COD removal = 85%–95%	Lin et al. (2001)
Ozonation + ceramic membrane + BAC	MBR effluent of a petrochemical plant (COD = 60–200 mg L^{-1}, BOD$_5$ = 10–55 mg L^{-1}, TDS = 3000–7000 mg L^{-1}, pH = 6.5–8.0)	Size = 1.5 mm	Flow = 10 m^3/d, Bed height = 2 m, Bed contact time = 30 min	COD reduced to 17.9 mg L^{-1}	Hu et al. (2020)
Catalytic functionalization oxidation reactor (CFOR) + Clarifier + Heterogeneous AC Fenton catalytic oxidation (HAFCO) reactor + Fenton AC catalytic oxidation (FACCO) reactor + Fluidized immobilized cell carbon oxidation (FICCO)	Refinery spent caustic (pH = 13.4, COD = 20150 mg L^{-1}, BOD$_5$; COD = 0.21, TDS = 54.5 g L^{-1}, Sulfide = 4686 mg L^{-1})	AC (Rice husk, 400°C, 4 h, H$_3$PO$_4$, 800°C, 1 h)	HRT = 24 h, Dose = 30 g L^{-1}	BOD$_5$/COD increased to 0.32 by FACCO, and reduced to 0.03 by FICCO. Removal efficiency COD = 95.3%, BOD$_5$ = 99.2%, TN = 67.3%, S^{2-} = 100%	Natarajan et al. (2020)

the treatment of petroleum refinery wastewater (El-Naas et al., 2014). In the treatment sequence, the GAC column could operate continuously for >24 h to meet the effluent discharge standards for COD and phenols. A pilot plant at the continuous flow of 1 m^3 h had been constructed according to this laboratory-scale model. The removal rate of COD and phenols was recorded at 96% and 100%, respectively, which was close to the laboratory findings. Despite being a mere polisher, AC adsorption contributed to approximately 40% of the overall COD removal.

The performance of the RWTP in the Emirates National Oil Company (ENOC), United Arab Emirates (Al Hashemi et al., 2014), with four parallel SBRs, each with a load of 20–80 m^3, as the secondary treatment, subsequently by GAC polishing, has been adopted, reported a 13% enhancement of the COD removal. Similarly, a three-stage pilot-scale system has been built for the decontamination of crude oil leakages and spills of the groundwater near the Tehran oil refinery company (TORC) (Aliyar Zanjani et al., 2015). The aforementioned three stages involve a DAF unit, activated sludge bioreactor (ASBR), and GAC packed column, and recorded the TPH and COD removal of 98%.

Today, RO is the most economically feasible technology for water purification with ultralow salinity. The problem of membrane fouling could be alleviated by a pretreatment procedure, notably UF and GAC column adsorption, which may significantly remove these organic matters from the surface. The benefits of integrating PAC into the UF process have been affirmed (Zhang et al., 2005). The flux declining trend was arrested, due to the adsorptive uptake of organic matters by the PAC, however, backwashing for 30 s is necessary at every 5 h for flux restoration.

GAC was commonly applied in the treatment of RO concentrate for the reduction of organic content to the COD discharge limit within 100–150 mg L^{-1} (Wong, 2012). A pilot-scale MBR + GAC combination was adopted for the purification of RO concentrate from the oil impounding basin, and the RO concentrate was recycled after the treatment process (Widiasa, Susanto, & Susanto, 2014b). A hybrid process of PAC accumulative countercurrent two-stage adsorption-microfiltration was designed for the water reclamation of biologically treated wastewater (Zhao et al., 2013). With a two-stage RO process, the RO concentrate from the first RO unit was treated with PAC adsorbent in a suspension mode, and the microfiltration module. The module effluent was sent to the second RO unit to achieve an overall water recovery of 91%, which was much higher than the 70%–80% recovery in a single-stage RO system in the petroleum and petrochemical industries. However, the gradual deposition of PAC particles on the module external surface would slowly lead to a reduction of the overall performance.

Regeneration and reusability

When the adsorbents reach saturation, regeneration remains a preferred strategy over the replacement for the waste disposal liability. Thermal regeneration in hearth by adopting steam as the oxidant in the furnaces or rotary kilns at above 900°C, is an established practice in industrial applications. The major drawbacks of thermal regeneration are carbon loss, properties alteration, high energy consumption, and the generation of hazardous gases. It was estimated that 5%–10% of the carbon mass was lost, and this accounts for approximately 20%–40% of the regeneration cost (de Abreu Domingos & da Fonseca, 2018). Moreover, the additional cost is incurred as the spent AC needs to be transported to the designated reactivation centers of a third-party service provider.

Chemical regeneration is the most widely applied alternative to conventional steam regeneration. It desorbs or decomposes the adsorbates using chemical treatment, mostly at mild temperature, and promises minimal tampering of the AC properties, and negligible carbon loss. The reactivity of the stripping agent and the solubility of the adsorbate are the two major factors dictating the effectiveness of the regeneration process. Phenol is best desorbed from the AC surface by NaOH$_{(aq)}$, mainly due to the formation of sodium phenolate ions, that are readily dissolved in the aqueous stripping solution. Using 0.1 M NaOH$_{(aq)}$, Anirudhan, Sreekumari, and Bringle (2009) and Mojoudi et al. (2019) have managed to restore 80%–90% of the phenol uptake capacity for up to 4 adsorption-desorption cycles. However, El-Naas et al. (2010b) reported that ethanol showed a better desorption efficiency for phenol in the petroleum wastewater, with 86% capacity revival at the fourth cycle, as compared to 35% for NaOH$_{(aq)}$.

The feasibility of water as a cheaper and safer replacement for steam and chemical reactivation had been examined, but no satisfactory outcome was acquired. The desorption of benzene from GAC via 3-day deionized water washing yielded only 50% of regeneration (Flores-Chaparro et al., 2016). Similarly, hot water (80°C–90°C) washing for phenol desorption has been attempted and the adsorption capacity was recorded at 30%–40% after the fourth regeneration run (Anirudhan et al., 2009; El-Naas et al., 2010b). With the aid of ultrasonication, petroleum ether has been adopted as the regenerating solvent for motor oil desorption, recording a 74% of recovery at the sixth regeneration cycle (Raj & Joy, 2015). A mixture of ethanol-hexane has been further examined for the extraction of BTX from the salty wastewater, and 83% of effectiveness was found after five regenerating cycles (Abdel-Aziz et al., 2019), while the stripping of naphthenic acid was best accomplished by methanol, with 85% of recovery at the fourth regenerating cycle (Singh et al., 2020). Degradation of the adsorbed organic contaminants at the adsorbent surface using microbial metabolism is biological regeneration (bioregeneration). Under this regeneration scheme, bacteria were fed with the organic contaminants as the carbon source in the presence of oxygen. An offline bioregeneration scheme has been attempted by Roccaro, Lombardo, and Vagliasindi (2015) using oilfield-produced water, reported 57% and 50% of benzene and toluene regenerative capacity, respectively. The suboptimal regeneration was grossly ascribed to biological fouling and structural modification of the carbon materials.

Perspective and outlook

The petroleum refinery and petrochemical industrial wastewaters are extremely diverse in terms of volume, composition, characteristics, and specific stages it was produced. Generally, these wastewaters are oily and hypersaline, with high contents of refractory organics and total dissolved solids, and low biodegradability. Among all, biological treatment remains the most feasible option for the large-scale treatment of petroleum and petrochemical wastewaters. The current approach may engage different available technology to enhance the biodegradability for the subsequent microbial breakdown or strengthen the biodegradation process. In this regard, the identification of the specific characteristics and the target treatment is required for the treatment plant design, and recycling of the wastewater appears to be a growing trend.

Due to the exorbitant materials cost, the GAC adsorption column is seldom installed in the primary treatment stage. It serves as a refining step after the biological treatment, and for the successive accomplishment in the treatment plant, the cost of CAC needs to be significantly reduced, and the exploration of natural biomass feedstock should be considered. Adsorptive reactivity, a higher degree of customization for textural development and chemical functionalization, a correlation between the preparation parameters and the textural characteristics, and postsynthetic nitrogenation or sulfuration procedures are some of the significant perspectives to be highlighted for the better handling of refinery wastewater or the major constituents.

One category of pollutants in petroleum and petrochemical wastewater that has been overlooked is heavy metals. The most toxic heavy metals, namely Pb, Cd, Hg, Ni, and As may be present in the petroleum industrial effluents. Research has shown that the modification of AC could make excellent adsorptive sequestration of heavy metals from the aqueous medium, and several industrial wastewaters (Ugwu & Agunwamba, 2020). However, it was still unclear about the metal adsorption performance of AC in the oily-saline medium of the petroleum and petrochemical wastewater, especially when considering the variable speciations of heavy metals in the ionic and organometallic forms.

Catalytic oxidation is another powerful treatment method that attracted esthetic attention, although the performance might reduce after several repetitive sessions (Kakavandi & Babaei, 2016). Leaching of the active metal from the AC support has been cited to be the primary reason, while the deposition of intermediates onto the catalyst surface could have blocked the pores, or reacted irreversibly with the catalytic metal. In such cases, a regeneration protocol needs to be further developed, and impregnation procedures should be reviewed and revised innovatively to minimize the leaching risk. Beyond the laboratory-scale operation, a pilot-scale setup of the catalytic oxidation, preferably in continuous operation mode, would be highly desirable for a closer insight into the practical potential of this advanced technology.

Other than AC itself, the layout of the treatment train should be deeply optimized in terms of the treatment sequence, reactor configurations, and process conditions. Specifically, the possible operating modes of GAC and BAC filter columns, including up-flow, downflow, packed bed, fluidized bed, and anaerobic variations. Additionally, the single, hybrid, or combination of AC treatment, notably the aeration or ozonation rate, hydraulic retention time, carbon dose, or nutrients supply remains another component to be verified, and lately, the preparation of GAC via microwave irradiation makes a promising pursuit, in light of the good microwave absorbance of AC.

The aforementioned treatment scheme can be fortified with the techno-economic analysis, and a detailed cost analysis along with the integrated treatment train has been presented by Wong (2012), Hu et al. (2020), and Jothinathan et al. (2021). The economic viability of the proposed treatment strategy has been analyzed in terms of the unit cost, long-term performance of the hybrid processes, and the robustness toward the alteration of the influent quality (Kakavandi & Babaei, 2016). Within this framework, the modeling and simulation of the process units are a viable technique for a predictive insight into the performance in the face of possible load changes for different systems.

The microbial decomposition of organic pollutants could be exploited for the generation of bioenergy. The microbial fuel cell is a bioelectrochemical system that harnesses the electrons and protons produced during microbial metabolism to generate bioelectricity. Recently, the feasibility of simultaneous microbial organics degradation and power generation by GAC-integrated microbial fuel cells (GAC-MFCs) have been investigated (Fazli et al., 2019; Hejazi et al., 2019). Although much is to be improved, the power density produced was appreciable up to 4.0 W m^2 as compared to AC-free setups and other sources of industrial wastewaters (Slate, Whitehead, Brownson, & Banks, 2019), indicating the potential of GAC-MFCs and petroleum or petrochemical wastewater for green energy derivation.

Despite the huge potential of AC, the growing research on the applications of AC is degrading in comparison to other advanced nanomaterials, notably metal-organic frameworks (MOFs). While the ardent pursuit of commercialization keeps propelling the development of MOFs to greater heights, the well-established commercialization of AC seems to be holding back the ground-breaking developments of AC. In this perspective, multiple directions that are awaiting further exploration have been underlined. The potential of AC is still far away from the full realization for the treatment of industrial wastewater and its saturation point, and hence in-depth research within this area is deeply desired.

Conclusions

AC, the father of adsorbents, is crucial to environmental preservation, air, and water purification, driven by the high specific surface area, large pore volume, excellent adsorptive reactivity, and chemical stability. AC can be prepared industrially via established production protocols, while the surface properties are readily tunable by manipulating the synthesis parameters or by postsynthetic chemical modification. Generally, the abundant presence of oxygen-containing organic functional groups may elevate the surface acidity, hydrophilicity, and polarity of the AC sample. AC is produced by physical activation, chemical activation, or physiochemical activation. Non-carbon heteroatoms can be introduced to the carbon surface by relevant chemical treatments for specific applications.

For the treatment of petroleum refinery and petrochemical wastewater, AC can serve as an adsorbent, a catalyst, catalyst support, and a growth medium for the microorganisms. AC can be integrated into the operating units for primary, secondary, or tertiary treatment. The addition of AC into the DAF unit, sequencing batch reactor, and ozonation reactor are typical examples of AC integration units. The AC-integrated hybrid processes and the GAC adsorption columns have been proposed in multistage treatment schemes, and experimentally verified for COD removal using laboratory- or pilot-scale systems. Similarly, GAC column adsorption is a valuable pretreatment option to the RO modules for the successive removal of residual organics, which might induce membrane fouling for overall performance enhancement.

References

Abdel-Aziz, M. A., Younis, S. A., Moustafa, Y. M., & Khalil, M. M. H. (2019). Synthesis of recyclable carbon/lignin biocomposite sorbent for in-situ uptake of BTX contaminants from wastewater. *Journal of Environmental Management, 233*, 459−470. Available from https://doi.org/10.1016/j.jenvman.2018.12.044.

Ahmadi, M., Kakavandi, B., Jaafarzadeh, N., & Akbar Babaei, A. (2017). Catalytic ozonation of high saline petrochemical wastewater using PAC@FeIIFe2IIIO4: Optimization, mechanisms and biodegradability studies. *Separation and Purification Technology, 177*, 293−303. Available from https://doi.org/10.1016/j.seppur.2017.01.008.

Ahmed, M. E., Al-Dhafeeri, A., & Mydlarczyk, A. (2018). Predominance of attached vs suspended growth in a mixed-growth, continuous-flow biological reactor treating primary-treated petrochemical wastewater. *The Arabian Journal for Science and Engineering, 44*, 4111−4117. Available from https://doi.org/10.1007/s13369-018-3315-y.

Al Hashemi, W., Maraqa, M. A., Rao, M. V., & Hossain Md, M. (2014). Characterization and removal of phenolic compounds from condensate-oil refinery wastewater. *Desalination and Water Treatment*, 660−671. Available from https://doi.org/10.1080/19443994.2014.884472.

Aleghafouri, A., Hasanzadeh, N., Mahdyarfar, M., SeifKordi, A., Mahdavi, S. M., & Zoghi, A. T. (2015). Experimental and theoretical study on BTEX removal from aqueous solution of diethanolamine using activated carbon adsorption. *Journal of Natural Gas Science and Engineering, 22*, 618−624. Available from https://doi.org/10.1016/j.jngse.2015.01.010.

Aliyar Zanjani, M. H., Vafaie, F., Mirbagheri, S. A., & Hadipour, V. (2015). Investigation of petroleum-contaminated groundwater remediation using multistage pilot system: Physical and biological approach. *Desalin. Water Treat, 57*, 9679−9689.

Anirudhan, T. S., Sreekumari, S. S., & Bringle, C. D. (2009). Removal of phenols from water and petroleum industry refinery effluents by activated carbon obtained from coconut coir pith. *Adsorption, 15*(5−6), 439−451. Available from https://doi.org/10.1007/s10450-009-9193-6.

Ayinla, R. T., Dennis, J. O., Zaid, H. M., Sanusi, Y. K., Usman, F., & Adebayo, L. L. (2019). A review of technical advances of recent palm bio-waste conversion to activated carbon for energy storage. *Journal of Cleaner Production, 229*, 1427−1442. Available from https://doi.org/10.1016/j.jclepro.2019.04.116.

Barton, T. J., Bull, L. M., Klemperer, W. G., Loy, D. A., McEnaney, B., Misono, M., ... Yaghi, O. M. (1999). Tailored porous materials. *Chemistry of Materials, 11*(10), 2633−2656. Available from https://doi.org/10.1021/cm9805929.

Cecen, F. (2011). Water and wastewater treatment: Historical perspective of activated carbon adsorption and its integration with biological processes. In F. Cecen, & O. Aktas (Eds.), *Activated carbon for water and wastewater treatment: Integration of adsorption and biological treatment* (1, pp. 1−11). Wiley.

Chen, C., Chen, H., Guo, X., Guo, S., & Yan, G. (2014). Advanced ozone treatment of heavy oil refining wastewater by activated carbon supported iron oxide. *Journal of Industrial and Engineering Chemistry, 20*(5), 2782−2791. Available from https://doi.org/10.1016/j.jiec.2013.11.007.

Chen, C., Wei, L., Guo, X., Guo, S., & Yan, G. (2014). Investigation of heavy oil refinery wastewater treatment by integrated ozone and activated carbon -supported manganese oxides. *Fuel Processing Technology, 124*, 165−173. Available from https://doi.org/10.1016/j.fuproc.2014.02.024.

Chen, C., Yan, X., Xu, Y. Y., Yoza, B. A., Wang, X., Kou, Y., ... Li, Q. X. (2019). Activated petroleum waste sludge biochar for efficient catalytic ozonation of refinery wastewater. *Science of the Total Environment, 651*, 2631−2640. Available from https://doi.org/10.1016/j.scitotenv.2018.10.131.

Cheng, H. H., & Hsieh, C. C. (2013). Removal of aromatic volatile organic compounds in the sequencing batch reactor of petroleum refinery wastewater treatment plant. *Clean - Soil, Air, Water, 41*(8), 765−772. Available from https://doi.org/10.1002/clen.201100112.

Cunningham, W., Felch, C., Smith, D., & Vollstedt, T. (2013). *Conference proceedings 86th annual water environment federation technical exhibition and conference. Water environment federation technical exhibition and conference (WEFTEC)*.

Danish, M., & Ahmad, T. (2018). A review on utilization of wood biomass as a sustainable precursor for activated carbon production and application. *Renewable and Sustainable Energy Reviews, 87*, 1−21. Available from https://doi.org/10.1016/j.rser.2018.02.003.

de Abreu Domingos, R., & da Fonseca, F. V. (2018). Evaluation of adsorbent and ion exchange resins for removal of organic matter from petroleum refinery wastewaters aiming to increase water reuse. *Journal of Environmental Management, 214*, 362−369. Available from https://doi.org/10.1016/j.jenvman.2018.03.022.

El-Naas, M. H., Alhaija, M. A., & Al-Zuhair, S. (2014). Evaluation of a three-step process for the treatment of petroleum refinery wastewater. *Journal of Environmental Chemical Engineering, 2*(1), 56−62. Available from https://doi.org/10.1016/j.jece.2013.11.024.

El-Naas, M. H., Alhaija, M. A., & Al-Zuhair, S. (2017). Evaluation of an activated carbon packed bed for the adsorption of phenols from petroleum refinery wastewater. *Environmental Science and Pollution Research, 24*(8), 7511−7520. Available from https://doi.org/10.1007/s11356-017-8469-8.

El-Naas, M. H., Al-Zuhair, S., & Alhaija, M. A. (2010a). Reduction of COD in refinery wastewater through adsorption on date-pit activated carbon. *Journal of Hazardous Materials, 173*(1−3), 750−757. Available from https://doi.org/10.1016/j.jhazmat.2009.09.002.

El-Naas, M. H., Al-Zuhair, S., & Alhaija, M. A. (2010b). Removal of phenol from petroleum refinery wastewater through adsorption on date-pit activated carbon. *Chemical Engineering Journal, 162*(3), 997−1005. Available from https://doi.org/10.1016/j.cej.2010.07.007.

Fazli, N., Mutamim, N. S. A., Shem, C. Y., & Rahim, S. A. (2019). Bioelectrochemical cell (BeCC) integrated with granular activated carbon (GAC) in treating spent caustic wastewater. *Journal of the Taiwan Institute of Chemical Engineers, 104*, 114−122. Available from https://doi.org/10.1016/j.jtice.2019.08.019.

Flores-Chaparro, C. E., Ruiz, L. F. C., Alfaro-De la Torre, M. C., & Rangel-Mendez, J. R. (2016). Soluble hydrocarbons uptake by porous carbonaceous adsorbents at different water ionic strength and temperature: Something to consider in oil spills. *Environmental Science and Pollution Research, 23*(11), 11014−11024. Available from https://doi.org/10.1007/s11356-016-6286-0.

Foo, K. Y., & Hameed, B. H. (2011). Preparation of activated carbon from date stones by microwave induced chemical activation: Application for methylene blue adsorption. *Chemical Engineering Journal, 170*, 338−341.

Gao, Y., Yue, Q., Gao, B., & Li, A. (2020). Insight into activated carbon from different kinds of chemical activating agents: A review. *Science of t+ + + he Total Environment, 746*. Available from https://doi.org/10.1016/j.scitotenv.2020.141094.

González-García, P. (2018). Activated carbon from lignocellulosics precursors: A review of the synthesis methods, characterization techniques and applications. *Renewable and Sustainable Energy Reviews, 82*, 1393−1414. Available from https://doi.org/10.1016/j.rser.2017.04.117.

Habeeb, O. A., Kanthasamy, R., Saber, S. E. M., & Olalere, O. A. (2020). Characterization of agriculture wastes based activated carbon for removal of hydrogen sulfide from petroleum refinery waste water. *Materials Today: Proceedings, 20*, 588−594. Available from https://doi.org/10.1016/j.matpr.2019.09.194.

Hami, M. L., Al-Hashimi, M. A., & Al-Doori, M. M. (2007). Effect of activated carbon on BOD and COD removal in a dissolved air flotation unit treating refinery wastewater. *Desalination, 216*(1−3), 116−122. Available from https://doi.org/10.1016/j.desal.2007.01.003.

Hariz, I. B., Ayni, F. A., & Monser, L. (2014). Removal of sulfur compounds from petroleum refinery wastewater through adsorption on modified activated carbon. *Water Science and Technology, 70*(8), 1376−1382. Available from https://doi.org/10.2166/wst.2014.384.

Heidarinejad, Z., Dehghani, M. H., Heidari, M., Javedan, G., Ali, I., & Sillanpää, M. (2020). Methods for preparation and activation of activated carbon: A review. *Environmental Chemistry Letters, 18*(2), 393−415. Available from https://doi.org/10.1007/s10311-019-00955-0.

Hejazi, F., Ghoreyshi, A. A., & Rahimnejad, M. (2019). Simultaneous phenol removal and electricity generation using a hybrid granular activated carbon adsorption-biodegradation process in a batch recycled tubular microbial fuel cell. *Biomass and Bioenergy, 104*, 105336. Available from https://doi.org/10.1016/j.biombioe.2019.105336.

Hu, J., Fu, W., Ni, F., Zhang, X., Yang, C., & Sang, J. (2020). An integrated process for the advanced treatment of hypersaline petrochemical wastewater: A pilot study. *Water Research, 182*. Available from https://doi.org/10.1016/j.watres.2020.116019.

Ji, H., Xie, W., Liu, W., Liu, X., & Zhao, D. (2020). Sorption of dispersed petroleum hydrocarbons by activated charcoals: Effects of oil dispersants. *Environmental Pollution, 256*. Available from https://doi.org/10.1016/j.envpol.2019.113416.

Jorfi, S., Pourfadakari, S., & Kakavandi, B. (2019). A new approach in sono-photocatalytic degradation of recalcitrant textile wastewater using MgO@Zeolite nanostructure under UVA irradiation. *Chemical Engineering Journal, 343*(2018), 95−107.

Jothinathan, L., Cai, Q. Q., Ong, S. L., & Hu, J. Y. (2021). Organics removal in high strength petrochemical wastewater with combined microbubble-catalytic ozonation process. *Chemosphere, 263*, 127980. Available from https://doi.org/10.1016/j.chemosphere.2020.127980.

Kakavandi, B., & Ahmadi, M. (2019). Efficient treatment of saline recalcitrant petrochemical wastewater using heterogeneous UV-assisted sono-Fenton process. *Ultrasonics Sonochemistry, 56*, 25−36. Available from https://doi.org/10.1016/j.ultsonch.2019.03.005.

Kakavandi, B., & Babaei, A. A. (2016). Heterogeneous Fenton-like oxidation of petrochemical wastewater using a magnetically separable catalyst (MNPs@C): Process optimization, reaction kinetics and degradation mechanisms. *RSC Advances, 6*(88), 84999−85011. Available from https://doi.org/10.1039/c6ra17624k.

Kulkarni, S. J., Tapre, R. W., Patil, S. V., & Sawarkar, M. B. (2013). Adsorption of phenol from wastewater in fluidized bed using coconut shell activated carbon. *Procedia Engineering, 51,* 300−307. Available from https://doi.org/10.1016/j.proeng.2013.01.040.

Kurniawan, T. A., Lo, W. H., & Sillanpää, M. E. (2011). Treatment of contaminated water laden with 4-chlorophenol using coconut shell waste-based activated carbon modified with chemical agents. *Separation Science and Technology, 46*(3), 460−472. Available from https://doi.org/10.1080/01496395.2010.512030.

Li, B., Lei, Z., Zhang, X., & Huang, Z. (2010). Adsorption of simple aromatics from aqueous solutions on modified activated carbon fibers. *Catalysis Today, 158*(3−4), 515−520. Available from https://doi.org/10.1016/j.cattod.2010.08.014.

Lin, C.-K., Tsai, T.-Y., Liu, J.-C., & Chen, M.-C. (2001). Enhanced biodegradation of petrochemical wastewater using ozonation and bac advanced treatment system. *Water Research, 35*(3), 699−704.

Lin, J. H., & Wang, S. B. (2017). An effective route to transform scrap tire carbons into highly-pure activated carbons with a high adsorption capacity of ethylene blue through thermal and chemical treatments. *Environmental Technology and Innovation, 8,* 17−27. Available from https://doi.org/10.1016/j.eti.2017.03.004.

Market, T. (2020). *Activated carbon market − Industry analysis, size, share, growth, trends, and forecast* (pp. 2020−2030).

Mojoudi, N., Mirghaffari, N., Soleimani, M., Shariatmadari, H., Belver, C., & Bedia, J. (2019). Phenol adsorption on high microporous activated carbons prepared from oily sludge: Equilibrium, kinetic and thermodynamic studies. *Scientific Reports, 9*(1). Available from https://doi.org/10.1038/s41598-019-55794-4.

Mosbah, M. B., Mechi, L., Khiari, R., & Moussaoui, Y. (2020). Current state of porous carbon for wastewater treatment. *Processes, 8,* 1651.

Muneron Mello, J. M., Brandão, H. L., Valério, A., de Souza, A. A. U., de Oliveira, D., da Silva, A., & e Souza, S. M. A. G. U. (2019). Biodegradation of BTEX compounds from petrochemical wastewater: Kinetic and toxicity. *Journal of Water Process Engineering, 32,* 100914. Available from https://doi.org/10.1016/j.jwpe.2019.100914.

Natarajan, P., Jayavel, K., Somasundaram, S., & Ganesan, S. (2020). Integrated catalytic oxidation and biological treatment of spent caustic wastewater discharged from petrochemical industries. *Energy, Ecology and Environment.* Available from https://doi.org/10.1007/s40974-020-00178-y.

Nazifa, T.H., Hadibarata, T., Yuniarto, A., Elshikh, M.S., & Syafiuddin, A. (2019). Equilibrium, kinetic and thermodynamic analysis petroleum oil adsorption from aqueous solution by magnetic activated carbon. *IOP Conference Series: Materials Science and Engineering, 495,* 012060.

Norman, L. M., & CHA, C. Y. (1995). Production of activated carbon from coal chars using microwave energy. *Chemical Engineering Communications, 140,* 87−110.

Oghenejoboh, K. M., Otuagoma, S. O., & Ohimor, E. O. (2016). Application of cassava peels activated carbon in the treatment of oil refinery wastewater - A comparative analysis. *Journal of Ecological Engineering, 17*(2), 52−58. Available from https://doi.org/10.12911/22998993/62290.

Qin, G., & Gong, D. (2014). Pretreatment of petroleum refinery wastewater by microwave-enhanced Fe0/GAC micro-electrolysis. *Desalination and Water Treatment, 52*(13−15), 2512−2518. Available from https://doi.org/10.1080/19443994.2013.811120.

Raj, K. G., & Joy, P. A. (2015). Coconut shell based activated carbon-iron oxide magnetic nanocomposite for fast and efficient removal of oil spills. *Journal of Environmental Chemical Engineering, 3*(3), 2068−2075. Available from https://doi.org/10.1016/j.jece.2015.04.028.

Rashidi, N. A., & Yusup, S. (2017). A review on recent technological advancement in the activated carbon production from oil palm wastes. *Chemical Engineering Journal, 314,* 277−290. Available from https://doi.org/10.1016/j.cej.2016.11.059.

Razmi, R., Ramavandi, B., Ardjmand, M., & Heydarinasab, A. (2019). Efficient phenol removal from petrochemical wastewater using biochar-La/ultrasonic/persulphate system: Characteristics, reusability, and kinetic study. *Environmental Technology (United Kingdom), 40*(7), 822−834. Available from https://doi.org/10.1080/09593330.2017.1408694.

Roccaro, P., Lombardo, G., & Vagliasindi, F. G. A. (2015). Offline bioregeneration of spent activated carbon loaded with real produced water and its adsorption capacity for benzene and toluene. *Desalination and Water Treatment, 55*(3), 756−766. Available from https://doi.org/10.1080/19443994.2014.964328.

Sayyahzadeh, A. H., Ganjidoust, H., & Ayati, B. (2016). MBBR system performance improvement for petroleum hydrocarbon removal using modified media with activated carbon. *Water Science and Technology, 73*(9), 2275−2283. Available from https://doi.org/10.2166/wst.2016.013.

Schneider, E. E., Cerqueira, A. C. F. P., & Dezotti, M. (2011). MBBR evaluation for oil refinery wastewater treatment, with postozonation and BAC, for wastewater reuse. *Water Science and Technology, 63*(1), 143−148. Available from https://doi.org/10.2166/wst.2011.024.

Seyed Hosseini, N., & Fatemi, S. (2013). Experimental study and adsorption modeling of COD reduction by activated carbon for wastewater treatment of oil refinery. *Iranian Journal of Chemistry and Chemical Engineering, 32*(3), 81−89. Available from http://www.ijcce.ac.ir/IJCCE/index.htm.

Shafeeyan, M. S., Daud, W. M. A. W., Houshmand, A., & Shamiri, A. (2010). A review on surface modification of activated carbon for carbon dioxide adsorption. *Journal of Analytical and Applied Pyrolysis, 89*(2), 143−151. Available from https://doi.org/10.1016/j.jaap.2010.07.006.

Shokry, H., Elkady, M., & Salama, E. (2020). Eco-friendly magnetic activated carbon nano-hybrid for facile oil spills separation. *Scientific Reports, 10*(1). Available from https://doi.org/10.1038/s41598-020-67231-y.

Singh, R., Naik, D. V., Dutta, R. K., & Kanaujia, P. K. (2020). Biochars for the removal of naphthenic acids from water: A prospective approach towards remediation of petroleum refinery wastewater. *Journal of Cleaner Production, 266.* Available from https://doi.org/10.1016/j.jclepro.2020.121986.

Slate, A. J., Whitehead, K. A., Brownson, D. A. C., & Banks, C. E. (2019). Microbial fuel cells: An overview of current technology. *Renewable and Sustainable Energy Reviews, 101,* 60−81. Available from https://doi.org/10.1016/j.rser.2018.09.044.

Souza, B. M., Cerqueira, A. C., Sant'Anna, G. L., & Dezotti, M. (2011). Oil-refinery wastewater treatment aiming reuse by advanced oxidation processes (AOPs) combined with biological activated carbon (BAC). *Ozone: Science and Engineering, 33*(5), 403−409. Available from https://doi.org/10.1080/01919512.2011.604606.

Sun, Y., Zhang, Y., & Quan, X. (2008). Treatment of petroleum refinery wastewater by microwave-assisted catalytic wet air oxidation under low temperature and low pressure. *Separation and Purification Technology, 62*(3), 565–570. Available from https://doi.org/10.1016/j.seppur.2008.02.027.

Susanto, H., Samsudin, A. M., Rokhati, N., & Widiasa, I. N. (2014). Immobilization of glucose oxidase on chitosan-based porous composite membranes and their potential use in biosensors. *Enzyme and Microbial Technology, 52*(6-7), 386–392.

Tafreshi, N., Sharifnia, S., & Moradi Dehaghi, S. (2019). Photocatalytic treatment of a multicomponent petrochemical wastewater by floatable ZnO/Oak charcoal composite: Optimization of operating parameters. *Journal of Environmental Chemical Engineering, 7*, 103397. Available from https://doi.org/10.1016/j.jece.2019.103397.

Ugwu, E. I., & Agunwamba, J. C. (2020). A review on the applicability of activated carbon derived from plant biomass in adsorption of chromium, copper, and zinc from industrial wastewater. *Environmental Monitoring and Assessment, 192*.

Verma, S., Prasad, B., & Mishra, I. M. (2014). Adsorption kinetics and thermodynamics of COD removal of acid pre-treated petrochemical wastewater by using granular activated carbon. *Separation Science and Technology (Philadelphia), 49*(7), 1067–1075. Available from https://doi.org/10.1080/01496395.2013.870200.

Wang, B., Shui, Y., Ren, H., & He, M. (2017). Research of combined adsorption-coagulation process in treating petroleum refinery effluent. *Environmental Technology (United Kingdom), 38*(4), 456–466. Available from https://doi.org/10.1080/09593330.2016.1197319.

Widiasa, I. N., Susanto, A. A., & Susanto, H. (2014). Performance of an integrated membrane pilot plant for wastewater reuse: Case study of oil refinery plant in Indonesia. *Desalination and Water Treatment, 52*(40–42), 7443–7449. Available from https://doi.org/10.1080/19443994.2013.830759.

Wong, J. M. (2012). The history and present status of the world's first major membrane-based water reuse system in the petrochemical industry. In *85th annual water environment federation technical exhibition & conference (WEFTEC)*.

Xiong, W., Chen, N., Feng, C., Liu, Y., Ma, N., Deng, J., ... Gao, Y. (2019). Ozonation catalyzed by iron- and/or manganese-supported granular activated carbons for the treatment of phenol. *Environmental Science and Pollution Research, 26*(20), 21022–21033. Available from https://doi.org/10.1007/s11356-019-05304-w.

Yang, X., Wan, Y., Zheng, Y., He, F., Yu, Z., Huang, J., ... Gao, B. (2019). Surface functional groups of carbon-based adsorbents and their roles in the removal of heavy metals from aqueous solutions: A critical review. *Chemical Engineering Journal, 366*, 608–621. Available from https://doi.org/10.1016/j.cej.2019.02.119.

Younker, J. M., & Walsh, M. E. (2015). Impact of salinity and dispersed oil on adsorption of dissolved aromatic hydrocarbons by activated carbon and organoclay. *Journal of Hazardous Materials, 299*, 562–569. Available from https://doi.org/10.1016/j.jhazmat.2015.07.063.

Zhang, J. C., Wang, Y. H., Song, L. F., Hu, J. Y., Ong, S. L., Ng, W. J., & Lee, L. Y. (2005). Feasibility investigation of refinery wastewater treatment by combination of PACs and coagulant with ultrafiltration. *Desalination, 174*(3), 247–256. Available from https://doi.org/10.1016/j.desal.2004.09.014.

Zhao, C., Gu, P., & Zhang, G. (2013). A hybrid process of powdered activated carbon countercurrent two-stage adsorption and microfiltration for petrochemical RO concentrate treatment. *Desalination, 330*, 9–15. Available from https://doi.org/10.1016/j.desal.2013.09.010.

Chapter 14

Life-cycle assessment and cost-benefit analysis of petroleum industry wastewater treatment

Yuh Nien Chow and Keng Yuen Foo
River Engineering and Urban Drainage Research Center (REDAC), Engineering Campus, Universiti Sains Malaysia, Penang, Malaysia

Introduction

The petroleum industry, generally encompasses four major sectors of (1) exploration, development, and production; (2) hydrocarbon processing (refineries and petrochemical plants); (3) storage, transportation, and distribution; and (4) retail or marketing, is responding to the rising global demand of 3.5 billion tons of crude oil, and 2.5 giga m^3 of natural gas and derivatives annually. Seawater, groundwater, and/or surface water play multiple roles for the upstream and downstream operations, as these sources could provide cooling water, process water, and desalinated water for the boiler feed, and potable water for the onshore or off-shore platforms (Adham, Hussain, Minier-Matar, Janson, & Sharma, 2018). This water could be polluted at every stage of its processing steps, in the form of oil spill contaminated water, cooling water, re-injected water, and treatment chemicals applied for oil recovery; and along the refinery process (Munirasu, Haija, & Banat, 2016). Containing a wide range of organic and inorganic components, including aliphatic and aromatic hydrocarbons, emulsified oil and greases, heavy metals, phenolic compounds, total dissolved solids, and sodium, according to the location, geology, and age of the oil and gas field, these complex mixtures are normally discharged to the local environment or injected into shallow or deep aquifers, resulting in the pollution of watercourses and soil media.

Petroleum wastewater management has been a significant agenda, and the treatment technique could be briefly classified into several stages, mainly: (1) oil and grease removal for downstream treatment; (2) biological treatment; (3) membrane separation and adsorptive treatment; and (4) dissolved pollutants removal for water reuse Fig. 14.1. The general sequence of the petroleum refinery effluent treatment system is depicted in Fig. 14.2, and the selection for appropriate treatment technologies or their combination should comply with the stipulated environmental regulatory standards for the desired end-use quality within the economical feasibility and environmental sustainability context (Arroyo & Molinos-Senante, 2018).

Within this context, life-cycle assessment (LCA), remains a process to evaluate the environmental burdens associated with a product, process, or activity by identification and quantification of energy, materials, and wastes; to assess the impact of energy and material release to the environment, and to verify and evaluate the opportunities for improvements, is a well-established quantitative multicriteria approach for the development of efficient environmental policies. LCA can be applied in macro-scale and micro-scale areas, specifically the public sector, individual organizations, eco-design, and product engineering. Within the petroleum wastewater treatment industries, LCA has been integrated with the economic analysis, as a fundamental tool to examine the practicability of a plant design. From an economic point of view, an in-depth economic analysis could provide viable information for the management to ascertain the possible options, that are of particular significance to be taken into account (Goffi, Trojan, de Lima, Lizot, & Thesari, 2018). This chapter outlines an overview of the LCA and economic analysis with respect to petroleum wastewater treatment, with respect to the historical development, unique features, methodological framework, and impact assessment outcomes. The possible challenges and the available research gaps within this area are evaluated.

Petroleum Industry Wastewater. DOI: https://doi.org/10.1016/B978-0-323-85884-7.00005-9

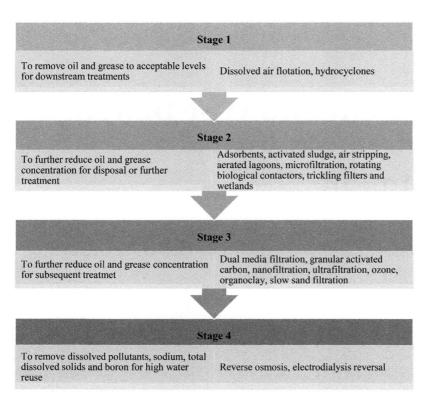

FIGURE 14.1 Categorization of treatment processes for petroleum wastewater.

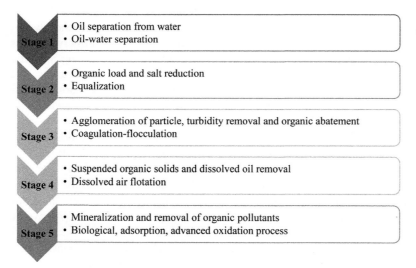

FIGURE 14.2 Schematic diagram of general sequence of petroleum refinery effluent treatment processes.

An overview on life-cycle assessment

The concept of LCA originates from the "net energy analysis" studies, that were firstly published in the 1970s. From 1974 to 1985, the concept has further evolved, with the inclusion of waste and emissions in the assessment, through the quantification of materials and energy consumption. However, the complex environmental issues have received esthetic attention from environmental scientists and policymakers for the need for a more comprehensive approach. In 1990, the Society for Environmental Toxicology and Chemistry (SETAC) took the initiative to develop a general methodology of LCA studies, and similar work has been carried out by the International Organization for Standardization (ISO) on developing the principles and guidelines of LCA methodology. A general consensus on the methodological framework has been established with the collective and continuous efforts from different organizations, specifically the United

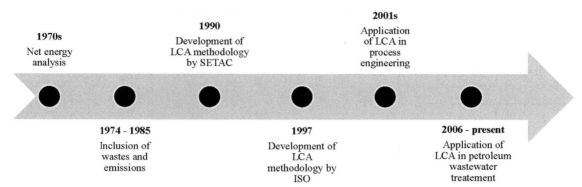

FIGURE 14.3 Historical development of life-cycle assessment.

States Environmental Protection Agency (USEPA), SETAC-Europe, United Nations Environment Program (UNEP), Center of Excellence in Cleaner Production (Australia), The Center for Environmental Strategy (UK), LCA Center (Denmark) and Global Alliance of LCA Centers (Azapagic, 1999; Parra-Saldivar, Bilal, & Iqbal, 2020). Thereafter, the application of LCA has been widened from product to processes or services assessments. In 2001, LCA has been adopted in the engineering design systems, and in 2006, the application in petroleum wastewater treatment has been successfully developed (Fig. 14.3).

The methodology was traditionally designed to the (1) assessment of environmental impacts, (2) comparison of process routes in the production of substitutable products or processes, and (3) comparison of alternative ways for the delivery of a specific function. The unique feature of LCA is the focus on products or processes in a life-cycle perspective to avoid problem-shifting: (1) from one phase of the life-cycle to another, (2) from one region to another, or (3) from one environmental problem to another. The main advantage of LCA over other, site-specific methods for environmental analysis, lies in broadening the system boundaries to include all burdens and impacts in the life cycle of a product or a process. LCA could be conducted as a (1) "cradle-to-gate" approach, in which the study ends at the gate of the factory, and end of life of the manufactured product is not considered; (2) "gate to gate," in which the system boundaries of the LCA end at the manufacturing gate, and do not consider the whole life cycle; or (3) "cradle-to-grave" approach, that includes the whole life cycle.

Life-cycle phases

The development of a complete LCA implies four major phases: goal and scope definition, life-cycle inventory analysis, life-cycle impact assessment, and interpretation. The related subcomponents of these phases are summarized in Fig. 14.4.

Goal and scope definition

In the goal and scope definition phase, the reasons to conduct the study, definition of the system, intended applications, and targeted audiences are clearly defined, with the descriptions of the functional units and system boundaries. The functional units generally referred to as the quantitative measurement of the performance of the goods or services provided in the system, serve as the primary purpose to ensure the compatibility of the LCA results. By providing a reference to which the inputs and outputs are related, functional units could ensure the comparison has been conducted on a common basis (International Organization for Standardization [ISO], 2006).

The system boundaries determine the inclusion of the unit processes within LCA, and in setting the system boundaries, it is useful to distinguish the foreground and background systems. The foreground system is defined as the set of processes affected directly by the study delivering a functional unit specified in the goal and scope definition, while the background system refers to the subsystems that supply energy and materials to the foreground system. Differentiation between the foreground and background systems is important to decide on the type of data to be applied. In general, the foreground system is described by specific process data, while the background is normally represented by data for a mix of a set of mixes of different technologies or processes (Azapagic, 1999).

System boundaries can be divided into three major types: (1) between the technical system and the environment, (2) between significant and insignificant processes, and (3) between the technological system under that study and other

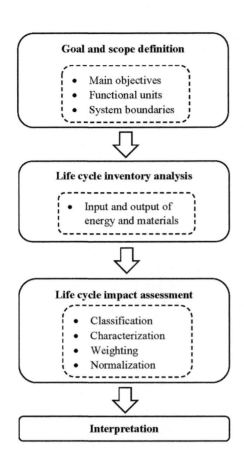

technological systems. Under some special conditions, time and geographical limits are identified as the system boundaries. In the LCA of the petroleum treatment process, the system was modeled independently of a specific geographic location for the exclusion of the transportation of the materials to a specific site. Additionally, the facility decommissioning was excluded since it could vary greatly according to the local conditions and regulations (Vlasopoulos, Memon, Butler, & Murphy, 2006). With the system boundary between the technical system and the environment, LCA should include the entire life cycle, while the inputs should ideally be traced back to raw materials, that are originated from nature. The system boundary between significant and insignificant processes is difficult since the insignificant data could not be identified in advance. A general approach that can be included is data verification, the importance of data checking, life-cycle inventory analysis, and life-cycle impact assessment in the iterative loops until the required precision has been achieved (Finnveden et al., 2009).

Within the LCA for petroleum wastewater treatment, the operation of the wastewater treatment plant is part of the system boundaries. However, only one study considered the construction phase (Vlasopoulos et al., 2006). Most of the available studies did not consider the construction phase, except for the evaluation of the extensive treatment technology with a high amount of materials transported or consumed during the operation, particularly related to constructed wetland or pond systems. Sludge treatment is usually adopted in the system boundaries, owing to the significant contribution to the overall impacts. However, the exception is generally observed in the nonconventional technology that does not produce sludge, and in Turkey, the sludge disposal including transportation of excess sludge generation, carbon dioxide emission, and fuel energy was listed within the system boundary, while other minor emissions were excluded (Muñoz et al., 2020).

Life-cycle inventory analysis

The life-cycle inventory phase encompasses the inputs and outputs of energy and materials in and out of the system boundaries, according to the defined functional units. Additionally, the environmental burdens are quantified according to the resources consumption and generation of air, water, and solid waste. The life-cycle inventory analysis is essential in LCA, since it affects the life-cycle impact assessment phase, and it is of utmost importance for the interpretation phase and conclusions. A clear definition and description of the life-cycle inventory analysis is crucial to ensure the reproducibility of the study, and should clearly describe the primary and background data and their sources.

The major problem affecting the reproducibility of life-cycle inventory analysis arises from a common procedure during the wastewater treatment inventories, which is the miscellany of data sources (Corominas, Flores-Alsina, Snip, & Vanrolleghem, 2012). Within the petroleum wastewater treatment, the data to synthesize the auxiliary materials were applied, due to the missing background data (corrosion inhibitor, antichlor, desulfurized, catalysts, and chemicals). Additionally, the background data from the Europe database rather than from the Eastern region were applied. Whilst the influent compositions from an oil refinery was referred to, the electrical energy supply is modeled from the secondary data by tailoring the database to the respective country conditions, leading to the data misinterpretation of local offshore extraction, natural gas transport, and final products (Liu et al., 2020; Sakamoto, Ronquim, Martins Seckler, & Kulay, 2019).

Life-cycle impact assessment and interpretation

This phase is aimed at the understanding and evaluation on the magnitude and significance of the potential environmental impacts of the tested system by providing additional information assess from the inventory analysis (International Organization for Standardization [ISO], 2006). In other terms, life-cycle impact assessment interprets the inventory results into the respective potential impacts, with respect to the areas of protection by LCA, in particular human health, natural environment, resources, and to some extent, man-made environment (Finnveden et al., 2009). These impacts on the areas of protection are modeled on the relationships between interventions in the form of resource extractions, emissions, land and water use, and the detrimental impacts to the environment. Life-cycle impact assessment applies a holistic perspective on environmental impacts and attempts to model any impact from the product system, which can be expected to damage one or more areas of protection. This phase addresses not only the toxic impacts that are available in the chemical risk assessment, but also the accumulative impacts associated with the emissions of air pollutants (global warming, stratospheric ozone depletion, acidification, photochemical ozone, and smog formation), waterborne discharges (eutrophication and oxygen depletion), and from different forms of land and water use, noise and radiation, or renewable and nonrenewable resources. In the LCA of the petroleum industry, the major evaluated impacts are global warming, aquatic eutrophication, freshwater, marine ecotoxicity, abiotic depletion, acidification, photooxidant formation, freshwater consumption, particulate matter pollution, and nonrenewable energy demand.

The life-cycle impact assessment phase constitutes classification and characterization. Classification involves the identification of the categories of environmental impacts that is of relevance to the study. This is conducted by assigning the emissions from the inventory to these impact categories according to the substances' ability to contribute to different environmental problems. For the characterization step, the relevant characterization models are selected, and the impact of each emission is modeled quantitatively according to the environmental mechanism, and expressed as an impact score in a unit common to all contributions within the impact category (International Organization for Standardization [ISO], 2006). In petroleum wastewater treatment, impact assessment for global warming is expressed as the CO_2-equivalents (eq.) to global warming potentials for a 100-year horizon as the characterization factor, marine ecotoxicity is expressed as 1,4-dichlorobenzene eq. emissions to seawater, and aquatic eutrophication is generally represented by the nitrate-eq. or phosphate-eq. emissions to water (Hauschild & Potting, 2005; Huijbregts, 2016). Characterization allows the summation of the contributions from different emissions and resource extractions within each impact category, to be translated into a profile of environmental impact scores.

The final step of the impact assessment is grouping or weighting of the environmental impact categories and resource consumptions, to reflect the relative importance they are assigned in the study. Weighting may be necessary when trade-off situations occur, in particular where improvements for one impact score are obtained at the expense of another impact score. The weighting step is the most normative part of the method, with its application of preferences and stakeholder values in a ranking, grouping, or quantitative weighting of the impact categories. There is no objective method to perform weighting and hence, no correct set of ranks or weighting factors. Whilst normalization expresses the relative magnitudes of the impact scores and resource consumptions, weighting expresses their relative significance according to the goal of the study. In the interpretation phase, the results from the previous phases are evaluated in relation to the goal and scope for the conclusions and recommendations (International Organization for Standardization [ISO], 2006). To facilitate the interpretation of the results, normalization, where the results from the characterization are related to the reference values is conducted. Selection of impact categories, classification, and characterization are mandatory in impact assessment, while normalization and weighting are usually optional (ISO, 2006).

Life-cycle impact assessment of petroleum wastewater treatment

Global warming

Global warming represents an adverse effect on human health and the ecosystem, while climate change is related to the emission of greenhouse gases. The available LCA on the petroleum wastewater treatment reported that the magnitude of the global warming impact is an interplay between the energy requirement, thermal demand, electricity consumption, and greenhouse gases emissions of the treatment activities. The emissions of carbon dioxide could be originated from the decomposition of submerged biomass as a consequence of dam formation, burning fuel in the thermoelectric plants, chlorofluorocarbon (CFC-113) volatilization, and calcium carbonate calcination during lime production. The global warming impact, however, could be effectively reduced by several modifications of the treatment systems. The application of waste heat in the oil refining plants as the source of thermal energy for the evaporation and catalytic cracking processes, and substitution of multieffect distillation by the steam recompression, coupling with the exclusive supply of electricity as the energy demand for the evaporation unit, have successfully reduced the global warming performance index at the highest rate (Sakamoto et al., 2019). The latter prevails to be a practical application since the waste heat system is subordinate to the operations of refinery sectors. These findings are well-corroborated with the previous findings that the global warming impacts are more sensitive to the intensity oscillation of thermal demands than to the electrical requirements of the treatment systems (Pintilie, Torres, Teodosiu, & Castells, 2016). Similarly, (Muñoz et al., 2020) reported that reclaimed petroleum refinery wastewater reuse for boiler feedwater could completely offset the global warming impact, by the benefits from steam substitution, and to a lower extent by those associated with substituted freshwater treatment by ion exchange. This is mainly due to the warmer nature of the reverse osmosis effluent, which contributes to an estimated saving of 79 MJ steam per m^3 wastewater.

Aquatic eutrophication

Aquatic eutrophication manifests as the excessive growth of phytoplankton driven by overabundant nitrogen, phosphorus, and other nutrient supplies, that deteriorate the water quality related to the bacterial contamination in the eutrophicated water reservoirs. Wastewater reclamation in the petroleum refineries has shown a great contribution to aquatic eutrophication, primarily due to the direct emission of the leading pollutants, specifically the nitrogen compounds and phosphate, from the wastewater treatment plants to the ocean (Liu et al., 2020; Muñoz et al., 2020). Eutrophication could be originated from the waste generated during the petroleum refining process, with 28% from the upstream in the catalytic cracking unit (MEPC, 2003). According to the existing LCA studies on petroleum wastewater treatment plants, aquatic eutrophication is among the most relevant impact to be evaluated.

Interestingly, the impact of aquatic eutrophication could be diluted by the reclaimed petroleum refinery wastewater reuse strategy for the boiler and cooling systems (Muñoz et al., 2020). This finding supports the earlier LCA conducted by (Vlasopoulos et al., 2006), which reported that hybrid hydrocyclones, microfiltration, and organoclay treatment processes is the most favorable option for petroleum wastewater remediation to support the cooling system feed reuse, with the least aquatic eutrophication impact.

Freshwater and marine ecotoxicity

Freshwater ecotoxicity refers to the impact on freshwater ecosystems as a result of emissions of toxic substances to air, water, and soil, while marine ecotoxicity is defined as the impacts of toxic substances on marine ecosystems. The ecotoxicity implications associated with the petroleum wastewater treatment are mainly driven by the (1) final effluent from the crystallization and desupersaturation units during the water recovery process, (2) gangue minerals generated from the coal mining; and (3) nickel emissions from the catalytic cracking process crystallization. The disposal of the dissolved inorganic matter in the refinery wastewater, specifically barium chloride, could induce a series of adverse effects to the environment, flora and fauna, and human health via the soil-aquifer reservoirs pathway. The released barium cations are highly mobile, in the waterways, to be bioaccumulated in the soil, and translocated into the plants. The phytotoxicity of barium is manifested by the symptoms of leaf wilting, growth inhibition of roots, shoots, and leaf, disturbance in potassium transportation from epidermal cells to guard cells, subsequently to the overall growth rate retardation (Llugany, Poschenrieder, & Barceló, 2000; Nogueira, deMelo, Fonseca, Marques, & He, 2010; Suwa et al., 2008). Meanwhile, the toxicity of barium in animals is depicted by gastrointestinal perforation and peritonitis, and in the human body, barium is related to cardiovascular and kidney diseases, metabolic disruption, neurological and mental disorders (Kravchenko, Darrah, Miller, Lyerly, & Vengosh, 2014).

The utilization of coal as the energy source of the Brazilian matrix (BR grid), is another important contributory precursor to freshwater ecotoxicity, due to the tailings discarded from the mining and ashes generated from the thermoelectric plants (Ronquim, Sakamoto, Mierzwa, Kulay, & Seckler, 2020). This environmental load has been reported to be associated with the high electrical demand. In the petroleum refinery, the impact on freshwater ecotoxicity was primarily attributable to the nickel and bromine release from the catalytic cracking, catalytic reforming, and wastewater treatment process. However, the environmental impact of freshwater ecotoxicity could be efficiently reduced by establishing fine catalyst material balance for the catalytic cracking and reforming units, taking targeted measures from raw materials replacement, and reaction control to the end fuel gas treatment (Liu et al., 2020).

Petroleum industry wastewater-related marine ecotoxicity impact has been mainly associated with the emissions of the aromatic hydrocarbons which contain more than 21 carbon atoms (Muñoz et al., 2020). The toxicity-related impact category for LCA remains uncertain at the inventory and impact assessment level, mainly due to the substantial effects contributed by the selection of the hydrocarbon indicator, specifically anthracene, that has been recognized as the representative for hydrocarbon fraction according to risk assessment guidelines. With reference to the ecotoxicity model developed by Huijbregts (2016), pyrene and phenanthrene could be the representative compounds from the same source, but they might demonstrate substantially different characterization factors in the marine ecotoxicity, which is one order of magnitude above or below that of anthracene. On the contrary, the wastewater reuse in the firewater, cooling systems, and boilers have shown positive outcomes in the reduction of marine ecotoxicity, as less amount of the hazardous hydrocarbons were discharged to the sea (Muñoz et al., 2020).

Freshwater consumption

Wastewater management remains a significant challenge to the petroleum refineries and petrochemicals plants, and the escalating demand of petroleum products at pace with the world population growth and technological development requires a sustainable water management strategy for the reduction of freshwater consumption, and remediation of environmental impacts. Within this context, (Ronquim et al., 2020) have reported that freshwater consumption was reduced up to 50% and 36%, respectively, with the application of desupersaturation, and a hybrid desupersaturation-softening precipitation as reverse osmosis pretreatment for the zero liquid discharge in oil refineries. The comparison between these modified treatment systems demonstrated that higher water consumption in the softening precipitation was mainly attributed to a large amount of cooling water requirement during the production of chloride from hydrochloric acid via electrolysis. The water consumption was associated closely with the energy demand, and with the application of decarbonized water for a thermoelectric generation. Nonetheless, the reuse of reclaimed refinery wastewater has successfully achieved a life-cycle freshwater saving of approximately 1 m^3 freshwater per m^3 of refinery wastewater, taking into account the entire supply chain on the affected activities (Arroyo & Molinos-Senante, 2018; Muñoz et al., 2020). However, when the application was adopted, the price of energy consumption resulted in a net increase, which is not compensated by the avoided energy consumption associated with freshwater usage.

Acidification, photooxidant formation, abiotic depletion, and particulate matter pollution

Acidifying substances induce a wide range of impacts on soil, groundwater, surface water, organisms, ecosystems, and materials, while photooxidant formation, represents the formation of reactive substances (mainly ozone), that are injurious to humans health and ecosystems. Abiotic depletion is concerned with the protection of human welfare, and this impact category is generally referred to as the consumption of nonbiological resources, notably fossil fuels, minerals, metals, and water. With the LCA conducted on the petroleum water treatment process, electrodialysis reversal has contributed to the acidification and photooxidant formation impact categories, in which a majority of these impacts could be observed during the construction phase. Specifically, the main precursor, platinum consumed in the electrode manufacturing process, was responsible for 99% of the construction phase impacts. Meanwhile, activated sludge which exhibited the highest levels of environmental impact with respect to acidification, photooxidant formation, and abiotic depletion, was related to the highest operational energy consumption (0.920 kWh m^3) (Vlasopoulos et al., 2006). The LCA findings further revealed that the consumption of plastic materials, particularly glass-reinforced plastic, polyvinyl chloride, and high-density polyethylene for the construction of wetlands, trickling filters, rotating biological contactors, sand filtration, and dual media filtration technology were responsible for the relatively higher environmental impacts.

Particulate matter is a complex mixture of extremely small particles, and particle pollution can be made up of several components, including acids, organic chemicals, metals, soil, or dust particles. In terms of particulate matter pollution, the treatments of petroleum refinery wastewater have been associated with substantial emissions of fine particulate matter from the power plants, owing to the higher electricity demand, that is mainly supplied by coal. The pollution impact was exacerbated by the reclaimed water reuse for boiler, cooling and fire water purposes, as higher electricity demand is required, and this rising demand is insufficient to compensate with the electricity required for the substituted freshwater treatment process (Muñoz et al., 2020).

Energy demand

Energy demand is a common impact category for LCA, as the major rationale behind the energy usage impact category is the limited energy resources for sustainability. Energy demand is often assessed together with other impact categories, notably global warming potential, acidification potential, and toxicity potential to shed light on the different types of environmental impacts. Energy demand indicators are a good proxy indicator for the environmental impacts in general, and these include nonrenewable energy requirements, energy consumption, cumulative primary energy demand, cumulative fossil energy demand, primary energy consumption, energy balance, and net energy balance.

This was evidenced by the LCA findings in petroleum wastewater treatment, which revealed that activated sludge construction exhibited the highest level of environmental impacts, mainly attributed to the highest energy consumption (0.920 kWh m^3), while microfiltration, with lower energy consumption (0.200 kWh m^3) during the operation phase, exhibited a lower environmental impact (Vlasopoulos et al., 2006). Meanwhile, (Ronquim et al., 2020) reported that the pretreatment of reverse osmosis concentrate with desupersaturation, and a combination of both desupersaturation and softening precipitation resulted in 56% and 60% reduction of energy demand. The pretreatment method, coupled with the exclusive use of steam recompression to replace multieffect distillation, which dismissed the use of natural gas, but solely supplied by electricity, has reduced the total energy demand in the refining treatment plant drastically (Sakamoto et al., 2019). From the perspective of petroleum refinery wastewater reuse as cooling or firewater, higher energy demand was reported, due to the electricity consumption by the exhaustive treatment to reach the desired quality. On the contrary, a low energy demand impact was observed as boiler feed water, leading to a net energy saving of 49 MJ per m^3 wastewater (Muñoz et al., 2020).

Challenges and research gaps for life-cycle assessment of water treatment technology

Despite the current development of the LCA for wastewater treatment, several challenges remain to be unaddressed:

- Ignorance of detailed information related to the influent, effluent, and sludge production throughout the treatment processes.
- Incomplete database on the inventory analysis and information on auxiliary materials.
- Geographical and temporal distribution among the developing countries has not been widely explored.
- Environmental impacts with regards to construction, operation, and disposal phases.

Economic efficiency analysis

In designing a new treatment process, the decisions are mainly impacted by economic factors. In the simplest term, the feasibility of a treatment process is primarily governed by the technical requirements, project operation system, and economic analysis. Contrasting to the well-established literature on the economic analysis for wastewater treatment, the cost-benefit analysis for oil and petroleum industry wastewater treatment is still in the infancy of development.

The two major costs involved in the water treatment processes are (1) capital expenditure (CAPEX) and (2) operating expenditure (OPEX), with CAPEX and OPEX as the nonlinear functions of the treatment capacity. Due to this nonlinear nature of the correlation, the treatment capacity within the acceptable range is effective, and exceeding the threshold limit may increase the cost in a nonlinear or exponential manner. A summary of the CAPEX and OPEX with respect to different treatment technology for the petroleum wastewater treatment industry has been provided in Table 14.1. For the membrane separation treatment unit, the CAPEX ranged between US$18,464 and US$44,102,563, while the OPEX cost within US$4551/year to US$112,924,358/year. Among all, ceramic membrane unit and microfiltration treatment systems have recorded the highest CAPEX and OPEX. On the contrary, the photolysis treatment process, recorded a higher OPEX value as compared to CAPEX, while a combination of coagulation, flotation and precipitation treatment showed the CAPEX and OPEX range from US$30,576 to US$194,662, and from US$7,277/year to US$18,743/year respectively. The CAPEX value of the biological treatment systems, particularly activated sludge,

TABLE 14.1 A summary of the capital expenditure and operating expenditure for different petroleum wastewater treatment technology.

Type of treatment	Treatment technology	Capital expenditure (CAPEX) (USD)	Operating expenditure (OPEX) (USD/year)	Reference
Physicochemical treatment	Nanofiltration (NF)	472,500–44,102,563	299,855–112,924,358	Moser, Ricci, Alvim, Cerqueira, and Amaral (2018), Santos, Crespo, Santos, and Velizarov (2016)
	Ceramic membrane unit (CM)	445,945	89,399	Salehi, Madaeni, Shamsabadi, and Laki (2014)
	Microfiltration (MF)	18,464	4,551	Kiss, Keszthelyi-Szabó, Hodúr, and László (2014)
	Submerged MF-Reverse osmosis (RO)	0.28–0.32[a]	0.45–0.52[a]	Ni, Lin, Chang, and Lin (2017)
	Filtrate	22,932	7338	Li, Zhong, Duan, and Song (2011)
	RO	2,330	5445	Ronquim et al. (2020)
	RO-Desupersaturation	1,280	3720	
	RO-Desupersaturation-Softening precipitation	1,067	3455	
	UV/H$_2$O$_2$ photolysis	369,653–525,967	424,948–1,550,520	Moser, Ricci, Alvim et al. (2018), Moser, Ricci, Reis et al. (2018)
	UV/H$_2$O$_2$ photolysis-NF	375,852	237,308	Moser, Ricci, Alvim et al. (2018)
	Coagulation	30,576	16,817	Li et al. (2011)
	Coagulation-flotation	64,057	7277	
	Precipitation	194,662	18,743	
Biological treatment	Activated sludge system	5,389,323	530,614	Li et al. (2011), Tellez, Nirmalakhandan, and Gardea-Torresdey (2002)
	Biodegradable activated sludge	382,199	716,791	
	Traditional biological filter	147,529	15,288	
	Biological aerated filters	1,161,884	135,054	
	Anaerobic-aerobic process	2,211,326	164,193	
	Biological contact oxidation process	3,354,681	948,877	

biological aerated filters, anaerobic-aerobic process, and biological contact oxidation process are generally higher than the physicochemical treatment technology, with the estimated cost of US$147,529 to US$5,389,323, whilst the corresponding OPEX fell within the range of US$15,288/year to US$948,877/year.

TABLE 14.2 Classification of major components in capital expenditure.

Components of capital expenditure (CAPEX)	
Direct capital costs	**Indirect capital costs**
Main operating system	Engineering and supervision
Installation	Contractor's fees
Instrument and controlling devices	Construction expenses
Building, yard, and auxiliary	Contingency
Land	

Capital expenditure

The CAPEX of a treatment process, generally known as the total investment cost, including the costs for land acquisition, construction, equipment installation for the process units, and the costs of design, can be further categorized into direct capital and indirect capital costs. The costs for the main operating system, installation, instrument and controlling devices, building, yard, auxiliary, and land constitute the direct capital costs; while the indirect capital costs include engineering and supervision fees, contractor's fees, construction expenses, and contingency (Table 14.2).

Within the petroleum industry wastewater treatment, the main operating system expenditure constitutes approximately 38.69%−57% of the total CAPEX. Being the primary contributor to the total CAPEX, the main operating system in membrane separation treatment technology refers to the membrane system required for the operation, including the ultrafiltration, nanofiltration, microfiltration, and ceramic membrane unit, with the investment cost related to the materials, filtration element, filtration area, housing capacity, volumetric flow, and design lifetime of the operating system. The membrane unit cost for microfiltration and ultrafiltration are very similar, with an estimated cost of US$120/$m^2$ (Vas-Vincze, 2011), while nanofiltration membrane costs within US$50/$m^2$ to US$1580/$m^2$ (Moser et al., 2018; Santos et al., 2016). CAPEX for the photolysis treatment process is mainly affected by the cost of the UV plant and UV reactor, specifically the surface area, illumination area, number of lamps, yearly operation duration, UV energy, and total volume of water loaded into the plant. Meanwhile, for the desalination plant with zero liquid discharge, the main operating system cost comprises a combination of the expenditures for reverse osmosis system, precipitation, desupersaturation, evaporation, and crystallization units (Ronquim et al., 2020).

The contributory percentage of the remaining direct capital costs to CAPEX of petroleum industrial wastewater treatment is according to the order: installation > building, yard and auxiliary > land > instrument and controlling devices, with the percentage ranging from 3.87% to 15.48%, 0% to 5.81%, 0% to 2.32%, and 0% to 2.32%, respectively. The installation cost applies for the expenditure on piping, water supply structure, feed tank, and valves; while the instrument and controlling devices included the pump, thermometer, manometer, rotameter, controller, and other auxiliary pieces of equipment. Specifically, the expenditure of building and land acquisition was not considered under several circumstances due to the assumption that the treatment system would be built within the refinery facility (Ronquim et al., 2020; Santos et al., 2016).

The largest contribution of the indirect capital costs to CAPEX the X has been ascribed to the expenditure on construction, while the remaining indirect costs have been identified in the descending order of engineering and supervision (11.50%−11.61%) > contingency (6.41%−7.41%) > contractor's fee (3.21%−3.52%). The contribution of different expenditures to the CAPEX of petroleum wastewater treatment is summarized in Fig. 14.5.

Operating expenditure

OPEX encompasses the cost that enables the operation of the wastewater treatment unit, specifically energy, chemicals, operation and performance, maintenance and replacement, labor, electricity, amortization, and deprecation of a system (Fig. 14.6).

The contribution of different components to OPEX in the petroleum wastewater treatment industry is displayed in Fig. 14.7. Generally, OPEX could be expressed in the descending order: depreciation > chemicals > operation and performance > amortization > energy > electricity > maintenance > labor.

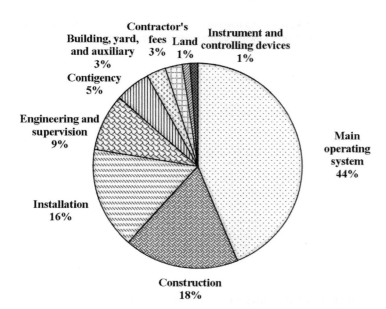

FIGURE 14.5 Contribution of different cost components to capital expenditure of petroleum wastewater treatment.

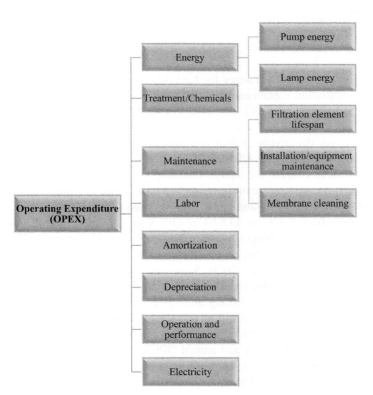

FIGURE 14.6 Major components of operating expenditure.

Energy, operation, and performance cost

Energy has been included as a new cost factor in petroleum engineering economics by the environmental regulatory body. This industry is absorbing this cost while revising the oilfield technology for environmental impact control. Energy consumption is primarily classified into the pump and lamp energy to run water pumps, air pumps, evaporators, and high pressure feed, in the membrane separation process, advanced oxidation process, and desalination unit. Specifically, the energy usage is dependent on the treatment process design, performance ratio, either the treatment facility is single-purpose, dual-purpose, or multipurpose (Wade, 1993). The estimation of the energy consumption could

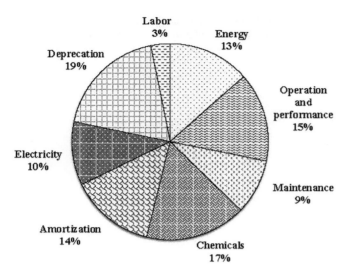

be achieved by the instantaneous measurement of a treatment plant by using an energy measuring device, or according to the equation (Xu, Lin, & Zhang, 2016), given by

$$Q_f = \frac{J \cdot A}{RR} \tag{14.1}$$

$$E_f = \frac{\Delta P \cdot Q_f}{\eta} \tag{14.2}$$

where Q_f refers to the nanofiltration feed flow (L/h), E_f is denoted as the energy for feed pump (kW/h), J is the average nanofiltration permeate flux (L/h m^2), A is the total nanofiltration membrane area (m^2), RR is the water recovery rate (%), ΔP is the applied pressure (bar), and η is the feed pump efficiency (%).

The energy index and energy cost were ascertained as 117.65 kWh m^3 and US$0.17 kWh for the nanofiltration system (Santos et al., 2016), while the pump and lamp energy for the advanced oxidation process has been estimated to be US$0.014 m^3 and US$0.06 m^3, respectively, at the energy price fixed at US$0.04/kWh according to the energy tariff paid by the mining company in Brazil (Moser et al., 2018). In overall, the energy consumption ranged from 4.67% to 77.65% of OPEX, in the descending order: advanced oxidation process > nanofiltration > preozonation > ceramic membrane unit > microfiltration > combination of advanced oxidation process and membrane filtration (Table 14.3).

Similarly, the operation and performance cost may range from 8.12% to 43.87% of OPEX, in the order: microfiltration < ceramic membrane unit < preozonation < advanced oxidation process < nanofiltration < desalination unit. The highest cost reported in the desalination unit for oil refineries and nanofiltration treatment was related to the final transport, hazardous residues handling, and disposal activities (Ronquim et al., 2020; Santos et al., 2016). However, the main OPEX for the nanofiltration treatment system could be significantly reduced by routing the concentrated caustic to pH adjustment circuits in the crude desalinating units, with regular monitoring and systematic maintenance control (Santos et al., 2016). Despite the high operation and performance cost, nanofiltration membrane treatment for the oil refinery spent caustic remains as a very competitive technology option in comparison to the wet air oxidation, that operates under high temperature and high energy consumption, with the CAPEX and annual OPEX at 9 million euros and 1 million euros, respectively. Comparatively, the current nanofiltration system has recorded a similar OPEX, but a lower (−10%) CAPEX, with the removal efficiency for polar oil and grease at 99.95%.

Maintenance cost

The maintenance cost consists of the expenditure on installation, equipment maintenance, filtration element replacement, and membrane cleaning. The per capita cost for maintenance and repair decreases with the rising capacity of the treatment plant. Although the maintenance and repair cost for the general water treatment system is usually considered at a lower percentage of 1%−3% with respect to CAPEX, in the petroleum wastewater treatment systems, it constitutes approximately 4%−40% of CAPEX, and 0.65%−22.43% of OPEX in the order: advanced oxidation process < desalination unit < combination of the advanced oxidation process and membrane filtration < nanofiltration < tubular nanofiltration < microfiltration < ceramic membrane unit < preozonation (Table 14.4).

TABLE 14.3 Costs of energy, operation, and performance for different treatment processes of the petroleum wastewater treatment system.

Treatment technology	Energy cost (% of OPEX)	Operation and performance cost (% of OPEX)	Reference
Tubular nanofiltration	26.85	42.65	Santos et al. (2016)
Nanofiltration	4.67	–	Moser, Ricci, Alvim et al. (2018)
Microfiltration	16.23	8.12	Kiss et al. (2014)
Ceramic membrane unit	19.95	9.98	Salehi et al. (2014)
Advanced oxidation process (AOP)	15.26–77.65	37.93	Moser, Ricci, Alvim et al. (2018), Moser, Ricci, Reis et al. (2018)
Combination of AOP and membrane filtration	4.80	–	Moser, Ricci, Alvim et al. (2018)
Pre-ozonation	22.43	11.21	Kiss et al. (2014)
Desalination unit	–	43.87	Ronquim et al. (2020)

TABLE 14.4 Maintenance and repair costs for the petroleum wastewater treatment processes.

Treatment technology	Maintenance (% of OPEX)	Reference
Tubular nanofiltration	15.62	Santos et al. (2016)
Nanofiltration	7.60	Moser, Ricci, Alvim et al. (2018)
Microfiltration	16.23	Kiss et al. (2014)
Ceramic membrane unit	19.95	Salehi et al. (2014)
Advanced oxidation process (AOP)	0.65–1.18	Moser, Ricci, Alvim et al. (2018), Moser, Ricci, Reis et al. (2018)
Combination of AOP and membrane filtration	7.38	Moser, Ricci, Reis et al. (2018)
Pre-ozonation	22.43	Kiss et al. (2014)
Desalination unit	1.85	Ronquim et al. (2020)

For membrane separation treatment, the quality of the membranes, replacement rate, cleaning frequency, and operation efficiency are primarily governed by the scale and degree of the treatment process (Ettouney, El-Dessouky, Faibish, & Gowin, 2002). The expenditure on membrane replacement could range from 5% in the low-salinity water treatment plants feeds to 20% in the high-salinity plants feeds. The typical replacement cost is estimated to be 1.30 ¢/m^3 of permeate for moderately sized plants, and 3.10–5.30 ¢/m^3 of permeate for seawater desalination (Bhojwani, Topolski, Mukherjee, Sengupta, & El-Halwagi, 2019). In the nanofiltration treatment of petroleum refinery wastewater and spent caustic of oil refinery effluent, at the membrane lifespan range within 1.50–5 years, the membrane replacement cost constitutes 9.64%–14.88% of OPEX. For the cost estimation of the chemical cleaning agent, it could be assumed that the cleaning frequency has been scheduled at once a week for an hour. The applied cleaning agent was 0.20% (w/w) hydrochloric acid solution, with an estimated price of US$7.10 L. The volume of this cleaning solution is generally dependent on the nanofiltration modules, and the predicted volume of the feed and return pipes (Table 14.5).

TABLE 14.5 Costs of membrane replacement and cleaning in membrane separation of petroleum wastewater.

Treatment technology	Membrane replacement (% of OPEX)	Membrane cleaning (% of OPEX)	Reference
Tubular nanofiltration	14.88	–	Santos et al. (2016)
Nanofiltration	9.64	1.46	Moser, Ricci, Alvim et al. (2018)
Microfiltration	–	12.17	Kiss et al. (2014)
Ceramic membrane unit	–	–	Salehi et al. (2014)
Combination of AOP and membrane filtration	9.60	1.48	Moser, Ricci, Reis et al. (2018)

Chemicals, labor and electrical costs

Chemicals are required for the treatment operation, cleaning, pretreating and posttreating water in the wastewater treatment facilities, and the consumption is independent of the capacity and the performance ratio. The cost for the chemicals and reagents may vary according to the proximity to the manufacturing facilities and global market, and therefore it is recognized as the nonfixed OPEX. The contribution of chemicals cost to OPEX ranged from 21.67% to 36.28% (Moser, Ricci, Alvim, et al., 2018; Moser, Ricci, Reis, et al., 2018). Labor cost, known as the employment expenditure for the regular maintenance, operations, and cleaning work, is generally site-specific and dependent on the ownership of the treatment plants, type of technology to be handled, and level of skills and professionality. The labor cost contributes approximately 12.17%−16.82% of OPEX and 3% of CAPEX. Generally, 15.70%−19.48% of OPEX was attributed to the electricity cost, which is usually location-specific. Some of the desalination plants have integrated power generation with water treatment to reduce the energy requirements by recycling the waste heat as renewable energy (Bhojwani et al., 2019).

Amortization and depreciation costs

Amortization cost of the main equipment used in the treatment process, is essential for the estimation of total costs, by taking into consideration of the operating costs and the total treated volume per year. The amortization cost (AC), defined as a function of investment rate (i_c, %) for the investment cost (IC) and design lifetime (DL, year), is given by:

$$AC = IC \cdot \left(\frac{i_c}{1 - [1 + i_c]^{-DL}} \right) \tag{14.3}$$

For the nanofiltration and advanced oxidation treatment for petroleum wastewater, the amortization cost has been reported to range between 9.48% and 39.21% of OPEX (Moser, Ricci, Alvim, et al., 2018; Moser, Ricci, Reis, et al., 2018). Contributing approximately 27.12%−35.15% to OPEX, the depreciation cost, estimated according to the Vebena model, is computed by

$$\text{Depreciation costs} = \left(\frac{1}{3}\right)\text{maintenance costs} + \left(\frac{1}{15}\right)\text{engineering and supervision} + \left(\frac{1}{5}\right)\text{membrane system} \tag{14.4}$$

Break-even point and payback period

In essence, life-cycle analysis considers the costs and benefits to the point where these values are balanced, instead of consideration over the entire life of a project. Break-even point (BEP) in wastewater treatment is defined as the position where a treatment facility could recover its cost through the profit that it generates. This economic parameter represents the economical potential of a treatment process and is estimated without taking into consideration of OPEX, given by:

$$\text{Break} - \text{evenpoint}(\%) = \left\{ \frac{\text{Fixed capitals}}{(\text{Total profit} - \text{operating capitals})} \right\} \tag{14.5}$$

TABLE 14.6 Classification for break-even point on the economic feasibility of a treatment process.

Classification	Description
BEP < 40%	Project is economically feasible.
40% < BEP < 55%	A more detailed study is required.
BEP > 55%	Project is economically rejected.

This economic feasibility assessment is generally carried out for the determination of the positive economic benefits of the treatment system. According to Salehi et al. (2014), the BEP for the ceramic membrane unit applied for coke-contaminated petrochemical wastewater was estimated to be 3%, with a total profit deduced from the yearly saving of filter coalescers at US$265,500, indicating that ceramic membrane unit implementation is economically competent according to the BEP indicator (Table 14.6).

The payback period, a dynamic economic indicator of the duration before the investment has paid for itself, is the first-cut assessment that could be computed by:

$$\text{Payback period} = \frac{\text{CAPEX}}{(\text{CAPEX} + \text{Net present value}) \times \text{lifetime}} \tag{14.6}$$

$$\text{Cashflow}_{\text{accumulated}} = \sum_{i=1}^{\text{Lifetime}} \text{Cashflow}_i \tag{14.7}$$

$$\text{Cashflow}_i = \frac{\text{EBIAT}_i + \left(\frac{\text{CAPEX}}{\text{Lifetime}}\right)}{(1 + \text{updatefactor})^i} \tag{14.8}$$

$$\text{EBIAT} = \left(\text{Revenues} - \text{OPEX} - \left[\frac{\text{CAPEX}}{\text{Lifetime}}\right]\right) \times (1 - 0.33) \tag{14.9}$$

which Net Present Value is defined as the accumulated cash flow at the end of the lifetime of the installation (year); i is denoted as each year between the first year after the investment, and the estimated years of a lifetime for the plant; Earnings Before Interest and After Taxes (EBIAT) refers to the estimated taxes according to the available refinery investment scenarios. With a taxes assumption of 33% at an annual update factor of 11.90% for a nanofiltration plant, Santos et al. (2016) reported that the payback time was 1.10 years, with an investment ratio of exceeding 355%. According to the sensitive analysis, an increment or decrement of 15% in the OPEX value would result in the change of payback period to the range from 0.91 to 1.38 years, while an increase of more than 57% in OPEX, would compromise the overall treatment performance. The key challenges which can be derived from the economic analysis for petroleum wastewater treatment are listed as: Limited studies have been carried out on the economic analysis for petroleum wastewater treatment, particularly on the biological treatment processes. As the nature of some parameters in CAPEX and OPEX, that are site-specific or dependent on the market price, the comparison on the expenditures could be hardly achieved.

Concluding remarks

In summary, LCA on petroleum wastewater treatment process has been associated with a series of environmental impacts, contributing to the (1) global warming, primary and cumulative energy demand, and freshwater consumption, primarily related to the greenhouse gases emission and energy consumption throughout the treatment process; (2) aquatic eutrophication and marine and freshwater ecotoxicity, owing to the consumption and release of toxic metallic or inorganic components; and (3) acidification, abiotic depletion, photooxidant formation, and particulate matter pollution that could be attributed to the usage of plastic materials, glass-reinforced plastic, polyvinyl chloride, and high-density polyethylene during the construction of wastewater treatment plants. Meanwhile, the economic analysis on petroleum wastewater treatment indicated that the cost for the main operating system is the major contributory factor to the capital expenditure while the costs on deprecation, chemicals, operation, amortization, energy, and electricity have been contributing a large portion to the operating expenditure.

The synergies between different industrial sectors, coupled with LCA strategies deliver highly efficient, precise, reliable, and sustainable production processes, with a minimal environmental load. A properly designed LCA procedure could provide comprehensive information to producers, consumers, and policymakers, or legislative authorities to further broaden the practical applications of LCA, minimize environmental insecurity, and alleviate the related human health risks. LCA and economic analysis could reduce the environmental load and economic burden of the petroleum industry wastewater treatment processes, by opting for alternative systems. As a whole, these assessments would contribute to the process safety and environmental security, with the additional benefits of optimal treatment, cost, minimal contamination, and water reusability. Future work gearing toward updating the regional life-cycle database, the inclusion of construction phase, and in-depth LCA and economic analysis on biological treatment processes for petroleum industry wastewater would be beneficial to the green and sustainable future highlighted by UN and the Sustainable Development Goals (SDG).

References

Adham, S., Hussain, A., Minier-Matar, J., Janson, A., & Sharma, R. (2018). Membrane applications and opportunities for water management in the oil & gas industry. *Desalination, 440*, 2−17. Available from https://doi.org/10.1016/j.desal.2018.01.030.

Arroyo, P., & Molinos-Senante, M. (2018). Selecting appropriate wastewater treatment technologies using a choosing-by-advantages approach. *Science of the Total Environment, 625*, 819−827. Available from https://doi.org/10.1016/j.scitotenv.2017.12.331.

Azapagic, A. (1999). Life cycle assessment and its application to process selection, design and optimisation. *Chemical Engineering Journal, 73*(1), 1−21. Available from https://doi.org/10.1016/S1385-8947(99)00042-X.

Bhojwani, S., Topolski, K., Mukherjee, R., Sengupta, D., & El-Halwagi, M. M. (2019). Technology review and data analysis for cost assessment of water treatment systems. *Science of the Total Environment, 651*, 2749−2761. Available from https://doi.org/10.1016/j.scitotenv.2018.09.363.

Corominas, L., Flores-Alsina, X., Snip, L., & Vanrolleghem, P. A. (2012). Comparison of different modeling approaches to better evaluate greenhouse gas emissions from whole wastewater treatment plants. *Biotechnology and Bioengineering, 109*(11), 2854−2863. Available from https://doi.org/10.1002/bit.24544.

Ettouney, H. M., El-Dessouky, H. T., Faibish, R. S., & Gowin, P. J. (2002). Evaluating the economics of desalination. *Chemical Engineering Progress, 98*(12), 32−39.

Finnveden, G., Hauschild, M. Z., Ekvall, T., Guinée, J., Heijungs, R., Hellweg, S., ... Suh, S. (2009). Recent developments in life cycle assessment. *Journal of Environmental Management, 91*(1), 1−21. Available from https://doi.org/10.1016/j.jenvman.2009.06.018.

Goffi, A. S., Trojan, F., de Lima, J. D., Lizot, M., & Thesari, S. S. (2018). Economic feasibility for selecting wastewater treatment systems. *Water Science and Technology, 78*(12), 2518−2531. Available from https://doi.org/10.2166/wst.2019.012.

Hauschild, M., & Potting, J. (2005). Spatial differentiation in life cycle impact assessment-The EDIP2003 methodology. *Environmental News, 80*, 1−195.

Huijbregts, M.A.J. (2016). ReCiPe 2016 a harmonized life cycle impact assessment method at midpoint and endpoint level report i: Characterization. The Netherlands. P.O. Box, 1.

International Organisation for Standardisation [ISO]. (2006). *Environmental management - Life cycle assessment - Principles and framework* (p. 14040) International Organization for Standardization. ISO.

ISO. (2006). *Environmental management − Life cycle assessment − Requirements and guidelines* (p. 14044) ISO.

Kiss, Z. L., Keszthelyi-Szabó, G., Hodúr, C., & László, Z. (2014). Economic evaluation for combinated membrane and AOPs wastewater treatment methods. *Annals of the Faculty of Engineering Hunedoara, 12*(4).

Kravchenko, J., Darrah, T. H., Miller, R. K., Lyerly, H. K., & Vengosh, A. (2014). A review of the health impacts of barium from natural and anthropogenic exposure. *Environmental Geochemistry and Health, 36*(4), 797−814. Available from https://doi.org/10.1007/s10653-014-9622-7.

Li, C., Zhong, S., Duan, L., & Song, Y. (2011). Evaluation of petrochemical wastewater treatment technologies in liaoning province of China. *Procedia Environmental Sciences, 10*(C), 2798−2802. Available from https://doi.org/10.1016/j.proenv.2011.09.434.

Liu, Y., Lu, S., Yan, X., Gao, S., Cui, X., & Cui, Z. (2020). Life cycle assessment of petroleum refining process: A case study in China. *Journal of Cleaner Production, 256*. Available from https://doi.org/10.1016/j.jclepro.2020.120422.

Llugany, M., Poschenrieder, C., & Barceló, J. (2000). Assessment of barium toxicity in bush beans. *Archives of Environmental Contamination and Toxicology, 39*(4), 440−444. Available from https://doi.org/10.1007/s002440010125.

MEPC. (2003). *Cleaner production standard petroleum refinery industry*. Ministry of Environment Protection of the People's Republic of China.

Moser, P. B., Ricci, B. C., Alvim, C. B., Cerqueira, A. C. F., & Amaral, M. C. S. (2018). Removal of organic matter of electrodialysis reversal brine from a petroleum refinery wastewater reclamation plant by UV and UV/H202 process. *Journal of Environmental Science and Health - Part A Toxic/Hazardous Substances and Environmental Engineering, 53*(5), 430−435. Available from https://doi.org/10.1080/10934529.2017.1409580.

Moser, P. B., Ricci, B. C., Reis, B. G., Neta, L. S. F., Cerqueira, A. C., & Amaral, M. C. S. (2018). Effect of MBR-H_2O_2/UV Hybrid pretreatment on nanofiltration performance for the treatment of petroleum refinery wastewater. *Separation and Purification Technology, 192*, 176−184. Available from https://doi.org/10.1016/j.seppur.2017.09.070.

Munirasu, S., Haija, M. A., & Banat, F. (2016). Use of membrane technology for oil field and refinery produced water treatment - A review. *Process Safety and Environmental Protection, 100*, 183−202. Available from https://doi.org/10.1016/j.psep.2016.01.010.

Muñoz, I., Aktürk, A. S., Ayyıldız, Ö., Çağlar, Ö., Meabe, E., Contreras, S., . . . Jiménez-Banzo, A. (2020). Life cycle assessment of wastewater reclamation in a petroleum refinery in Turkey. *Journal of Cleaner Production, 268*. Available from https://doi.org/10.1016/j.jclepro.2020.121967.

Ni, C. H., Lin, Y. C., Chang, C. Y., & Lin, J. C. T. (2017). Recycling of biological secondary effluents in petrochemical industry using submerged microfiltration and reverse osmosis—pilot study and economic evaluation. *Desalination and Water Treatment, 96*, 45−54. Available from https://doi.org/10.5004/dwt.2017.21463.

Nogueira, T. A. R., deMelo, W. J., Fonseca, I. M., Marques, M. O., & He, Z. (2010). Barium uptake by maize plants as affected by sewage sludge in a long-term field study. *Journal of Hazardous Materials, 181*(1−3), 1148−1157. Available from https://doi.org/10.1016/j.jhazmat.2010.05.138.

Parra-Saldivar, R., Bilal, M., & Iqbal, H. M. N. (2020). Life cycle assessment in wastewater treatment technology. *Current Opinion in Environmental Science and Health, 13*, 80−84. Available from https://doi.org/10.1016/j.coesh.2019.12.003.

Pintilie, L., Torres, C. M., Teodosiu, C., & Castells, F. (2016). Urban wastewater reclamation for industrial reuse: An LCA case study. *Journal of Cleaner Production, 139*, 1−14. Available from https://doi.org/10.1016/j.jclepro.2016.07.209.

Ronquim, F. M., Sakamoto, H. M., Mierzwa, J. C., Kulay, L., & Seckler, M. M. (2020). Eco-efficiency analysis of desalination by precipitation integrated with reverse osmosis for zero liquid discharge in oil refineries. *Journal of Cleaner Production, 250*. Available from https://doi.org/10.1016/j.jclepro.2019.119547.

Sakamoto, H., Ronquim, F. M., Martins Seckler, M., & Kulay, L. (2019). Environmental performance of effluent conditioning systems for reuse in oil refining plants: A case study in Brazil. *Energies, 12*(2). Available from https://doi.org/10.3390/en12020326.

Salehi, E., Madaeni, S. S., Shamsabadi, A. A., & Laki, S. (2014). Applicability of ceramic membrane filters in pretreatment of coke-contaminated petrochemical wastewater: Economic feasibility study. *Ceramics International, 40*(3), 4805−4810. Available from https://doi.org/10.1016/j.ceramint.2013.09.029.

Santos, B., Crespo, J. G., Santos, M. A., & Velizarov, S. (2016). Oil refinery hazardous effluents minimization by membrane filtration: An on-site pilot plant study. *Journal of Environmental Management, 181*, 762−769. Available from https://doi.org/10.1016/j.jenvman.2016.07.027.

Suwa, R., Jayachandran, K., Nguyen, N. T., Boulenouar, A., Fujita, K., & Saneoka, H. (2008). Barium toxicity effects in soybean plants. *Archives of Environmental Contamination and Toxicology, 55*(3), 397−403. Available from https://doi.org/10.1007/s00244-008-9132-7.

Tellez, G. T., Nirmalakhandan, N., & Gardea-Torresdey, J. L. (2002). Performance evaluation of an activated sludge system for removing petroleum hydrocarbons from oilfield produced water. *Advances in Environmental Research, 6*(4), 455−470. Available from https://doi.org/10.1016/S1093-0191(01)00073-9.

Vas-Vincze, I. (2011). *Using integrated membrane processes for producing healthy semifinish products.* Doctoral dissertation.

Vlasopoulos, N., Memon, F. A., Butler, D., & Murphy, R. (2006). Life cycle assessment of wastewater treatment technologies treating petroleum process waters. *Science of the Total Environment, 367*(1), 58−70. Available from https://doi.org/10.1016/j.scitotenv.2006.03.007.

Wade, N. M. (1993). Technical and economic evaluation of distillation and reverse osmosis desalination processes. *Desalination, 93*(1−3), 343−363. Available from https://doi.org/10.1016/0011-9164(93)80113-2.

Xu, Y., Lin, Z., & Zhang, H. (2016). Mineralization of sucralose by UV-based advanced oxidation processes: UV/PDS vs UV/H$_2$O$_2$. *Chemical Engineering Journal, 285*, 392−401. Available from https://doi.org/10.1016/j.cej.2015.09.091.

Chapter 15

Sustainability of wastewater treatment

Naim Rashid, Snigdhendubala Pradhan and Hamish R. Mackey

Division of Sustainable Development, College of Science and Engineering, Hamad Bin Khalifa University, Qatar Foundation, Doha, Qatar

Background

Petroleum industries generate a large volume of wastewater during drilling, oil refinery, and other industrial processes. According to one estimate, globally 33 million barrels per day of effluent are generated from petroleum industries (Diya'Uddeen, Daud, & Abdul Aziz, 2011). Petroleum industry wastewater (PIW) is characterized by various pollutants such as hydrocarbons, ammonia, sulfides, aromatic organics such as phenols, and refractory pollutants. These pollutants have many health and ecological implications (Fakhru'l-Razi et al., 2009; Li, Yan, Xiang, & Hong, 2006; Mohammadi et al., 2020). Thus, PIW needs to undergo several treatment processes before discharging it into the environment. Several physical, chemical, and biological techniques such as coagulation, flotation, chemical oxidation, adsorption, aerobic and anaerobic biodegradation, bioelectrochemical cells (BECs), phytoremediation, and membrane filtration have been developed for the treatment of PIW. However, each of these techniques faces its own limitations, which may include high capital cost, maintenance, low treatment efficiency, chemical and material inputs, and technical complexities. These limitations reduce the sustainability of PIW treatment and must be carefully assessed for the selection of suitable techniques.

The sustainability of PIW treatment can be enhanced by the implementation of resource recovery from the PIW in the form of water, energy, metals, nutrients, and carbon (Iddya et al., 2020; Wei, Kazemi, Zhang, & Wolfe, 2020). These materials can potentially be reused for various purposes including irrigation, agricultural amendments, fuel and heating, and as chemical feedstocks in various industries. A resource recovery approach can also improve energy, water, and environmental footprints (Tian et al., 2020). While these technologies usually come at increased capital and operational costs, there is also strong potential for them to result in economic returns from reduced maintenance in other parts of the plant or process, as well as providing a potential revenue stream. Thus, the integration of PIW treatment with recycling and reuse can provide economic benefits in addition to environmental sustainability.

This chapter outlines sustainability concepts and their application to PIW treatment. At first, the characteristics of various PIW are discussed and the general environmental and human health risks of PIW. The concept of sustainability is then introduced before describing various physicochemical and biological treatment technologies about their technology potential, recent developments, and limitations. Resource recovery potential from the PIW with a particular focus on water, organic material, inorganic material, and nutrients is explored. A perspective is provided to set future research directions to enhance the sustainability of PIW treatment processes while reviewing relevant sustainability assessments for PIW treatment.

Characteristics

Petroleum industries produce a significant amount of wastewater through various processes. Oily wastes and hydrocarbon-contaminated water sources can also result from leaks during storage, distribution, war and political crises, and natural disasters (Ossai, Ahmed, Hassan, & Hamid, 2020). The quality and characteristics of the PIW significantly depend on the particular geological structures from the source reservoir, the age of the field, the quality of the gas/oil, and the processing techniques utilized.

PIW is mainly characterized by suspended and emulsified oil, aliphatic, and aromatic compounds including benzene, toluene, ethylbenzene, and xylenes (BTEX) grease, ammonia, sulfide, and other organic compounds (Jain, Majumder,

Petroleum Industry Wastewater. DOI: https://doi.org/10.1016/B978-0-323-85884-7.00008-4

Ghosal, & Gupta, 2020; Varjani, Gnansounou, & Pandey, 2017; Varjani, Joshi, Srivastava, Ngo, & Guo, 2020). Other organic compounds may include carboxylic acids including fatty acids, alkanes, ketones, esters, sulfonates, amides, wax, and tars (Ossai et al., 2020). Many of the organic compounds in PIW are refractory or recalcitrant (Jain et al., 2020; Ossai et al., 2020; Riham et al., 2020). The specific composition of PIW depends on its origin and the type of petroleum product it is associated with. PIW can be classified as (1) oil and gas produced water (2) oil processing water (3) gas processing water and (4) fuel synthesis wastewater, such as process water from the gas-to-liquids (GTL) process.

Oil and gas produced water generated during extraction contains a variety of pollutants including organic compounds such as phenols, volatiles, polycyclic aromatic hydrocarbons (PAHs); dissolved solids (DS) and suspended solids (SS); heavy metals such as cadmium (Cd), lead (Pb), chromium (Cr) and iron (Fe)); alkali and alkali earth metals including barium (Ba), sodium (Na), magnesium (Mg) and calcium (Ca); anions such as chloride (Cl^-), carbonate (HCO_3^-) and sulfate (SO_4^{2-}); and dissolved gases (oxygen, hydrogen sulfide (H_2S), and carbon dioxide (CO_2)) (Liang, Ning, Liao, & Yuan, 2018; Lu, Liu, Lu, Jiang, & Wan, 2010; Nasiri, Jafari, & Parniankhoy, 2017). Oil processing involves several chemical processes such as cracking, distillation, polymerization, and coking, which generate a large volume of wastewater. This wastewater contains oil and grease, phenol, PAHs, benzene, H_2S, ammonia (NH_3), metals, cyanides (CN^-), and SS (Moreno, Farahbakhshazad, & Morrison, 2002; Singh et al., 2017). Produced water from gas processing contains some contaminants including Br^-, Cl^-, benzene, Fe, Ca, toluene, Na, potassium (K), surfactants, DS, and SS (Liang et al., 2018; Sengupta, Nawaz, & Beaudry, 2015). Fuel synthesis wastewater will depend on the particular synthesis method used and end product. For GTL, natural gas is converted into liquid through the Fischer–Tropsch (FT) reaction, which results in a large volume of wastewater. GTL wastewater is characterized by alcohols, volatile fatty acids, ketones, alkanes, metals, low concentrations of chloride and sulfate, and dissolved gases such as hydrogen sulfide and carbon dioxide (Riham et al., 2020). The characteristics of PIW differ across the petroleum industries due to differences in daily flow, processes, composition, and handling of water.

Impact on environmental and human health

Petroleum hydrocarbons pose a threat to both aquatic and terrestrial environments and impose human and ecological health risks (Ossai et al., 2020; Singh et al., 2017). In aquatic environments, oil and hydrocarbons in untreated or partially treated PIW discharge form an immiscible thin surface layer on water bodies, which decreases light permeation leading to photosynthesis inhibition and a decrease in dissolved oxygen (DO) level (Jain et al., 2020). Low DO level can affect the health of aquatic species. PIW contains harmful contaminants such as xylenes, metals, NH_3, sulfides, and PAHs (Chiha, Hamdaoui, Ahmedchekkat, & Pétrier, 2010; Ku Ishak & Abdalla Ayoub, 2019). The presence of these contaminants, in particular, PAHs, poses chronic and acute toxicity risks to aquatic organisms. Aromatic compounds, being lipid-soluble and hydrophobic, can accumulate in aquatic species causing various health issues when consumed by higher trophic organisms including humans. Similarly, the positive charge of most metals species binds strongly to negatively charged cell surfaces, typically causing chronic toxicity issues. The presence of NH_3, nitrate (NO_3^-), and nitrite (NO_2^-) can cause a reduction in the DO level or pose direct toxicity, particularly in the case of NH_3, to fish. Similarly, PIW contains SO_4^{2-}, which can generate H_2S in hypoxic freshwater environments. H_2S is highly toxic for aquatic health (Singh et al., 2017).

PIW is also toxic to humans, with the short and long-term health effects depending on the PIW composition and pollutant concentrations, characteristics, exposure level, and exposure time. PIW exposure can lead to short-term and long-term health effects. Short-term effects include cough, eye irritation, headache, skin irritation and nausea. Long-term health effects include carcinogenicity, mutagenicity, cardiotoxicity, neurotoxicity, depression, and endocrine toxicity (Ossai et al., 2020; Wang et al., 2018). In addition to less soluble PAHs, PIW contains a large number of organic contaminants that are readily water-soluble. Humans when exposed to these contaminants through various routes can cause serious health effects. The intake of benzene is reported to cause liver and kidney failure, xylene damages the nervous system, and phenol impairs human vision. Phenol can react with human skin and mucous membranes forming insoluble protein, which damages the viability of the cells.

The effects of PIW contaminants on plant health include reduced soil fertility and can stunt growth. Its presence lowers the water and nutrients intake by the plants leading to photosystem damage and attenuated growth. Plants exposed to PIW show less resistance to pest attack and suffer coagulation of proteins and tissue damage (Varjani et al., 2020). PIW can also cause sodicity and salinity issues in the soil. High sodicity can deflocculate clay that lowers the water and air permeability. A change in soil properties can cause a loss of plant diversity (Pichtel, 2016).

Sustainability

Sustainability is a triple bottom line (TBL) concept, which involves the evaluation of a technology based on its environmental, economic, and social competitiveness (Kizilet, Veral, & Ćemanović, 2017; Mohseni, Abdollahi, & Siadat, 2019; Slaper, 2011). The primary goal is the selection of technologies that will enable continued future use of resources with minimal impact on these three pillars. TBL incorporates an economic and societal agreement to underline the impact of technology with recent developments and suggests the way for improvement. The TBL dimensions are commonly known as the 3Ps: people, planet, and profit (Kizilet et al., 2017; Mohseni et al., 2019; Slaper, 2011). Finding a common unit for the 3Ps is challenging; therefore, TBL calculation is complex. TBL can be calculated in terms of an index but concerns exist regarding the appropriate weighting of each component in the index. For instance, is an equal weighting of all Ps reasonable, and who or what should drive the weighting criterion? Hence, there is no standard method of TBL calculation. TBL provides a general framework whereby different considerations and metrics can be incorporated in decision-making processes (Kizilet et al., 2017) by the stakeholders, depending on the type and scope of the project. For example, the Genuine Progress Indicator (GPI) is comprised of 25 economic, social, and environmental variables. Similarly, Minnesota and Environmental Quality Board developed a progress indicator of 42 variables to gauge the progress of a project (Kizilet et al., 2017; Mohseni et al., 2019; Slaper, 2011). The indicators should be applied at different stages of the process design to maximize process efficiency and improve sustainability (Argoti, Orjuela, & Narváez, 2019).

In TBL, economic variables are important to determine. Economic evaluations should be holistic over the lifetime of the infrastructure and facility service life, considering both capital and operational costs, as well as salvage, dismantling, and disposal costs. The entire life cycle of a process would require consideration of various factors at different stages with the level of information available. In a truly sustainable economic assessment, the economic analysis should extend beyond the direct process and consider indirect economic costs associated with other services, resource use, business opportunities, and environmental rehabilitation. However, these marginal costs can be difficult to determine. To determine these costs, the economic evaluation should incorporate the level of influence of raw materials, production, and consumption prices both at a local and global scale. A complete economic evaluation can be performed by deploying a distinct framework with the ability to incorporate additional dimensions according to the nature of a project. Economic sustainability is highly contextualized, thus its basis should be updated regularly to incorporate new information (Argoti et al., 2019).

Social measures would include the public acceptability of various technologies and their impact on social well-being, health and wellness, and life quality. For instance, while water reuse of PIW in the municipal sector may be economically and environmentally beneficial, if public acceptance is not strong then the system is likely to fail. Such examples exist with municipal wastewater reuse projects during the Australian Millennium drought, where infrastructure was built, only to not be utilized due to significant public resistance (Heberger, 2011; Hurlimann & Dolnicar, 2010). Similarly, social evaluation should consider all stakeholders and representative groups. Although PIW treatment facilities are usually well separated from populations, still the impact of PIW facilities on the infrastructure development, transportation, employment, capital investment, and disturbance in cultural/religious and recreational events, educational facilities, and other related businesses (real estate, hospitality, tourism) should be taken into consideration.

Environmental sustainability evaluates both immediate and long-term harm to natural resources, to preserve resources for future generations. The environmental sustainability of PIW should evaluate the impact on the aquatic community, marine health, food chain, micropollutants, energy demand, land use, resource use, waste generation, and air pollution (Li et al., 2006). Traditionally in wastewater treatment, the focus has been on the removal of toxic chemicals and limiting eutrophication potential. Recently, the wastewater treatment systems are transitioning to resource recovery to support the circular economy. Life cycle assessment (LCA) is widely used to assess the environmental impact of traditional wastewater treatment systems (WWTS) and emerging resource recovery options (Kizilet et al., 2017; Shi & Guest, 2020). LCA provides a global perspective including the direct and indirect impact of WWTS on environmental sustainability (Corominas et al., 2020). It can be a useful decision-support tool for future strategic planning and technology development. The implementation of LCA is complex due to the involvement of many steps in the treatment process and their interdependency. The ISO standard (ISO 14040) (Byrne, Lohman, Cook, Peters, & Guest, 2017; Corominas et al., 2020) provides only a general guideline to conduct LCA, although there is no comprehensive document guiding LCA specifically for wastewater treatment processes, such as which functional unit should be used (organic strength, wastewater volume, specific pollutant mass). Impact assessment within LCA improves the interpretability of life cycle data by translating raw data into a small set of impact categories. It should be noted that LCA is a tool to assess global environmental impacts. Local environmental and socioeconomical impacts are generally better assessed by environmental impact assessment (EIA) studies (Liebsch, 2021; Morero, Rodriguez, & Campanella, 2015). EIA usually addresses the risks, issues, and neighborhood disturbances including noise, odor, landform aesthetics, land valuation, and transportation impacts. Recent studies have demonstrated the integration of LCA with EIA to adopt a synergetic approach for decision support (Larrey-Lassalle et al., 2017; Tukker, 2000).

Treatment technologies

Several technologies can be employed for the treatment of PIW. The choice of treatment technology depends on many factors including physical, chemical, and biological properties, nature of the pollutant, resources, target level of pollutant removal, and cost-effectiveness (Ossai et al., 2020). The major purpose of any treatment is to remove, extract, and transform the pollutant into a relatively less toxic and reactive form. Regardless of the efficiency of treatment technology, its large-scale application should be evaluated within the framework of sustainability. The following sections provide an overview of different technologies being applied for the treatment of PIW as described in Fig. 15.1. A summary of various techniques is provided in Table 15.1.

Physicochemical treatment

Coagulation

Coagulation is a widely adopted technique for the treatment of various wastewaters. It is a simple and efficient treatment method. Coagulation involves adding chemicals to reduce the surface charge of the pollutant particles, causing them to aggregate and separate from the medium. The aggregation of the particles first requires flash mixing to disperse the coagulant followed by slow mixing for particle agglomeration and floc formation. Coagulation is mainly applied for the removal of SS from the wastewaters.

Although widely applied, and relatively effective, coagulation has moderate energy inputs for flash mixing requirements. Moreover, it involves the addition of a high dosage of chemicals as coagulants. The synthesis and supply of these coagulants are neither cost-effective nor environmentally friendly. The process also generates large volumes of chemically-laden sludge, which poses a threat to the environment and has limited disposal options with little to no reuse applicability. Moreover, the sludge needs to be dewatered before final disposal, which is an energy-intensive process, increasing the carbon footprint.

The main advancements with coagulation are in the area of the coagulant chemical, to improve its effectiveness, or reduce either or both of its cost or environmental burden. The most commonly used coagulants for PIW wastewater

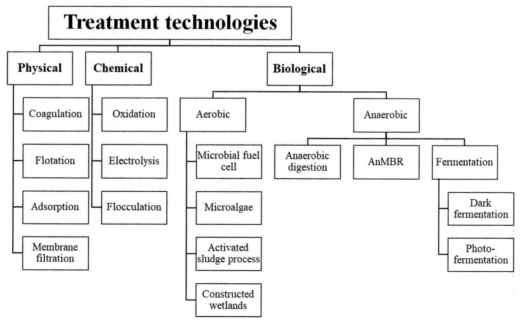

FIGURE 15.1 Various treatment technologies for petroleum industry wastewater.

TABLE 15.1 Summary of considerations related to different treatment technologies.

Different technologies	Major benefits	Drawbacks of the process	Waste generation	Reuse of materials that go into the process	Energy consumption
Coagulation and flocculation (Dąbrowski, Podkościelny, Hubicki, & Barczak, 2005; Derwent, Jenkin, & Saunders, 1996)	Simple method integrated with the physicochemical process, the reasonable capital cost of a wide range of commercial coagulants, rapid and efficient removal of insoluble contaminants, SS, TOC, BOD and COD, good sludge settling and dewatering characteristics, bacterial inactivation capability	Requires one-time use of chemicals (coagulants, flocculants, aid chemicals), increased hazardous sludge volume generation, high installation cost	–	No	Moderate
Biodegradation (Lu et al., 2010)	Very effective for highly biodegradable wastewaters, removal of soluble and many suspended pollutants, economically attractive and well accepted, efficiently eliminates biodegradable organic matter, NH_3, Fe, color, BOD, SS and emergent contaminants. One of most effective treatment methods for moderate-high concentration wastewaters	Slow process with large footprint, uncontrolled degradation products and production of sludge, poor decolorization, low removal for recalcitrant pollutants, possible inhibition and instability, likely to require nutrient addition for PIW	C_2H_4O, C_3H_8O and C_3H_6O will be generated during the removal of hydrocarbons	No	Moderate
Adsorption or filtration by commercial activated carbons or alumina, sand mixed materials and silica gel, etc. (Esmaeili & Saremnia, 2018)	Technologically simple and adaptable to many treatment formats, a wide range of commercial products available, effective removal of wide variety of contaminants which can be done with specificity. Adsorption can remove COD, color, nutrients and metals. Filtration can remove SS and low concentrations of COD and color.	Relatively high investment of activated carbon and alumina, the requirement of several types of adsorbents to remove different pollutants, chemical derivatization to improve their adsorption capacity, rapid saturation and clogging issues, regeneration is expensive and involves chemicals or energy. Disposal without regeneration just moves pollutant from water to elsewhere	Spent adsorbent	Yes	Low
Chemical precipitation (Crini & Lichtfouse, 2019; Mohammadi et al., 2020)	Simple physicochemical process, economically advantageous and efficient, adapted to high pollutant loads, very efficient for metals elimination, not metal selective, significant reduction of COD.	Consumption of CaO, $FeCl_3$, NaOH, require frequent monitoring of physicochemical properties of the effluent, ineffective to remove PAH, BTEX and other organics as well as trace metal ions, the additional requirement of oxidation for complexed metals, high cost to handle and manage the precipitate disposal	High sludge production	–	Moderate

(Continued)

TABLE 15.1 (Continued)

Different technologies	Major benefits	Drawbacks of the process	Waste generation	Reuse of materials that go into the process	Energy consumption
Phytoremediation (Zouboulis & Avranas, 2000)	Economically efficient, good esthetic and ecologically sustainable, carbon sequestration, wildlife habitat	Slow treatment with very large land requirement, potential for increased bioavailability of contaminants and incomplete degradation, large evaporation losses if reuse is intended, extensive maintenance	–	Yes	Very low
Dissolved air flotation (Petre, Piché, Normandin, & Larachi, 2007)	Low reactant consumption and no production of byproducts, cost-effective for removal of emulsified and small suspended contaminants from highly concentrated wastewater	Lower removal efficiency, removed contaminants must subsequently be disposed of, potential for loss of volatile contaminants which may require air treatment	–	No	High
Solvent extraction (Crini & Lichtfouse, 2019)	Used for large scale of operations with a load of contaminants, simple control and monitoring of the process, easy to perform extraction or stripping, economically viable when both solute concentrations and wastewater flow rates are high, efficient for metal removal, separate C_6H_5OH and a good alternative for H_3PO_4 recovery	High equipment cost, uneconomic for trace contaminants, use of a large volume of organic extractants and toxic solvents, entrainment of phases giving poor effluent quality, emulsification of phase with poor separation and fire risk from use of organic solvents	VOC emissions	–	High
Chemical oxidation (Petre et al., 2007)	Natural process, operates rapidly, cause oxidation of other impurities and minimum waste production, increases biodegradability of product, no sludge production, possibility of water recycle disinfection, rapid and complete removal of PAH	Economically nonsustainable, H_2O_2 is highly corrosive, high explosion risk and require special handling	Generates sludge	Yes	High
Fenton oxidation (Tsai, Sah, & Kao, 2010)	Simple and efficient process	Acidic and used chemicals are very expensive	CO_2; Cl^- and Fe^{3+}	No	High
Ozonation (Alcántara-Garduño, Okuda, Nishijima, & Okada, 2008)	Effective and fast removal of contaminants	Low solubility of ozone in water, ozone scavengers commonly found in environment and incomplete oxidation	Secondary organic aerosols	–	High
Wet air oxidation (Petre et al., 2007)	For treatment of spent caustic (very high H_2S and often organic compounds) from refining process, equipment is highly resistant to corrosion	Nonspecific oxidation and safety concerns associated with high pressure and high temperature process	Halogenated subproducts	–	High

NH_3, ammonia; Fe, iron; SS, suspended solids; TOC, total organic carbon; BOD, biochemical oxygen demand; COD, chemical oxygen demand; C_2H_4O, acetaldehyde; C_3H_8O, propanol; C_3H_6O, acetone; CaO, lime; $FeCl_3$, iron chloride, $NaOH$, sodium hydroxide; H_3PO_4, phosphoric acid; H_2O_2, hydrogen peroxide; H_2S, hydrogen sulfide; CO_2, carbon dioxide; C_8H_{10}, ethylbenzene; PAH, polyaromatic hydrocarbons; Cl^-, chloride; Fe^{3+}, ferric ion.

treatment are poly-aluminum chloride, poly-aluminum zinc silicate chloride, aluminum sulfate ($Al_2(SO_4)_3$), poly-silicate aluminum ferric sulfate, ferric chloride ($FeCl_3$), and chitosan. Among these, chitosan is most effective due to its high charge density, relatively low dosage requirement, and biodegradable nature (Zeng, Yang, Zhang, & Pu, 2007). Electro-coagulation is an advanced form of coagulation. In this technique, the breakage of oil emulsion is carried out by electric current and then the SS is agglomerated by coagulation. It is an efficient technique to remove SS from the PIW but requires higher energy inputs (Al Hawli, Benamor, & Hawari, 2019; Crini & Lichtfouse, 2019). Recent research attention has been focused on exploring cost-effective and environmentally friendly biomaterials as coagulants. In this regard, biocoagulants such as starch/cellulose-based material and polysaccharides showed promising results (Zhao et al., 2020; Zouboulis & Avranas, 2000) but are expensive to use.

Flotation

Flotation relies on the introduction of microair bubbles in the PIW that agglomerate oil particles and rise to form a scum layer on the water surface. This layer is separated from the aqueous medium. Flotation produces relatively less sludge and provides a high separation efficiency for oil droplets and colloidal solids (Yu, Han, & He, 2017). The separation efficiency depends on the particle size, airflow rate, hydraulic retention time, and surface characteristics of the particles. Flotation can be carried out by different methods such as dissolved air flotation, dispersed air flotation, and electro-flotation. In dissolved air flotation, a pressurized air stream is contacted with the main flow or a return flow, which when entering the main vessel under ambient temperature results in the formation of fine bubbles. Air bubbles generated attach with the suspended particles and raise them to the surface. In contrast, dispersed air flotation involves the injection of nonpressurized air into the wastewater. Dispersed air-flotation generates larger bubbles than dissolved air flotation (Uduman, Qi, Danquah, Forde, & Hoadley, 2010), exhibiting lower flotation efficiency in general, but is efficient in removing free oil. Recently, electro-flotation is gaining widespread attention. It presents high removal efficiency due to the formation of uniform and tiny bubbles of hydrogen and oxygen during the electrolysis process. The small bubble size provides a large surface area for particle attachment that facilitates flotation (Santiago Santos et al., 2018). Flotation was found an effective technique for the treatment of concentrated oily wastewaters ($14 \, g \, L^{-1}$ of oil) (Yu et al., 2017). The treatment efficiency of flotation is often enhanced by the use of surfactants (Painmanakul, Sastaravet, Lersjintanakarn, & Khaodhiar, 2010). The process treatment efficiency of flotation largely depends on pH, gas flow rate, and flocculants dosage. The most commonly used flocculants are alum, polyethyleneimine, poly dimethylaminoethyl methacrylate, ferric sulfate, ferric chloride, and chitosan (Brian & Zongli, 2019).

Flotation is an effective technique for PIW treatment. It is cheap, simple in operation, and compact with short residence time. However, all methods have high-energy consumption and can drive off volatile compounds prevalent in PIW requiring extensive air extraction and treatment systems (Yalcinkaya, Boyraz, Maryska, & Kucerova, 2020). Flotation also requires the addition of coagulants or surfactants to improve its removal efficiency (Alwared & Faraj, 2015), with similar drawbacks and considerations as mentioned previously under coagulation.

Adsorption

Adsorption is a simple, safe, and typically cost-effective technique for removing organics and inorganic ions, including metals, from PIW. Several adsorbents including organo-clay, zeolite, resins, activated carbon, nutshells, and biochar can be used for PIW treatment (Dąbrowski et al., 2005; Fakhru'l-Razi et al., 2009; Singh et al., 2017). The pollutants are removed through their attachment to the surface of the adsorbent by physical or chemical adhesion (Yalcinkaya et al., 2020). In oil adsorption, oil particles coalesce, then move into the pore spaces and onto the surface of the adsorbent through diffusion or interception (Esmaeili & Saremnia, 2018; Kundu & Mishra, 2019). Adsorption systems typically rely on either fixed bed design, where the wastewater is passed through the bed and adsorption is allowed to occur, or in a completely mixed batch adsorber system.

Ion exchange is a similar, but different, technique to adsorption that relies on the same general process layout of creating contact of the wastewater with a solid media for ion exchange to occur. In ion exchange, there is only an exchange of the ions between the liquid and solid phase, whereas in adsorption ions are removed from the liquid phase. Ion exchange resins have been used for the treatment of PIW. Both naturally occurring, and synthetic ions exchangers can be used. The most commonly used ion exchangers for PIW treatment include zeolites, clays, or materials such as polystyrene, granular activated carbon, and granular polypropylene loaded with a particular exchange ion. The removal efficiency of the resins can be improved by surface modification or by using a combination of different resins. Ion exchange resins show high performance to recover metals from the wastewater (Claudia & Vassilis, 2015).

The selection of an appropriate adsorbent/ion-exchange material is crucial to demonstrate the sustainability of the adsorption technique. The key criteria for the selection of an adsorbent should be based on (1) adsorption material with low inputs/emissions from its preparation, (2) high adsorption/exchange capacity to reduce media use between regenerations, (3) regeneration capacity to ensure the longevity of use before final disposal, (4) rapid adsorption/exchange kinetics to minimizing retention time and associated pumping/mixing costs, and (5) high selectivity in the case where specific pollutants are being targeted. The material should be abundantly available at a low cost and with low energy and chemical requirements in its preparation. Organic materials, particularly those derived from waste materials, are desirable to reduce production impacts, preparation costs, and potentially sequester carbon. Adsorption capacity is a key parameter to select an adsorbent. Synthetic and hybrid adsorbents often show higher adsorption capacity than natural materials. However, taking into consideration the material used in the synthesis process, processing cost, and safety risks, these factors may lead to reduced environmental and economic sustainability. It is also important to consider the equilibrium capacity at the design effluent concentration and the use of multistage adsorption systems, as these can have significant impacts on the mass of adsorbent required (Alyasi, Mackey, Loganathan, & McKay, 2020). Regeneration is another key property to demonstrate the sustainability of an adsorbent and should be considered in conjunction with adsorption capacity at equilibrium to determine the overall quantity of pollutant an adsorbent can remove during its lifetime. Regeneration involves the recycling of the adsorbent and recovering adsorbate from the aqueous medium. Reuse of an adsorbent can display many economic, environmental, and social benefits. However, it should be noted that regeneration is carried out by thermal or chemical treatment and the performance of the adsorbent will deteriorate with each subsequent cycle of regeneration. The use of chemicals, heat, and mixing pose safety issues, environmental emissions, and economic burdens that must also be considered against those of the adsorbent production and disposal (Kyzas & Kostoglou, 2014). A trade-off exists between these selection parameters. Thus, to select an appropriate adsorbent optimization studies should be conducted to achieve a good agreement between different parameters resulting in less environmental burden, cost-effectiveness, and high social benefits.

Both inorganic and organic adsorbents have been used for the treatment of PIW. Organic adsorbents show high removal efficiency for organic pollutants, are cheap, readily available, and can be made from waste products. For instance, walnut and other nutshells are popular and have been commercially used for many years, particularly for oil separation from production water. Organo-clay is another natural adsorbent that has been reported as an effective adsorbent for the removal of free hydrocarbons from PIW (Fakhru'l-Razi et al., 2009). Hasan, Chong, and Setiabudi (2019) reported the use of silica-rich rice husk for the removal of heavy metals, BOD, and COD from PIW. Zeolites are often used for the removal of dissolved organic carbon from PIW and can be regenerated by acid backwash.

Various studies have shown extremely high adsorption capacities and kinetics with good regeneration for dissolved and colloidal oil using various nanoadsorbents such as graphene oxides and carbon nanotubes (Diraki, Mackey, Mckay, & Abdala, 2019; Kazemi, Bahramifar, Heydari, & Olsen, 2019; Mohammadi et al., 2020). However, such adsorbents are costly to produce and require further demonstration of economic feasibility. Activated carbon is an effective adsorbent for the treatment of PIW. However, commercial activated carbon is quite expensive and saturates rapidly (De Gisi, Lofrano, Grassi, & Notarnicola, 2016). Nevertheless, it is one of the most sustainable treatment methods for PIW through LCA (Vlasopoulos, Memon, Butler, & Murphy, 2006), particularly as a second-stage treatment step. Renewable materials originating from natural resources provide alternative adsorbents that are relatively inexpensive and environmentally sustainable. Cellulose-based materials are abundantly available and widely distributed in plants and several marine species. They are highly efficient, nontoxic, and cost-effective. However, while their regeneration potential is high, the long-term stability of the adsorbent due to the biodegradable nature of cellulose must be considered (Peng et al., 2020; Syed, 2015).

Membrane technology

Membrane technology is an effective method of PIW treatment offering high separation efficiency, limited chemical addition, and compact treatment (Padaki et al., 2015). Its primary drawbacks include moderate to high energy use and a concentrated aqueous retentate that must be disposed of. In this process, a membrane serves as a barrier between two phases, and selected pollutants are retained on the membrane surface. Different types of membranes can be used for the treatment of PIW. Membrane types classified in order of decreasing pore size are microfiltration (MF), ultrafiltration (UF), nanofiltration (NF), and reverse osmosis (RO) (De Gisi et al., 2016). UF membranes are one of the most commonly employed, due to their high efficiency to remove oil, as well as relatively low energy and chemical additive requirements (He & Jiang, 2008). Salahi, Gheshlaghi, Mohammadi, and Madaeni (2010) employed UF in a cross-flow filtration mode for the treatment of oily wastewater. They achieved 94% removal of SS and 97% removal of oil content with a permeation flux of $96\,L\,m^{-2}\,h^{-1}$. Similar performance was achieved by Bilstad and Espedal (1996) for the treatment of oilfield wastewater using UF membranes. They achieved 96% removal efficiency of total hydrocarbons, 54% removal of benzene, toluene, and 95% removal of heavy metals (Cu, Zn) with a permeate recovery of 96% and flux of 96 L m^{-2} h^{-1} at a transmembrane pressure of 6–10 bars (Bilstad & Espedal, 1996). He and Jiang (2008) used a UF membrane system equipped with polyvinylidene fluoride membrane modified with nanosized alumina particles, achieving 90% COD removal and 98% TOC removal.

Ceramic membranes are frequently favored for PIW treatment in contrast to polymer membranes. Ceramic membranes have chemical and thermal stability and give a high flux. They are resistant to microbial degradation, harsh chemical environments (Bader, 2007), and show less pressure drop during operation (Brian & Zongli, 2019). They show relatively less fouling than polymeric membranes, and thus, require less cleaning and have a long life span. However, the manufacturing cost of the ceramic membrane is still high, which is a major obstacle to their market expansion. A few studies have emphasized the use of an integrated membrane system with other techniques for the effective treatment of oily wastewater and to reduce membrane fouling. For example, an RO system combined with dissolved gas flotation was successfully used for PIW treatment (Tsang & Martin, 2004). Similarly, Çakmakce, Kayaalp, and Koyuncu (2008) used the combination of NF and RO membrane after adsorption. However, this system demanded frequent regeneration of absorbers, which was not cost-effective.

The major issue in using membrane technology is fouling due to the deposition of particles on the membrane surface or within the pores. Oily wastewater contains hydrophobic and viscous molecules that adhere to most membrane material surfaces. The deposition of these molecules decreases the volumetric flux and performance efficiency, requiring higher energy input, more frequent cleaning or backwashing, and the increased retentate. These factors reduce the economic and environmental sustainability of the membrane process.

Several strategies have been proposed to deal with the fouling issue. Surface modification is one of the most effective methods to control fouling. Surface modification controls the pore size distribution and shape, reduces deposition of foulants on the membrane surface, and improves hydrophilicity of the membrane, and thus, increases flux. The surface modification also promotes self-cleaning of the membrane. Surface modification is frequently achieved through either coating the membrane or doping nanofillers into mixed matrix composite membranes during casting. Surface coating is often done with nanoparticles such as titanium oxide, iron oxide, zirconium dioxide, or hydrophilic polymers including polyethylene, and polyelectrolytes (Yalcinkaya et al., 2020). They improve the hydrophilicity of the membrane and facilitate the cleaning process.

The inclusion of nanofillers during membrane casting provides a blend of properties between the original polymer and nanofiller. At the right blending ratio to achieve good dispersion, mechanical properties can be significantly enhanced, reducing compression of surface pores during operation and altering surface charge and pore opening structure at the membrane surface. Excellent enhancements can be achieved. For instance, Sali, Mackey, and Abdala (2019) achieved a 2.7 × increase in flux through doping only 0.2% graphene oxide in a polyethersulfone membrane while simultaneously achieving a slight improvement in diesel rejection. The same study also highlighted the importance of the nanofiller production route, showing significant differences in the performance of the mixed matrix membrane based on the graphene oxide preparation method. Such findings highlight how small changes in membrane fabrication could influence significantly the chemicals and energy used in membrane fabrication as well as their resulting treatment system efficiency, which both need to be considered in a sustainability assessment.

Regular hydraulic backwashing assists in removing reversible fouling. However, irreversible fouling occurs and over time leads to overall flux deterioration, requiring extensive chemical cleaning. Depending on the membrane type, configuration, and PIW characteristics will determine how regularly such cleaning is necessary. Chemical dosing requirements for either pretreatment (pH adjustment, antiscalant, etc.) or cleaning processes of membranes can contribute significantly to environmental burdens in a variety of impact categories, such as marine ecotoxicity, ozone depletion, and abiotic resource depletion (Al-Kaabi, Al-Sulaiti, Al-Ansari, & Mackey, 2021). Cleaning is typically carried out using dilute hydrochloric acid or sulfuric acid solutions, mild hypochlorite, or hydrogen peroxide, all of which are detrimental to polymeric membranes in the long term. Several studies have proposed the use of sonication to overcome this, but high energy requirements make its large-scale application infeasible (Padaki et al., 2015).

Several membrane technologies have been developed for the treatment of PIW. However, selecting appropriate technology is critical. Selection should be based on integrated criteria of environmental and economic sustainability. RO is the required method for removal of salinity from the water but requires high energy inputs to overcome osmotic gradients, with high pretreatment requirements, including chemical inputs, and concentrated brine that must be disposed of. Membrane distillation is also an effective method of wastewater treatment (Roy & Ragunath, 2018). It requires less energy and produces high-quality water but it provides less flux and less resistance to high concentration brine. NF operates at a much lower pressure than the RO, generates high water flux, and the initial investment is also less. MF and UF are relatively less efficient in pollutant removal, but are less energy-intensive and give high water recovery. There is a trade-off between permeate water quality and energy and chemical footprints which must be determined based on the desired water quality. The sustainability of any membrane system can be judged based on the metrics. Basic indicators of the metrics are flexibility, modularity, recovery, product quality, brine production, size, operating conditions, and chemical usage (Figoli & Criscuoli, 2017).

Overall, membranes provide an effective treatment system but can be further improved through modification of fabrication techniques to improve flux, reduce fouling, and through the development of innovative and green chemistry techniques for foulant cleaning. Over recent years significant developments have been made to improve membrane performance. It was estimated that the energy consumption for the RO membrane decreased from 12 to 2 kW m^{-3} in the last three decades due to an improvement in membrane properties, operational parameters, and reduction in membrane fouling (Roy & Ragunath, 2018).

Oxidation

Oxidation is a promising technique to remove organic and some inorganic contaminants from PIW, as well as undesirable microbes. It involves the addition of an oxidizing agent, which decomposes the contaminants. This method results in less sludge generation than the other conventional methods; however, its operating cost is very high and it can cause toxicity in downstream processing of the wastewater due to the generation of by-products (Mohammadi et al., 2020). The most commonly used oxidizing agents are hydrogen peroxide, ozone, permanganate, and chlorine dioxide. Fenton oxidation is an effective technique for the treatment of PIW. It is an aqueous solution consisting of hydrogen peroxide and ferrous ions (Mohammadi et al., 2020; Tsai et al., 2010). In this reaction, hydroxyl radicals are produced with high oxidative power. These hydroxyl radicals are then used for the degradation of organic matter. Hydrogen peroxide can further break into environmentally safe species such as H_2O and O_2. It is an efficient, simple, and safe method of PIW treatment (Min, Changyong, & Yuexi, 2020). In addition to Fenton's reaction, catalytic oxidation is also used for the generation of free radicals. Metals such as Mn, Co, and Fe react with oxygen; free radicals are generated using electron transformation from metals to ozone or oxygen. Several factors such as catalyst type, catalyst loading rate, and supporting material affect the efficiency of catalytic oxidation (Tian et al., 2020). It was found that the degradation of organic material increased with an increase in catalyst loading rate (Xiao et al., 2018). However, a catalytic loading rate beyond a certain level reduced the exposure opportunity of a catalyst and specific organic pollutants (Mohammadi et al., 2020).

Oxidation is an efficient method of PIW treatment. However, there are several disadvantages associated with this method. The operational cost of an oxidation process is relatively high due to the constant supply of energy and chemicals, where the environmental impacts are dependent on energy and chemicals used (Petre et al., 2007). Chemical oxidants can generate by-products and sludge. Many of the by-products are toxic to humans and the ecosystem at large. Moreover, there are many safety risks to storing oxidants. Given these issues associated with oxidation techniques, its sustainability for large-scale application is not demonstrated yet and is a technology more suited to the removal of trace recalcitrant pollutants.

Various emerging techniques have been used for PIW or associated petrochemical wastewaters. In a study, it found that electrochemical oxidation was suitable for the removal of COD and TOC from the petrochemical wastewater (Wei et al., 2020) where oxidation time and current densities were the most influencing parameters for the removal performance. Fu, Wu, Zhou, Zuo, and Song (2019) evaluated the combination of different oxidation techniques such as only ozone, ozone with hydrogen peroxide, and ozone with a catalyst for the removal of organic matter from the petrochemical wastewater. It found that these techniques were effective for the removal of organic carbon and degradation of recalcitrant organic pollutants. The addition of hydrogen peroxide in ozone increased the transformation of hydrophobic components into hydrophilic components. Ozone with the catalyst showed the highest degradability of petrochemical wastewater and total organic carbon removal followed by ozone with hydrogen peroxide (Derwent et al., 1996).

An integrated system of photo-electro-catalysis and ozonation was found effective for the removal of color, turbidity, inorganic carbon, and organic carbon from oil-produced wastewater (Alcántara-Garduño et al., 2008; De Brito et al., 2019). The main areas of development for oxidation technology are in the identification of new oxidants and catalysts, where the latter can provide significant enhancements with existing oxidation processes.

Biological treatment

Biological treatment involves the use of microorganisms to degrade or bioaccumulate the pollutants through biochemical processes. It is a promising technique for PIW treatment with a high potential for energy and resource recovery.

The ubiquitous presence of the microbes in nature, their diversity, and their adaptation ability provide a wide range of opportunities for bioremediation. A variety of microorganisms exist that can be applied including algae, bacteria, fungi, and yeast. The majority of these organisms use hydrocarbons as substrates and convert toxic chemicals into less harmful compounds such as water, carbon dioxide, and inorganic compounds (Varjani et al., 2017). In some circumstances, the often recalcitrant or inhibitory nature of certain organics or other components (e.g., metals, salinity) that may be present in PIW may require bioaugmentation of specific organisms to the natural mixed cultures typically employed. In such cases, exogenic and/or genetically modified microorganisms are employed to remove the pollutants. For example, (Covino et al., 2015) employed autochthonous fungi as an exogenic microbe isolated from hydrocarbon-contaminated soil to degrade hydrocarbons. They achieved a removal efficiency of 79%.

Biological treatment using microorganisms can be carried out under aerobic and anaerobic conditions. In addition, phytoremediation—using plants can also be applied. All three processes present their advantage and disadvantages, which are described in the following sections.

Aerobic treatment

This is a widely adopted technology for the treatment of various wastewaters due to its ease of operation, high pollutants removal efficiency, and high microbial growth rate. In this treatment process, oxygen is supplied as a terminal electron acceptor and cosubstrate to degrade the contaminants. Several compounds such as BTEX, PAHs, phenol, refractory pollutants can be degraded by aerobic treatment (Singh et al., 2017). In aerobic treatment, the most commonly used techniques are biological aerated filters (BAFs) and the activated sludge process (ASP). The choice of treatment technique depends on the composition and nature of the pollutants, as some contaminants are more resistant to bioremediation in comparison to others. For example, long-chain organic compounds show more resistance to bioremediation than aliphatic compounds (Ossai et al., 2020).

BAFs have been useful for the treatment of PIW. BAFs are a submerged fixed-film type bioreactor with depth filtration to remove suspended solids as well as dissolved pollutants through microbial treatment. BAFs have been demonstrated to remove 81% COD from PIW (Wei et al., 2020). BAFs are efficient, compact, and easy to operate. BAFs show reasonably efficient mass transfer of oxygen to the biofilm in the reactor. The long sludge retention time of BAF also helps to retain nitrifying microbes and other sensitive organisms (Pramanik, Fatihah, Shahrom, & Ahmed, 2012). The performance of BAF depends on their configuration, packing media type, surface area, and the fate of microorganisms to develop a biofilm. BAFs can be a promising choice in built-up areas that have limited space. However, its cost is still high due to continuous air supply and periodic backwashing requirements. The requirements for backwashing can be reduced by selecting a suitable packing media and optimizing the operational parameters. Optimization studies might govern a reduction in power consumption to warrant the sustainability of BAFs.

The treatment of PIW is often carried out using the ASP. ASP offers high treatment efficiency, removing more than 90% COD (Mohammadi et al., 2020). In this process, the microorganisms form bioflocs due to cell aggregation that enables them to be separated from the liquid stream by gravity. However, this requires significant additional space compared to BAFs for a settling device. ASP also typically runs at lower sludge retention times. This requires less oxygen but produces greater quantities of biological sludge which must be handled. Sludge dewatering and drying result in large energy and hence carbon footprints and often have large land requirements (Motlagh & Goel, 2014). The cost of ASP can be reduced to some extent by utilizing the sludge for energy recovery, such as biogas. Significant opportunities exist for process optimization to reduce environmental impacts and operating costs, particularly when considering sludge handling, treatment, and aeration requirements plant-wide (Georgios, Dimitrios, Eleni, Alexandros, & Elisavet, 2018).

Some limitations exist for all aerobic growth treatment systems. The primary disadvantage is the supply of dissolved oxygen to the liquid, which is an energy-intensive and inefficient process due to the low solubility of oxygen. A further limitation is that nutrients (nitrogen, phosphorus, etc.) or trace elements (Mg, Ca, K, Fe, Mn, Mo, Cu, Zn, etc.) are frequently limiting for biological growth and therefore may need to be added to sustain the process. The supply of these nutrients comes with a significant environmental burden, for instance, most trace elements and phosphorus are mined, contributing to abiotic depletion, while nitrogen is typically derived from ammonia salts produced via the energy-intensive Haber-Bosch process. One way to balance the PIW wastewater composition for aerobic treatment is to combine it with other wastewaters, which are rich in nutrients (Morée, Beusen, Bouwman, & Willems, 2013).

Algae can be used for the treatment of PIW. Algae can bioaccumulate and degrade hydrocarbons under autotrophic, heterotrophic, and mixotrophic conditions. While not aerobic organisms, they can be grown under aerobic conditions utilizing CO_2 and producing oxygen. Genera suitable for PIW treatment include *Chlorella*, *Chlamydomonas*, and *Phoridium*. Different studies have demonstrated the use of microalgae in an open pond or closed system for PIW treatment. Microalgae-based treatment systems have shown high removal efficiency of COD from the PIW demonstrating it as a potential alternative to traditional oxic treatment (Fu, Wu, Zhou, Zuo, Song, et al., 2019). However, there are many issues that need to be addressed to demonstrate their large-scale application. These include affordable systems allowing high average light intensity and maintaining the desired microalgae species. To overcome this second challenge, since sterilization is not amenable to PIW treatment, extremophile microalgae have been tested for wastewater treatment that can thrive in environments with minimal competition from other organisms (Henkanatte-Gedera et al., 2017). Such conditions can be found with PIW, particularly produced water that may be highly acidic or saline.

Recently, the concept of coculture for the treatment of PIW has received significant research attention. In this system, bacteria and microalgae are grown together. Bacteria oxidize organic matter and produce CO_2, which in turn is used by microalgae to develop their biomass (Lutzu & Dunford, 2019). This system offers competitive advantages over the monoculture system by reducing aeration/CO_2 supply requirements. A study displayed >90% removal of COD from the PIW by employing a coculture system (Mujtaba, Rizwan, Kim, & Lee, 2018). The coculture system also showed higher pollutants degradation of alkyl aromatic compounds than the monoculture (Jones et al., 2008). Coculture systems for PIW have been tested at a lab-scale only, but have been used at pilot or full scale for other wastewaters. The effectiveness of these systems to reduce aeration requirements is not fully understood as the relative algal fraction of the microbial community depends on biomass concentrations, light penetration, and the particular algal species present as to whether they can grow heterotrophically during dark (night) periods. Such systems are also highly influenced by the season (temperature and daylight hours).

Anaerobic treatment

PIW wastewater treatment can be carried out under anaerobic conditions. In comparison to aerobic treatment, anaerobic treatment presents many advantages. It allows the opportunity for energy recovery, requires less energy input and capital investment, and produces less sludge volume (Kong, Li, Xue, Yang, & Li, 2019). Anaerobic treatment can be used to recover resources such as biomethane, biohydrogen, and volatile fatty acids (VFA) (Kong et al., 2019). Furthermore, due to the low biomass growth rates and yield of anaerobic respiration, fewer nutrients and trace elements need to be supplied for nutrient imbalanced PIW.

Anaerobic digestion (AD) is a set of biochemical processes that result in the production of biogas, containing methane, CO_2, and hydrogen. Major steps in AD involve hydrolysis, acidogenesis, acetogenesis, and methanogenesis. In hydrolysis, long-chain organic molecules are converted into smaller molecules, called monomers. These monomers are further degraded by microorganisms into VFA and this process is called acidogenesis. Depending on the microbial composition and biochemical route, VFA is further degraded into acetic acid, methane, hydrogen, and carbon dioxide. Acetogens are responsible to convert VFA into acetic acid, whereas methanogens convert acetic acid or H_2 and CO_2 into methane. This process is highly sensitive to pH, temperature, and inhibitory compounds including undissociated VFA due to the slow growing nature of these organisms. The methanogenesis step is typically rate-limiting which can result in the buildup of volatile fatty acids, resulting in a positive feedback failure loop. This also causes a buildup of hydrogen, inhibiting hydrogenotrophic methanogenesis (methane from hydrogen). Typically the process is run at mesophilic ($\sim 35°C$) or thermophilic ($\sim 55°C$) conditions due to the relatively slow rates of the anaerobic processes. If the AD process is deliberately interrupted before methanogenesis a fermentation broth containing VFA is produced. VFA are extremely biodegradable and can be used to produce a wide variety of bioresources including polyhydroxyalkanoates (PHA), bioelectricity, biodiesel, and single-cell protein.

Energy can be recovered in the form of biohydrogen and biogas through the AD system (Ali, Elreedy, Ibrahim, Fujii, & Tawfik, 2019). Biogas recovery from the wastewater can be upgraded to natural gas suitable for network injection, or combusted after desulfurization for onsite combined heat and power, providing an overall energy-positive treatment process. The biogas yield and the quality highly depend on the feed composition and operational parameters, the former which can be challenging. Moreover, the optimization of operational parameters such as carbon to nitrogen ratio, pH, temperature, and microbial diversity is a complex process. An imbalance in operational parameters may result in impurities in the biogas leading to the low calorific value, as well as corrosion of the digester, or could lead to failure of the digestion process (loss of methane production). Unintentional biogas production, soluble methane in the reactor effluent or poorly sealed treatment systems can lead to the emissions of various greenhouse gases including carbon dioxide, carbon monoxide, sulfur dioxide, methane, and nitrous oxide, the most prominent and influential being methane (Paolini et al., 2018). Safe and economical handling of digestate (sludge after anaerobic treatment) is another concern to implement biogas technology at a large scale. The digestate can contain toxic heavy metals and a high concentration of nitrogen and phosphorus, which can pose a negative impact on the environment (Paolini et al., 2018).

Anaerobic membrane bioreactors (AnMBR) are an emerging technology that aims at ambient temperature anaerobic processes with high-quality effluent, achieving both goals through the use of a membrane module embedded in the anaerobic digester. The membrane serves as a physical barrier to retain sludge and facilitates short hydraulic retention time and prolonged sludge retention time (Smith, Stadler, Love, Skerlos, & Raskin, 2012). In a study, AnMBR showed a 100% removal of oil and grease and 97% COD removal. Currently, several pilot-scale studies have demonstrated the use of AnMBR for the treatment of various wastewaters (Vinardell et al., 2020). It was viewed that this technology can make the wastewater treatment process energy neutral or even energy positive. However, membrane fouling is more severe in anaerobic conditions and is the current technology bottleneck (Kong et al., 2019). It can be controlled by cleaning or backwashing but it is expensive and involves significant chemicals and energy.

BECs are another emerging technology tested for the treatment of PIW. In a BEC degradation of organic contaminants occurs at the anode and generates electrons, which are guided toward the cathode via an external circuit that creates current. They take on two forms of relevance for PIW treatment—microbial fuel cells (MFC) and microbial electrolysis cells (MEC). In MFC the product is electricity. PIW is fed into an anode, oxygen is typically supplied at the cathode, and the system is connected to an electric circuit. In MEC the product is some chemical, using the chemical energy from the organic oxidation at the anode to offset or reduce the energy input required across the circuit for the chemical production to occur at the cathode. Both systems provide energy savings, either directly or indirectly, and have low sludge production (Fakhru'l-Razi et al., 2009; Li et al., 2006; Mohammadi et al., 2020). A study demonstrated the use of hydrocarbon wastewater in an MFC for the removal of phenol (23 mg L^{-1}) and generated current (1.9 mA) simultaneously. In one study, constructed wetlands were integrated with MFC reactors for the treatment of oily wastewater. An oil removal of 95%, COD removal of 75%, and power density of 3868 W m^{-3} were achieved (Yang, Wu, Liu, Zhang, & Liang, 2016). However, BEC faces significant issues for full-scale deployment. The systems are prohibitively expensive due to the materials such as electrodes, catalysts, wires, and separators, whose manufacture also poses many environmental issues (Li, Yu, & He, 2014). The long-term performance also suffers from membrane fouling, clogging of electrodes, and difficulty in maintaining an active electrochemical community. Due to these issues, the BEC technology is far from the level of commercialization. Advancement in reactor engineering, microbial ecology, and material synthesis are necessary for the sustainable application of BEC (Nealson, 2017).

Phytoremediation

Phytoremediation is another technique to treat PIW utilizing plants to remove contaminants from the wastewater through volatilization, degradation, and adsorption mechanisms (Epps, 2006; Li et al., 2012; McCutcheon & Schnoor, 2004). The plants are associated with the soil environment and soil microorganisms create a distinct habitat around the roots, which enables the degradation of the pollutants. Plant roots provide oxygen to the microorganisms in the rhizosphere, which enhances the degradation of hydrocarbons since aerobic degradation is more rapid than anaerobic degradation. The plants can uptake pollutants through the root, which are then bioaccumulated in the root tissues or transpired into the atmosphere (Yavari, Malakahmad, & Sapari, 2016). Several plants have been identified for the remediation of PIW. For instance, *Impatiens balsamina* was shown to remove 65% hydrocarbons from wastewater of 10,000 mg kg^{-1} hydrocarbons in four months (Cai, Zhou, Peng, & Li, 2010). Similarly, *Canna indica* removed 80% of BTEX in 21 days only. Suitable plants have large root surface areas and a dense rhizosphere that propagates microbial degradation (Boonsaner, Borrirukwisitsak, & Boonsaner, 2011). The efficiency of phytoremediation can be enhanced by inoculating plant species with hydrocarbon-degrading bacteria (Khan, 2006; Sudarsan, Annadurai, Subramani, & George, 2016). They improve a plant's tolerance to hydrocarbon pollutants by providing various enzymes. These bacteria are reported to improve plant growth, protect against various diseases, and enhance removal efficiency (Buyan, Shah, & JaeSeong, 2014). Phytoremediation has been demonstrated at the field scale for PIW and shows much promise.

In Oman, the Nimr Water Treatment plant consists of a 480 ha constructed wetland treating 175,000 m^3 day^{-1} of produced water from an oilfield using *Typha*, *Cyperus*, and *Juncus* which has gone through three stages of expansion (NIMR, 2019, 2020). The plant has hydrocyclone pretreatment and achieves an effluent with less than 0.5 mg L^{-1} of hydrocarbons from an average influent concentration of 300 mg L^{-1}. The effluent is used for saline irrigation of corn and cotton (Stefanakis, Al-Hadrami, & Prigent, 2017), while the wetland biomass is also harvested for biofuel production. Overall, the system uses only 0.06 kWh m^{-3} and the establishment of the wetland contributes significantly to the countries greenhouse gas reduction targets as well as increasing biodiversity in the arid environment (Breuer, Thaker, Headley, & Al Sharji, 2012). A US EPA-2000 report indicated the total cost for phytoremediation of hydrocarbon wastewater on 1 ha was $2500–$15,000, which was much lower than microbial remediation ($7500–$20,000) (Epps, 2006). The cost of phytoremediation would largely depend on the plant species employed and land value, the latter which does not appear to be included in the costs given previously.

Some limitations impede the widespread uptake of phytoremediation systems. These systems are less efficient for concentrated wastewaters and require a large area and long treatment duration. The efficiency of phytoremediation is very sensitive to the temperature and PIW is often deficient in nitrogen and phosphorus—requiring the addition of these nutrients. However, there are many associated benefits with the phytoremediation technique, which can enhance its sustainability. It can improve air quality by taking up carbon from the atmosphere. It would promote biodiversity, restoration of ecosystems, enhance the esthetic value of land, and reduce soil erosion.

Sustainability of petroleum industry wastewater treatment

It is well known that the petroleum industry has a largely negative environmental impact and positive economic impact amongst various sectors. Significant changes in country-level legislature, as well as global agreements such as the Paris Accord, place increasing requirements on the operations of the petroleum industry to develop ever-more sustainable solutions to their product production, which also includes the management of PIW. Traditionally PIW management has focused purely on treatment, to meet regulatory discharge standards at minimum cost. These environmental and economic constraints will remain the primary drivers for PIW treatment selection, but growing awareness of the public toward climate change and other environmental stresses will drive social pressure for "greener" treatment systems. At the same time, competition from renewable energy sources will require more economic oil and gas refining, which includes more economic treatment systems, to maintain competitive pricing. The overall sustainability of PIW can be improved by reducing the use of nonrenewable resources, the risk of failure and cost, minimizing energy inputs, waste generation, and chemical inputs; and implementing resource recovery/recycling. Fig. 15.2 represents the major sustainability indicators of PIW treatment.

A key aspect of any wastewater treatment process, including PIW treatment, is that a single treatment unit is rarely sufficient to achieve the desired water quality. In most cases the most optimal treatment systems comprise multiple treatment units focused on removing specific pollutants within different concentration ranges. Given the high variability in PIW characteristics depending on the reservoir and its life-cycle stage, the processing techniques used, as well as different local economic, supply chain, and environmental regulation considerations, there is also no optimum treatment

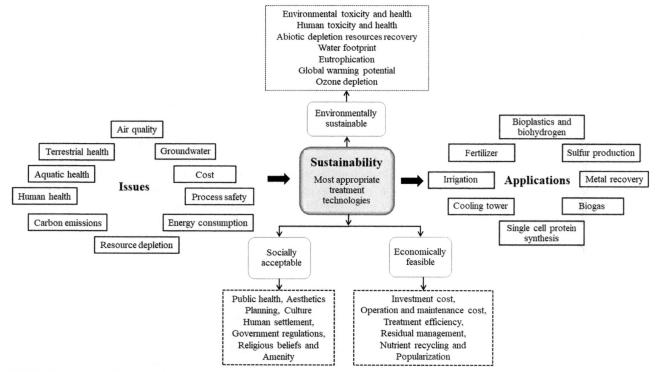

FIGURE 15.2   Sustainability approaches toward petroleum industry wastewater.

process for PIW that can be applied to all cases. It is therefore important in any sustainability assessment to provide whole plant modeling and evaluation, looking at inputs, emissions, and products from across the entire plant, including both liquid and solid stream treatment.

Energy for pumping, mixing, aeration, and other treatment equipment is an integral input for all treatment trains. It is therefore a critical consideration in sustainability assessment since different countries rely on vastly different energy mixes for electricity generation. These different energy sources can then have a significant impact on both the economic and environmental sustainability of a process. Regarding economics, price volatility is heavily impacted by many factors, including the renewable energy sources utilized. Renewable energy sources can either increase or decrease price volatility depending on the renewable energy technology, the degree of reliance on them and other energy sources in the grid, volatility time-span under consideration, and location/weather conditions (Johnson & Oliver, 2016; Rintamäki, Siddiqui, & Salo, 2017). In general, renewable energy sources are much more volatile in their electricity production, due to their strong reliance on weather conditions (e.g., sunlight, wind, etc.). However, because they have a low marginal cost for production, when conditions are favorable they tend to be the first technologies to contribute to the power demand with more conventional fossil fuels adding on-demand generation.

For environmental sustainability, the electricity grid mix also has a significant impact. Renewable technologies are generally considered green and environmentally friendly. However, it is important to note that each technology has different types of environmental impacts associated with its electricity generation, often arising from the manufacturing stage in the case of renewables. For instance, Al-Kaabi et al. (2021) demonstrated in a study on water desalination that photovoltaics are a very low impact with regards to the commonly considered global warming potential; they have a very high environmental burden compared to certain fossil fuel electricity sources for ozone depletion potential and abiotic depletion. Moreover, natural gas, despite being a fossil fuel, has lower impacts across a broad range of environmental impact categories than photovoltaics or heavy fuel oil, while wind power was the most low-impact energy source overall of the four energy sources evaluated. Photovoltaics have significant impacts associated with the materials that go into their production—requiring high energy, chemicals, and often the use of rare heavy metals that also pose issues with end-of-life disposal. Concentrated solar systems require large land areas and a large water footprint for cooling purposes. Hydropower systems result in large land and vegetation loss, impacts on fish migration, and may have variable impacts on global warming potential or water losses. However, they can also provide unforeseen benefits such as flood protection, increased irrigation opportunities, and recreational value. A detailed review of various potential impacts from a wide range of renewable energy sources can be found in Abbasi and Abbasi (2000).

Despite the significant role of energy in GWP from the wastewater treatment process itself, the overall contribution of wastewater treatment in the oil refinery process is estimated at around 0.40% (Li, Han, & Elgowainy, 2016). Given the high toxicity associated with certain hydrocarbons and metals, aquatic ecotoxicity impacts are potential of greater concern related to their disposal. For instance, Liu et al. (2020) performed a whole refinery LCA study for a medium-scale inland refinery in northern China. They found that wastewater treatment contributed significantly to freshwater eutrophication and freshwater ecotoxicity due to phosphate and nickel concentrations in the effluent, respectively. In most other impact categories, except ozone depletion, the impacts from the wastewater treatment process at that refinery were minimal and it shows local discharge regulations and degree of treatment are important considerations. Higher levels of the treatment process will reduce eutrophication and aquatic ecotoxicity but will require higher energy and chemical inputs.

It is also important to consider the countries from which these chemicals are acquired, as this can result in vastly different transportation distances and technologies used for the chemical products based on local legislation, which both define the overall impacts associated with the chemical (Al-Kaabi & Mackey, 2019) . For instance, the chlor-alkali process used for both sodium hydroxide and hypochlorite production, two commonly used chemicals in water treatment, have vastly different impacts depending on whether modern technology or older technology is used, particularly with relation to marine ecotoxicity from mercury release (Crook & Mousavi, 2016). In addition, many processes such as neutralization, coagulation, scaling minimization, and disinfection can be conducted with different chemicals, each which has significant differences in the environmental impacts associated with both their production and disposal (Al-Kaabi et al., 2021).

Existing life cycle assessments for petroleum industry wastewater treatment

A limited number of studies currently exist on the treatment of PIW. The earliest study was by Vlasopoulos et al. (2006), who undertook a comprehensive study of 20 different treatment technologies for the treatment of water from oil and gas extraction. The study divided treatment technologies into five different stages of treatment and considered over

600 possible combinations to achieve water quality suitable for seven different agricultural end uses and two industrial end uses. About 10 impact categories were assessed and a European-wide energy mix was used. For stage 1 treatment processes, DAF performed slightly better than hydrocyclones, mainly due to its lower energy usage. Activated sludge was the highest overall impact from stage 2 technologies while adsorbents were the lowest, with constructed wetlands, trickling filters, rotating biological contactors, and microfiltration all considered low impact. The impact was closely associated with energy use during the operational phase. Ozone oxidation was the worst-performing of the tertiary treatment units while dual media filters were the best overall. For stage 4 technologies, only reverse osmosis and electrodialysis reversal were considered. Reverse osmosis performed significantly better due to lower energy usage, while the platinum used in electrodialysis reversal electrodes also presented significant construction phase impacts. Depending on the end-use quality required, the optimal treatment train differed, and in certain instances, a longer treatment train was more environmentally optimal (MF was also considered as a combined stage 1 and 2 treatment).

Another important finding from the study was the relative contributions of the construction phase versus the operational phase. For instance, certain technologies such as constructed wetlands and slow sand filters had significant construction phase impacts due to the large amounts of polymer-based materials used in the construction and low operational phase energy use. It is important to note that the findings from this study are dependent on the produced water quality considered, the energy mix, and the functional unit, which was the quantity of water treated. For instance, constructed wetlands were in several optimal treatment trains, but were assumed to lose 40% water through evaporation. If recovered water had been a significant goal in the environmental analysis this would lead to a different finding. The increased contribution of renewables would also have a very significant impact on the assessment given the strong linkage of energy use to the overall performance of most technologies.

Muñoz et al. (2020) evaluated an integrated set of process units for a petroleum refinery to achieve water quality suitable for three different levels of reuse, in addition to existing ocean disposal quality. These reuse applications were firefighting water, cooling water, and boiler feedwater, while the process units consisted of DAF, activated sludge, and ceramic membrane ultrafiltration as the base for all reuse scenarios with additional treatment units of catalytic wet air-oxidation, advanced oxidation (ozone + hydrogen peroxide), and reverse osmosis (single/double pass), which were dependent on the fluctuating influent water quality and required effluent quality using a decision support system. They found for reuse as boiler feedwater, the additional global warming potential (GWP) caused by the more extensive treatment was offset by the reduced heating requirements of the refinery effluent since the reuse water is warmer than the natural water source. The other two water reuse options had higher GWP than the disposal route. Reuse was advantageous for reducing marine ecotoxicity, eutrophication, and freshwater footprint, but had higher particulate matter (air quality/human toxicity) than direct disposal due to the higher energy requirements. The role of RO to meet the desired water quality also influenced results, as this increased the proportion of discharge to the ocean as brine and reduced the overall recovery. An evaluation of the European grid mix was also conducted. Impact categories associated primarily with discharge to the ocean remained relatively unaffected (eutrophication, marine ecotoxicity, water footprint) but GWP, particulate matter, and nonrenewable energy demand all decreased significantly, albeit with similar relative performance. It should be noted that findings are very location-dependent. For instance, when considering reuse scenarios in the countries comprising the Gulf Cooperation Council (GCC), a major oil and gas producing region, desalination is the primary form of water supply. In such cases, reuse, unlike the Turkey case, is strongly desirable due to its lower impacts, particularly with relation to GWP.

Transportation is a key consideration when assessing offshore-produced water treatment. Torp (2014) assessed the handling of produced water on an offshore platform in Norway. She considered direct injection; two offshore treatment processes comprising (1) decanter + centrifuge + cartridge filter and (2) flocculation + DAF; and an onshore treatment process comprising decanter + flocculation + biological treatment with thermomechanical cutting cleaner for solids handling. In her assessment, offshore treatment systems performed significantly better than the onshore system, which in turn was better than direct injection. The transportation was a key differentiating element, as the average distance from the platform to onshore was 163 km with a further 271 km onshore transport from the port base to the treatment facility. Transportation therefore comprised at least 20% in any individual impact category. Recovery was also a key aspect of the overall impacts, with light oil recovered from the sludge treatment providing negative (offsetting) impacts. However, the study also highlights the importance of process-dependent choices. For instance, disposal of oil not recovered/sold was done by hazardous waste incineration and had a GWP contribution of roughly 70% in the onshore treatment case, where transport comprised almost all the remaining GWP contribution. With more efficient incineration processes with thermal recovery, such impacts may be significantly reduced. Lastly, when assessing the relative contribution of different treatment units for the onshore system, it was determined that flocculation-DAF was the major contributor to all categories, namely due to the chemical inputs required in this process.

Recovery of various resources from petroleum industry wastewater

Water

Recovery of treated PIW plays a vital role in sustainable water management for various applications such as various industrial processes, and potentially agricultural irrigation, aquaculture, and any nonhuman contact utilization. Additionally treated wastewater reuse in petroleum industries is the most convenient alternative for reducing freshwater consumption for different processes. Recycling treated wastewater in cooling towers, steam generation systems and firefighting can save 40%–45% of total freshwater use in refineries (Chen, Wu, Qi, Ding, & Zhao, 2020; Fu et al., 2019; William, Lindgren, Noling, & Dennis, 2003). Chen, Lu, Jiao, Wang, and Chang (2013) reported that reuse wastewater is priced at $0.784 m^{-3} which represents only one to three-quarters of desalinated water cost. Additionally, Pearce (2008) reported that energy required to treat wastewater using conventional activated sludge systems post-treated with MF/UF and RO is 0.8–1.2 kWh m^{-3} while the energy required to desalinate seawater could be around 4 kWh m^{-3}.

Water pinch analysis is an important part of any chemical process facility to achieve efficiency and can realize significant benefits within a refinery. For instance, water pinch analysis was used to minimize freshwater requirements in a Tehran oil refinery (Mohammadnejad, Bidhendi, & Mehrdadi, 2011). In this case, 26% of water was recovered after removing suspended solids, hardness, and COD from the wastewater and could be reused in cooling towers in the oil refinery. Separate studies, also in Tehran, have also shown a common approach of water reuse in cooling towers (Hashemi et al., 2020; Manouchehr & Mishana, 2008; Singh, 2019). This end-use required reductions of 95% TDS, 100% total hydrocarbons, 93% Cl$^-$, and 97% SiO$_2$ achieved primarily by RO. There are also many opportunities for reusing oily wastewater in steam boilers or recycling it in injection wells to enhance crude oil exploitation. The China Department of Petroleum recommends maximum limits of 2 mg L^{-1} oil and suspended solids for reuse in boiler inlets, and 10 mg L^{-1} of oil and 5 mg L^{-1} of suspended solids for recycling into injection wells (Moreno et al., 2002; Singh et al., 2017). The (US EPA, 2012) requires less than 30 mg L^{-1} of BOD and TSS, 200 fecal coliforms100 mL^{-1}, and 1 mg-Cl$_2$ residue L^{-1}. US EPA guidelines for boiler water makeup vary depending on the boiler drum operating pressure, but require maximum levels of 3000 mg L^{-1} for TDS, 0.1 mg L^{-1} for total iron, 0.05 mg L^{-1} for copper, 0.3 mg $^{-1}$ CaCO$_3$ for hardness, and 1 mg L^{-1} for nonvolatile TOC and oil for the least stringent case (US EPA, 2012). Due to the low inorganic contamination of fuel synthesis wastewater, this water is highly amenable for reuse applications. ORYX GTL, one of the gas to liquid technology operators in Qatar, reuses 80% (approximately 1.38 million m^3) of its treated water in 2012 for landscaping irrigation, cooling tower make-up, and fire extinction (Jasim, Saththasivam, Loganathan, Ogunbiyi, & Sarp, 2016). It was found that the treated wastewater reuse plan would have a range of positive economic impacts. For example, the market value for recovered wastewater is set between $0.24 (€0.22) and $0.31 (€0.28) m^{-3} for agricultural uses, while the estimated operation and maintenance cost for producing high-quality recycled water is $0.16 (€0.14) m^{-3} (Smith et al., 2012). Therefore, recovery of treated PIW by the most appropriate treatment technology is economically feasible.

Organics, inorganics, and nutrients

PIW contains a wide variety of organic compounds, which contain high energy content and offer chances for carbon recovery. Some PIW streams, as well as produced water, may also contain nutrients such as nitrogen, phosphorus, and sulfur; phenols; suspended solids, and dissolved solids with certain refinery streams also including various metal ions that may arise from catalysts used in the process. The types of wastewater stream generated by the petroleum refining industry can be classified into desalter water, cooling tower spent catalyst, spent caustic, water used for flushing during maintenance and shut down, sour water, and other residuals (Mustapha, 2018). Various petroleum refining unit operations like desalting of crude, atmospheric distillation, catalytic cracking, hydrocracking, thermal cracking, alkylation, polymerization, hydrotreating, lube and asphalt, and isomerization, etc., produce high salt wastewater, desalter sludge, oily sour water, or quench water with vastly different characteristics (US EPA, 2019). Recovery may be more efficient and economic for specific compounds by focusing on streams with high concentrations of the target resource prior to the combination for whole plant treatment. Therefore, characterization of these different streams and evaluation of treatment-recovery options that may involve more than one treatment process on the refinery site should be evaluated.

Table 15.2 summarizes various treatment technologies providing resource recovery options of organics, nutrients, and metals. For organic material, hydrocarbons may be able to be recovered directly through physical separation processes such as DAF, creating an oily sludge. Oily sludge is high in energy with a typical calorific content of over 3900 kcal kg^{-1} (Hou et al., 2013). In oily sludge storage facilities in the Niger Delta area, a recovery of about 68% hydrocarbon oil was obtained, out of which 87% were aromatic, using a solvent extraction process (Hu, Li, & Hou,

TABLE 15.2 Recovery of organics, inorganics, nutrients, and heavy metals from petroleum industry wastewater by different treatment technologies.

Different treatment technologies	Recovery	Remark
Algal treatment in hydraulic fracturing technology (Lutzu & Dunford, 2019)	Growing algae and harvesting the algal biomass	Commercialization and energy-efficient compared to other conventional wastewater treatment technologies but high water loss
Sequencing batch reactor (SBR) (Mannina, Presti, Montiel-Jarillo, & Suárez-Ojeda, 2019)	$74 \pm 8\%$ PHA recovery with 100% purity	Cheap and environmentally friendly manner
Bioelectrochemical (Kuntke et al., 2011)	$NH_3(g)$ recovery in the form of $(NH_4)_2SO_4$	–
Electrodialysis (Vaneeckhaute et al., 2017)	80%–83% NH_3 recovery from high strength wastewater	
Combination of solvent extraction and freeze-thaw (Hu et al., 2015)	40% oil recovery rate	–
Electrically conducting gas-stripping membrane (Iddya et al., 2020)	68.8 ± 8.0 g-N m^2 day^{-1} of NH_3 was recovered, with an energy consumption of 7.1 ± 1.1 kWh kg-N	–
Ammonia stripping (an integrated scheme involving ultrafiltration, reverse osmosis, and cold stripping air stripping in packed bed tower) (Sengupta et al., 2015)	Recovered 71% of NH_4HCO_3 from high strength wastewater	High commercial viability process optimized for pH and temperature
Emulsification assisted by ultrasonic probe (Chiha et al., 2010)	95% Cu recovery	Higher ultrasonic power consumption
Cyclonic gas stripping (Wang et al., 2018)	96% of oil with 3100 t a^{-1} of recovery and 647 t a^{-1} of high-activity catalyst recovery	Direct economic benefits of this process as high as 37.28 million CNY year^{-1}
Corrugated plate separator (Arthur et al., 2005)	Oil recovery from emulsions or water with high oil content prior to discharge	–
Constructed wetlands (Agarry, Oghenejoboh, Latinwo, & Owabor, 2020; Li et al., 2012; Moreno et al., 2002; Sudarsan et al., 2016)	90 to 95% of BOD, COD, TSS, 65 to 90% of phenol, 90% of NH_3, 93% TPH, 67% of TN, 54%–71% TP, 90.4% oil and grease, 94% Cd, 92.5% Pb, 93% Cr, 94.8% Fe, 92.2% Ni, and 57.7% Cl^- removed by *Typha latifolia, Pragmites, Acorus calamus* and *Eichhornia crassipes* plants including *Acinetobacter, Rhizobium,* and *Rhodobacter* bacteria development.	Robust to high flow variations but large land requirement.

2015; Taiwo & Otolorin, 2009). 39% to 88% crude oil can be recovered by solvent extraction. However, the recovered oil has a high sulfur content and some carbon residue impurities (Hui et al., 2020). A variety of other methods also exist, such as surfactant enhanced oil recovery, freeze-thaw, microwave treatment, electrokinetic methods, centrifugation, pyrolysis, and ultrasound (Arthur, Bruce, Langhus, & Patel, 2005; Hui et al., 2020). Alternatively, biological treatment processes allow the conversion of soluble or emulsified hydrocarbon into biogas using AD, or into biomass that can be utilized for energy generation in a thermochemical conversion such as pyrolysis, gasification, or incineration following sludge dewatering and drying. The conversion to sludge and then subsequently energy is not preferential due to large energy requirements for drying and loss of chemical energy to biological catabolic processes.

A further carbon recovery route is through sludge or biochar—the former produced from AD and the latter by thermochemical conversion (both of which produce some energy also). These products can then be applied to land to improve agricultural productivity. However, risks with contaminants present in PIW sludge and biochar exist, which may prevent their application to land in many countries. Other biotechnology routes also exist, to produce metabolites such as polyhydroxyalkanotes, volatile fatty acids, or other chemical feedstocks or materials for industrial or commercial use (Mannina et al., 2019). However, such processes have not been applied at scale to PIW and their economics may be limited by the more recalcitrant nature of the organic compounds in PIW, making it a less favorable feedstock for such processes (Kuntke et al., 2011).

Nutrients can be recovered directly, typically through processes such as adsorption or precipitation. In the case of phosphorus and sulfur, biological processes can also be applied (Singh & Borthakur, 2018; Vaneeckhaute et al., 2017) such as enhanced biological phosphorus removal and sulfide oxidation/sulfate reduction processes to produce elemental sulfur such as with the Thiopaq process. Sulfur is the most commonly recovered nutrient due to its prevalence and negative impacts in various PIW streams. For instance, petroleum refinery wastewater contains a high concentration of sulfate ($15-1332$ mg L^{-1}) (El-Naas, Al-Zuhair, Al-Lobaney, & Makhlouf, 2009). Metals can also be recovered primarily through adsorption/ion exchange processes, or by chemically or bioinduced precipitation. Metals, due to both their common toxicity and often high value, should be considered as a priority for recovery in PIW.

CO$_2$ capture and emission reduction

Li et al. (2016) reported that GHG emissions from petroleum industries comprise about 1.57% of global GHG emissions (49 Gt CO$_{2e}$). According to USEPA (2009), petroleum refineries wastewater contributes 25,000 mtCO$_{2e}$ per annum, primarily as CH$_4$ and N$_2$O. Therefore, CO$_2$ capture from PIW treatment technology could provide small contributions to mitigate climate change.

Phototrophic systems involve the use of microalgae, which perform oxygenic photosynthesis using CO$_2$ as a substrate, converting it to biomass that can then be processed into biofuels or biogas. Jacob-Lopes and Franco (2013) developed an integrated system of biotransformation of CO$_2$ in oil refineries wastewater that contains 92% VOCs to cultivate *Aphanothece microscopica* in a bubble column photobioreactor. The system showed an optimum specific growth rate of 1.4 day^{-1} with a maximum CO$_2$ elimination capacity of 22.9 mg L^{-1} min^{-1}. Each g of CO$_2$ converted produces 0.75 g of O$_2$, with a small fraction (3.64%) of the CO$_2$ converted to fixed biomass form. Additionally 0.08 g lipid L^{-1} day^{-1} biodiesel was produced in this process. However, phototrophic systems require a large land area and energy for mixing and subsequent biomass processing can offset advantages from the photosynthetic carbon sequestration. Microalgae also cannot perform photosynthesis and organic carbon biodegradation simultaneously.

Constructed wetlands are another form of the phototrophic treatment system. Due to their almost zero energy input required for treatment, they provide a significant GHG emission reduction in comparison to other methods of treatment. For instance, the Nimr treatment wetlands estimated energy reductions at 98% compared to deep well injection, equating to a 3.9 kg CO$_{2e}$ m^{-3} reductions in GHG emissions. Carbon sequestration in the biomass of wetlands is less certain. In general, wetlands are considered carbon-sequestering when considering methane decay rates in the atmosphere and sufficiently long timelines (Mitsch et al., 2013). However, this is highly dependent on the relative growth of plant biomass and fluxes of methane, which are influenced by root submergence and temperature. Therefore, in constructed wetlands, which are submerged with high organic concentrations, methane fluxes could be significant (Agarry et al., 2020).

MEC technology provides another route for GHG reduction. MFCs can generate electricity that may offset overall GHG emissions from a PIW treatment process. However, MEC systems provide carbon sequestration through electrosynthesis of carbon dioxide to larger organic molecules. Currently, there are no studies on electrosynthesis from PIW and the high cost and existing development issues make their potential in the future somewhat limited.

Future perspective

Despite the current transition to renewable energy and fuel sources, global demand for petroleum continues to increase, and with it the need for wastewater treatment from the industry. Stringent global standards and increasing freshwater scarcity have necessitated new and existing facilities to achieve higher levels in wastewater treatment to recover more water for reuse. These factors, coupled with a more intensified interest in resource recovery and circular economy provide strong drivers for enhancement of existing treatment processes and development of new ways of PIW treatment.

With regards to water reuse, it is expected that membrane technology will continue to receive significant development and focus, primarily in the area of nanomaterial enhancements to reduce fouling and improve membrane rigidity and performance. Such developments will be key for both ceramic and polymeric membranes, both of which are important to the PIW sector. Advanced oxidation processes, particularly combined with catalysis or photocatalysis, are also expected to see increased development to remove recalcitrant and toxic contaminants. Zero liquid discharge facilities have received growing attention and development, coupled with the need for increased water reuse. Such facilities are likely to continue growing, particularly in inland areas. However, such facilities suffer from high energy inputs. Reduction of energy through increased plant integration and development of solar-based technologies is likely, while the concept of zero liquid discharge may come under increased scrutiny with increasing LCA-based assessments.

Bioremediation is another area of significant future interest. The high organic content of most PIW streams provides opportunities for carbon capture and chemical energy capture, which can be most readily achieved through biological processes. Many bioprocesses have potential with PIW, such as polyhydroxyalkanoate production, microalgae treatment for biodiesel, biohydrogen production, etc. These processes can also work for nutrient and metal recovery. However, these processes usually are limited by current cost-effectiveness, often in part due to support services such as solids separation or gas purification. Anaerobic MBR is one technology that is rapidly maturing and highly suited to high strength nature of PIW from refineries where water reuse and energy recovery are targets.

Increased focus on metal recovery, even at low concentrations, is expected, due to both their toxicity and value. Ion exchange processes are likely to remain dominant for this purpose, with cheaper and more effective ion-exchange media and adsorbents being developed, particularly those that are biobased or waste-derived. With the growth of renewable energy sources, energy-demanding techniques such as electrodialysis may also become viable.

Historically environmental performance of wastewater systems has focused on the removal of highly toxic organic compounds common to PIW, as well as less significant concentrations of nutrients and pathogens. More recently, energy reduction and greenhouse gas emissions have received focus. With the emergence of life cycle analysis, a much broader approach to environmental performance is now expected, also considering offsite impacts from chemical and material production and disposal. With increased studies on various PIW treatment facilities, it is anticipated that these studies will drive processes with reduced inputs and byproducts, in addition to resource recovery previously mentioned. Care is needed to focus on relevant impacts that are substantial from PIW treatment, through the use of normalization. These analyses are likely to promote new treatment trains and processes in the future, along with industrial symbiosis.

Many powerful process modeling software exists, for both the general chemical industry and for water treatment. These tools which are typically used in process system design can aid significantly in systems optimization to reduce energy and chemical inputs. To better leverage these tools to understand overall process sustainability, they can be integrated with life cycle assessment tools to evaluate various system design points and configurations from a holistic environmental perspective. This is an emerging area and some relevant examples exist within the municipal sector (Bisinella de Faria, Spérandio, Ahmadi, & Tiruta-Barna, 2015). Such methods are not yet reported for PIW treatment and this is an area of significant opportunity for further environmental improvement.

Conclusions

PIW contains toxic pollutants, which must be removed before entering into the effluent stream to safeguard human and environmental health. Several techniques have been developed for the treatment of PIW. For large-scale application, every treatment technology should be evaluated in the context of sustainability, which is based on a triple bottom line concept involving environmental, social, and, economic competitiveness. It is emphasized that the sustainability of PIW treatment technologies can be established by integrating treatment with resource recovery. Resource recovery in the form of energy, nutrients, metals, and water can provide an additional gain in the treatment process. Several physico-chemical and biological techniques have been discussed in this study. Among them, adsorption, membrane filtration, phytoremediation, and AnMBR hold the strong potential of resource recovery, with relatively less environmental and economic burden. However, each suffers from certain drawbacks that detract from their overall sustainability. There is therefore a need to evaluate technology integration holistically and in-depth, based on context-specific factors, to provide a good balance between resource recovery, treatment efficiency, cost-effectiveness, and environmental emissions to ensure the sustainability of PIW treatment.

Acknowledgments

The authors would like to thank Hamad Bin Khalifa University (HBKU), under Qatar Foundation (QF), for supporting the research. Any opinions, findings, and conclusions, or recommendations expressed in this material are those of the author(s) and do not necessarily reflect the views of HBKU or QF.

References

Abbasi, S. A., & Abbasi, N. (2000). The likely adverse environmental impacts of renewable energy sources. *Applied Energy*, *65*(1−4), 121−144. Available from https://doi.org/10.1016/S0306-2619(99)00077-X.

Agarry, S. E., Oghenejoboh, K. M., Latinwo, G. K., & Owabor, C. N. (2020). Biotreatment of petroleum refinery wastewater in vertical surface-flow constructed wetland vegetated with *Eichhornia crassipes*: Lab-scale experimental and kinetic modelling. *Environmental Technology (United Kingdom)*, *41*(14), 1793−1813. Available from https://doi.org/10.1080/09593330.2018.1549106.

Al Hawli, B., Benamor, A., & Hawari, A. A. (2019). A hybrid electro-coagulation/forward osmosis system for treatment of produced water. *Chemical Engineering and Processing - Process Intensification*, 143. Available from https://doi.org/10.1016/j.cep.2019.107621.

Alcántara-Garduño, M. E., Okuda, T., Nishijima, W., & Okada, M. (2008). Ozonation of trichloroethylene in acetic acid solution with soluble and solid humic acid. *Journal of Hazardous Materials*, 160(2−3), 662−667. Available from https://doi.org/10.1016/j.jhazmat.2008.03.106.

Ali, M., Elreedy, A., Ibrahim, M. G., Fujii, M., & Tawfik, A. (2019). Hydrogen and methane bio-production and microbial community dynamics in a multiphase anaerobic reactor treating saline industrial wastewater. *Energy Conversion and Management*, 186, 1−14. Available from https://doi.org/10.1016/j.enconman.2019.02.060.

Al-Kaabi, A., Al-Sulaiti, H., Al-Ansari, T., & Mackey, H. R. (2021). Assessment of water quality variations on pretreatment and environmental impacts of SWRO desalination. *Desalination*, 500, 114831. Available from https://doi.org/10.1016/j.desal.2020.114831.

Al-Kaabi, Abdulrahman, H, & Mackey, Hamish, R (2019). Environmental assessment of intake alternatives for seawater reverse osmosis in the Arabian Gulf. *Journal of Environmental Management*, 242, 22−30. Available from https://doi.org/10.1016/j.jenvman.2019.04.051.

Alwared, A., & Faraj, N. (2015). Coagulation − Flotation process for removing oil from wastewater using sawdust + bentonite. *Journal of Engineering*, 21, 62−76.

Alyasi, H., Mackey, H. R., Loganathan, K., & McKay, G. (2020). Adsorbent minimisation in a two-stage batch adsorber for cadmium removal. *Journal of Industrial and Engineering Chemistry*, 81, 153−160. Available from https://doi.org/10.1016/j.jiec.2019.09.003.

Argoti, A., Orjuela, A., & Narváez, P. C. (2019). Challenges and opportunities in assessing sustainability during chemical process design. *Current Opinion in Chemical Engineering*, 26, 96−103. Available from https://doi.org/10.1016/j.coche.2019.09.003.

Arthur, J., Bruce, P., Langhus, G., & Patel, C. (2005). *Technical summary of oil & gas produced water treatment technologies*. ALL Consulting, LLC.

Bader, M. S. H. (2007). Seawater vs produced water in oil-fields water injection operations. *Desalination*, 208(1−3), 159−168. Available from https://doi.org/10.1016/j.desal.2006.05.024.

Bilstad, T., & Espedal, E. (1996). Membrane separation of produced water. *Water Science and Technology*, 34(9), 239−246. Available from https://doi.org/10.1016/S0273-1223(96)00810-4.

Bisinella de Faria, A. B., Spérandio, M., Ahmadi, A., & Tiruta-Barna, L. (2015). Evaluation of new alternatives in wastewater treatment plants based on dynamic modelling and life cycle assessment (DM-LCA). *Water Research*, 84, 99−111. Available from https://doi.org/10.1016/j.watres.2015.06.048.

Boonsaner, M., Borrirukwisitsak, S., & Boonsaner, A. (2011). Phytoremediation of BTEX contaminated soil by Canna × generalis. *Ecotoxicology and Environmental Safety*, 74(6), 1700−1707. Available from https://doi.org/10.1016/j.ecoenv.2011.04.011.

Breuer, R., Headley, T.R., Thaker, Y. I., & Al Sharji, B. (2012). The first year operation of the Nimr water treatment plant in Oman-sustainable produced water management using wetlands. In *Society of Petroleum Engineering Journal. International conference on health, safety and environment in oil and gas exploration and production*, Perth, Australia, September 11−13. https://doi.org/10.2118/156427-MS.

Brian, B., & Zongli, X. (2019). The use of polymers in the flotation treatment of wastewater. *Processes*, 7(6), 374. Available from https://doi.org/10.3390/pr7060374.

Buyan, C., Shah, S., & JaeSeong, R. (2014). Bioaugmented phytoremediation: A strategy for reclamation of diesel oil-contaminated soils. *International Journal of Agriculture and Biology*, 16, 624−628.

Byrne, D. M., Lohman, H. A. C., Cook, S. M., Peters, G. M., & Guest, J. S. (2017). Life cycle assessment (LCA) of urban water infrastructure: Emerging approaches to balance objectives and inform comprehensive decision-making. *Environmental Science: Water Research and Technology*, 3(6), 1002−1014. Available from https://doi.org/10.1039/c7ew00175d.

Cai, Z., Zhou, Q., Peng, S., & Li, K. (2010). Promoted biodegradation and microbiological effects of petroleum hydrocarbons by *Impatiens balsamina* L. with strong endurance. *Journal of Hazardous Materials*, 183(1−3), 731−737. Available from https://doi.org/10.1016/j.jhazmat.2010.07.087.

Çakmakce, M., Kayaalp, N., & Koyuncu, I. (2008). Desalination of produced water from oil production fields by membrane processes. *Desalination*, 222(1−3), 176−186. Available from https://doi.org/10.1016/j.desal.2007.01.147.

Chen, Q., Wu, W., Qi, D., Ding, Y., & Zhao, Z. (2020). Review on microaeration-based anaerobic digestion: State of the art, challenges, and prospectives. *Science of the Total Environment*, 710. Available from https://doi.org/10.1016/j.scitotenv.2019.136388.

Chen, W., Lu, S., Jiao, W., Wang, M., & Chang, A. C. (2013). Reclaimed water: A safe irrigation water source? *Environmental Development*, 8(1), 74−83. Available from https://doi.org/10.1016/j.envdev.2013.04.003.

Chiha, M., Hamdaoui, O., Ahmedchekkat, F., & Pétrier, C. (2010). Study on ultrasonically assisted emulsification and recovery of copper(II) from wastewater using an emulsion liquid membrane process. *Ultrasonics Sonochemistry*, 17(2), 318−325. Available from https://doi.org/10.1016/j.ultsonch.2009.09.001.

Claudia, C., & Vassilis, I. (2015). Elsevier BV, pp. 425−498. https://doi.org/10.1016/b978-0-12-384746-1.00010-0.

Corominas, L., Byrne, D. M., Guest, J. S., Hospido, A., Roux, P., Shaw, A., & Short, M. D. (2020). The application of life cycle assessment (LCA) to wastewater treatment: A best practice guide and critical review. *Water Research*, 184. Available from https://doi.org/10.1016/j.watres.2020.116058.

Covino, S., D'Annibale, A., Stazi, S. R., Cajthaml, T., Čvančarová, M., Stella, T., & Petruccioli, M. (2015). Assessment of degradation potential of aliphatic hydrocarbons by autochthonous filamentous fungi from a historically polluted clay soil. *Science of the Total Environment*, 505, 545−554. Available from https://doi.org/10.1016/j.scitotenv.2014.10.027.

Crini, G., & Lichtfouse, E. (2019). Advantages and disadvantages of techniques used for wastewater treatment. *Environmental Chemistry Letters*, 17(1), 145−155. Available from https://doi.org/10.1007/s10311-018-0785-9.

Crook, J., & Mousavi, A. (2016). The chlor-alkali process: A review of history and pollution. *Environmental Forensics*, 17(3), 211−217. Available from https://doi.org/10.1080/15275922.2016.1177755.

Dąbrowski, A., Podkościelny, P., Hubicki, Z., & Barczak, M. (2005). Adsorption of phenolic compounds by activated carbon—A critical review. *Chemosphere*, 1049–1070. Available from https://doi.org/10.1016/j.chemosphere.2004.09.067.

De Brito, J. F., Bessegato, G. G., De Toledo, E., Souza, P. R. F., Viana, T. S., De Oliveira, D. P., . . . Zanoni, M. V. B. (2019). Combination of photo-electrocatalysis and ozonation as a good strategy for organics oxidation and decreased toxicity in oil-produced water. *Journal of the Electrochemical Society*, *166*(5), H3231–H3238. Available from https://doi.org/10.1149/2.0331905jes.

De Gisi, S., Lofrano, G., Grassi, M., & Notarnicola, M. (2016). Characteristics and adsorption capacities of low-cost sorbents for wastewater treatment: A review. *Sustainable Materials and Technologies*, *9*, 10–40. Available from https://doi.org/10.1016/j.susmat.2016.06.002.

Derwent, R. G., Jenkin, M. E., & Saunders, S. M. (1996). Photochemical ozone creation potentials for a large number of reactive hydrocarbons under European conditions. *Atmospheric Environment*, *30*(2), 181–199. Available from https://doi.org/10.1016/1352-2310(95)00303-G.

Diraki, A., Mackey, H. R., Mckay, G., & Abdala, A. (2019). Removal of emulsified and dissolved diesel oil from high salinity wastewater by adsorption onto graphene oxide. *Journal of Environmental Chemical Engineering*, *7*(3). Available from https://doi.org/10.1016/j.jece.2019.103106.

Diya'Uddeen, B. H., Daud, W. M. A. W., & Abdul Aziz, A. R. (2011). Treatment technologies for petroleum refinery effluents: A review. *Process Safety and Environmental Protection*, *89*(2), 95–105. Available from https://doi.org/10.1016/j.psep.2010.11.003.

El-Naas, M. H., Al-Zuhair, S., Al-Lobaney, A., & Makhlouf, S. (2009). Assessment of electrocoagulation for the treatment of petroleum refinery wastewater. *Journal of Environmental Management*, *91*(1), 180–185. Available from https://doi.org/10.1016/j.jenvman.2009.08.003.

Epps, A. (2006). Phytoremediation of petroleum hydrocarbons. *Environmental Protection Agency*.

Esmaeili, A., & Saremnia, B. (2018). Comparison study of adsorption and nanofiltration methods for removal of total petroleum hydrocarbons from oil-field wastewater. *Journal of Petroleum Science and Engineering*, *171*, 403–413. Available from https://doi.org/10.1016/j.petrol.2018.07.058.

Fakhru'l-Razi, A., Pendashteh, A., Abdullah, L. C., Biak, D. R. A., Madaeni, S. S., & Abidin, Z. Z. (2009). Review of technologies for oil and gas produced water treatment. *Journal of Hazardous Materials*, *170*(2–3), 530–551. Available from https://doi.org/10.1016/j.jhazmat.2009.05.044.

Figoli, A., & Criscuoli, A. (2017). *Sustainable membrane technology for water and wastewater treatment*.

Fu, L., Wu, C., Zhou, Y., Zuo, J., & Song, G. (2019). Effects of residual ozone on the performance of microorganisms treating petrochemical wastewater. *Environmental Science and Pollution Research*, *26*(26), 27505–27515. Available from https://doi.org/10.1007/s11356-019-05956-8.

Fu, L., Wu, C., Zhou, Y., Zuo, J., Song, G., & Tan, Y. (2019). Ozonation reactivity characteristics of dissolved organic matter in secondary petrochemical wastewater by single ozone, ozone/H_2O_2, and ozone/catalyst. *Chemosphere*, *233*, 34–43. Available from https://doi.org/10.1016/j.chemosphere.2019.05.207.

Georgios, S., Dimitrios, T., Eleni, T., Alexandros, K., & Elisavet, A. (2018). Innovative approach on aerobic activated sludge process towards more sustainable wastewater treatment. *Proceedings*, 645. Available from https://doi.org/10.3390/proceedings2110645.

Hasan, R., Chong, C. C., & Setiabudi, H. D. (2019). Synthesis of KCC-1 using rice husk ash for Pb removal from aqueous solution and petrochemical wastewater. *Bulletin of Chemical Reaction Engineering & Catalysis*, *14*(1), 196–204. Available from https://doi.org/10.9767/bcrec.14.1.3619.196-204.

Hashemi, F., Hashemi, H., Shahbazi, M., Dehghani, M., Hoseini, M., & Shafeie, A. (2020). Reclamation of real oil refinery effluent as makeup water in cooling towers using ultrafiltration, ion exchange and multioxidant disinfectant. *Water Resources and Industry*, 23. Available from https://doi.org/10.1016/j.wri.2019.100123.

He, Y., & Jiang, Z. W. (2008). Technology review: Treating oilfield wastewater. *Filtration and Separation*, *45*(5), 14–16. Available from https://doi.org/10.1016/S0015-1882(08)70174-5.

Heberger, M. (2011). Australia's millennium drought: Impacts and responses. In Gleick, P. H. (ed.), *The world's water: The biennial report on freshwater resources. in Australia's millennium drought: Impacts and responses* (pp. 97–125). Island Press/Center for Resource Economics, Washington, DC.

Henkanatte-Gedera, S. M., Selvaratnam, T., Karbakhshravari, M., Myint, M., Nirmalakhandan, N., Van Voorhies, W., & Lammers, P. J. (2017). Removal of dissolved organic carbon and nutrients from urban wastewaters by *Galdieria sulphuraria*: Laboratory to field scale demonstration. *Algal Research*, *24*, 450–456. Available from https://doi.org/10.1016/j.algal.2016.08.001.

Hou, B., Xie, S. X., Chen, M., Jin, Y., Hao, D., & Wang, R. S. (2013). The treatment of refinery heavy oil sludge. *Petroleum Science and Technology*, *31*(5), 458–464. Available from https://doi.org/10.1080/10916466.2012.708083.

Hu, G., Li, J., & Hou, H. (2015). A combination of solvent extraction and freeze thaw for oil recovery from petroleum refinery wastewater treatment pond sludge. *Journal of Hazardous Materials*, *283*, 832–840. Available from https://doi.org/10.1016/j.jhazmat.2014.10.028.

Hui, K., Tang, J., Lu, H., Xi, B., Qu, C., & Li, J. (2020). Status and prospect of oil recovery from oily sludge: A review. *Arabian Journal of Chemistry*, *13*(8), 6523–6543. Available from https://doi.org/10.1016/j.arabjc.2020.06.009.

Hurlimann, A., & Dolnicar, S. (2010). When public opposition defeats alternative water projects – The case of Toowoomba Australia. *Water Research*, *44*(1), 287–297. Available from https://doi.org/10.1016/j.watres.2009.09.020.

Iddya, A., Hou, D., Khor, C. M., Ren, Z., Tester, J., Posmanik, R., . . . Jassby, D. (2020). Efficient ammonia recovery from wastewater using electrically conducting gas stripping membranes. *Environmental Science: Nano*, *7*(6), 1759–1771. Available from https://doi.org/10.1039/c9en01303b.

Jacob-Lopes, E., & Franco, T. T. (2013). From oil refinery to microalgal biorefinery. *Journal of CO_2 Utilization*, *2*, 1–7. Available from https://doi.org/10.1016/j.jcou.2013.06.001.

Jain, M., Majumder, A., Ghosal, P. S., & Gupta, A. K. (2020). A review on treatment of petroleum refinery and petrochemical plant wastewater: A special emphasis on constructed wetlands. *Journal of Environmental Management*, 272. Available from https://doi.org/10.1016/j.jenvman.2020.111057.

Jasim, S. Y., Saththasivam, J., Loganathan, K., Ogunbiyi, O. O., & Sarp, S. (2016). Reuse of treated sewage effluent (TSE) in Qatar. *Journal of Water Process Engineering*, *11*, 174–182. Available from https://doi.org/10.1016/j.jwpe.2016.05.003.

Johnson, E. P., & Oliver, M. E. (2016). Renewable energy and wholesale electricity price variability. *International Association for Energy Economics: Bergen*, 25−26.

Jones, D. M., Head, I. M., Gray, N. D., Adams, J. J., Rowan, A. K., Aitken, C. M., ... Larter, S. R. (2008). Crude-oil biodegradation via methanogenesis in subsurface petroleum reservoirs. *Nature*, 176−180. Available from https://doi.org/10.1038/nature06484.

Kazemi, A., Bahramifar, N., Heydari, A., & Olsen, S. I. (2019). Synthesis and sustainable assessment of thiol-functionalization of magnetic graphene oxide and superparamagnetic Fe_3O_4@SiO_2 for Hg(II) removal from aqueous solution and petrochemical wastewater. *Journal of the Taiwan Institute of Chemical Engineers*, 95, 78−93. Available from https://doi.org/10.1016/j.jtice.2018.10.002.

Khan, A. G. (2006). Mycorrhizoremediation−an enhanced form of phytoremediation. *Journal of Zhejiang University. Science. B.*, 7(7), 503−514. Available from https://doi.org/10.1631/jzus.2006.B0503.

Kizilet, A., Veral, M., & Ćemanović, A. (2017). *The use of membrane processes to promote sustainable environmental protection practices* (Vol. 2, pp. 17−22).

Kong, Z., Li, L., Xue, Y., Yang, M., & Li, Y. Y. (2019). Challenges and prospects for the anaerobic treatment of chemical-industrial organic wastewater: A review. *Journal of Cleaner Production*, 231, 913−927. Available from https://doi.org/10.1016/j.jclepro.2019.05.233.

Ku Ishak, K. E. H., & Abdalla Ayoub, M. (2019). Removal of oil from polymer-produced water by using flotation process and statistical modelling. *Journal of Petroleum Exploration and Production Technology*, 9(4), 2927−2932. Available from https://doi.org/10.1007/s13202-019-0686-x.

Kundu, P., & Mishra, I. M. (2019). Treatment and reclamation of hydrocarbon-bearing oily wastewater as a hazardous pollutant by different processes and technologies: A state-of-the-art review. *Reviews in Chemical Engineering*, 35(1), 73−108. Available from https://doi.org/10.1515/revce-2017-0025.

Kuntke, P., Geleji, M., Bruning, H., Zeeman, G., Hamelers, H. V. M., & Buisman, C. J. N. (2011). Effects of ammonium concentration and charge exchange on ammonium recovery from high strength wastewater using a microbial fuel cell. *Bioresource Technology*, 102(6), 4376−4382. Available from https://doi.org/10.1016/j.biortech.2010.12.085.

Kyzas, G. Z., & Kostoglou, M. (2014). Green adsorbents for wastewaters: A critical review. *Materials*, 7(1), 333−364. Available from https://doi.org/10.3390/ma7010333.

Larrey-Lassalle, P., Catel, L., Roux, P., Rosenbaum, R. K., Lopez-Ferber, M., Junqua, G., & Loiseau, E. (2017). An innovative implementation of LCA within the EIA procedure: Lessons learned from two wastewater Treatment Plant case studies. *Environmental Impact Assessment Review*, 63, 95−106. Available from https://doi.org/10.1016/j.eiar.2016.12.004.

Li, H., Hao, H., Yang, X., Xiang, L., Zhao, F., Jiang, H., & He, Z. (2012). Purification of refinery wastewater by different perennial grasses growing in a floating bed. *Journal of Plant Nutrition*, 35(1), 93−110. Available from https://doi.org/10.1080/01904167.2012.631670.

Li, Q., Han, J., & Elgowainy, A. (2016). *Industrial wastewater treatment in GREET® model: Energy intensity, water loss, direct greenhouse gas emissions, and biogas generation potential.*

Li, W. W., Yu, H. Q., & He, Z. (2014). Towards sustainable wastewater treatment by using microbial fuel cells-centered technologies. *Energy and Environmental Science*, 7(3), 911−924. Available from https://doi.org/10.1039/c3ee43106a.

Li, Y. S., Yan, L., Xiang, C. B., & Hong, L. J. (2006). Treatment of oily wastewater by organic-inorganic composite tubular ultrafiltration (UF) membranes. *Desalination*, 196(1−3), 76−83. Available from https://doi.org/10.1016/j.desal.2005.11.021.

Liang, Y., Ning, Y., Liao, L., & Yuan, B. (2018). *Special focus on produced water in oil and gas fields: Origin, management, and reinjection practice. Formation damage during improved oil recovery: Fundamentals and applications* (pp. 515−586). Elsevier. Available from https://doi.org/10.1016/B978-0-12-813782-6.00014-2.

Liebsch, T. (2021). *Environmental impact assessment (EIA) − How is it different from LCA?* Ecochain. https://ecochain.com/knowledge/environmental-impact-assessment-eia-how-is-it-different-from-lca/.

Liu, Y., Lu, S., Yan, X., Gao, S., Cui, X., & Cui, Z. (2020). Life cycle assessment of petroleum refining process: A case study in China. *Journal of Cleaner Production*, 256. Available from https://doi.org/10.1016/j.jclepro.2020.120422.

Lu, Y., Liu, J., Lu, B., Jiang, A., & Wan, C. (2010). Study on the removal of indoor VOCs using biotechnology. *Journal of Hazardous Materials*, 182(1−3), 204−209. Available from https://doi.org/10.1016/j.jhazmat.2010.06.016.

Lutzu, G. A., & Dunford, N. T. (2019). Algal treatment of wastewater generated during oil and gas production using hydraulic fracturing technology. *Environmental Technology (United Kingdom)*, 40(8), 1027−1034. Available from https://doi.org/10.1080/09593330.2017.1415983.

Mannina, G., Presti, D., Montiel-Jarillo, G., & Suárez-Ojeda, M. E. (2019). Bioplastic recovery from wastewater: A new protocol for polyhydroxyalkanoates (PHA) extraction from mixed microbial cultures. *Bioresource Technology*, 282, 361−369. Available from https://doi.org/10.1016/j.biortech.2019.03.037.

Manouchehr, N., & Mishana, J. (2008). Reuse of refinery treated wastewater in cooling towers. *Iranian Journal of Chemistry and Chemical Engineering*, 27(4), 1−7.

McCutcheon, S., & Schnoor, J. (2004). *Overview of phytotransformation and control of wastes* (pp. 1−58).

Min, X., Changyong, W., & Yuexi, Z. (2020). Advancements in the Fenton process for wastewater treatment. *IntechOpen*. Available from https://doi.org/10.5772/intechopen.90256.

Mitsch, W. J., Bernal, B., Nahlik, A. M., Mander, U., Zhang, L., Anderson, C. J., ... Brix, H. (2013). Wetlands, carbon, and climate change. *Landscape Ecology*, 28(4), 583−597. Available from https://doi.org/10.1007/s10980-012-9758-8.

Mohammadi, L., Rahdar, A., Bazrafshan, E., Dahmardeh, H., Susan, M. A. B. H., & Kyzas, G. Z. (2020). Petroleum hydrocarbon removal from wastewaters: A review. *Processes*, 8(4). Available from https://doi.org/10.3390/PR8040447.

Mohammadnejad, S., Bidhendi, G. R. N., & Mehrdadi, N. (2011). Water pinch analysis in oil refinery using regeneration reuse and recycling consideration. *Desalination*, 265(1−3), 255−265. Available from https://doi.org/10.1016/j.desal.2010.07.059.

Mohseni, M., Abdollahi, A., & Siadat, S. H. (2019). Sustainable supply chain management practices in petrochemical industry using interpretive structural modeling. *International Journal of Information Systems and Supply Chain Management, 12*(1), 22−50. Available from https://doi.org/10.4018/IJISSCM.2019010102.

Morée, A. L., Beusen, A. H. W., Bouwman, A. F., & Willems, W. J. (2013). Exploring global nitrogen and phosphorus flows in urban wastes during the twentieth century. *Global Biogeochemical Cycles, 27*(3), 836−846. Available from https://doi.org/10.1002/gbc.20072.

Moreno, C., Farahbakhshazad, N., & Morrison, G. M. (2002). Ammonia removal from oil refinery effluent in vertical up flow macrophyte column systems. *Water, Air, and Soil Pollution, 135*(1−4), 237−247. Available from https://doi.org/10.1023/A:1014753817216.

Morero, B., Rodriguez, M. B., & Campanella, E. A. (2015). Environmental impact assessment as a complement of life cycle assessment. Case study: Upgrading of biogas. *Bioresource Technology, 190*, 402−407. Available from https://doi.org/10.1016/j.biortech.2015.04.091.

Motlagh, A. M., & Goel, R. K. (2014). *Sustainability of activated sludge processes. Water reclamation and sustainability* (pp. 391−414). Elsevier Inc. Available from https://doi.org/10.1016/B978-0-12-411645-0.00016-X.

Mujtaba, G., Rizwan, M., Kim, G., & Lee, K. (2018). Removal of nutrients and COD through coculturing activated sludge and immobilized *Chlorella vulgaris*. *Chemical Engineering Journal, 343*, 155−162. Available from https://doi.org/10.1016/j.cej.2018.03.007.

Muñoz, I., Aktürk, A. S., Ayyıldız, Ö., Çağlar, Ö., Meabe, E., Contreras, S., . . . Jiménez-Banzo, A. (2020). Life cycle assessment of wastewater reclamation in a petroleum refinery in Turkey. *Journal of Cleaner Production, 268*. Available from https://doi.org/10.1016/j.jclepro.2020.121967.

Mustapha, H. (2018). *Treatment of petroleum refinery wastewater with constructed wetlands*.

Nasiri, M., Jafari, I., & Parniankhoy, B. (2017). Oil and gas produced water management: A review of treatment technologies, challenges, and opportunities. *Chemical Engineering Communications, 204*(8), 990−1005. Available from https://doi.org/10.1080/00986445.2017.1330747.

Nealson, K. H. (2017). Bioelectricity (electromicrobiology) and sustainability. *Microbial Biotechnology, 10*(5), 1114−1119. Available from https://doi.org/10.1111/1751-7915.12834.

NIMR, O. (2019). *Expansion for Oman's flagship industrial constructed wetland*. International Water Association. Available from https://www.thesourcemagazine.org/expansion-for-omans-flagship-industrial-constructed-wetland/.

NIMR, O. (2020). *International ecological engineering society*. https://iees.ch/lighthouse-projects/nimr-water-treatment-plant/.

Ossai, I. C., Ahmed, A., Hassan, A., & Hamid, F. S. (2020). Remediation of soil and water contaminated with petroleum hydrocarbon: A review. *Environmental Technology and Innovation, 17*. Available from https://doi.org/10.1016/j.eti.2019.100526.

Padaki, M., Surya Murali, R., Abdullah, M. S., Misdan, N., Moslehyani, A., Kassim, M. A., . . . Ismail, A. F. (2015). Membrane technology enhancement in oil-water separation. A review. *Desalination, 357*, 197−207. Available from https://doi.org/10.1016/j.desal.2014.11.023.

Painmanakul, P., Sastaravet, P., Lersjintanakarn, S., & Khaodhiar, S. (2010). Effect of bubble hydrodynamic and chemical dosage on treatment of oily wastewater by induced air flotation (IAF) process. *Chemical Engineering Research and Design, 88*(5−6), 693−702. Available from https://doi.org/10.1016/j.cherd.2009.10.009.

Paolini, V., Petracchini, F., Segreto, M., Tomassetti, L., Naja, N., & Cecinato, A. (2018). Environmental impact of biogas: A short review of current knowledge. *Journal of Environmental Science and Health - Part A Toxic/Hazardous Substances and Environmental Engineering, 53*(10), 899−906. Available from https://doi.org/10.1080/10934529.2018.1459076.

Pearce, G. K. (2008). UF/MF pretreatment to RO in seawater and wastewater reuse applications: A comparison of energy costs. *Desalination, 222*(1−3), 66−73. Available from https://doi.org/10.1016/j.desal.2007.05.029.

Peng, B., Yao, Z., Wang, X., Crombeen, M., Sweeney, D. G., & Tam, K. C. (2020). Cellulose-based materials in wastewater treatment of petroleum industry. *Green Energy and Environment, 5*(1), 37−49. Available from https://doi.org/10.1016/j.gee.2019.09.003.

Petre, C. F., Piché, S., Normandin, A., & Larachi, F. (2007). Advances in chemical oxidation of total reduced sulfur from Kraft mills atmospheric effluents. *International Journal of Chemical Reactor Engineering*. Available from https://doi.org/10.2202/1542-6580.1574.

Pichtel, J. (2016). Oil and gas production wastewater: Soil contamination and pollution prevention. *Applied and Environmental Soil Science, 2016*. Available from https://doi.org/10.1155/2016/2707989.

Pramanik, B. K., Fatihah, S., Shahrom, Z., & Ahmed, E. (2012). Biological aerated filters (BAFs) for carbon and nitrogen removal: A review. *Journal of Engineering Science and Technology, 7*(4), 428−446. Available from http://jestec.taylors.edu.my/Vol%207%20Issue%204%20June%2012/Vol_7_4_428-446_%20PRAMANIK%20BIPLOB.pdf.

Riham, S., H., E.-N. M., M., V. L. M. C., Abdelbaki, B., Fatima, A.-N., & Udeogu, O. (2020). Biotechnology for gas-to-liquid (GTL) wastewater treatment: A review. *Water, 2126*. Available from https://doi.org/10.3390/w12082126.

Rintamäki, T., Siddiqui, A. S., & Salo, A. (2017). Does renewable energy generation decrease the volatility of electricity prices? An analysis of Denmark and Germany. *Energy Economics, 62*, 270−282. Available from https://doi.org/10.1016/j.eneco.2016.12.019.

Roy, S., & Ragunath, S. (2018). Emerging membrane technologies for water and energy sustainability: Future prospects, constraints and challenges. *Energies, 11*(11). Available from https://doi.org/10.3390/en11112997.

Salahi, A., Gheshlaghi, A., Mohammadi, T., & Madaeni, S. S. (2010). Experimental performance evaluation of polymeric membranes for treatment of an industrial oily wastewater. *Desalination, 262*(1−3), 235−242. Available from https://doi.org/10.1016/j.desal.2010.06.021.

Sali, S., Mackey, H. R., & Abdala, A. A. (2019). Effect of graphene oxide synthesis method on properties and performance of polysulfone-graphene oxide mixed matrix membranes. *Nanomaterials, 9*(5). Available from https://doi.org/10.3390/nano9050769.

Santiago Santos, G., de, O., de Salles Pupo, M. M., Vasconcelos, V. M., Barrios Eguiluz, K. I., & Salazar Banda, G. R. (2018). *Electrochemical water and wastewater treatment* (pp. 77−118). Butterworth-Heinemann. Available from https://doi.org/10.1016/B978-0-12-813160-2.00004-3.

Sengupta, S., Nawaz, T., & Beaudry, J. (2015). Nitrogen and phosphorus recovery from wastewater. *Current Pollution Reports, 1*(3), 155−166. Available from https://doi.org/10.1007/s40726-015-0013-1.

Shi, R., & Guest, J. S. (2020). BioSTEAM-LCA: An integrated modeling framework for agile life cycle assessment of biorefineries under uncertainty. *ACS Sustainable Chemistry and Engineering, 8*(51), 18903–18914. Available from https://doi.org/10.1021/acssuschemeng.0c05998.

Singh, P., & Borthakur, A. (2018). A review on biodegradation and photocatalytic degradation of organic pollutants: A bibliometric and comparative analysis. *Journal of Cleaner Production, 196*, 1669–1680. Available from https://doi.org/10.1016/j.jclepro.2018.05.289.

Singh, P., Jain, R., Srivastava, N., Borthakur, A., Pal, D. B., Singh, R., ... Mishra, P. K. (2017). Current and emerging trends in bioremediation of petrochemical waste: A review. *Critical Reviews in Environmental Science and Technology, 47*(3), 155–201. Available from https://doi.org/10.1080/10643389.2017.1318616.

Singh, S. (2019). *Treatment and recycling of wastewater from oil refinery/petroleum industry. Advances in biological treatment of industrial wastewater and their recycling for a sustainable future* (pp. 303–332). Springer.

Slaper, T. (2011). The triple bottom line: What is it and how does it work? Indiana business review. *Indiana Business Review, 86*(1), 4–8.

Smith, A. L., Stadler, L. B., Love, N. G., Skerlos, S. J., & Raskin, L. (2012). Perspectives on anaerobic membrane bioreactor treatment of domestic wastewater: A critical review. *Bioresource Technology, 122*, 149–159. Available from https://doi.org/10.1016/j.biortech.2012.04.055.

Stefanakis, A., Al-Hadrami, A., & Prigent, S. (2017). Reuse of oilfield produced water treated in a Constructed Wetland for saline irrigation under desert climate. In *7th international symposium on wetland pollutant dynamics and control (wetpol)*.

Sudarsan, J. S., Annadurai, R., Subramani, S., & George, R. B. (2016). Petrochemical wastewater treatment using constructed wetland technique. *Pollution Research, 35*(4), 727–732. Available from http://www.envirobiotechjournals.com/journal_details.php?jid = 4.

Syed, S. (2015). Approach of cost-effective adsorbents for oil removal from oily water. *Critical Reviews in Environmental Science and Technology*, 1916–1945. Available from https://doi.org/10.1080/10643389.2014.1001143.

Taiwo, E. A., & Otolorin, J. A. (2009). Oil recovery from petroleum sludge by solvent extraction. *Petroleum Science and Technology, 27*(8), 836–844. Available from https://doi.org/10.1080/10916460802455582.

Tian, X., Song, Y., Shen, Z., Zhou, Y., Wang, K., Jin, X., ... Liu, T. (2020). A comprehensive review on toxic petrochemical wastewater pretreatment and advanced treatment. *Journal of Cleaner Production, 245*. Available from https://doi.org/10.1016/j.jclepro.2019.118692.

Torp, A. (2014). *Life cycle assessment of wastewater treatment for oil and gas operations*.

Tsai, T. T., Sah, J., & Kao, C. M. (2010). Application of iron electrode corrosion enhanced electrokinetic-Fenton oxidation to remediate diesel contaminated soils: A laboratory feasibility study. *Journal of Hydrology, 380*(1–2), 4–13. Available from https://doi.org/10.1016/j.jhydrol.2009.09.010.

Tsang, P.B., & Martin, C.J. (2004). Economic evaluation of treating oilfield produced water for potable use. In *SPE international thermal operations and heavy oil symposium proceedings* (pp. 305–320). Society of Petroleum Engineers. https://doi.org/10.2118/86948-ms.

Tukker, A. (2000). Life cycle assessment as a tool in environmental impact assessment. *Environmental Impact Assessment Review, 20*(4), 435–456. Available from https://doi.org/10.1016/S0195-9255(99)00045-1.

Uduman, N., Qi, Y., Danquah, M. K., Forde, G. M., & Hoadley, A. (2010). Dewatering of microalgal cultures: A major bottleneck to algae-based fuels. *Journal of Renewable and Sustainable Energy, 2*(1). Available from https://doi.org/10.1063/1.3294480.

USEPA. (2009). Technical support document for wastewater treatment: Proposed rule for mandatory reporting of greenhouse gases. *Climate change division office of atmospheric programs*.

US EPA. (2012). *Guidelines for water reuse*.

US EPA. (2019). *Detailed study of the petroleum refining category – 2019 report*. Pennsylvania Avenue.

Vaneeckhaute, C., Lebuf, V., Michels, E., Belia, E., Vanrolleghem, P. A., Tack, F. M. G., & Meers, E. (2017). Nutrient recovery from digestate: Systematic technology review and product classification. *Waste and Biomass Valorization, 8*(1), 21–40. Available from https://doi.org/10.1007/s12649-016-9642-x.

Varjani, S., Joshi, R., Srivastava, V. K., Ngo, H. H., & Guo, W. (2020). Treatment of wastewater from petroleum industry: Current practices and perspectives. *Environmental Science and Pollution Research, 27*(22), 27172–27180. Available from https://doi.org/10.1007/s11356-019-04725-x.

Varjani, S. J., Gnansounou, E., & Pandey, A. (2017). Comprehensive review on toxicity of persistent organic pollutants from petroleum refinery waste and their degradation by microorganisms. *Chemosphere, 188*, 280–291. Available from https://doi.org/10.1016/j.chemosphere.2017.09.005.

Vinardell, S., Astals, S., M., P., Cardete, M. A., Fernández, I., Mata-Alvarez, J., & Dosta, J. (2020). Advances in anaerobic membrane bioreactor technology for municipal wastewater treatment: A 2020 updated review. *Renewable and Sustainable Energy Reviews*, 109936. Available from https://doi.org/10.1016/j.rser.2020.109936.

Vlasopoulos, N., Memon, F. A., Butler, D., & Murphy, R. (2006). Life cycle assessment of wastewater treatment technologies treating petroleum process waters. *Science of the Total Environment, 367*(1), 58–70. Available from https://doi.org/10.1016/j.scitotenv.2006.03.007.

Wang, H., Fu, P., Li, J., Huang, Y., Zhao, Y., Jiang, L., ... Huang, C. (2018). Separation-and-recovery technology for organic waste liquid with a high concentration of inorganic particles. *Engineering, 4*(3), 406–415. Available from https://doi.org/10.1016/j.eng.2018.05.014.

Wei, X., Kazemi, M., Zhang, S., & Wolfe, F. A. (2020). Petrochemical wastewater and produced water: Treatment technology and resource recovery. *Water Environment Research, 92*(10), 1695–1700. Available from https://doi.org/10.1002/wer.1424.

William, Byers, Lindgren, Glen, Noling, Calvin, & Dennis, Peters (2003). *Chapter 5: Water use in industries of the future. Industrial Water Management: A Systems Approach* (Second, pp. 5-1–5-68). American Institute of Chemical Engineers.

Xiao, H., Wu, J., Wang, X., Wang, J., Mo, S., Fu, M., Chen, L., & Ye, D. (2018). Ozone-enhanced deep catalytic oxidation of toluene over a platinum-ceria-supported BEA zeolite catalyst. *Molecular Catalysis, 460*, 7–15. Available from https://doi.org/10.1016/j.mcat.2018.09.005.

Yalcinkaya, F., Boyraz, E., Maryska, J., & Kucerova, K. (2020). A review on membrane technology and chemical surface modification for the oily wastewater treatment. *Materials, 13*(2). Available from https://doi.org/10.3390/ma13020493.

Yang, Q., Wu, Z., Liu, L., Zhang, F., & Liang, S. (2016). Treatment of oil wastewater and electricity generation by integrating constructed wetland with microbial fuel cell. *Materials*, *9*(11). Available from https://doi.org/10.3390/ma9110885.

Yavari, S., Malakahmad, A., & Sapari, N. B. (2016). Effects of production conditions on yield and physicochemical properties of biochars produced from rice husk and oil palm empty fruit bunches. *Environmental Science and Pollution Research*, *23*(18), 17928−17940. Available from https://doi.org/10.1007/s11356-016-6943-3.

Yu, L., Han, M., & He, F. (2017). A review of treating oily wastewater. *Arabian Journal of Chemistry*, *10*, S1913−S1922. Available from https://doi.org/10.1016/j.arabjc.2013.07.020.

Zeng, Y., Yang, C., Zhang, J., & Pu, W. (2007). Feasibility investigation of oily wastewater treatment by combination of zinc and PAM in coagulation/flocculation. *Journal of Hazardous Materials*, *147*(3), 991−996. Available from https://doi.org/10.1016/j.jhazmat.2007.01.129.

Zhao, C., Zhou, J., Yan, Y., Yang, L., Xing, G., Li, H., . . . Zheng, H. (2020). Application of coagulation/flocculation in oily wastewater treatment: A review. *Science of the Total Environment*. Available from https://doi.org/10.1016/j.scitotenv.2020.142795.

Zouboulis, A. I., & Avranas, A. (2000). Treatment of oil-in-water emulsions by coagulation and dissolved-air flotation. *Colloids and Surfaces A: Physicochemical and Engineering Aspects*, *172*(1−3), 153−161. Available from https://doi.org/10.1016/S0927-7757(00)00561-6.

Chapter 16

Circular economy in petroleum industries: implementing Water Closed Loop System

Mohammad-Hossein Sarrafzadeh

UNESCO Chair on Water Reuse, School of Chemical Engineering, College of Engineering, University of Tehran, Tehran, Iran

Industrial water and wastewater

Implementation of sustainable development goals (SDGs) in a country is bound to the development and sustainability of the industries so that all SDGs are directly or indirectly influenced by industrial development (Griggs et al., 2013). A wide and various set of economic and political indices including employment index, gross domestic product, environmental indices such as carbon emission and Human Development Index are completely dependent on the rotation of the wheels of industries and it is obvious that the human societies are highly sensitive to the changes in the industrial sector. During recent years, the environmental impacts of industries in terms of CO_2 emission and climate change have been highlighted and the attention are attracted more than before on evaluating the performance of industries in the integrated framework of sustainable development with a special look toward their destructive effects on the environment and ecosystem. One of the most important natural elements which are in direct linkage with industries is water.

Water is a crucial material in the productivity of any industrial activity where it can be consumed in a various range of activities from process applications to utility functions in cooling and energy generation (Bavar, Sarrafzadeh, Asgharnejad, & Norouzi-Firouz, 2018). About 20% of global freshwater demand is being consumed in the industrial sector which makes about 920 km^3 in 2019. However, there is a direct connection between the level of industrialization and development in a country with the industrial water consumption, so that about 55% of total water abstraction in Europe is being consumed in industrial sectors which are due to the number of active industries in this area (Flörke et al., 2013). According to the statistical analyses, it is predicted that water demand for the industry by 2050 will increase everywhere around the world. Water demand for the industry will increase by 800% in Africa, where present industry usage is negligible. Industrial water consumption in Asia will also increase by 250% by 2050 which means a significant growth in water abstraction from the natural resources within forthcoming decades (Boretti & Rosa, 2019).

Considering the fact that global water resources are strictly under pressure and the available freshwater for human activities has been reduced to its lowest levels in the last 50 years, the threat of shut-down due to lack of water overshadows industries more than any time. Among three water-consuming sectors (agricultural, industrial and municipal), in a water scarcity condition, industries are normally considered as the last priority by water supply organizations. Therefore, the industries need to find a sustainable solution for managing their water consumption.

The industrial water management pyramid (see Fig. 16.1) is the approach that has been proposed for water consumption management in industries (Mohammadnejad et al., 2012; Norling, Wood-Black, Masciangioli, & Roundtable, 2004; Ng, Foo, & Tan, 2009).

Physiochemical analyses (qualitative and quantitative analyses of water and greywater) and mathematical tools such as optimization, pinch and integration analysis, fuzzy logic, and resource allocation are the main tools that are essential for efficient application of reuse approach in an industrial unit.

The third priority in industrial water consumption management is "recycle" which is also known as "regeneration reuse" and "recovery" in different terminologies (Mohammadnejad, Bidhendi, & Mehrdadi, 2011). The development of this approach depends on the progress and developments in treatment technologies. Generally, recycling is based on the

Petroleum Industry Wastewater. DOI: https://doi.org/10.1016/B978-0-323-85884-7.00002-3

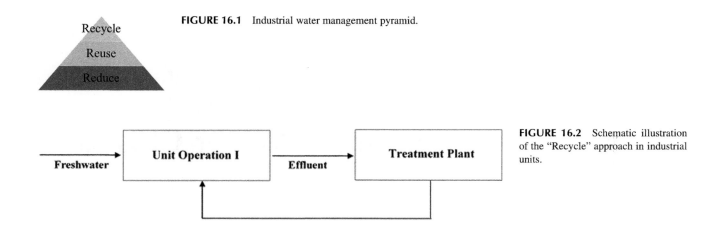

FIGURE 16.1 Industrial water management pyramid.

FIGURE 16.2 Schematic illustration of the "Recycle" approach in industrial units.

high-efficiency treatment of a waste stream (effluent) with the objective of maximum recovery of the materials with added value (water, raw materials, etc.) to be reused in the same unit operation or any other operations or applications (Zbontar & Glavic, 2000) (Fig. 16.2).

Advancements in physiochemical methods of wastewater treatment and resource recovery are the key step in the development of recycling approaches. Recycling not only decreases the need for fresh natural resources in the industrial systems but also minimizes the emission of contaminants that are being discharged to the environment in the form of effluent. Zero liquid discharge (ZLD) is the ideal case of the recycling approach in which the waste emission from the unit is zero and all of the waste streams are recovered within the system. ZLD is one of the fundaments of a new concept called "circular economy" which has attracted attention within the framework of SDGs and will be comprehensively investigated in this chapter.

Freshwater shortage especially during the recent decade has raised concern among the decision-makers about the sustainability of human societies regarding the water supply. Therefore, noticeable efforts have been carried out for finding alternative solutions which have led to the concept of unconventional water resources. Rainwater, seawater, saline water, and wastewater (used water) are among the unconventional water resources that have drawn attention to themselves as sustainable alternatives to conventional water resources (surface water, underground resources, etc.) in different consuming sectors (Smakhtin et al., 2001).

Among all unconventional water resources, wastewaters have found a special place in the industrial sector since it is always available and can be considered as a reliable resource in all seasons. Moreover, industrial wastewaters usually carry a significant amount of different high-value materials (e.g., a part of raw material, products, or byproducts of a manufacturing unit) which can be separated and recycled in the process or considered as a value-added product. In this approach which is known as "resource recovery," wastewater is considered as a "resource" instead of "waste" which is a fundamental approach in circular economy studies and will be thoroughly discussed in this chapter. The significance of wastewater as a collection of resources has caused rapid growth in the scientific efforts for developing technologies of wastewater management and treatment.

Despite the ongoing efforts for developing wastewater treatment technologies, more than 80% of all the globally-produced wastewaters are being discharged into the environment, mostly without any treatment which not only causes destructive health problems for humans and the ecosystem but also wastes an enormous opportunity for an economic enhancement (Association, 2018). The total global water and wastewater treatment market value is about US$425 billion which is divided into two big categories of utilities and technologies. About 42% of this market is related to efficient and sustainable technologies and services which makes it a very suitable platform for research and innovation (Royan, 2012). Developing efficient technologies of water and wastewater management remarkably helps to grow this market and decrease the amount of untreated wastewater discharge to the environment which subsequently leads to enhanced environmental protection, higher productivity of industrial sectors, and efficient implementation of SDGs.

Water and wastewater in oil and gas industries

The oil and gas sector is one of the largest industries in the world which is in charge of producing energy for different activities. Any change in the global oil and gas market can initiate revolutionary changes and it is not an exaggeration to claim it is the most significant industry in sustainability studies in a nation that has direct and indirect impacts on all SDGs. Processes in this industry are categorized into three main groups of upstream, midstream, and downstream (Colson, 1999). Upstream processes include all activities that are conducted for exploration and extraction of crude oil and gas from reservoirs, while midstream processes are the ones dealing with oil and gas transportation from wells to refineries by pipeline, rail, tankers, etc. Downstream processes contain the activities performed on crude oil and gas in a refinery to produce value-added products such as gasoline, jet fuel, diesel, LPG, and CNG (2013).

Water is widely consumed in downstream and upstream processes as a critical agent for cooling, washing, and direct use in the processes (see Fig. 16.3).

Water consumption in downstream processes largely depends on the products that are being produced in the refinery and numerous studies have been carried out to evaluate the specific water requirements in this category. Previous studies have shown that processing 1 gal of crude oil in a refinery in the US needs 1−1.9 gal of water while another study suggested 1.1 L of water consumption (including freshwater, reused and recycled water) per liter of processed crude oil and the freshwater of 0.53−1.38 L per liter of crude oil (Sun, Elgowainy, Wang, Han, & Henderson, 2018). Gasoline production normally is the biggest water-consuming unit of an oil refinery with consumption of 0.6−0.71 gal water gal^{-1} gasoline. Jet fuel production consumes the least amount of water with 0.09 gal water gal^{-1} fuel (Sun et al., 2018). Water consumption in gas processing refineries also varies in the range of 0−2 gal $MMBtu^{-1}$; in other words, about 0.268 L of water is needed for processing 1 m^3 of natural gas (Mielke, Anadon, & Narayanamurti, 2010). Global oil refinery throughput is about 83 million barrels per day since 2019, which means an average amount of 13 million m^3 of water is being globally consumed per day in oil refineries. This amount is about 4.4 million m^3 per day for gas processing in which 586 billion cubic feet of natural gas is being processed per day. A simple comparison reveals that the whole water that is being consumed in oil and gas refineries all over the world in one day equals the annual water consumption of 14,750 US citizens.

On the other hand, a noticeable amount of wastewater is being generated in these industries annually which is a potential unconventional water resource for nonprocess and utility functions including cooling and washing. It is estimated that an annual amount of 848 billion gallons of wastewater is being generated during conventional onshore oil and gas industries (Lewis, 2012). A significant portion of these wastewaters have resulted from cooling functions, that is, thermal contamination (high temperature) is the only challenge in reusing these waters and the other qualitative specifications are similar to the freshwater. Combination of these high-quality wastewaters with other waste streams that are highly polluted with COD, TDS, TSS, and nondecomposable compounds such as aromatics, olefins, and paraffins is not a rational approach, since wastes a valuable water resource with high reuse potential. Therefore, it is suggested to switch to an internal approach regarding wastewater categorization instead of a cumulative approach in which an industrial unit is considered as a black box with a single water inlet and wastewater outlet Fig. 16.4.

The internal approach provides a powerful tool for the categorization and evaluation of the interconnection of different sectors regarding water consumption and wastewater generation in an industrial unit. It remarkably facilitates holistic management of water consumption in the industrial unit and also provides the capability of monitoring the flow of contaminants and resources within the plant. This holistic look toward the water and resources network in an industrial unit has resulted in an integrated system called the WCLS which is the main focus of this chapter and is defined as an executive lever for the implementation of circular economy principles in industries.

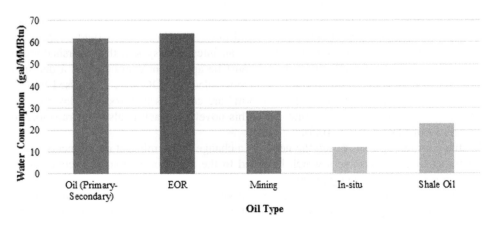

FIGURE 16.3 Water consumption in upstream phase of oil and gas industry.

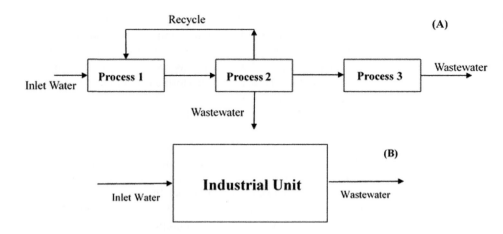

FIGURE 16.4 (A) Internal versus (B) cumulative look toward water and wastewater network in an industrial unit.

FIGURE 16.5 The linkage between technological progress and circular economy in the framework of 5R approach.

Principles of circular economy

The circular economy is an economic system in which the ultimate objective is eliminating wastes by reusing them instead of consuming natural resources. It is an alternative approach to the linear economy in which the lean is on extracting from natural resources for supplying the required raw materials (Kirchherr, Reike, & Hekkert, 2017). In this economic system, the waste materials are looked at as a resource, not an undesired leftover and the main objective is to maximize recovery of different materials with added values from the wastes. A circular economy has also a special place in the UN sustainable development program and SDG 12 comprehensively discusses this concept under the title of "sustainable production patterns" (Schroeder, Anggraeni, & Weber, 2019). However, implementation of the objectives of circular economy is directly bound to technological development and innovation in different fields, especially engineering and manufacturing. It is important to bear in mind that SDG is an integrated system that overshadows a process in all stages. In other words, the objective of minimizing wastes must be applied in every step of a decision from problem-definition to providing the solutions. This sort of integrated look ruled by SDGs has progressed the 3R approach into its evolutionary version of 5R in which "Refuse" and "Reform" are added as new Rs that play roles in the very early stages of problem-definition and building up the first idea. This novel approach is also generalizable to the concept of circular economy as a result of SDG 12 (Fig. 16.5).

SDG 12 and its focus on sustainable consumption has raised the need for a change in the mindset of engineers, manufacturers, decision, and policymakers as well as consumers which have led to the inclusion of two new concepts of refuse and reform in the integrated system of resource management.

Refuse and reform are the mechanisms that will be activated in the prefeasibility study (PFS) and feasibility study (FS) stages of an economic activity where the problem is being defined for the first time and its requirements, necessities, procedure, and consequences are being evaluated. As a matter of fact, refuse and reform are the responses to the "Why" question while conventional 3Rs answer to the "How" questions. These mechanisms are the early levels of decision-making according to the SDGs integrated system in which we ask three main questions:

1. Why this decision must be made?
2. What will be its essentials and consequences?
3. Is it worth?

The main scope of circular economy is deciding about the type, amount, and procedure of consuming resources and waste management. Water is a valuable resource in all industrial activities that can be the focus of circular economy to minimize freshwater requirement and wastewater generation through maximizing reuse and recycling. Water closed-loop system (WCLS) is the sustainable solution that is proposed by UNESCO Chair on Water Reuse based on the logic of circular economy with the primary goal of water-saving. It also results in the recovery of value-added by-products within the framework of resource recovery.

Water closed-loop system; theories and concept

Considering the ongoing critical situation of freshwater resources in both quantitative and qualitative aspects, the value of this substance has even got increased in different sections of human activities, especially industry and it is rational to consider water as a precious input in industrial units. Therefore, unlike previous decades, considering the costs related to water supply and wastewater treatment and discharge is inevitable in the economic analysis of industrial units. This is even more vital in water-consuming industries such as oil and gas in which the daily water consumption is as much as a small town.

Like other industrial inputs, water can also be managed based on the principle of the circular economy under the topic of ZLD. However, the circular economy is a more generalized concept than ZLD which not only focuses on 5R, but also evaluates the potential of resource recovery. WCLS is an integrated system that is proposed and developed by UNESCO Chair on Water Reuse for implementation of circular economy as a theoretical context for water and wastewater management in industrial units. The ultimate objective of this system is minimizing water consumption and wastewater generation in an industrial unit according to the topics of SDGs 6 and 12 using theoretical and technological tools such as pinch analysis and process integration methods, water and wastewater treatment technologies, and qualitative-quantitative analysis of the water and wastewater streams (Fig. 16.6).

The main difference of WCLS with previous methods of water consumption management is in three levels of problem definition, decision making, and practical solutions as Table 16.1.

The fundamental difference is in the problem definition stage where our approach toward the existing challenge is being determined. In WCLS, the focus is on water network synthesis within the whole units which shows the distribution of water all through the industrial unit. This stage is named as "water audit" in which the input-output relations

FIGURE 16.6 The concepts employed in water closed-loop system.

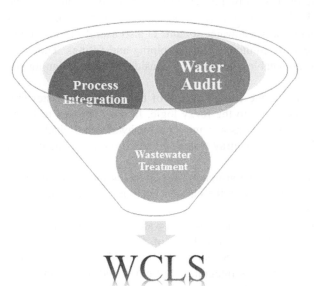

TABLE 16.1 Water closed-loop system versus previous methods of industrial water consumption and wastewater generation management.

Method	Problem definition	Decision making	Practical solutions
WCLS	Internal approach	Fuzzy logic (white-gray-black)	5Rs approach
Conventional methods	Cumulative approach	Absolute logic (freshwater–wastewater)	3Rs approach

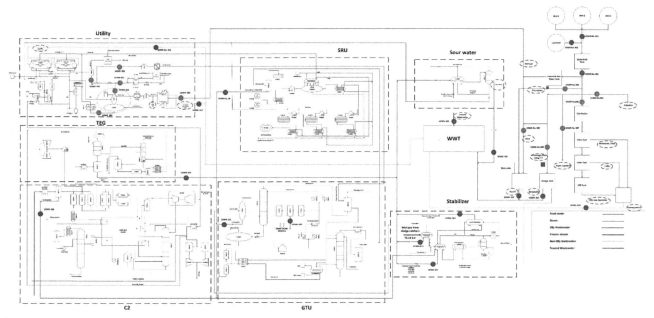

FIGURE 16.7 Water and wastewater network in a natural gas refinery.

between different operations is determined. The ultimate objective of a water audit is to define the characteristics of all water and wastewater streams from quantitative, qualitative, and local aspects.

The index of quantitative measurement is usually volumetric or mass flow rate while the qualitative indices can be each qualitative parameter of water (e.g., TDS, TSS, COD, BOD, etc.) or a combination of numerous parameters depending on the process characteristics. Localization of the water and wastewater streams is also useful in the synthesis of the water distribution network in order to maximize reuse with minimum costs of piping and fittings. The costs related to piping and water transfer between two units may be noticeable and must be considered in redesigning the water allocation plan (see Fig. 16.7).

Furthermore, the results of the water audit are necessary for the is the categorization of the water and wastewater streams based on their potential for reuse, recycle and recovery according to the fuzzy logic. In our approach to fuzzy logic in WCLS, we consider the qualitative specifications of water streams as a range with freshwater and black water being the boundaries and any specific type of wastewater lays in between as gray water and based on its characteristics (see Fig. 16.8) (Yager & Zadeh, 2012).

When the problem is fully defined (i.e., water audit is carried out), decision-making for choosing the optimum solutions must be initiated. The core policy in choosing the solutions is dictated by the circular economy in which the main focus is on minimizing wastes and maximizing reuse and recovery. However, a water audit also makes it possible to evaluate the possibility of "Reform" and "Refuse" as two practical solutions for managing water consumption problems in industrial units. It is a great opportunity that WCLS provides in the PFS and FS phases of a water management project. Apart from reform and refuse, the mechanism of problem-solving in WCLS follows the hierarchy of water resource management (Fig. 16.1). Mathematical models and the principles of process integration are employed for designing or retrofitting the water network to minimize freshwater consumption through maximizing reuse and recovery.

FIGURE 16.8 Water categorization based on its qualitative specification according to fuzzy logic.

Process integration is a set of operations being conducted to retrofit the process to minimize the requirement for water, energy, or other inputs with focusing on the solutions provided by the circular economy (El-Halwagi, 2006). Pinch analysis and mathematical optimization are two powerful methods of process integration for the management of inputs (e.g., water) consumption in industrial units (Beninca, Trierweiler, & Secchi, 2011). Pinch is an inherent characteristic of any thermodynamic system which defines the minimum freshwater requirement possible to reach by direct and regeneration reuse (Foo, 2009). It is a start point for planning water allocation and designing water networks in industrial units. On the other hand, mathematical optimization is the methodology of network design and retrofit using optimization algorithms and single or multiple objective functions (Bavar et al., 2018). However, it is not the focus of this chapter to study different methods of water network synthesis using pinch analysis or mathematical optimization, since numerous references comprehensively cover the fundaments of process integration and can be referred to for further information (Foo, 2009; Kemp, 2011).

In WCLS, we are focused on choosing the most appropriate method of process integration to achieve the simplest retrofitted network with maximum water reuse and minimum freshwater requirement. For this purpose, the knowledge of process engineering (chemical engineering) and wastewater treatment needs to be employed during the process integration step to provide the economically optimum solution that is also operationally feasible. There are different methods of solving the problem of water network synthesis and process integration including water source diagram (WSD), water cascade analysis, and bridge analysis which will be employed in WCLS based on the necessities of the process (Bonhivers, Svensson, Berntsson, & Stuart, 2014; de Souza et al., 2009; Francisco, Mirre, Calixto, Pessoa, & Queiroz, 2018; Manan, Tan, & Foo, 2004). The significant advantage of merging process engineering and wastewater treatment knowledge with process integration analyses is providing vital information for probable wastewater treatment plants including the objective contaminant, desired treatment efficiency, and the method of treatment which are valuable information for minimizing the process costs. It is one of the key differences of WCLS with other methods of industrial water consumption management.

Case study in oil industry

In this section, the results of applying WCLS on reducing water consumption and wastewater generation in an Iranian oil refinery with the capacity of 40,000 barrel of crude oil per day is studied. Considering the significance of oil industries and the role they play in the implementation of SDGs in a country like Iran whose economy is based on fossil fuels, refuse and reform seem to be infeasible in the studied oil refinery and WCLS must look for alternative options of reducing water consumption and wastewater generation through reuse and recycle.

In this sector, after performing a water audit and synthesizing the water network of the refinery, water pinch analysis is employed to retrofit the network for reducing freshwater consumption by two approaches of direct reuse and regeneration reuse. The results of these two approaches are compared with each other at the end.

Water audit

Water and steam network

A water audit is the first practical step that we followed in this project. In the water audit stage, the process flow diagram with a focus on the distribution of water and wastewater within the whole unit is synthesized. Moreover, the flow rate of water in each line is determined (Fig. 16.9).

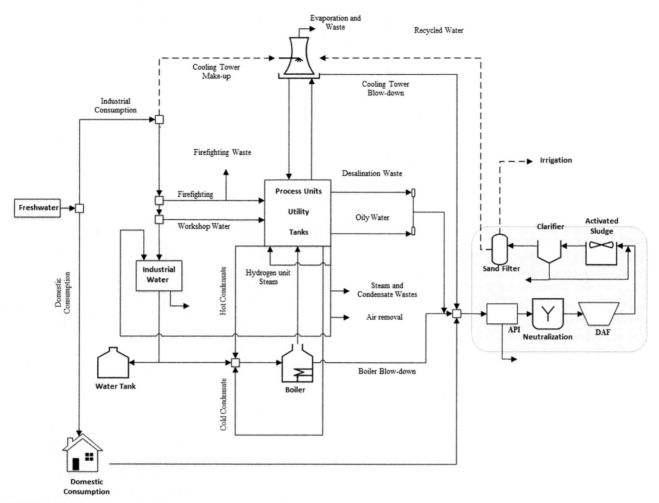

FIGURE 16.9 Current network of water-steam-wastewater in the studied oil refinery obtained from water audit stage in water closed-loop system.

The freshwater consumed in this refinery is 177 ton h^{-1} which is being supplied from three wells. This amount of water is being consumed in two different sectors of industrial and domestic usages. The domestic usages are related to the houses of the personnel living adjacent to the refinery or irrigation of the green built environment.

However, the major part of the water consumption is related to industrial usages which consume about 166 ton h^{-1} (about 94% of the total water consumption) mainly as make-up water of the cooling towers, firefighting unit water, and workshop and industrial water production unit. Cooling towers provide the required water for cooling rotary devices, water coolers, pumps, compressors, and heat exchangers. In this unit, 4700 ton h^{-1} of the cooling water loses its temperature from 48°C to 29°C in a closed cycle. In the cooling towers, a part of the temperature of the water will be discharged to the environment through evaporation, therefore, a certain amount of water is being evaporated from the cooling towers which needs to be made up. 45.3 ton h^{-1} is the amount of evaporation in the cooling tower system of the studied oil refinery. Furthermore, blowing the air by the fans into the cooling towers causes carrying out water drops which are estimated to be between 0.1% and 0.5% of the circulated water (4.7 ton h^{-1}). Moreover, since water evaporation increases the salinity of the water and the probability of sedimentation and corrosion, consequently, a blow-downstream of 15 ton h^{-1} will be discharged out of the tower to keep the salinity in the desirable range. This stream is sent to the wastewater treatment unit. To make these lost waters up, a 65 ton h^{-1} of freshwater is being fed to the cooling towers.

The water of the firefighting unit is another sector in which industrial water is being consumed. This water is being used for the protection of different units against fire. Although it is not a permanent consumption, in case of existence it will designate an enormous amount of water to itself. The main portion of this water will also be discharged to the wastewater treatment system and a part will be wasted during consumption. The average consumption of firefighting water is 21 ton h^{-1}; 3 ton h^{-1} of it will be wasted during consumption and the remaining will be transferred to the treatment system.

7 ton h^{-1} is the water requirement of workshop consumption which is being consumed for washing equipment, site washing, and similar usages. The effluent of these usages will be discharged to the wastewater treatment system of the refinery.

Industrial water production has the highest share in the industrial sector of water consumption. Industrial water production means water degasification, removing hardness and silicon from water. In the studied refinery, the hot lime process is applied for generating industrial water. In this method, water will be fed to a reaction tower with 7 ton h^{-1} of low-pressure steam and $Ca(OH)_2$. In temperatures higher than $100°C$, Ca and Mg bicarbonates will be removed from water in the form of sediments. To remove the remaining suspended solids, the treated water will pass through carbon filters and ion exchange beds. The filters and ion exchange beds need to be regenerated once in 24 h to be functionalized. This regeneration process needs 0.5 ton h^{-1} of treated water which will completely be discharged as wastewater. Wastewater of the reaction tower and the waste of steam to the atmosphere is 5.5 ton h^{-1} totally. A cumulative amount of 6 ton h^{-1} is the waste of industrial water unit which has a significant TDS and cannot be entered into the wastewater treatment system due to its destructive impacts on the performance of the biological treatment unit. This wastewater is being discharged into evaporation lagoons currently. 74.5 ton h^{-1} of industrial water is generated in this refinery. A portion of this water can be reserved in tanks for emergency consumptions. The remaining will be added to 32 ton h^{-1} of hot condensate water and 18.5 ton h^{-1} of cold condensate water and transferred to the boilers for producing steam.

Cold condensate water is the outlet stream of the compressor of the steam turbines which is being used for electricity generation in the refinery. Hot condensate is the stream that is being recycled from different units of the refinery in which heat exchange between the steam and a process or water stream occurs.

Steam production in the boilers of the studied refinery is 114 ton h^{-1} which is being used for utility consumptions of the refinery. Also, 11 ton h^{-1} of wastewater is being drained out of the boilers in the form of blow-down. 18 ton h^{-1} of the steam is consumed in different processes which is mainly turned into sour water. This sour water after treatment in sour water strippers is consumed in an oil desalination unit which produces 18 ton h^{-1} of wastewater. The wastewater will be discharged to the wastewater treatment system. From the remaining amount of the steam (96 ton h^{-1}), 7 ton h^{-1} is used for removing air from the feed water of the boilers and drained to the atmosphere at last. 32 and 18.5 ton h^{-1} are recovered as hot and cold condensate, respectively which are fed to the boilers (45% of steam recovery). An approximate amount of 28.5 ton h^{-1} is also wasted as steam or inconsumable condensates within the network Table 16.2.

As it is obvious, a water audit provides the requisite information for conducting water management plans, since the "water profile" of the refinery (including water consumption, wastewater generation, water distribution in different sectors, and water consumption bottlenecks) is completely determined. Now, the authority to act is wider for choosing more flexible methods of process integration and water reuse.

The studied refiner has a previously designed wastewater treatment system which means regeneration-reuse solutions may also be taken into account in water management scenarios. In the industrial units with a previously-built wastewater treatment system (like the studied oil refinery), it is rational to design the regeneration reuse scenario based on the potentials and removal efficiency of the existing treatment system.

Wastewater treatment system

The wastewater treatment system in the studied refinery is composed of an API separator, equalizer lagoon, dissolved air floatation (DAF), biological treatment with activated sludge, clarifiers, sludge removal and sand filtration.

TABLE 16.2 Summary of water consumption in different units of the oil refinery.

Unit	Water consumption (ton h^{-1})
Cooling towers make-up	65
Firefighting	21
Workshop consumption	7
Industrial water generation	74.5
Domestic consumption	9.5

71 ton h^{-1} of the industrial wastewater of the refinery is entered to the API separator after being collected from different units. The amount of the oily compounds that is being separated in API is reported to be 1% of the total wastewater (0.71 ton h^{-1}). The remaining is mixed with the domestic wastewater (81.29 ton h^{-1} cumulatively) and will be fed to DAF unit for suspended solids removal. The outlet of DAF has a noticeable organic load in form of COD, therefore it must be fed to a biological treatment unit with activated sludge for organics removal. The outlet of the activated sludge has an acceptable organic load and after sludge and suspended solid removal in sand filters will be consumed for green-built irrigation in the oil refinery.

Fuzzy categorization

The arrangement of the unit operations in the treatment system is a combination of physicochemical and biological methods which indicates an acceptable efficiency in dissolved and suspended solids (TDS and TSS) removal as well as organic loads including COD and BOD. This flexibility in the removal efficiency of the treatment system is a great advantage that increases the freedom in choosing the limiting contaminant and enhances the recovery potential of the whole process.

In the fuzzy categorization step, water-consuming units (sinks) and the qualitative properties that they accept are determined. Moreover, the qualitative specifications of the wastewater streams (sources) that are being reused for supplying the sinks are defined. The minimum accepting threshold of each source is a key parameter that must be defined for each water-consuming unit in the refinery.

As it was mentioned, domestic consumptions, firefighting, workshop water, industrial water production, and cooling tower makeup are the five sectors that consume freshwater in the oil refinery. The first step is to choose the contaminant(s) that confines reuse or recycle more than others which are known as limiting contaminant(s). Limiting contaminants must be chosen based on the operational conditions and the equipment design. The following points are the limiting conditions that must be taken into account in choosing limiting contaminant:

- An organic load may produce slurry in the walls of the boiler and result in a reduction in heat transfer;
- The organic load may lead to biofilm formation which results in an increased rate of corrosion;
- Suspended solid may increase the probability of fouling in the pipeline and the filters;
- Existence of dissolved solid increases sedimentation and corrosion in the pipes and equipment;
- Oily compounds are usually of high concentration in refinery wastewater which normally is toxic and needs to be considered;
- H$_2$S is a very important contaminant that can cause hazardous damages to the health of the personnel;
- NH$_3$ is an abundant compound in industrial wastewaters such as oil refineries and is important in the design of the biological treatment plant.

These constraints are imposed on the process integration procedure by the unit operations and types of equipment that existed in the studied refinery. It is obvious that any changes in the process or equipment may result in changes in the constraints, consequently.

Therefore and considering the above-mentioned conditions, COD, TSS, TDS, Oil, H$_2$S, and NH$_3$ are chosen as the limiting contaminants in this study. Since the arrangement of the wastewater treatment system is in a way that it is capable of removing all of these contaminants, the regeneration reuse scenario can also be considered as a powerful option with lower uncertainty Table 16.3.

The values presented in Table 16.3.

On the other hand, the qualitative specifications and the flow rates of the wastewater streams, as the sources with which the sinks must be supplied, need to be determined. Boiler blow-down, cooling tower blow-down, wastewater of the oil desalination, sanitary effluent, and oily water are the wastewater streams that were diagnosed in the refinery and are shown in Table 16.4.

The final step of fuzzy categorization is an analysis of the wastewater treatment plant efficiency. In order to have a more realistic vision about the regenerative ruse scenario, the data of removal efficiency in the existing treatment system is necessary, since it is more favorable to use the existing facilities for recycling wastewater instead of design and construction of a new system Table 16.5.

Water reuse

As it was explained before, for applying reuse scenarios, process integration methods such as water pinch analysis and mathematical optimization are employed. The objective of these methods is to retrofit the water network (Fig. 16.10) in a way that water consumption and wastewater generation reduces through water reuse. However, it is important to recall that in reuse, the focus is on consuming water again without any kind of treatment.

TABLE 16.3 Minimum accepting threshold and the required flowrate of each sink in the oil refinery.

							Sink
0	0	441	0	5	0	11	Domestic
0	0	441	0	5	5	21	Firefighting
0	0	441	0	5	5	73	Industrial water production
0	0	441	0	5	5	7	Workshop
0	0	500	1	15	40	65	Cooling tower make-up

TABLE 16.4 Quantitative and qualitative specifications of the wastewater streams in the studied oil refinery.

							Wastewater
0	0	2370	0	3	1	11	Boiler blow-down
0	0	1650	1	7	0	15	Cooling tower blow-down
4	61	1452	9.3	7	374	18	Desalination wastewater
33	10.2	1009	6.5	157	106	11	Sanitary wastewater
622	165	1350	325	95	802	27	Oily water
0	0	441	0	5	0		Freshwater

TABLE 16.5 Performance of the existing wastewater treatment system in the studied oil refinery.

Irrigation standard (Fipps, 2003)	Treated wastewater	Parameter (ppm)
200	40	COD
100	15	TSS
10	1	Oil
2000	1488	TDS
—	0	H_2S
—	0	NH_3

The focus of this chapter is not on explaining the procedure of pinch analysis and the reader can refer to the supplementary references that have been introduced above. In this section, the results of the process integration is only reported and the impacts it had on reducing water consumption and wastewater generation of the refinery. However, for the complete procedure, you can refer to our similar research in a corn processing industrial unit (Bavar et al., 2018).

Water pinch analysis proposes that the boiler and cooling tower blow-down wastewaters can be reused as the cooling tower make-up. Applying these modifications as described in figure 12, freshwater water consumption and wastewater generation are reduced to 126.27 and 31.27 ton h^{-1}, respectively. Initial amounts of freshwater consumption and wastewater generation in the studied refinery were 177 and 8 ton h^{-1}, respectively. In other words, applying the reuse scenario has resulted in a 28.66% and 61.87% reduction in freshwater requirement and wastewater generation of the refinery, respectively. It is essential to bear in mind that these reductions are obtained without any treatment.

Wastewater treatment and regeneration reuse

It is mentioned before that the difference between direct reuse and regeneration reuse in WCLS is in applying wastewater treatment methods in the regeneration reuse approach. Therefore, it is necessary to have a deep knowledge of wastewater treatment techniques for the efficient application of WCLS. In the case of the studied refinery, it is more rational

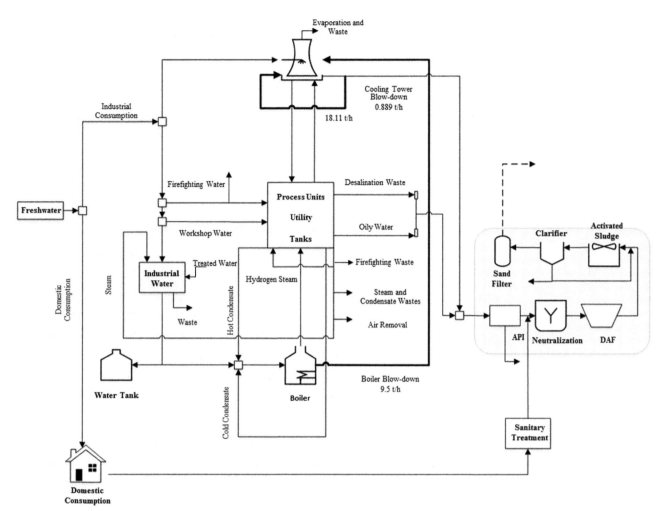

FIGURE 16.10   Retrofitted water—steam—wastewater network with direct reuse scenario.

to use the potential of the existing treatment system instead of constructing a new one. Therefore, the retrofitted water network based on the regeneration reuse scenario is as Fig. 16.11.

Concluding all, by applying WCLS in the oil refinery, water consumption and wastewater generation is reduced to its half; it means about 88 ton h^{-1} of freshwater which is equal to the average hourly water consumption of about 610 persons in a city. This case study clearly shows the potential of WCLS in water consumption management in industrial units.

Conclusion and recommendations

Oil and gas industries are among the greatest water consumers that necessitate applying sustainable methods of water consumption management. The circular economy has generated new solutions in the field of water management which are based on reuse, recycle and recovery. WCLS is an integrated approach for water consumption management in water-consuming industries (e.g., oil and gas industries) which is based on the novel approach of 5R which is governed by the circular economy. WCLS is a holistic algorithm for minimizing water consumption and wastewater generation through maximizing water reuse and recycling. WCLS makes a bridge between fundamental concepts such as water management, process integration, and fuzzy logic with the technical knowledge of wastewater treatment and process engineering and provides a practical and feasible context for implementing the objectives of circular economy regarding industrial water and wastewater management. Flexibility in this systematic approach is its advantage over other methods of industrial water management. In other words, WCLS has the potential of evolution in its different stages. Theoretical and technical advancement in methods of process integration (e.g., pinch analysis, mathematical optimization, bridge

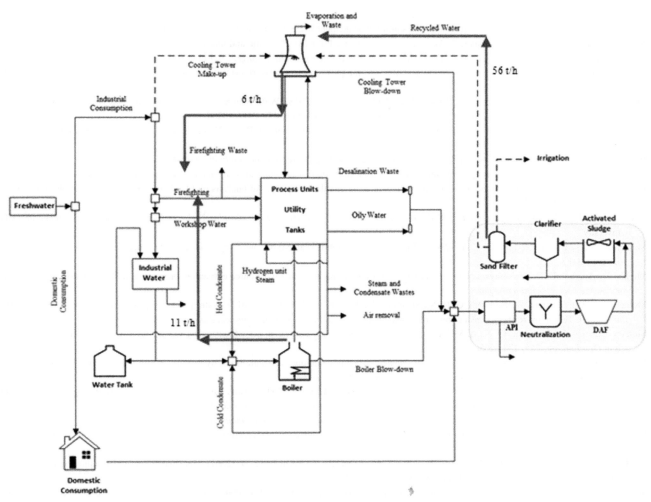

FIGURE 16.11 The water network after applying water closed-loop system with recycling. Retrofitted water network of the refinery based on the recycle scenario.

methodology, etc.) or wastewater treatment methods can be applied to WCLS as well because it is highly flexible for being coupled with modern methodologies. Therefore, WCLS is a constantly progressive method of water consumption management whose aim is to minimize freshwater consumption and wastewater generation in the industrial unit. On the other hand, WCLS can be applied either for the design or retrofitting of the process. In the initial design step, WCLS can open new doors under the topic of water-sensitive designing in industrial sectors. In retrofitting, it can be an efficient instruction for changing existing industrial units toward more sustainable processes.

References

Bavar, M., Sarrafzadeh, M.-H., Asgharnejad, H., & Norouzi-Firouz, H. (2018). Water management methods in food industry: Corn refinery as a case study. *Journal of Food Engineering, 238,* 78−84.

Beninca, M., Trierweiler, J. O., & Secchi, A. R. (2011). Heat integration of an Olefins Plant: Pinch analysis and mathematical optimization working together. *Brazilian Journal of Chemical Engineering, 28,* 101−116.

Bonhivers, J.-C., Svensson, E., Berntsson, T., & Stuart, P. (2014). Comparison between pinch analysis and bridge analysis to retrofit the heat exchanger network of a Kraft pulp mill. *Applied Thermal Engineering, 70,* 369−379.

Boretti, A., & Rosa, L. (2019). Reassessing the projections of the world water development report. *NPJ Clean Water, 2,* 1−6.

Colson, J. (1999). Upstream, midstream, downstream-the valuation of royalties on federal oil and gas leases. *University of Colorado Law School, 70,* 563.

de Souza, A. U., Forgiarini, E., Brandao, H., Xavier, M., Pessoa, F., & Souza, S. G. U. (2009). Application of water source diagram (WSD) method for the reduction of water consumption in petroleum refineries. *Resources, Conservation and Recycling, 53,* 149−154.

El-Halwagi, M. M. (2006). *Process integration.* Elsevier.

Fipps, G. (2003). Irrigation water quality standards and salinity management strategies. *Texas FARMER Collection.*

Flörke, M., Kynast, E., Bärlund, I., Eisner, S., Wimmer, F., & Alcamo, J. (2013). Domestic and industrial water uses of the past 60 years as a mirror of socio-economic development: A global simulation study. *Global Environmental Change, 23,* 144–156.

Foo, D. C. Y. (2009). State-of-the-art review of pinch analysis techniques for water network synthesis. *Industrial & Engineering Chemistry Research, 48,* 5125–5159.

Francisco, F. S., Mirre, R. C., Calixto, E. E., Pessoa, F. L., & Queiroz, E. M. (2018). Water sources diagram method in systems with multiple contaminants in fixed flowrate and fixed load processes. *Journal of Cleaner Production, 172,* 3186–3200.

Griggs, D., Stafford-Smith, M., Gaffney, O., Rockström, J., Öhman, M. C., Shyamsundar, P., . . . Noble, I. (2013). Sustainable development goals for people and planet. *Nature, 495,* 305–307.

International Water Association. (2018). *The reuse opportunity, cities seizing the reuse opportunity in a circular economy.*

Kemp, I. C. (2011). *Pinch analysis and process integration: A user guide on process integration for the efficient use of energy.* Elsevier.

Kirchherr, J., Reike, D., & Hekkert, M. (2017). Conceptualizing the circular economy: An analysis of 114 definitions. *Resources, Conservation and Recycling, 127,* 221–232.

Lewis, A. (2012). *Wastewater generation and disposal from natural gas wells in Pennsylvania.* Nicholas School of the Environment of Duke University.

Manan, Z. A., Tan, Y. L., & Foo, D. C. Y. (2004). Targeting the minimum water flow rate using water cascade analysis technique. *AIChE Journal, 50,* 3169–3183.

Mielke, E., Anadon, L. D., & Narayanamurti, V. (2010). Water consumption of energy resource extraction, processing, and conversion. *Belfer Center for Science and International Affairs.*

Mohammadnejad, S., Ataei, A., Bidhendi, G. R. N., Mehrdadi, N., Ebadati, F., & Lotfi, F. (2012). Water pinch analysis for water and wastewater minimization in Tehran oil refinery considering three contaminants. *Environmental Monitoring and Assessment, 184,* 2709–2728.

Mohammadnejad, S., Bidhendi, G. N., & Mehrdadi, N. (2011). Water pinch analysis in oil refinery using regeneration reuse and recycling consideration. *Desalination, 265,* 255–265.

Ng, D. K. S., Foo, D. C. Y., & Tan, R. R. (2009). Automated targeting technique for single-impurity resource conservation networks. *Part 1: Direct Reuse/Recycle, Industrial & Engineering Chemistry Research, 48,* 7637–7646.

Norling, P., Wood-Black, F., Masciangioli, T. M., & Roundtable, N. R. C. C. S. (2004). *Water and sustainable development: opportunities for the chemical sciences: a workshop report to the chemical sciences roundtable.* National Academies Press.

Royan, F. (2012). *Sustainable water treatment technologies in the 2020 global water market, mountain view.* Frost & Sullivan.

Schroeder, P., Anggraeni, K., & Weber, U. (2019). The relevance of circular economy practices to the sustainable development goals. *Journal of Industrial Ecology, 23,* 77–95.

Smakhtin, V., Ashton, P., Batchelor, A., Meyer, R., Murray, E., Barta, B., . . . Terblanche, D. (2001). Unconventional water supply options in South Africa: a review of possible solutions. *Water International, 26,* 314–334.

Sun, P., Elgowainy, A., Wang, M., Han, J., & Henderson, R. J. (2018). Estimation of US refinery water consumption and allocation to refinery products. *Fuel, 221,* 542–557.

Yager, R. R., & Zadeh, L. A. (2012). An introduction to fuzzy logic applications in intelligent systems. *Springer Science & Business Media.*

Zbontar, L., & Glavic, P. (2000). Total site: Wastewater minimization: Wastewater reuse and regeneration reuse. *Resources, Conservation and Recycling, 30,* 261–275.

Index

9780323858847